Exploring Nanotechnology for a Sustainable Economy

Nanotechnology has been recognized for its transformative potential across various industries for decades, sparking widespread interest among researchers and innovators seeking to explore its capabilities for sustainable solutions. It remains one of the most dynamic and extensively studied fields in modern science. Nanotechnology involves the manipulation of matter at the nanoscale, where unique properties of materials emerge due to their size and structure. These nanomaterials play a pivotal role in advancing applications in energy, healthcare, environmental remediation, and sustainable economic development. In recent years, green nanotechnology, particularly the eco-friendly synthesis of nanomaterials, has gained significant attention for its ability to minimize environmental impact while maximizing efficiency. Nanomaterials exhibit remarkable properties such as enhanced strength, conductivity, and reactivity that make them indispensable in addressing global challenges such as climate change, water scarcity, and healthcare innovation, positioning nanotechnology as a cornerstone of sustainable progress.

The book delves into the structural and functional properties of nanomaterials, which are critical for their applications in clean energy production, environmental remediation, water treatment, and healthcare. It highlights their environmental significance, including their use in mitigating climate change, enhancing construction materials, and developing fire-retardant solutions. The fundamentals of nanotechnology are thoroughly explored, alongside cutting-edge developments in synthesis, characterization methods, and applications, covering chemical, optical, thermal, and structural properties. Additionally, the book addresses the environmental fate, health hazards, and regulatory frameworks of nanomaterials, offering a balanced perspective on their deployment. With contributions from leading experts, this book serves as an authoritative resource for researchers, policymakers, and industry professionals to drive sustainable economic growth while addressing pressing global challenges.

Sustainability: Contributions through Science and Technology
Series Editors: Thomas P. Umile, Ph.D, Villanova University, Pennsylvania, USA
William M. Nelson, US Army ERDC, USA

Preface to the Series

Sustainability is rapidly moving from the wings to center stage. Overconsumption of non-renewable and renewable resources, as well as the concomitant production of waste has brought the world to a crossroads. Green chemistry, along with other green sciences technologies, must play a leading role in bringing about a sustainable society. The Sustainability: Contributions through Science and Technology series focuses on the role science can play in developing technologies that lessen our environmental impact. This highly interdisciplinary series discusses significant and timely topics ranging from energy research to the implementation of sustainable technologies. Our intention is for scientists from a variety of disciplines to provide contributions that recognize how the development of green technologies affects the triple bottom line (society, economic, and environment). The series will be of interest to academics, researchers, professionals, business leaders, policy makers, and students, as well as individuals who want to know the basics of the science and technology of sustainability.

Michael C. Cann

Green Chemistry for Environmental Sustainability
Edited by Sanjay Kumar Sharma, Ackmez Mudhoo, 2010

Microwave Heating as a Tool for Sustainable Chemistry
Edited by Nicholas E. Leadbeater, 2010

Green Organic Chemistry in Lecture and Laboratory
Edited by Andrew P. Dicks, 2011

A Novel Green Treatment for Textiles: Plasma Treatment as a Sustainable Technology
C. W. Kan, 2014

Environmentally Friendly Syntheses Using Ionic Liquids
Edited by Jairton Dupont, Toshiyuki Itoh, Pedro Lozano, Sanjay V. Malhotra, 2015

Catalysis for Sustainability: Goals, Challenges, and Impacts
Edited by Thomas P. Umile, 2015

Nanocellulose and Sustainability: Production, Properties, Applications, and Case Studies
Edited by Koon-Yang Lee, 2017

Sustainability of Biomass through Bio-based Chemistry
Edited by Valentin Popa, 2021

Nanotechnologies in Green Chemistry and Environmental Sustainability, 2022

Towards Sustainability in the Wine Industry by Valorization of Waste Products: Bioactive Extracts
Edited by Patricia Joyce Pamela Zorro Mateus and Siby Inés Garcés Polo, 2023

Climate Change and Carbon Recycling: Surface Chemistry Applications,
K.S. Birdi

Exploring Nanotechnology for a Sustainable Economy

Edited by
Bimal Krishna Banik, Prashant Singh,
and Garima Pandey

CRC Press
Taylor & Francis Group
Boca Raton London New York

CRC Press is an imprint of the
Taylor & Francis Group, an **informa** business

First edition published 2026
by CRC Press
2385 NW Executive Center Drive, Suite 320, Boca Raton FL 33431

and by CRC Press
4 Park Square, Milton Park, Abingdon, Oxon, OX14 4RN

CRC Press is an imprint of Taylor & Francis Group, LLC

© 2026 Bimal Krishna Banik, Prashant Singh, and Garima Pandey

ISBN: 9781041053736 (hbk)
ISBN: 9781041053798 (pbk)
ISBN: 9781003632498 (ebk)

DOI: 10.1201/9781003632498

Typeset in Times
by Apex CoVantage, LLC

Contents

Preface...vii

Organization of the Book..ix

About the Editors xi

List of Contributors...xiii

1   Assessing the Business Potential of Nanotechnology 1
    Ravi Kumar, Pashupati Pratap Neelratan, Abhay Singh Rana, Shivom,
    Mahender Pal, and Sanjeev Kumar Sharma

2   Nanomaterials for Green and Clean Energy Production and Storage....... 39
    Shailesh S. Ankalgi, Aviraj M. Teli, Dattakumar S. Mhamane,
    and Mayur S. Ankalgi

3   Life Cycle Thinking of Nanotechnology 64
    Himanshi Goel and Kumar Rakesh Ranjan

4   Current Status and Future Prospects of Nanotechnology in the
    Healthcare Industry..81
    Vishakha Rathi, Umesh Kumar, Garima Rathi,
    and Pritipriyambada Baisakh

5   Nanotechnology in Environmental Remediation 93
    Monika Verma, Ruchi Bharti, and Renu Sharma

6   Application and Prospects of Nanotechnology for Water Treatment........ 125
    Bibhudendu Behera, Bhupender Sahu, Shalini Rai, Siya Sharma,
    and Umesh Kumar

7   Eco-Friendly Alchemy: Harnessing Nature's Power for the Green
    Synthesis of Nanomaterials... 151
    Priya Kaushik, Ruchi Bharti, Renu Sharma, and Monika Verma

8   Environmental Fate, Transport, and Health Hazards of
    Nanomaterials 174
    Vijay Kumar Vishvakarma and Gyanendra Kumar

9   Nanomaterials for the Construction Industry 193
    Vaibhav Sharma and Piyush Gupta

10 Nanomaterials as Fire Retardants..211
    Seema

11 Nanotechnology: A Sustainable Solution for Climate
    Change Mitigation 236
    Bhawani Shankar

12 Applications of Nanotechnology for Eco-Friendly and Sustainable
    Economic Development 249
    Surbhi Dhadda, Nikita Choudhary, and Sanjay K. Meena

13 Advancing Sustainable Development Goals through
    Nanotechnology 260
    Himakshi Adhikari, Ishika Gupta, Sudhir Upadhyaya, Himanshi Goel,
    and Kumar Rakesh Ranjan

14 Policy and Regulatory Issues in Achieving a Green Economy through
    Nanotechnology 281
    Ikechukwu P. Ejidike, Opeyemi N. Avoseh, Adeniyi A. Adejare, Taiwo A.
    Akande, Toyib S. Oyewole, Folajimi T. Avoseh, Chinwendu B. Sunday,
    and Mercy O. Bamigboye

15 Nanotechnology and Its Role in Building a Sustainable Economy 304
    Upendra Kumar Mishra, Satyendra Pratap Singh, S. Gaurav,
    and Vishal Singh Chandel

Index 339

Preface

In recent decades, the rapid advancement of nanotechnology has catalyzed transformative changes across diverse sectors from energy and healthcare to environmental remediation and industrial development. As nations and industries strive toward sustainable development and green economies, the promise of nanotechnology stands out not merely as a scientific innovation but as a pivotal enabler of long-term economic, environmental, and social progress.

This book, *Exploring Nanotechnology for a Sustainable Economy*, is a collective effort by researchers, academicians, and professionals to bridge the gap between cutting-edge nanoscience and its real-world applications in achieving sustainable goals. The chapters presented herein explore the multifaceted role of nanomaterials and nanotechnological innovations in areas such as green energy production, water purification, eco-friendly manufacturing, healthcare, construction, climate change mitigation, and regulatory policy.

Each contribution is grounded in rigorous scientific inquiry while also contextualizing the broader implications for industry, governance, and the environment.

Chapter 1 starts with an insightful exploration of the business potential of nanotechnology, highlighting its commercialization and market integration. Subsequent chapters delve into technical domains such as energy storage, green synthesis of nanomaterials, life cycle analysis, and the environmental fate of nanoparticles. The book also addresses cross-cutting themes such as policy frameworks and sustainable development goals that are essential to the responsible and ethical adoption of nanotechnology. This integrated vision is important for fostering a future where technological advancement harmonizes with environmental stewardship and social equity.

This book will serve as a valuable resource for researchers, students, policymakers, and industry professionals who are committed to advancing sustainability through science and innovation. May it inspire new ideas, responsible action, and collaborative endeavors toward a greener and more sustainable world.

Organization of the Book

The book is organized into 15 chapters. A brief description of each of the chapters follows:

Chapter 1 focuses on analyzing the commercial viability and market opportunities of nanotechnology applications across industries.

Chapter 2 explores the role of nanomaterials in clean energy production and storage, emphasizing sustainable energy solutions.

Chapter 3 discusses the environmental and economic impacts of nanotechnology through life cycle assessment frameworks.

Chapter 4 examines current applications and future prospects of nanotechnology in advancing healthcare innovations.

Chapter 5 highlights nanotechnology potential in addressing environmental pollution and ecosystem restoration.

Chapter 6 focuses on nanotechnology-driven solutions for efficient and sustainable water purification and treatment.

Chapter 7 explores eco-friendly methods for synthesizing nanomaterials using nature-inspired processes.

Chapter 8 investigates the fate, transport, and potential health risks associated with nanomaterials in the environment.

Chapter 9 discusses the use of nanomaterials to enhance durability and sustainability in the construction industry.

Chapter 10 explores nanotechnology applications in developing advanced fire-retardant materials.

Chapter 11 examines nanotechnology's role in reducing environmental impacts and combating climate change.

Chapter 12 highlights nanotechnology's contributions to eco-friendly and economically viable development strategies.

Chapter 13 aligns nanotechnology innovations with the Global Sustainable Development Goals (SDGs).

Chapter 14 addresses policy challenges and regulatory considerations for integrating nanotechnology into a green economy.

Chapter 15 explores how nanotechnology can drive sustainable economic growth and innovation.

About the Editors

Bimal Krishna Banik, Ph.D., is a professor in the Deanship of Research at Prince Mohammed Bin Fahd University, Saudi Arabia. He earned his doctorate from the Indian Association for the Cultivation of Science, Calcutta, and conducted postdoctoral research at Case Western Reserve University and Stevens Institute of Technology, USA. Previously, he served as a tenured full professor, the first president's endowed professor in science and engineering, and the vice president of research and education development at the University of Texas. His research focuses on synthetic organic chemistry and chemical biology, particularly in cancer, antibiotics, hormones, catalysis, and natural products. As the principal investigator, he secured $7.25 million in grants from the NIH and NCI. He has over 700 publications, 530 presentation abstracts, and has edited/authored 23 books. His work, cited over 10,000 times, ranks him in the top 2% of scientists globally. Dr. Banik is a prolific leader in the scientific community, serving as the editor-in-chief of 12 journals, a reviewer for 93 journals, and an editorial board member for 26. He has organized over 20 ACS symposia, chaired numerous conferences, and introduced approximately 300 speakers. His accolades include the Indian Chemical Society's Lifetime Achievement Award, the Mahatma Gandhi Pravasi Honor Medal, and the University of Texas Board of Regents' Outstanding Teaching Award. Dr. Banik has delivered keynote lectures in 35 countries and hosted distinguished figures, including Nobel Laureates and US Senators.

Prashant Singh completed an M.Sc. in chemistry from the Indian Institute of Technology, Delhi, India. He joined the research group of the late Dr. N. N. Ghosh at Dr. B. R. Ambedkar Centre for Biomedical Research, University of Delhi, India, and received his Ph.D. He is currently a professor of chemistry at Atma Ram Sanatan Dharma (ARSD) College, University of Delhi, Delhi, India. He has published about 190 research articles and reviews in journals of international repute, along with chapters. His h-index is 37, i-10-index is 131, and citations of his around 5,000. He has been awarded the UGC postdoctoral research fellowship and was also offered a post-doctoral fellowship from North-West University, South Africa. Seven students have received their Ph.D. under his guidance, and a few others are currently working with him.

Garima Pandey is Professor and Head of the Department of Chemistry at SRM Institute of Science and Technology (SRMIST), Delhi-NCR Campus. With a Ph.D. in chemistry from CCS University, she has over two decades of academic and research experience, specializing in nanotechnology, green chemistry, electrochemical sensing, and environmental sustainability. Dr. Pandey serves as an academic editor for *PLOS ONE* and is on the editorial board of *Scientific Reports*. She is also a reviewer for high-impact journals such as *Discover Sustainability, Environmental Monitoring* *and Assessment*, and *Discover Materials*. She has published more than 50 research articles and reviews in reputed international journals and contributed over 25 book chapters. She has authored and edited multiple books and holds three patents, including innovations in hydrogen production and plastic waste pyrolysis. Dr. Pandey has been featured in the prestigious list of the Top 2% Most-Cited Scientists worldwide for two consecutive years. She has also received several accolades including, the Best Researcher and Best Faculty Awards at SRMIST. A dynamic academic leader and mentor, she is actively involved in research guidance, administrative roles, and academic policy development.

Contributors

Adeniyi A. Adejare
Department of Chemistry
Faculty of Science
Lagos State University
Ojo, Nigeria

Himakshi Adhikari
Department of Chemistry
Amity Institute of Applied Sciences
Amity University
Noida, Uttar Pradesh, India

Taiwo A. Akande
Department of Chemistry
Faculty of Science
Lagos State University
Ojo, Nigeria

Mayur S. Ankalgi
Department of Physics
Raje Ramrao College
Jath, India

Shailesh S. Ankalgi
Department of Chemistry
Sangmeshwar College
Solapur, Maharashtra, India

Folajimi T. Avoseh
Institute of Chemical and Biotechnology
Vaal University of Technology
Southern Gauteng Science and
    Technology Park
Sebokeng, South Africa

Opeyemi N. Avoseh
Department of Chemistry
Faculty of Science
Lagos State University
Ojo, Nigeria

Pritipriyambada Baisakh
Department of Biochemistry and
    Biotechnology
Annamalai University
Annamalai Nagar, Chidambaram
Cuddalore, Tamil Nadu, India

Mercy O. Bamigboye
Department of Industrial Chemistry
Faculty of Physical Sciences
University of Ilorin
Ilorin, Nigeria

Bibhudendu Behera
Department of Zoology
Central University of Jammu
Jammu and Kashmir, India

Ruchi Bharti
Department of Chemistry
UIS, Chandigarh University
Gharuan, Mohali, Punjab, India

Vishal Singh Chandel
Department of Applied Science and
    Humanities
Rajkiya Engineering College
Ambedkar Nagar, Uttar Pradesh, India

Nikita Choudhary
Department of Chemistry
Faculty of Basic and Applied Sciences
Vivekananda Global University
Jagatpura, Jaipur (Raj), India

Surbhi Dhadda
Department of Chemistry
Faculty of Basic and Applied Sciences
Vivekananda Global University
Jagatpura, Jaipur (Raj), India

Ikechukwu P. Ejidike
Biological and Analytical Application
    of Synthetic and Nanomaterials
    Research (BAASNMR) Group
Department of Chemical Sciences
Faculty of Science
Anchor University
Lagos, Nigeria
and
Department of Chemistry
College of Science, Engineering, and
    Technology
University of South Africa
Florida Park, South Africa

S. Gaurav
Department of Physics
Amity Institute of Applied
    Sciences
Amity University
Uttar Pradesh, Noida, India

Himanshi Goel
Department of Chemistry
Amity Institute of Applied Sciences
Amity University
Uttar Pradesh, India

Ishika Gupta
School of Material Science and
    Technology, IIT BHU
Varanasi, Uttar Pradesh, India

Piyush Gupta
Department of Chemistry
Faculty of Engineering and
    Technology
SRM Institute of Science and
    Technology, Delhi-NCR Campus
Ghaziabad, Uttar Pradesh, India

Priya Kaushik
Department of Chemistry
UIS, Chandigarh University
Gharuan, Mohali, Punjab, India

Gyanendra Kumar
Department of Chemistry
Swamishraddhanand College
University of Delhi
Delhi, India

Ravi Kumar
Biomaterials and Sensor
    Laboratory
Department of Physics
Ch. Charan Singh University
Meerut, Uttar Pradesh, India

Umesh Kumar
Chandigarh University
NH5, Chandigarh-Ludhiana Highway
Gharuan, Mohali, Punjab, India

Sanjay K. Meena
Department of Chemistry
Faculty of Basic and Applied
    Sciences
Vivekananda Global University
Jagatpura, Jaipur (Raj), India

Dattakumar S. Mhamane
Department of Chemistry
Sangmeshwar College
Solapur, Maharashtra, India

Upendra Kumar Mishra
Department of Applied Science
Shri Ramswaroop Memorial
    College of Engineering and
    Management
Lucknow, Uttar Pradesh, India

Pashupati Pratap Neelratan
Biomaterials and Sensor
    Laboratory, Department
    of Physics
Ch. Charan Singh University
Meerut, Uttar Pradesh, India

Toyib S. Oyewole
Department of Chemistry
Faculty of Science
Lagos State University
Ojo, Nigeria

Mahender Pal
Biomaterials and Sensor
    Laboratory
Department of Physics
Ch. Charan Singh University
Meerut, Uttar Pradesh, India

Shalini Rai
Centre for Molecular Biology
Central University of Jammu
Jammu and Kashmir, India

Abhay Singh Rana
Biomaterials and Sensor
    Laboratory
Department of Physics
Ch. Charan Singh University
Meerut, Uttar Pradesh, India

Kumar Rakesh Ranjan
Department of Chemistry
Amity Institute of Applied Sciences
Amity University
Noida, Uttar Pradesh, India

Garima Rathi
Delhi Public School
Uttar Pradesh, Ghaziabad, India

Vishakha Rathi
Forensic Science Laboratory
Government of NCT Delhi
Sector 14, Rohini, Delhi

Bhupender Sahu
Centre for Molecular Biology
Central University of Jammu
Samba, Jammu and Kashmir, India

Seema
Department of Chemistry
Shivaji College
University of Delhi, Delhi, India

Bhawani Shankar
Department of Chemistry
Deshbandhu College
University of Delhi, Delhi, India

Renu Sharma
Department of Chemistry
UIS, Chandigarh University
Mohali, Punjab, India

Sanjeev Kumar Sharma
Biomaterials and Sensor Laboratory
Department of Physics
Ch. Charan Singh University
Meerut, Uttar Pradesh, India

Siya Sharma
Centre for Molecular Biology
Central University of Jammu
Samba, Jammu and Kashmir, India

Vaibhav Sharma
Department of Chemistry
SRM Institute of Science and
    Technology, Delhi-NCR Campus
Ghaziabad, Uttar Pradesh, India

Shivom
Biomaterials and Sensor
    Laboratory
Department of Physics
Ch. Charan Singh University
Meerut, Uttar Pradesh, India

Satyendra Pratap Singh
Department of Physics
Amity Institute of Applied Sciences
Amity University
Uttar Pradesh, Noida, India

Chinwendu B. Sunday
Department of Science Education
Faculty of Education
University of Nigeria
Nsukka, Nigeria

Aviraj M. Teli
Dongguk University-Seoul Campus,
    Nano-Optoelectronic Device
    Research Lab
Seoul, South Korea

Sudhir Upadhyaya
School of Material Science and
    Technology, IIT BHU
Varanasi, Uttar Pradesh, India

Monika Verma
Department of Chemistry
UIS, Chandigarh University
Mohali, Punjab, India

Vijay Kumar Vishvakarma
Delhi School of Skill Enhancement &
    Entrepreneurship Development
Department of Chemistry
Institution of Eminence
University of Delhi
Delhi, India

1

Assessing the Business Potential of Nanotechnology

Ravi Kumar, Pashupati Pratap Neelratan, Abhay Singh Rana, Shivom, Mahender Pal, and Sanjeev Kumar Sharma

1.1 Introduction

Nanotechnology (NT) has been considered a transformative breakthrough of the 21st century, defined by the National Nanotechnology Initiative (NNI) as the "understanding and control of matter at dimensions of roughly 1–100 nm," enabling novel applications in construction, healthcare diagnosis, textiles, semiconductor systems, energy harvesting and storage, and green/clean environment. According to the US Environmental Protection Agency (EPA), the matter is discussed in terms of contamination at the atomic and molecular levels with the unique properties of nanomaterials. According to the literature survey, nanocomposites or nanomaterials with advanced microscopes have opened new prospects across the industries, ranging from medicine to electronics, agriculture, and energy. To address the societal challenges of food, water, and health, it is predicted that 15% of global manufacturing will be considered, with many products available in the market, including paints, cosmetics, and electronics. The massive investments in research and development are fueling growth, depicted by a 19% CAGR (compound annual growth rate) between 2011 and 2014. The future of NT holds immense business opportunities in sectors such as fuel cells, solar cells, and consumer goods, although challenges remain in educating the public on safe usage. It is considered a disruptive innovation with the potential to solve critical global issues, including food security, energy, and environmental sustainability [1].

Key materials in NT synthesized by top-down and bottom-up approaches include carbon-based nanomaterials (NMs), such as graphene (Gr), carbon nanotubes (CNTs), and quantum dots (QDs); dendrimers; and metal and metal oxide nanoparticles (NPs) such as silver (Ag), gold (Au), titanium dioxide (TiO_2), and zinc oxide (ZnO). Two-dimensional (2D) materials, such as MXenes and molybdenum disulfide (MoS_2), are also gaining attention. Current trends focus on green synthesis methods, sustainable and biocompatible NMs, and multifunctional applications in the fields of nanomedicine, targeted drug delivery, nanoelectronics-based biosensing, flexible electronics, and energy storage (see Figure 1.1a). There is a growing interest in nano-enabled environmental remediation and agriculture. Emphasis is being placed on scalability, safety, and regulatory frameworks for successful commercialization and global adoption [2].

DOI: 10.1201/9781003632498-1

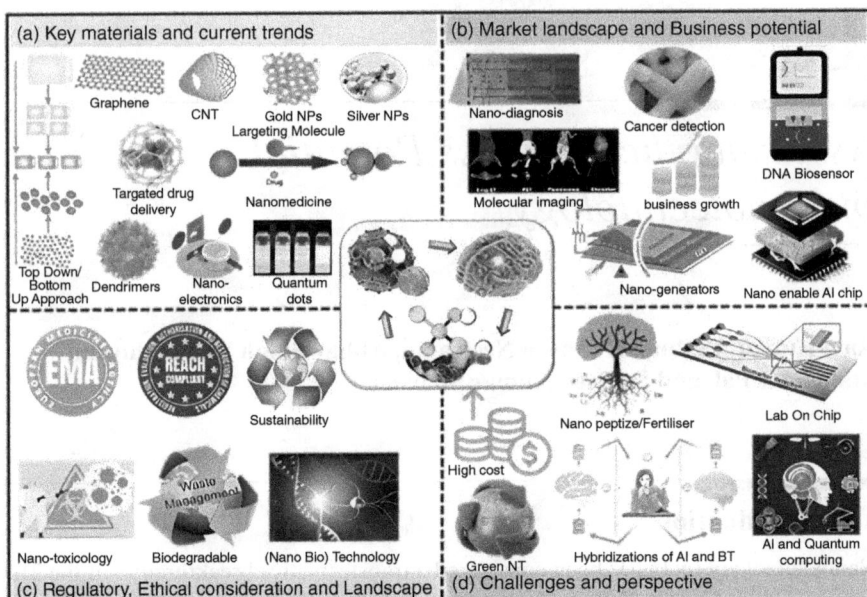

FIGURE 1.1 Assessment and business potential of NT, (a) key materials and current trends in NT, (b) market landscape and business potential of NT, (c) regulatory, ethical considerations, and landscape of NT, and (d) challenges and perspectives in NTs.

NT is rapidly growing, with a market valued at USD 1.1 trillion in 2023 and expected to reach USD 3 trillion by 2030. Key sectors driving growth include healthcare (nano-diagnostics, cancer detection, DNA biosensors, and molecular imaging), energy (nanogenerators), and nano-enabled AI chip and environmental remediation (see Figure 1.1b). The business potential is significant, particularly in personalized medicine, advanced diagnostics, renewable energy, agriculture (nano-pesticides and packaging), and consumer goods (cosmetics and textiles). As the technology advances, industries will benefit from increased efficiency, sustainability, and innovation. Companies investing in NT-based R&D and commercialization stand to capitalize on emerging markets and meet growing consumer demand for cutting-edge, eco-friendly products [3].

The regulatory landscape of NT is still developing, with agencies such as the European Medicines Agency (EMA) and Registration, Evaluation, Authorisation, and Restriction of Chemicals (REACH) beginning to address NM safety (nano-toxicology and nano-biotechnology), risk assessment (biodegradable), and labelling. However, clear, consistent regulations are still lacking (see Figure 1.1c), where the ethical concerns include potential health risks, privacy issues with nanosensors, environmental impact, and the "nano-divide" between developed and developing nations. Ethical considerations in the commercialization of NT are crucial, including concerns about equity, safety, and environmental sustainability. To ensure responsible innovation, frameworks such as Responsible Research and Innovation (RRI) are being developed to balance technological progress with public safety, inclusivity, and ethical standards [4].

NT faces several challenges, including scaling up NMs while maintaining their unique properties, ensuring safety due to potential health (green NT) and environmental risks (nano-peptizer/fertilizer), and navigating the lack of established regulatory frameworks. Economic barriers like high production costs and limited investment also impede progress. Despite these hurdles, NT holds immense potential across industries such as healthcare (lab-on-a-chip), energy, and electronics (AI and quantum computing (QM), hybridization of AI and biotechnology) (see Figure 1.1d). The development of scalable, cost-effective solutions, along with clearer safety regulations and ethical considerations, will drive its growth. Collaboration among academia, industry, and government is key to overcoming challenges and unlocking NT's transformative impact on society and the economy [5].

The technology bridges classical and quantum mechanics through mesoscopic systems, enabling innovations in medicine and manufacturing. Unlike traditional bulk production, NT mimics nature's bottom-up approach, building from atomic levels. While developed countries invest heavily in this field, developing nations struggle to keep pace due to economic challenges. A significant barrier to NT commercialization is the "Valley of Death," which represents the gap between academia and industry. While academia manages research in its initial phases and industry concentrates on advanced, high-profit technologies, the mid-TRL gap remains neglected. Due to their respective restricted funding, infrastructure, and technical knowledge, start-ups often find it challenging to thrive, resulting in high failure rates. The challenges of commercialization encompass the technical, biological, environmental, economic, and regulatory domains. Technical challenges comprise scalability, reproducibility, and a shortage of skilled personnel. Biological and environmental concerns pertain to the toxicity of NMs and their ecological effects. Challenges of an economic nature comprise a lack of investment and unequal access on a global scale. Regulatory challenges include uncertain policies and matters relevant to intellectual property [6].

NT is regarded as an emerging disruptive technology, capable of altering existing technological paradigms and fostering innovations. Scholars such as Bower and Christensen define disruptive technologies as generating new product performance attributes, requiring shifts in user behavior and industry standards. Governments worldwide have invested billions in NT research, restructuring institutions, and fostering interdisciplinary convergence with biotechnology, information technology, and cognitive sciences. Additionally, the chapter also discusses the societal implications of NT, categorized into social (environmental, health, and economic), ethical (technology misuse and social divides), and legal (intellectual property, privacy, and regulation) concerns. As a platform technology, NT has the potential to redefine multiple industries, presenting both opportunities and challenges for future research, policy, and application.

1.1.1 Business Scope of NT

NT has emerged as a revolutionary field of vast business potential in multiple industries, driving innovation, economic growth, and market expansion. Innovations in various fields, including healthcare, electronics, energy, textiles, agriculture, construction, water treatment, aerospace, and military, have been made available by the capacity to work with materials at the atomic and molecular level. As a result

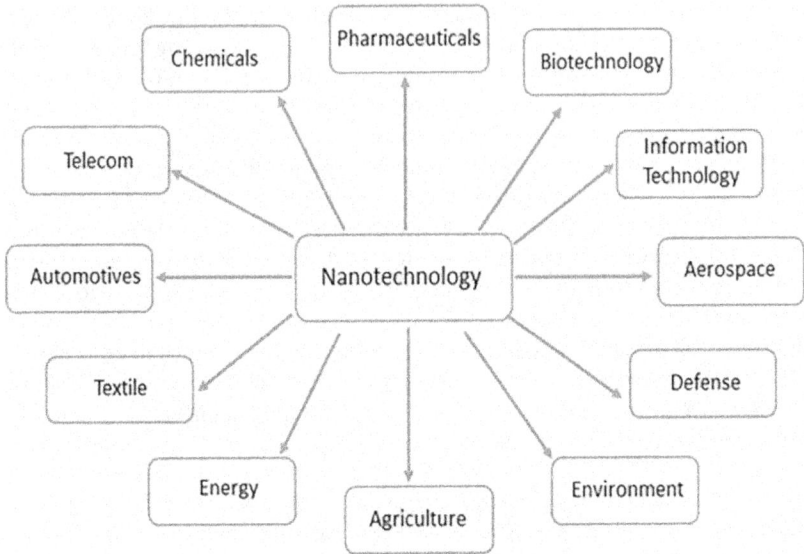

FIGURE 1.2   The business scope of NT in different fields. [11]

of growing international investment, NT is now the driving force behind upcoming technological developments, with businesses and academic institutions always looking for new uses [7].

In biotechnology, NT is reshaping drug delivery, diagnostics, and regenerative medicine. Nano-enabled drug carriers enhance treatment precision by targeting specific cells, reducing side effects, and improving therapeutic outcomes (see Figure 1.2) [8–11]. NPs are used in advanced imaging techniques and biosensors to improve disease detection and monitoring [11]. The pharmaceutical industry is increasingly adopting NT for the development of smart medicines, wearable health monitoring devices, and tissue engineering solutions [12].

NT significantly assists the telecom, information, and energy industries by facilitating shrinking, faster processing, and more energy efficiency. NMs such as graphene and CNTs are essential for the development of smaller, quicker, and more potent electrical components used in nanoelectronics, flexible displays, and QM [13, 14]. In the energy sector, NMs make better energy conversion and storage options possible, which improve the performance of solar cells, batteries, fuel cells, and supercapacitors. Nanostructured electrodes enhance the capacity and lifespan of lithium-ion batteries, while QDs and perovskite NMs have raised solar panel efficiency. These developments are drawing a lot of investment in green NT and accelerating the shift to sustainable energy sources [15].

The expansion of the consumer goods and textiles industries is further supported by involving sportswear, fashion, and military applications, all of which favor fabrics with nanocoatings since they give them antibacterial, water-repellent, and self-cleaning qualities [16, 17]. Furthermore, by improving performance and durability, NT has improved common consumer goods, including sunscreens, cosmetics, and household coatings [18]. NT is revolutionizing agriculture and the food industry by precision farming, nano-fertilizers, nano-pesticides, and smart sensors, enhancing crop yields and food safety [19–21]. In water and air purification, NT uses NMs to remove pollutants, promoting sustainability [22]. NT improves material strength, safety, and efficiency using CNTs and graphene in automotive, aerospace, and defense industries. It also advances security systems through nanosensors and nanorobotics [23].

Manufacturers of nanoscale instruments are expected to be the first major beneficiaries in the NT industry, with the US nanotech tools sector solely projected to grow by 30% annually. The industry's development is anticipated to follow a three-phase path: presently, NT is selectively applied in high-end sectors such as aerospace and automotive; by 2009, breakthroughs are expected to drive widespread use in electronics and IT; and after 2010, nanotech is predicted to become common in everyday products, particularly in healthcare and life sciences via nano-enabled drugs and devices. Despite its rapid growth, the field remains underrepresented in official statistics due to its novelty and interdisciplinary nature. Therefore, market estimates primarily come from private consultancies. Lux Research projects that by 2014, nanotech-based products will make up 15% of global manufacturing, generating $2.6 trillion in sales-comparable to the entire IT and telecom sectors. Similarly, SwissRe addresses that sales revenues from nanotech products have already reached substantial levels and are projected to grow exponentially.

1.1.2 Modern Industry Updated Systems

NT has significantly transformed the modern sector by incorporating smart manufacturing, cutting-edge materials, and AI-driven automation into several industries. Because of the rapid technological advancements, industries are employing NT to increase production processes' sustainability, longevity, efficiency, and precision. Upgraded systems in NT focus on intelligent monitoring systems, nanoscale manufacturing, and next-generation materials [24]. Nano-manufacturing systems are among the significant advances in contemporary NT since they enable an exact and large-scale manufacture of NMs. Molecular self-assembly, lithography, and atomic layer deposition (ALD) are some top-down and bottom-up techniques that have made it possible to create tailored NPs, nanocoatings, and ultra-thin films. These techniques are widely used in the semiconductor, medical, and defense industries, enabling the development of ultra-efficient chips, flexible electronics, and nanoscale sensors. 3D printing at the nanoscale or nano-printing is another emerging technology, allowing industries to manufacture customized nanostructures with extreme precision for biomedical implants, photonics, and aerospace components [25].

Advances in NT have produced high-speed transistors, flexible circuits, and QD displays in the electronics sector. Faster, more energy-efficient devices become possible by the use of 2D NMs such as graphene. The combination of neuromorphic computing and AI-powered nano-processors has fueled AI, robotics, and high-performance computing developments. Innovative nanomedicine powered by AI ensures accurate therapy with fewer adverse effects. While nanorobotics and imaging technologies facilitate non-invasive operations, regenerative medicines, and early illness detection, lab-on-a-chip devices provide quick diagnoses [26].

Modern NT systems assist the healthcare sector, especially in biosensing, nano-robotics, and targeted drug delivery. AI-driven drug delivery systems have emerged as an integral component of innovative nanomedicine, guaranteeing accurate drug administration at the cellular level while reducing adverse effects. Real-time illness detection and quick diagnostics are made possible by lab-on-a-chip (LOC) technology, which combines several laboratory operations onto a single nanoscale device. Furthermore, nanorobotics and nano-imaging technology have progressed in early cancer detection, regenerative medicine, and non-invasive surgery [27]. NT is transforming the energy sector with innovations such as hydrogen fuel cells, efficient perovskite cells, high-density batteries for devices and electric vehicles, and nano-engineered solar panels. Nano-photonics further enhances energy harvesting and light management in renewable energy sources. NT promotes environmental sustainability by enabling advanced water purification through nanofiltration and nanoadsorbents and improving air quality through photocatalytic materials. Therefore, NT is encouraging industrial processes that are more intelligent, environmentally sustainable, and productive.

1.1.3 Objectives of Assessment in Business

In the business sector, assessment is essential for boosting the effectiveness and efficiency of organizations. It primarily assists in assessing worker performance, highlighting both areas of strength and opportunities for development. This promotes focused training and development, raising morale among individuals and coordinating abilities with strategic objectives. Assessments use methods such as SWOT analysis and customer feedback to identify organizational strengths and shortcomings, and individual performance. These insights guide strategic planning, operational enhancements, and resource allocation decisions [28].

Developing exact requirements for performance to promote accountability is an essential objective. This promotes a culture where individuals own their responsibilities and work to fulfil established expectations. Regular assessments can improve performance management by identifying high achievers and taking proactive measures to resolve challenges [29]. Assessments give strategic planners the information they need to spot patterns, anticipate obstacles, and seize opportunities. Market and financial analyses assist companies in making wise financial decisions and maintaining their competitiveness. Assessments further encourage continual improvement by spotting inefficiencies and inspiring creativity, which is crucial for adjusting to a quickly changing environment. Finally, evaluations guarantee adherence to rules and industry norms, lowering legal risks and boosting stakeholder confidence. Therefore, business evaluations are essential for ongoing expansion, flexibility, and sustained success in a cutthroat industry.

1.2 Nanotechnology: Technological Revolutions

The manipulation of matter at the atomic and molecular level, typically within 1 and 100 nm, serves as NT a significant technological revolution. Since it has rendered previously inconceivable innovations possible, this branch of research and engineering has become a revolutionary force in several industries, including electronics, materials science, medicine, and energy. In 1959, Richard Feynman's lecture "There is Plenty of Room at the Bottom" introduced the idea of manipulating materials at the atomic and molecular level. He envisioned groundbreaking discoveries and emphasized the need for advanced, miniaturized instruments to control and study these nanoscale structures. This inspired a push for investment in nanoscale research due to its vast economic and societal potential, such as advancements in electronics, materials, defense, environment, and biotechnology. John Armstrong later predicted that NT would revolutionize the information age. By 1999, researchers agreed that NT would transform nearly all manufactured objects, sparking an industrial revolution. While the US leads in nanochemistry and biology, it lags in nanodevices and structural materials. Other nations, such as Japan and European countries, are building strong nanotech programs and specialized research centers [30].

While many products have benefited from nanoscale effects in recent years, the exact beginning of NT is hard to pinpoint. Because of some long-existing products, the hard drives have used the giant magnetoresistance (GMR) effect, which meets current nanotech definitions but was not labeled initially. Therefore, the line between "nanotechnology" and other technologies is often unclear. Beyond high-tech fields, nanobiotechnology is expected to transform life sciences, as biology naturally builds at the nanoscale through self-assembly. This efficient method could support sustainability, although its full potential remains unrealized now [31]. Despite its numerous benefits, the rapid advancement of NT also raises ethical and safety concerns. The potential risks of NMs, including toxicity and environmental impact, necessitate comprehensive research and regulation to ensure safe usage. As NT continues to evolve, it is crucial for policymakers, scientists, and industry leaders to collaborate in addressing these challenges while maximizing the benefits of this revolutionary technology [32]. In conclusion, NT stands as a hallmark of technological revolution, driving innovation across multiple sectors and reshaping the future of industries. Its applications in medicine, electronics, materials science, and energy highlight its transformative potential, offering solutions to some of the world's most pressing challenges. As research and development in this field continue to progress, the possibilities for NT are virtually limitless, promising a future where enhanced performance, efficiency, and sustainability are the norm.

1.2.1 Core Principles and Concepts

NT is a multidisciplinary field centered on manipulating matter at the nanoscale, typically defined as dimensions ranging from 1 to 100 nm. At this scale, materials exhibit unique physical, chemical, and biological properties that differ significantly from their bulk counterparts. One of the fundamental principles of NT is the significance of the nanoscale itself, where quantum mechanical effects become pronounced. Quantum mechanics plays a crucial role in determining the behavior of nanoscale materials, leading to phenomena such as quantum tunneling and size-dependent optical properties. This is exemplified by QDs, which can emit different light colors based on their size, enabling applications in advanced imaging and display technologies. Another key concept is the high surface area-to-volume ratio characteristic of NMs. As materials are reduced to the nanoscale, their surface area increases dramatically relative to their volume, enhancing reactivity and environmental interaction. This property is particularly advantageous in applications such as catalysis and drug delivery. NPs can act as highly effective catalysts or targeted drug carriers, improving the efficiency of chemical reactions and therapeutic outcomes.

Self-assembly is another fundamental principle in NT, referring to the spontaneous organization of molecules into structured arrangements without external direction. This process allows for the creation of complex nanostructures, such as nanospheres and nanowires, enabling precise control over material architecture and properties. Researchers utilize self-assembly techniques to develop innovative materials for drug delivery systems and nanoscale electronics [33]. NT also encompasses two primary fabrication approaches: bottom-up and top-down methods. Bottom-up approaches involve constructing NMs from atomic or molecular building blocks through chemical vapor deposition (CVD) and sol–gel synthesis processes. These techniques allow for precise control over material composition and structure. In contrast, top-down methods involve breaking down bulk materials into nanoscale components using techniques such as lithography and etching. While top-down approaches are often more suitable for mass production, bottom-up techniques enable the creation of more complex nanostructures.

The unique properties of NMs, such as increased strength, enhanced electrical conductivity, and altered optical characteristics, further highlight the potential of NT. CNTs exhibit exceptional tensile strength and conductivity, making them valuable in various applications, including electronics and composite materials. NT is inherently interdisciplinary, drawing knowledge and techniques from fields such as physics, chemistry, biology, and engineering. This collaborative nature fosters innovation and the development of novel solutions to complex challenges, from medical advancements to environmental sustainability. However, the rapid advancement of NT also raises important ethical and societal considerations, including potential health risks and environmental impacts. Addressing these concerns through responsible research and development practices is essential to ensure that the benefits of NT are realized while minimizing risks [35]. Therefore, the core principles and concepts of NT, including nanoscale definition, quantum mechanics, surface area-to-volume ratio, self-assembly, fabrication approaches, unique properties, interdisciplinary nature, and ethical considerations, are fundamental to understanding its transformative potential across various industries and applications.

1.2.2 Key Materials and Processes

NT encompasses a wide array of key materials and processes essential for developing and applying nanoscale structures across various fields. At the forefront of NT are NPs, which are typically defined as particles with dimensions of approximately 100 nm. These particles can be composed of metals, semiconductors, polymers, or carbon-based materials and exhibit unique properties that differ significantly from their bulk counterparts. Metal NPs, such as gold and silver, are known for their exceptional optical properties, making them invaluable in different applications such as medical imaging, drug delivery, and biosensing (see Figure 1.3a). Semiconductor NPs, commonly referred to as QDs, possess size-tunable electronic and optical properties, which enable their use in display technologies and photovoltaic cells (see Figure 1.3a). Polymer NPs serve as effective drug carriers, enhancing the solubility and bioavailability of therapeutic agents (see Figure 1.3a). Another category of materials

(a)

(b)

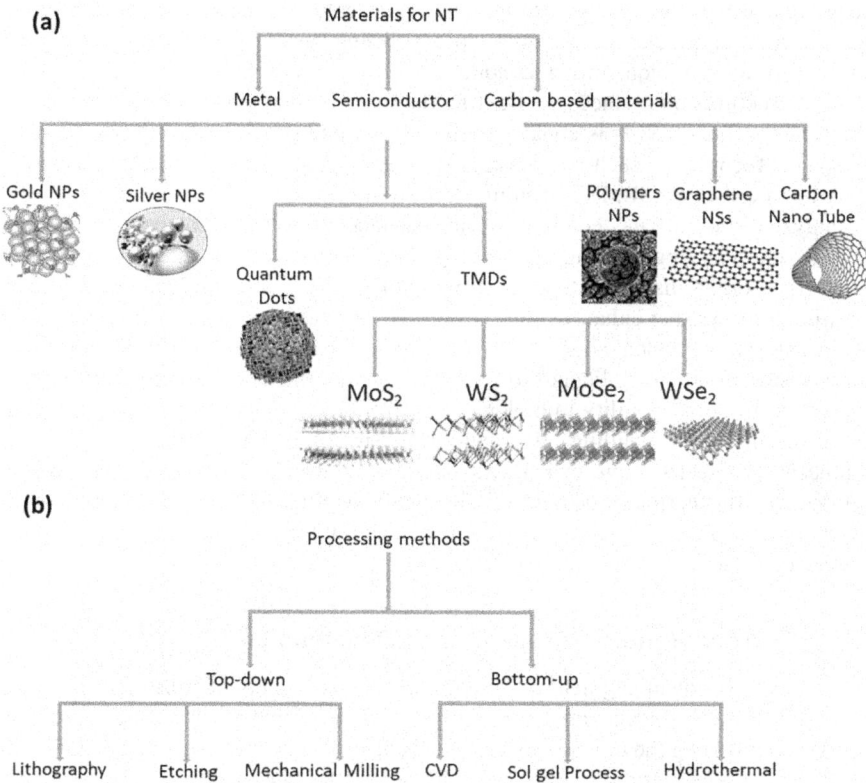

FIGURE 1.3 (a) Classification of materials used in NT, (b) NMs different processing techniques (bottom up and top down). [34]

in NT is nanocomposites, which combine nanoscale fillers with a matrix material to improve mechanical, thermal, and electrical properties. The incorporation of CNTs or graphene into polymers can significantly enhance their strength and conductivity, making these nanocomposites suitable for applications in aerospace, automotive, and electronics industries where high-performance materials are crucial [34].

Nanosheets (NSs) of Gr, CNT, and transition metal dichalcogenides (TMDs) (molybdenum disulfide (MoS_2), molybdenum diselenide ($MoSe_2$), tungsten disulfide (WS_2), and tungsten diselenide (WSe_2)) are 2D materials that have gained significant attention due to their remarkable properties. Graphene is renowned for its outstanding electrical, thermal, and mechanical characteristics, positioning it as a potential candidate for applications in flexible electronics, energy storage, and sensors [36]. The synthesis of NMs can be achieved through various methods, broadly classified into bottom-up and top-down approaches (see Figure 1.3b). Bottom-up approaches involve assembling nanostructures from molecular or atomic components through techniques of CVD, sol–gel processes, and hydrothermal synthesis. CVD is widely used to produce high-quality graphene by depositing carbon atoms onto a substrate at elevated temperatures, allowing for controlled material growth. While top-down approaches involve breaking down bulk materials into nanoscale components using techniques such as lithography, etching, and mechanical milling (see Figure 1.3b). Photolithography, a common technique in semiconductor fabrication, enables the precise patterning of nanostructures on surfaces, facilitating the mass production of nanoscale devices. Characterization of NMs is equally essential for understanding their properties and behavior. Techniques such as transmission electron microscopy (TEM), scanning electron microscopy (SEM), and atomic force microscopy (AFM) are commonly used to analyze the size, shape, composition, and surface properties of nanostructures [37].

TEM provides high-resolution imaging that reveals detailed internal structures, while SEM offers insights into surface morphology. The functionalization of NMs is a critical process that enhances their properties or enables specific interactions with other materials or biological systems. This can be achieved through chemical modifications, such as attaching ligands to NPs for targeted drug delivery or altering surface charges to improve stability in biological environments [38]. The key materials and processes in NT – including NPs, nanocomposites, and NSs – along with various synthesis, characterization, and functionalization methods, for instance, the sol–gel synthesis, hydrothermal processing, TEM/SEM imaging, XRD analysis, and surface functionalization, collectively provide the foundation for developing advanced nanostructured materials with tailored properties for diverse applications.

1.2.3 Current Trends in Research and Development

NT is a rapidly advancing field that continues to push the boundaries of science and engineering, with current trends in research and development reflecting a growing focus on harnessing the unique properties of NMs to tackle pressing challenges across various industries. One such area of NT research is nanomedicine, which emphasizes the use of nanoscale materials for medical applications, including targeted drug delivery, imaging, and diagnostics. Researchers are developing NPs that can encapsulate drugs and deliver them directly to specific cells, such as cancer cells, thereby enhancing therapeutic efficacy while minimizing side effects. This targeted approach often utilizes functionalized NPs that recognize specific biomarkers on the surface of diseased cells. Advancements in imaging techniques, including the use of QDs and magnetic NPs, are improving the sensitivity and specificity of diagnostic tools, allowing for earlier disease detection and more personalized treatment strategies. Another significant trend is the emphasis on sustainable NMs, which is driven by the need for environmentally friendly solutions. Researchers are exploring using renewable resources and green synthesis methods to create NMs with reduced environmental impact. Bio-inspired approaches that synthesize NPs using plant extracts or microbial processes are gaining traction, providing a sustainable alternative to traditional chemical synthesis. Furthermore, sustainable NMs are being developed for energy storage and conversion applications, such as batteries and supercapacitors, where they can enhance performance and efficiency while minimizing resource depletion. In electronics, the miniaturization of electronic components remains a significant trend, particularly in developing nanoscale transistors and components for next-generation electronic devices. As traditional silicon-based technology approaches its physical limits, researchers are investigating new materials, such as graphene and TMDs, which offer superior electrical conductivity, flexibility, and thermal management.

These materials pave the way for innovations in flexible electronics, wearable, and high-performance computing, as well as the integration of NT in QM, where nanoscale quantum bits (qubits) hold promise for faster and more powerful computation. Besides, NT is being increasingly applied to environmental challenges, including water purification, pollution remediation, and sustainable energy production. NMs are being developed for advanced filtration systems that efficiently remove contaminants from water, such as heavy metals, pathogens, and organic pollutants, by enhancing the adsorption capacity of filters. In pollution remediation, nanoscale zero-valent iron is being utilized to treat contaminated soil and groundwater, offering cost-effective and efficient solutions for environmental cleanup. NT is contributing to the development of next-generation solar cells, where NMs enhance light absorption and conversion efficiency, advancing renewable energy technologies [39, 40].

Therefore, the current trends in NT research and development highlight a multidisciplinary approach that integrates advancements in materials science, biology, and engineering to create innovative solutions addressing health, environmental, and technological challenges. As research in this field continues to evolve, it is poised to play a crucial role in shaping the future of various industries and improving the quality of life globally.

1.3 Market Landscape and Industry Applications

NT offers significant advantages by engineering materials at the nanoscale to achieve specific properties of increased strength, lightness, durability, conductivity, and reactivity. These enhancements have led to various real-world applications across various industries. It revolutionizes healthcare and biotechnology by enabling targeted drug delivery, early disease detection, and advanced imaging (see Figure 1.4) [41]. It improves vaccine development, tissue engineering, and antibacterial therapies. Meanwhile, nanosensors and carriers improve genetic engineering and diagnostics. Such developments are revolutionizing illness prevention, diagnosis, and individualized treatment plans by providing accurate, effective, and less intrusive treatments (see Figure 1.4) [42]. NT renders smaller, quicker, and more efficient devices, which propel developments in electronics, energy, and environmental solutions. For sustainable energy, it improves solar cells, batteries, and supercapacitors. By improving performance and using fewer resources, NMs help with waste treatment, pollution prevention, and water purification, providing environmentally responsible alternatives to global problems (see Figure 1.4) [43].

It enhances manufacturing and material science by creating stronger, lighter, and more durable materials. It improves surface properties such as conductivity, corrosion resistance, and self-cleaning abilities. Nanoscale innovations lead to advanced composites, innovative materials, and precise fabrication techniques, revolutionizing product performance across various industrial applications. In textiles, nanoscale treatments can add features such as stain and wrinkle resistance, antibacterial properties, and even ballistic protection. Nanoscale coatings improve the surfaces of eyeglasses and screens by making them water-repellent, self-cleaning, and scratch-resistant. Bright fabrics with embedded nanosensors are emerging for health monitoring and energy harvesting (see Figure 1.4) [44]. In energy and sustainability,

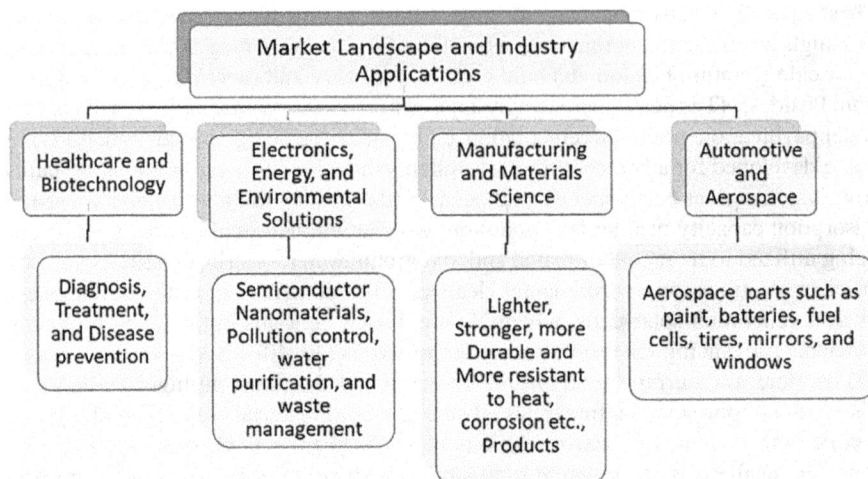

FIGURE 1.4　Market landscape and industrial applications of NT in different fields. [44]

nano-bioengineered enzymes help produce ethanol from plant waste, and cellulosic NMs are proving useful in construction and healthcare sectors. In transportation, NMs reduce weight and enhance performance in vehicles and sports equipment, while CNT offers thermal and electromagnetic management solutions in aerospace. Innovation in the automotive and aerospace industries boosts performance, safety, and efficiency. Nano-enhanced materials can increase strength and durability (see Figure 1.4) [45]. Nanocoatings improve resistance to corrosion, heat, and wear. In engines and fuel systems, nanotech enhances combustion and fuel efficiency. Besides, nanosensors enable real-time monitoring, contributing to smarter, safer, and more sustainable transportation systems. Industrial applications benefit from tougher coatings and longer-lasting lubricants, while NPs boost chemical reactions in petroleum refining processes. Household and personal care products also benefit from improved cleaners, coatings, sunscreens, and more. Overall, NT is revolutionizing materials science and delivering practical benefits in daily life.

1.3.1　Medical and Healthcare Applications

NT is significantly enhancing medicine via a specialized branch known as nanomedicine, which leverages nanoscale tools and materials to improve disease prevention, diagnosis, and treatment. By working at the same scale as many biological processes, nanomedicine offers exact and practical solutions. One promising application involves gold NPs, which are already being used commercially to detect specific sequences of nucleic acids and are being studied for their potential to treat cancer and other diseases. NT is also advancing diagnostic and imaging tools, allowing for earlier detection of illnesses and more personalized treatment plans, improving patient outcomes [46, 47]. In cardiovascular medicine, NT is being explored to diagnose and treat atherosclerosis. Scientists have developed NPs that mimic HDL, or "good" cholesterol, to help shrink arterial plaque. Similarly, advances in solid-state nanopore materials are offering next-generation gene sequencing technologies, which promise rapid, low-cost, and highly sensitive detection of single molecules with minimal sample preparation [48].

Cancer treatment is another key focus, with researchers designing NPs that can deliver medication directly to tumor cells while sparing healthy tissue, potentially reducing the harmful side effects of traditional chemotherapy. In regenerative medicine, NT is being applied to engineering bone and neural tissues. Scientists are developing materials that replicate the mineral structure of bone and can be used in dental restorations. Other research includes growing complex tissues for future organ transplants and using graphene nanoribbons to support spinal cord repair, as neurons have been shown to grow well on graphene surfaces [49–51]. Nanomedicine also plays a role in vaccine development. Researchers are exploring needle-free delivery methods and working toward creating a universal flu vaccine platform that would target multiple virus strains and reduce the resources needed for annual production [52]. Thus, NT is rapidly transforming the medical landscape, offering novel approaches to diagnosis, therapy, tissue regeneration, and vaccination. It holds the potential to revolutionize patient care in the years ahead.

1.3.2 Electronics and IT Applications

NT has significantly advanced the fields of computing and electronics, driving innovations that result in faster, smaller, more efficient, and portable systems capable of managing and storing vast amounts of data. One notable contribution of NT is the miniaturization of transistors. In the early 2000s, typical transistors measured between 130 and 250 nm. With continued innovation, Intel introduced a 14-nanometer transistor in 2014, followed by IBM's 7-nanometer version in 2015, and Lawrence Berkeley National Laboratory's breakthrough of a 1 nm transistor in 2016. These advancements are enabling devices to become increasingly compact and powerful, potentially allowing entire computer memories to be stored on a single chip [53]. Another transformative innovation is the development of magnetic random-access memory (MRAM), which is enabled by nanoscale magnetic tunnel junctions. MRAM allows for near-instant system boot-up and quickly saves data during a shutdown, supporting seamless resume–play features. This marks a significant shift in the speed and efficiency of data management [54].

Display technology has also significantly benefited from NT. QDs are now used in ultra-high-definition screens to enhance color quality and energy efficiency. In parallel, the rise of flexible and stretchable electronics, made possible by NMs such as graphene and cellulose, is expanding the application of electronics into new areas. These flexible electronics are being integrated into products such as wearable devices, medical tools, aerospace components, and the Internet of Things. Recently, semiconductor nanomembranes have been used in the displays of smartphones and e-readers, while wearable sensors and rollable electronic paper are becoming increasingly viable [55].

NT also plays a crucial role in producing various everyday electronic devices. Flash memory chips used in smartphones and thumb drives, ultra-responsive hearing aids, and antimicrobial coatings on keyboards and phone casings all utilize nanotech. Additionally, conductive inks containing NPs are being applied to create printed electronics for smart cards, RFID systems, and flexible displays. NP copper suspensions have emerged as a safer and more reliable alternative to traditional lead-based solder in manufacturing. These innovations reduce environmental hazards and improve the durability and performance of electronic assemblies. Therefore, NT continues to drive the evolution of electronics, making devices more intelligent, more adaptable, and more efficient across various industries [56]. Assessing the business potential requires analyzing cost–benefit ratios, scalability, environmental impact, and regulatory compliance. These advancements position NT as a critical enabler of next-generation industrial and environmental technologies.

1.3.3 Energy Applications and Environmental Remediation

NT plays a key role in improving traditional energy sources and advancing alternative energy solutions to meet growing global energy demands. It enhances fuel production efficiency through better catalysis, enables reduced fuel consumption in vehicles and power plants, and improves combustion efficiency. It is also being applied in oil and gas extraction, such as NT-enabled gas lift valves and NPs for detecting pipeline fractures. CNTs are being researched for use in "scrubbers" and membranes to capture carbon dioxide from power plant emissions [57]. In energy transmission, CNT wires are being developed to reduce resistance and minimize power loss in the electric grid. NT also enhances solar energy, with nanostructured solar cells offering higher efficiency and lower manufacturing costs. These cells could even be made in flexible, printable forms, potentially leading to "paintable" solar converters. Furthermore, new batteries being developed with NT promise faster charging, increased efficiency, lighter weight, and longer-lasting power [58].

Wind energy is benefiting from nanotech with the use of CNT epoxy in windmill blades, making them stronger, lighter, and more efficient at generating electricity. NT is also advancing energy harvesting through innovations in thin-film solar panels for electronic devices and piezoelectric nanowires in clothing to generate energy from light, friction, and body heat. Furthermore, research is ongoing to convert waste heat from various sources, involving computers and power plants, into usable electrical power [59]. NT is also driving energy efficiency across numerous sectors. It enables the development of more efficient lighting systems, lighter and stronger materials for vehicles, reduced energy consumption in electronics, and innovative coatings for glass that respond to light. These advances are contributing to cleaner, more efficient energy systems and helping to reduce environmental impact.

1.3.4 Future Transportation Benefits

NT is poised to revolutionize transportation by developing multifunctional materials for vehicles, aircraft, spacecraft, and ships that are lighter, safer, smarter, and more efficient. In automotive applications, nano-engineered materials improve parts of polymer nanocomposites, high-power batteries, thermoelectric materials, tires, sensors, and catalytic converters, contributing to cleaner exhaust and longer ranges. For infrastructure, NT enhances materials such as aluminum, steel, asphalt, and concrete, improving their durability and reducing life cycle costs. It also enables innovations such as self-repairing structures and energy-generating or transmitting components [60].

NT also offers continuous, cost-effective monitoring of transportation infrastructure. Nanoscale sensors can assess the integrity of bridges, tunnels, and roads, while nanoelectronics support smarter infrastructure systems that help drivers with tasks of lane positioning, collision avoidance, and route adjustments. Lightweight, high-strength NMs could also significantly reduce the weight of transportation vehicles. Reducing the weight of a commercial jet by 20% could lower fuel consumption by 15%, and advanced NMs could decrease the weight of launch vehicles by 63%, leading to energy savings and reduced launch costs. These advancements could revolutionize transportation, improving energy efficiency, safety, and mission success rates in aerospace [61].

1.4 Economic and Business Potential

The global NT market was valued at USD 3.69 billion in 2022 and is expected to grow rapidly, with a CAGR of 33.1% between 2023 and 2030 (see Figure 1.5). NT develops devices and systems at the nanoscale and has several applications in domains of chemistry, medicinal research, materials science, and mechanics. Its ability to create innovative materials with distinct properties has the potential to transform several sectors. Growing use in medical imaging and diagnostics, and ongoing technological developments, are significant factors propelling the market's expansion. Strong R&D spending and government backing are also essential. NT is employed in many industries, including manufacturing, transportation, agriculture, and pharmaceuticals, and it enables the creation of sophisticated, high-performing products at reduced costs. The market is expanding due to its expanding applications in electronics, healthcare, aerospace, and textiles. In addition, it supports the development of smaller and efficient transistors, memory devices, and sensors, contributing to innovations in electronics and computing.

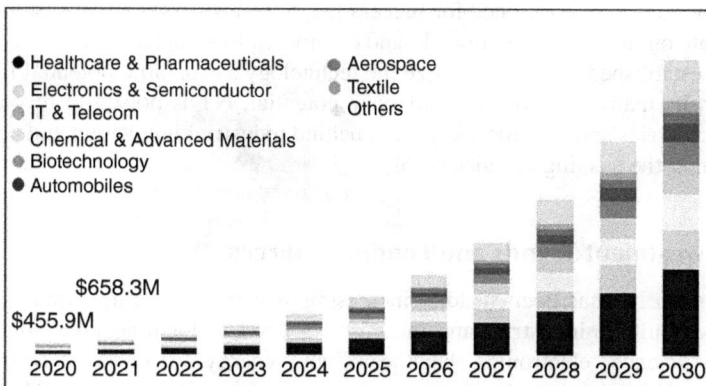

FIGURE 1.5 Asia Pacific nanotechnology market for different segments within the tenure of 2020–2030.
Reproduced with permission © GVR [62]

1.4.1 Market Size and Growth Projections

The manipulation of matter at the atomic and molecular level has emerged with profound implications across multiple industries, including electronics, healthcare, energy, the automobile industry, and agriculture [61, 63, 64]. With an expected market size of over $85 billion in 2023 and a projected CAGR of 14.5% from 2024 to 2030, the worldwide NT industry has been expanding at an exponential rate [65–67]. The growing use of NMs, nanoelectronics, and nanosensors is improving product efficiency and opening up new business opportunities, allowing quick expansion [68, 69]. Regionally, North America remains the dominant market due to substantial research and development (R&D) investments [70, 71], robust industrial adoption, and a favorable regulatory framework [72].

Europe is right behind, with growing government spending and strategic partnerships centered on advanced NMs and nanomedicine [73]. But due to industrial centers in China, Japan, and South Korea, as well as the expanding use of NT in consumer electronics and healthcare, the Asia-Pacific region is anticipated to have the quickest rate of market growth [74, 75]. As companies investigate uses of NT, emerging markets in the Middle East and Latin America are also gaining traction [76, 77]. Multiple factors are driving the growing business potential of NT, such as ongoing progress in material science, the escalating demand for smaller and more efficient electronic devices, expanded investments from both government and private sectors, and the increasing emphasis on eco-friendly and sustainable technologies [78, 79]. However, challenges such as high initial R&D costs, regulatory and ethical concerns, lack of industry-wide standardization, and limited awareness among businesses pose potential barriers to commercialization and widespread adoption [80].

Despite these obstacles, NT has a bright future, with projections of over $200 billion by 2030 [81]. New developments such as bio-NT, AI-driven NT, and nanomanufacturing are anticipated to spur further expansion and open up lucrative economic prospects [82]. Strategic alliances among academic institutions, governmental organizations, and private businesses will be essential to promoting research and commercialization, and startups and R&D projects concentrating on cutting-edge NT applications are well-positioned for success [83, 84]. Businesses and investors should concentrate on innovation, teamwork, and commercialization tactics as NT continues to upend established sectors to realize the technology's economic potential fully [85, 86]. With its many uses and revolutionary potential, NT is poised to revolutionize several industries and become a key force behind industrial innovation and economic expansion in the ensuing decades [6, 61].

1.4.2 Investment Trends and Funding Sources

Investment in NT has been steadily increasing, driven by its transformative potential across multiple industries and the growing demand for innovative solutions in materials science, electronics, healthcare, and energy [6]. Numerous financing sources, including government grants, private venture capital, business R&D investments, and academic research institutes, have contributed significantly to the NT sector [87]. Through national NT initiatives of the US NNI, the European Union's Horizon funding programs, and China's aggressive investments in nanoscience and nano-manufacturing, governments around the world are acknowledging the strategic significance of NT and are offering significant financial support [88–91]. These initiatives seek to improve commercialization prospects, expedite research, and fortify technological leadership in NT [6, 92]. Along with government assistance, venture capital firms and private equity investors are increasingly funding breakthrough research initiatives and enterprises in NT [93].

According to investment trends, there is a special emphasis on high-growth fields such as advanced NMs, nanomedicine, and nanoelectronics, where innovations might upend established sectors and yield substantial profits [18]. In an effort to incorporate nanoscale advances into their products and production processes, large multinational firms are also making significant investments in NT research, either through internal R&D or strategic acquisitions of promising startups [94, 95]. To decrease the gap between laboratory discoveries and commercial applications, academic and research organizations frequently work with industrial partners to make fundamental improvements in NT. Recently, a sharp increase in corporate expenditures in NT has been accounted, with leaders in the energy industry, pharmaceutical industries, and large technology businesses all dedicating considerable resources to nanotech R&D [88]. To accelerate discoveries and reduce time-to-market for novel NT applications, businesses are increasingly implementing open innovation models, in which they collaborate with startups, academic institutions, and other industry participants [96]. Furthermore, government financing initiatives frequently offer grants and incentives to promote private sector investments, which makes NT a desirable field for both existing businesses and early-stage entrepreneurs [6]. The increasing number of successful nanotech-based product launches, which show great market potential and revenue generation capabilities, is another factor driving private sector investment.

However, despite the optimistic investment environment, funding hurdles still exist due to the high expenses of NT research and the long timelines needed for commercialization [79]. Many NT ventures require substantial capital infusion to transition from research to mass production, allowing access to consistent funding, a crucial factor for success. A variety of finance strategies, such as impact investing, technology transfer programs, and public–private partnerships, are being investigated to overcome these issues and provide a long-lasting financial ecosystem for advances in NT [97]. As governments and private investors continue to recognize the vast economic and business potential of NT, funding opportunities are expected to expand, enabling further breakthroughs and accelerating the adoption of nanotech-based solutions across various industries. NT has the potential to become a significant technical field, propelling future industrial change and economic growth, provided it receives consistent investment and practical financing efforts [79, 98]. Increased cooperation among investors, academic institutions, and commercial businesses is anticipated as the sector develops, which will improve the scalability and commercialization of NT applications in international marketplaces [79, 92].

1.4.3 Commercialization Challenges and Opportunities

NT's commercialization presents significant challenges and immense opportunities, shaping its economic and business potential across multiple industries. The main obstacle in bringing NT products to market is the high cost of R&D, as NMs and nanofabrication processes require specialized equipment, highly skilled personnel, and extensive testing to ensure efficacy and safety [3, 79]. Another obstacle is regulatory barriers, which can cause delays in product clearance and market access since various nations have varied safety regulations and standards governing the use of NMs [99, 100]. Moreover, scaling up production from laboratory research to mass manufacturing remains a key challenge, as NMs behave differently at different scales, requiring new manufacturing techniques and infrastructure investment [56, 101]. However, despite these challenges, the opportunities for commercialization are vast. The increasing demand for NT-driven solutions in medicine, electronics, energy, and environmental applications provides a strong market pull for nanotech innovations. The commercialization gap can be closed by strategic alliances among government, business, and academia that offer financial resources, regulatory assistance, and technological know-how. Businesses that effectively use NM in their goods stand to benefit greatly from increased sustainability, efficiency, and performance [6]. Therefore, NT has the potential to transform a number of industries and open up profitable business possibilities for progressive companies and investors as improvements in nano-manufacturing technologies increase scalability and cost-effectiveness [102, 103].

1.5 Regulatory, Ethical Considerations, and Competitive Landscape

1.5.1 Global Regulatory Frameworks

A global regulatory framework is necessary as manipulation and use of materials at the nanoscale opens up several avenues for scientific progress and creativity, but poses issues because of the possible influence on safety, health, and the environment [104]. The key regulatory issues linked to NT include risk assessment and NMs toxicity, labelling of nano-products and protection of consumers, environmental impact assessment, worldwide harmonization, intellectual property and protection of patents, and ethical and societal issues [105, 106]. NMs exhibited distinct features compared to bulk versions, posing significant challenges due to which regulators encounter difficulties in determining the toxicity and possible dangers connected with these compounds [107]. It is critical to develop NM-specific standardized testing procedures and risk assessment frameworks. It is difficult to define and characterize ultrafine particles (UPS). Therefore, regulatory bodies often struggle to create precise definitions and measuring methodologies for identifying and distinguishing NPs [104].

The length, width, and surface features of NPs may have a substantial impact on their behavior and possible dangers. The proper labeling of NM-containing goods is critical for customer knowledge and choice. Regulators must decide if NM-infused items need particular labeling requirements as they might possess unanticipated environmental implications, having the potential to accumulate in ecosystems, affecting animals [108]. Furthermore, releasing NMs during the manufacture, usage, or disposal of nanoproducts might be problematic [109–111]. Environmental hazards must be addressed via regulatory frameworks and adequate monitoring and mitigation techniques. Efforts to create uniform standards and norms across nations via international harmonization may help to promote responsible NMs research and commercialization while promising global security and regulatory consistency

[112]. NT poses intellectual property issues owing to its transdisciplinary nature. Patent eligibility, originality, and non-obviousness requirements for NT-related innovations should all be addressed to foster innovation while preserving intellectual property rights [113, 114].

Ethical implications, such as privacy problems, possible abuse, and fair access, should be examined. Authorities must ensure that social principles and concerns are considered when making regulatory decisions [115]. Global regulatory bodies are actively tackling these concerns via cooperation, research projects, and the establishment of NT-specific rules and regulations. The objective is to promote the safe and responsible research, manufacturing, and use of NPs and NT-enabled products. These regulatory efforts are reflected in various visual representations that highlight the evolution and structure of global NT governance. Figure 1.6a delineates some major areas of NT regulation globally and global regulatory views. It emphasizes the importance of coordinated research, evaluation of safety, and standardization to counter the developing challenges as well as to enable safe development and utilization of NT across the world. Figure 1.6(b) documents the major milestones of NT development and regulation, from Feynman's lecture of 1959 to the current times. These consist of the naming of the term, significant inventions, institutional development, initial regulatory debate, and academic writings, revealing the gradual development of regulatory consciousness parallel to the development of NT. The pyramid shows the hierarchical structure of the regulatory framework, beginning with enabling legislation (acts and regulations) through to enforceable requirements such as licensing and regulatory documents. At the bottom, it encompasses guidance documentation such as staff review guides and info documents, providing clarity and advice for compliance within the formal system of law and regulation (see Figure 1.6c).

The European Union (EU) and the US, the world's major economic players, have taken quite different approaches in assessing the hazards of NT, and when and how to adopt legislation. The EU established an official definition of NMs that is integrated into various EU-wide legislation. In contrast, the US has not enacted federal NT law [116, 117]. NMs are strictly controlled in the EU; vertical regulations for NPs apply to individual items (chemicals and cosmetics), while horizontal regulations apply to specific substances, as it initially included NT-specific laws in the cosmetics regulation. In 2011, the European Commission approved an official description of NT, and nano-specific rules have been added to laws on plastic materials and goods that come into contact with food. The EU added nano-specific measures to the classification and labeling (CLP) regulation on hazardous chemicals and mixtures and the Biocidal Products Regulation to address overall issues. The US regulatory framework is market oriented, with the government serving as a facilitator and organizer of NT regulations [118]. The US weakly depends on nano-specific law and more on a broad set of state and federal laws. Congress has not taken any steps to update or analyze NPs-related legislation, leading to increased regulatory action via national and subnational agencies. In 2001, the US launched the first organized NT endeavor via the National NT Initiative (NNI), for which the President's fiscal 2015 budget includes more than $1.5 billion and a total of over $21 billion since then.

Despite the appearance of cooperation, the NNI is an umbrella organization with no budget or policy-making authority, with the private sector providing far more support for NT R&D [120]. In Japan, regulatory science attempts to obtain the optimal result for human wellness and society, as determined by suitable healthcare policies,

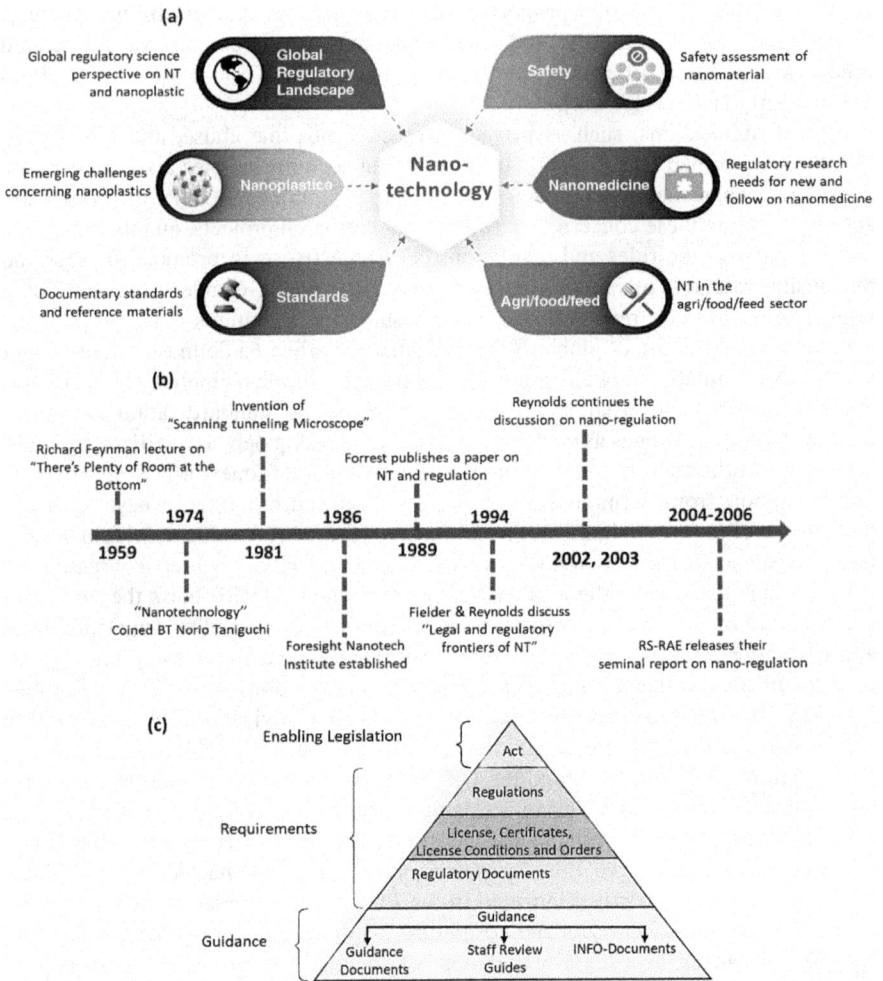

FIGURE 1.6　(a) Global regulatory frameworks. (b) Timeline of the nano-regulation debate. (c) Key elements of regulatory frameworks.
Reproduced with permission from © Elsevier [104] and © Research Gate [119]

with the goal of promptly translating research findings into practical implementations [121]. Japanese Ministry of Health, Labor and Welfare (MHLW), the Pharmaceutical Company and Medical Devices Agency (PMDA), the National Institute of Health-care Sciences (NIHS), and the Japanese Agency for Medical Research and Development (AMED) are responsible for regulatory science in this area [122–124]. Within this PMDA, a Regulatory Research Centre was formed in 2018 to promote creative approaches in advancing cures and technologies, including NT, emphasizing horizon scanning to help regulators keep up with emerging advances. Clinical data and electronic healthcare record information are being used to detect adverse occurrences and help product development, reviewing in large populations [104]. The Act on the Development of Healthcare Industries and Healthcare Technologies has increased regulatory research in science on cutting-edge pharmaceuticals [125]. In 2019, the Indian government released guidelines for the assessment of nanopharmaceuticals, covering the scientific justification to create such new drugs and their comparison with current medications to be achieved via both in vitro and in vivo research, with enhanced security, efficiency, and reduction in toxicology. Both the EU and the US have made significant efforts to collect and disseminate information regarding the scientific uncertainty, risks, and dangers of NMs in fast-globalizing markets. The EU and the US communicate with scientists, international networks, and private-sector organizations regularly when developing laws and regulations for NM-containing products. Despite their shared commitment to employing the greatest scientific data available, the EU and the US have chosen different methods to address these concerns [126].

1.5.2 Environmental and Health Concerns

Accelerating the development and use of NT provides significant economic and social advantages, as many NMs or nano-related items are currently in use in society; however, their influence on the environment, health, and security (EHS) is not well known or characterized. Interactions between NMs (derived from NT developments) and the human body, as well as the ecological system, have raised significant concerns [118]. The size of the NPs determines the amount that may be inhaled; particles with diameters below 100 nm are nanoscale, while particles smaller than 5 µm are classified as being part of the inhaled fraction for all types. Consequently, every aerosolized NP is considered to be a portion of the respirable fraction. Due to their volume and compact size, NPs have the potential to take off rapidly [127]. Since these free NPs may exist in large quantities, they are the subject of most contemporary research. The release of NPs into the surroundings, however, can also occur via friction on objects coated with or made of NPs via the product substrate, which is concerning. The absorption of NPs is seen as one such exposure mechanism. Although inhalation is the primary exposure route of concern, several situations and applications raise the risk of absorption and offer a potential entry point into the body [128]. The special chemical and physical characteristics of atoms at the nanoscale level, together with exposure to sunlight, are thought to boost the reactivity of NPs, enabling deeper penetration.

Diffusion is expected to allow the particles to enter the circulation after reaching the dermal layer, raising the same concerns about translocation as inhalation exposure [128]. Although it is not considered the main exposure route, intake of NPs is seen as a possible one since it can happen when other chemicals, often used in the workplace, are transferred from hand to mouth. Ingestion occurs when NPs are inhaled and then consumed by the mucociliary elevator's respiratory clearance functions [130]. A common occupational NM-handling environment is depicted in Figure 1.7, where workers could be exposed to engineered NMs (ENMs) through release, emission, and transportation. Even though application scenarios vary too much, exposure evaluation and management are essential in reducing effects on the production environment [129]. Proper risk management techniques, such as designing control systems for workplaces or providing safety gear for people, are crucial. It would also be instructive if future efforts were focused on assessing the effects of usage and application procedures,

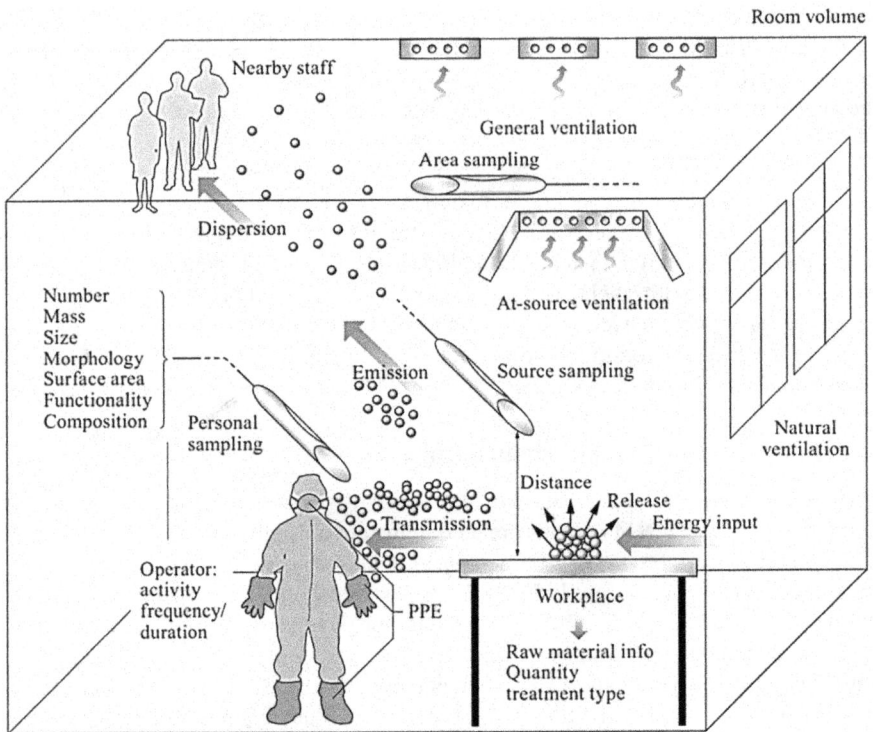

FIGURE 1.7 Familiar NM-handling environment.
Reproduced with permission from © Springer [129]

and the ultimate disposal of unrelated products. NT relies heavily on the NM manufacturing process, as the fact that mass manufacture of NMs typically involves pure and uncombined NMs strongly correlates with a large release potential. Workers in this setting are thus not merely at the highest risk of NT breakthroughs. Still, their workplaces also place a high value on developing and applying ethical norms [131].

1.5.3 Ethical Implications of NT

NT presents important ethical issues that the public, researchers, legislators, and nanotech entrepreneurs must face [4]. NMs react to organisms and the environment differently due to special characteristics. NMs can affect people, animals, or plants as toxic, hazardous, or allergenic. Studies have shown that CNTs, employed in a variety of applications such as medicine delivery, sensors, and batteries, can induce inflammation and lung damage [132]. The NT sector ought to make sure that it thoroughly evaluates the risks associated with the products, inspects them, and complies with all applicable safety and quality laws. Certain environmental issues, as contamination, global warming, and resource depletion, can be resolved by NT as it can be used to create catalysts, water filters, and effective solar cells [133]. On the other hand, it exacerbates already-existing environmental issues or causes new ones. Its impact on ecosystems and biodiversity can build up in the environment, posing additional risks.

The entire existence of these goods should be considered, and the environmental effects of their manufacture, usage, and disposal should be kept to a minimum. Development of new markets, industries, employment, and wealth are a few significant social and economic effects that NT may have [134]. People's well-being and quality of life may also be enhanced by NT, particularly among underdeveloped nations, where it could be utilized to supply renewable energy, affordable healthcare, and clean water, leading to decrease in new injustices, disputes, and disparities and raising the disparity between industrialized and emerging nations, or between the affluent and poor, thereby jeopardizing people's individuality, dignity, and human rights. Entrepreneurs in the sector should be conscious of the financial and social implications of goods and ensure that they are available, affordable, and advantageous to humanity. NT could question or alter some moral and ethical principles influencing how people behave and make decisions, facilitating the development of novel life forms, including synthetic cells, organs, and creatures [135]. Certain questions may be raised by these possibilities, including the significance and value of human life, bounds and constraints, and the rights and obligations of both human and non-human beings.

Nanotech entrepreneurs should consider the moral and ethical principles underpinning their products and engage in discussion and deliberation with a range of stakeholders and viewpoints. Employees in jobs involving NMs face ethical challenges related to employers and authorities, scientists identifying and communicating risks and hazards, workers accepting risk, the implementation of controls, the decision to participate in medical screening, and sufficient funding for the exposure control research. Identification and evaluation of dangers, autonomy, fairness, confidentiality, and respect for individuals are among the ethical concerns [136]. Despite prejudices and preferences, ethical judgments about workplace safety and health are based on factual scientific information, as scientific information is loaded with value. One scientific theory might be more subjective than another, but no theory can be said to be completely objective. The way that NT is portrayed has possible advantages and related dangers that form the basis of the ethical considerations. Finding the boundary between the required degree of safety and the risk that remains at a particular level of protection is crucial when knowledge of the dangers of NP is unclear.

Risk evaluations tend to be extremely political and subjective. Predictions are therefore loaded with values as they are diverse and dynamic, a single scenario for illustrating hazards, whose controls are insufficient. The ethical concerns will pertain only to the information base at a particular moment in time and to a particular production and usage scenario [62]. Providing high-quality goods such as food, clean water, a secure environment, and healthcare helps make the world a better place. Sustainable development should be prioritized when it comes to the distribution of goods and facilities produced by NTs. The use of NT in military applications must be limited to security and defense. The development of NPs raises ethical questions, leading to self-destructive scientific practices. Reliable sources should also serve as a foundation for any research done on NT. Development of these products must be founded on models integrating sustainable practices, including resource economy, green industrial processes, absence of hazardous materials, equitable worker compensation, and human rights workplace regulations [137].

1.5.4 Key Industry Players and Startups

Enabling revolutionary breakthroughs and enhancing product performance, NT is revolutionizing a number of industries. Leading countries such as Japan, Germany, the US, China, and France are at the forefront of research and commercialization in these areas. The US is a leader in nano-food, medicine, and advanced materials, while Japan is a leader in electronics, coatings, and energy storage; Germany and France both provide significant contributions to nano-food, supported by strong research institutes and EU projects; and China is emerging as new leader within nano-agriculture, with the most patents and substantial investments in related technology being driven by startups, especially in textile and health sectors. Technologies in healthcare, such as personalized medication distribution and diagnostic instruments, are revolutionizing treatment approaches. Smart fabrics with antimicrobial, UV-resistant, and energy-harvesting qualities are becoming more popular in the textile industry [6]. The US and Europe are important hubs for startups supporting green technologies, including biodegradable materials, renewable energy, and air and water purification. NT has commercialization hurdles, regulatory concerns, and public image difficulties despite its promise. It is necessary to address issues including environmental impact, cost-effectiveness, and safety. Collaboration among existing industrial players, upcoming businesses, and research groups is necessary to address these issues and fully exploit the potential of NT in a variety of industries [138].

1.5.5 Patents and Intellectual Property Trends

Using a techno-economic network paradigm, evaluating advancements in commercial applications and research, and assessing developments in NT across scientific, technological, and market dimensions. Patents are used to measure technological advancements, publications measuring scientific development, and the commercialization of NT goods to measure market expansion [136]. When examining these three pillars, it becomes clear that between 2000 and 2024, NT made significant strides in research and real-world applications. Research conducted in academia and industry facilities has largely driven the scientific component of NT innovation, with publications serving as a key gauge of scientific output. One important indicator is the number of publications, but evaluating their influence on the field requires an understanding of their quality, often shown by the frequency of citations. Commercialization of NT goods in the NT has been steadily growing over time, with notable increases seen in nations like China and India [40]. Patent filings demonstrate the capacity to translate scientific discoveries into technology applications and commercial commodities and are the best way to evaluate components of NT innovation (see Figure 1.8).

As patents facilitate research and commercialization, according to statistics from the European Patent Office, the number of NT patent families increased steadily between 2019 and 2024. The initial growth in patent filings was gradual until 2006, with a notable patent surge observed since 2019, mostly due to better partnerships among research facilities and global corporations. The rise of NT patents and publications reflects development, particularly in countries like China, the Republic of Korea, Japan, and the US. The number of articles regarding NT has increased dramatically [6]. By 2023, China will have more patents and publications than any other country in

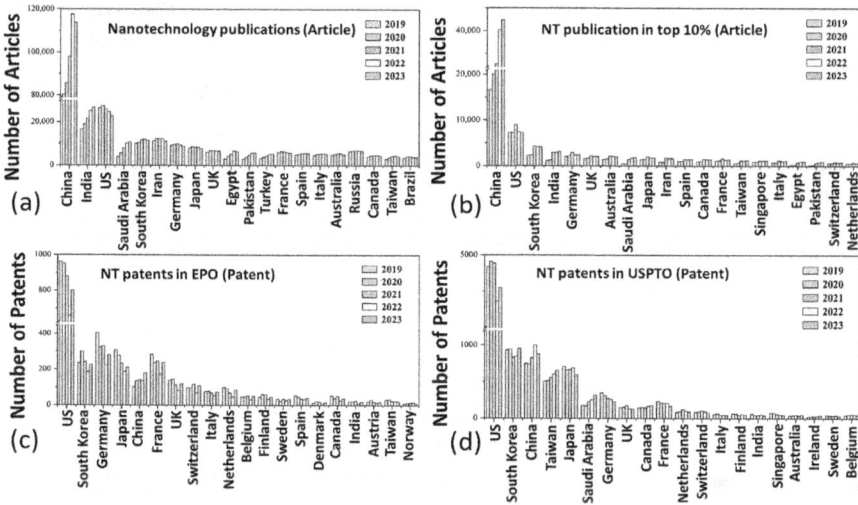

FIGURE 1.8 Global trends in NT research and innovation (2019–2023). This figure presents a comparative analysis of NT performance across four metrics from 2019 to 2023: (a) the total number of NT publications, (b) NT publications in the top 10% of cited articles, (c) NT patents filed in the European Patent Office (EPO), and (d) NT patents filed in the United States Patent and Trademark Office (USPTO). Each metric is broken down by country to highlight international contributions and trends in NT. **Reproduced with permission from © MDPI [6]**

the world, with about 40% of all patents related to NT. Figure 1.7 shows how China leads in NT publications and high-impact research, while the US excels in patent filings, particularly with the USPTO. South Korea and Japan also show strong performance in both publications and patents, highlighting China's academic dominance and the US's strength in commercialization and intellectual property. NT has experienced significant growth in research and commercialization, with China leading in publications and high-impact research, while the US excels in patent filings, reflecting a global innovation divide [126].

1.5.6 Regional and Global Market Dynamics

The global market for NT is expanding quickly, and its uses are found in many different industries, including production, agriculture, and medicine [120]. The significant economic effect of NT is demonstrated by the fact that the worldwide market is projected to reach approximately $183.7 billion by 2028, up from an anticipated $68.0 billion in 2023. The need for quicker and more effective technologies is driving the demand for nanotech-based systems, which is expected to keep driving market expansion [139]. There are notable regional variations in the creation and application of NT, leading companies in the NT industry, including China, Russia, Japan, the US, and Europe, which have distinct tactics and accomplishments that affect their competitive standing internationally. Europe's 18% market share demonstrates its strong position and dedication to NT advancement. To boost economic growth and competitiveness, NT is essential for improving industrial efficiency and developing new goods, being an important part of the Fourth Industrial Revolution, changing industries, and giving businesses new advantages in the global marketplace. Food, agriculture, and medicines are just a few of the industries that depend on the incorporation of nanotechnological advancements, enhancing industrial processes and providing sustainable solutions [40].

Notwithstanding its promise, NT's commercialization is beset by several obstacles, including legal restrictions, safety issues, and the requirement for a significant R&D expenditure. The effective transfer of nanotechnological discoveries from the laboratory to the market depends on overcoming these obstacles. Future approaches to optimizing NT's advantages include strengthening international collaboration, increasing regulatory frameworks, and encouraging multidisciplinary research. It has a crucial role in determining how regional and global markets develop in the future. Significant economic and technical improvements are anticipated due to its ongoing growth and integration across several industries, but obstacles must be strategically addressed.

1.6 Challenges and Perspectives

1.6.1 Technical and Manufacturing Limitations

NT has the potential to revolutionize a number of industries, including materials science, electronics, healthcare, and energy; however, its full-scale commercialization and business potential are still limited by a number of industrial and technological issues [61]. The intricacy of material management and nanoscale manufacturing is one such obstacle. Producing NMs with consistent size, shape, and functionality at a commercial scale remains difficult, as slight variations at the atomic level can significantly alter performance [56]. Another circumstance is scalability; methods such as CVD, sol–gel synthesis, or molecular self-assembly that perform well in the lab sometimes face problems with repeatability and cost-effectiveness when used in commercial production [140]. Furthermore, there are compatibility issues when integrating NMs into current production methods, especially in healthcare or electronics applications where high purity, safety, and dependability are crucial [141]. Product approval and market entry are further slowed down by the absence of established procedures and regulatory frameworks for testing, quality control, and environmental safety. Public mistrust and regulatory reluctance stem from the incomplete understanding of NM toxicity and lifespan analysis [118]. Startups and small firms are additionally hampered by the high initial costs associated with specialist equipment and cleanroom settings [142]. Collectively, these technical and manufacturing hurdles limit the pace at which NT innovations can transition from the lab to the market, thereby impacting the broader assessment of its immediate business viability and long-term commercial sustainability. Overcoming these challenges is essential to unlocking the full economic potential of NT [138].

1.6.2 Cost and Scalability Issues

Significant pricing and scalability issues still prevent NT from being widely commercialized and reaching its full business potential, despite its enormous promise across sectors [143, 144]. The high cost of manufacturing NMs and creating nanoscale devices is one of the most important problems. Advanced apparatus, highly regulated conditions, and costly raw materials are needed for processes such as electron-beam lithography, ALD, and precision chemical synthesis, all of which raise operating costs [145]. Furthermore, it is still far more expensive per unit of production than traditional technologies, which renders it unfeasible for mass-market applications across a wide range of industries [146]. Simultaneously, scaling continues to be a significant barrier since many laboratory-optimized nanofabrication techniques cannot be readily scaled to industrial levels without compromising performance or quality [147]. Problems, including aggregation, contamination, and variation in NP characteristics, frequently jeopardize the repeatability of nanostructures at scale. Large-scale production setup and maintenance costs are further increased by infrastructural constraints, such as the requirement for cleanrooms, specific handling procedures, and strict environmental controls [148, 149]. These obstacles may be prohibitive for small and medium-sized businesses (SMEs), which would restrict their ability to participate and innovate in the nanotech sector. Consequently, while many uses of NT have been shown to be scientifically feasible, their large-scale economic viability is still up for debate [79]. To fully realize the business potential of NT, cost reduction strategies and scalable manufacturing techniques must be developed and supported by public–private investments, regulatory clarity, and industrial partnerships [150].

1.6.3 Emerging Innovations in NT

Emerging innovations in NT are quickly pushing the limits of its economic viability and significantly increasing its potential for use in various industries. One such noteworthy development is the creation of green synthesis techniques for NPs, which employ microorganisms, plant extracts, or biodegradable polymers to provide economical and environmentally friendly substitutes for conventional chemical procedures [103]. Nano-enabled diagnostics and customized medication delivery systems employing liposomes and nanoshells are two medical advancements that are transforming personalized healthcare [151]. Similarly, in electronics, the use of CNTS and Gr is facilitating ultra-fast, flexible, and energy-efficient transistors and sensors, indicating a strong future for next-generation consumer devices [152]. Innovations are also being made in energy applications, as NMs enable greater charge storage and less energy loss, increasing the efficiency of solar panels, batteries, and supercapacitors. Furthermore, the combination of NT, artificial intelligence (AI), and machine learning is accelerating time-to-market, decreasing trial-and-error development, and enabling predictive production and better material design [153, 154]. Nanopesticides and nanofertilizers are becoming viable ways to increase agricultural crop output while reducing environmental damage. Significantly, advances in scalable nanomanufacturing methods, such as nanoimprint lithography and roll-to-roll printing, make it easier to create NMs in huge quantities at lower prices. These advancements in technology not only improve functionality and performance but also create new markets and income streams, establishing NT as a key component of future economic and industrial expansion [3, 155].

1.6.4 Strategies for Business Growth and Market Entry

Companies must implement well-thought-out growth and market entrance plans that meet the particular requirements of this high-tech industry if they are to fully realize the business potential of NT. Focusing on high-value applications and specialized markets, such as targeted medicine delivery, high-performance sensors, or nano-enhanced coatings, where NT provides clear benefits over traditional technologies and warrants greater cost, is an important tactic [79]. In order to get access to state-of-the-art knowledge, shared facilities, and cooperative growth prospects, startups and innovators should also give top priority to cooperative collaborations with academic institutions, research laboratories, and well-established industry players [156]. Early-stage businesses can be sustained and entrance obstacles reduced by utilizing government financing and support initiatives, such as R&D grants, incubation programs, and regulatory facilitation [157]. Developing a robust intellectual property (IP) portfolio is equally essential to protect innovations and attract investors. Businesses must prioritize regulatory compliance and standardization to expand their markets, ensuring their goods fulfil global safety and quality standards, especially in the delicate food and healthcare industries [158]. Adopting agile manufacturing techniques, including modular production and scalable synthesis methods, enables companies to adapt quickly to market demand without incurring unsustainable costs. Furthermore, successful market education and customer engagement methods may generate demand and trust by bridging the gap between technological innovation and consumer comprehension. Last but not least, adding sustainability and transparency to business plans may set nanotech companies apart in a global market that is becoming more environmentally conscientious, giving them a competitive edge and long-term success [159, 160].

1.6.5 Predictions for the Next Decade

Presently, NT is poised to become a cornerstone of global innovation, with its business potential expanding across virtually every industrial sector. NT is expected to have the greatest revolutionary influence on biomedical and healthcare applications, allowing regenerative medicine, tailored medication delivery, and precision diagnostics [161]. The distinction between consumer electronics and healthcare will become more hazy as wearable technology and smart implants benefit from NT advances, opening up new markets [162]. NMs are predicted to transform solar cells, supercapacitors, and battery technologies in the energy industry, resulting in more sustainable and effective energy sources. Nanoscale components will help the electronics industry by enabling quicker, smaller, and more energy-efficient devices, which will facilitate the development of sophisticated AI systems and QM [10, 163].

Furthermore, sustainable manufacturing practices and green NT will become more well-known, resolving environmental issues and supporting international climate goals. As scalable synthesis processes advance, mass manufacturing of NMs will become more viable from a business perspective, lowering prices and boosting commercial acceptance [164]. Governments will probably enact more stringent standards and regulatory frameworks, which will improve consumer confidence and market stability, even if they will make compliance more difficult [165]. Global investment in NT startups and R&D is expected to grow significantly, driven by its interdisciplinary nature and cross-sector utility. Therefore, the next decade will probably see NT evolve from a niche research domain into a mainstream economic driver, unlocking unprecedented business opportunities for early adopters and innovators worldwide [166].

1.7 Summary of Key Findings

NT is rapidly becoming an unparalleled technology with immense potential to revolutionize various sectors, including healthcare, energy, agriculture, and electronics. Thus, its commercial potential is significant, as controlling matter at the nanoscale permits the development of new materials and products with improved attributes of greater strength, conductivity, and reactivity, which can be used in numerous diverse fields. In the healthcare sector, NT promises to transform drug delivery systems, diagnostics, and imaging, providing targeted and personalized treatments that result in fewer side effects. In the energy domain, it could boost the efficiency of solar cells, develop new materials for various power sources, and improve energy storage systems.

Even though NT has a lot of potential, there are several obstacles in its commercialization process. Early-stage research at low technological readiness levels (TRLs) typically serves as the focus of academia, but interest wanes as the technology advances and becomes better known. Industry, however, is looking for high-TRL technology that offers quick profits. Many promising inventions are caught in this key gap, and they are unable to go from the lab to the market. Startups seeking to bridge this gap frequently encounter significant obstacles, such as inadequate infrastructure, a lack of technological expertise, and financial constraints. These obstacles restrict NT's overall commercial success by causing new businesses to fail at high rates. Technical issues, including scalability, repeatability, and guaranteeing the incorporation of NMs into industrial processes, impede commercialization.

Environmental and biological concerns present considerable challenges. There are possible hazards when NMs are added to consumer goods or manufacturing processes since their toxicity, biocompatibility, and ecological effects are still poorly understood. These uncertainties exacerbate public and regulatory worries about NT's safety and environmental sustainability. Low investment in translational research, poor access to technology across areas, and concerns about socioeconomic inequality are some of the economic obstacles to the commercialization of NT. The lack of precise rules and regulations for NT development, IP management, and product clearance further exacerbates regulatory issues.

Addressing these issues requires the collaboration among academics, industry, governments, and policymakers. Initiatives such as the EU's RRI framework attempt to minimize hazards and maximize possible advantages, ensuring that NT is developed properly and sustainably. Simpler regulatory paths and multidisciplinary cooperation will help NT realize its full economic potential.

REFERENCES

[1] Malanowski, N. and A. Zweck, Bridging the gap between foresight and market research: Integrating methods to assess the economic potential of nanotechnology. *Technological Forecasting and Social Change*, 2007. **74**(9): p. 1805–1822.

[2] Schulte, J., *Nanotechnology: Global strategies, industry trends and applications.* 2005, John Wiley & Sons.

[3] Golubev, S.S., et al., Nanotechnology market research: Development and prospects. *Revista Espacios*, 2018. **39**(36).

[4] Baran, A., Nanotechnology: Legal and ethical issues. *Ekonomia i Zarządzanie*, 2016. **8**(1).

[5] Mata, M., et al., Nanotechnology and sustainability-current status and future challenges, in *Life cycle analysis of nanoparticles. Risk, assessment, and sustainability.* 2015, DEStech Publications. p. 271–306.

[6] Khatoon, U.T. and A. Velidandi, An overview on the role of government initiatives in nanotechnology innovation for sustainable economic development and research progress. *Sustainability*, 2025. **17**(3): p. 1250.

[7] Naughtin, C., J. McLaughlin, and S. Hajkowicz, *Opportunities for growth: Driving forces creating economic opportunities for Queensland companies over the coming decade.* 2017, Brisbane, Australia: CSIRO.

[8] Sahu, T., et al., Nanotechnology based drug delivery system: Current strategies and emerging therapeutic potential for medical science. *Journal of Drug Delivery Science and Technology*, 2021. **63**: p. 102487.

[9] Mujawar, M.A., et al., Nano-enabled biosensing systems for intelligent healthcare: Towards COVID-19 management. *Materials Today Chemistry*, 2020. **17**: p. 100306.

[10] Passian, A. and N. Imam, Nanosystems, edge computing, and the next generation computing systems. *Sensors*, 2019. **19**(18): p. 4048.

[11] Jeyaraman, N., et al., Revolutionizing healthcare: The emerging role of quantum computing in enhancing medical technology and treatment. *Cureus*, 2024. **16**(8).

[12] Patel, S., R. Nanda, and S. Sahoo, Nanotechnology in healthcare: Applications and challenges. *Medicinal Chemistry*, 2015. **5**(12): p. 2161–0444.

[13] Hussein, H.S., The state of the art of nanomaterials and its applications in energy saving. *Bulletin of the National Research Centre*, 2023. **47**(1): p. 7.

[14] Naushad, M., R. Saravanan, and R. Kumar, *Nanomaterials for sustainable energy and environmental remediation.* 2020, Elsevier.

[15] Suberu, M.Y., M.W. Mustafa, and N. Bashir, Energy storage systems for renewable energy power sector integration and mitigation of intermittency. *Renewable and Sustainable Energy Reviews*, 2014. **35**: p. 499–514.

[16] Bratovcic, A. and J. Suljagic, Micro-and nano-encapsulation in food industry. *Croatian Journal of Food Science and Technology*, 2019. **11**(1): p. 113–121.

[17] Ayyaril, S.S., et al., Recent progress in micro and nano-encapsulation techniques for environmental applications: A review. *Results in Engineering*, 2023. **18**: p. 101094.

[18] Kaounides, L., H. Yu, and T. Harper, Nanotechnology innovation and applications in textiles industry: Current markets and future growth trends. *Materials Technology*, 2007. **22**(4): p. 209–237.

[19] Guo, H., X. Zhou, and Z. Liu, Advanced lightweight structural materials for automobiles: Properties, manipulation, and perspective. *Science of Advanced Materials*, 2024. **16**(5): p. 563–580.

[20] Nagaraju, S.B., et al., Lightweight and sustainable materials for aerospace applications, in *Lightweight and sustainable composite materials.* 2023, Elsevier. p. 157–178.

[21] Najam, A., Development of advanced materials for lightweight and fuel-efficient aircraft structures: A review. *Research Journal*, 2023. **1**(01): p. 1–9.

[22] Ashique, S., et al., Artificial intelligence integration with nanotechnology: A new frontier for sustainable and precision agriculture. *Current Nanoscience*, 2025. **21**(2): p. 242–273.

[23] Shafique, M. and X. Luo, Nanotechnology in transportation vehicles: An overview of its applications, environmental, health and safety concerns. *Materials*, 2019. **12**(15): p. 2493.

[24] Ramachandran Thampy Bindhu, V. and A. Kurumbadan Saseendran, *AI revolutionizing manufacturing: Cutting-edge advances: Redefining manufacturing with intelligent solutions*. 2024, Independent thesis Advanced level (degree of Master (One Year)) Halmstad University.

[25] Lancaster, C.A., *Nanomites to nano-coatings: Manipulation of plasmonic gold nanoparticle interfaces*. 2017, The University of Utah.

[26] Karst, J., et al., Enhancing maritime domain awareness through AI-enabled acoustic buoys for real-time detection and tracking of fast-moving vessels. *Sensors*, 2025. **25**(6): p. 1930.

[27] Solovchuk, D.R., Advances in AI-assisted biochip technology for biomedicine. *Biomedicine & Pharmacotherapy*, 2024. **177**: p. 116997.

[28] Rasheed, S., et al., Advances and challenges in portable optical biosensors for onsite detection and point-of-care diagnostics. *TrAC Trends in Analytical Chemistry*, 2024: p. 117640.

[29] Grimshaw, J., et al., How to combat a culture of excuses and promote accountability. *Strategy & Leadership*, 2006. **34**(5): p. 11–18.

[30] Yadugiri, V. and R. Malhotra, 'Plenty of room' – fifty years after the Feynman lecture. *Current Science* (00113891), 2010. **99**(7).

[31] Baird, D., A. Nordmann, and J. Schummer, *Discovering the nanoscale*. 2004, IOS Press.

[32] Wiesner, M.R., et al., *Assessing the risks of manufactured nanomaterials*. 2006, ACS Publications.

[33] Logothetidis, S., Nanotechnology: Principles and applications, in *Nanostructured materials and their applications*. 2011, Springer. p. 1–22.

[34] Naz, M.Y., et al., Synthesis and processing of nanomaterials, in *Solar cells: From materials to device technology*. 2020, Springer Nature Switzerland AG. p. 1–23.

[35] Arole, V. and S. Munde, Fabrication of nanomaterials by top-down and bottom-up approaches-an overview. *Journal of Materials Science*, 2014. **1**: p. 89–93.

[36] Murali, A. and B. Saraswathyamma, The role of engineered MoS2 and MoSe2 transition metal dichalcogenides in electrochemical sensors and batteries: A review. *Tungsten*, 2024: p. 1–28.

[37] Saravanan, P., R. Gopalan, and V. Chandrasekaran, Synthesis and characterisation of nanomaterials. *Defence Science Journal*, 2008. **58**(4): p. 504–516.

[38] Rizvi, M., H. Gerengi, and P. Gupta, Functionalization of nanomaterials: Synthesis and characterization, in *Functionalized nanomaterials for corrosion mitigation: Synthesis, characterization, and applications*. 2022, ACS Publications. p. 1–26.

[39] Dirote, E.V., *Trends in nanotechnology research*. 2004, Nova Publishers.

[40] Khan, W.S., E. Asmatulu, and R. Asmatulu, Nanotechnology emerging trends, markets and concerns, in *Nanotechnology safety*. 2025, Elsevier. p. 1–21.

[41] Eskandar, K., Revolutionizing biotechnology and bioengineering: Unleashing the power of innovation. *Journal of Applied Biotechnology & Bioengineering*, 2023. **10**(3): p. 81–88.

[42] Qin, L., et al., Prospects and challenges for the application of tissue engineering technologies in the treatment of bone infections. *Bone Research*, 2024. **12**(1): p. 28.

[43] Popli, S., R.K. Jha, and S. Jain, A survey on energy efficient narrowband internet of things (NBIoT): Architecture, application and challenges. *IEEE Access*, 2018. **7**: p. 16739–16776.

[44] Yetisen, A.K., et al., Nanotechnology in textiles. *ACS Nano*, 2016. **10**(3): p. 3042–3068.

[45] Kumari, R., et al., Nano-bioengineered sensing technologies for real-time monitoring of reactive oxygen species in in vitro and in vivo models. *Microchemical Journal*, 2022. **180**: p. 107615.

[46] Logothetidis, S., Nanomedicine: The medicine of tomorrow, in *Nanomedicine and nanobiotechnology*. 2011, Springer. p. 1–26.

[47] Joseph, T.M., et al., Nanoparticles: Taking a unique position in medicine. *Nanomaterials*, 2023. **13**(3): p. 574.

[48] Viljoen, A. and A.S. Wierzbicki, New approaches in the diagnosis of atherosclerosis and treatment of cardiovascular disease. *Recent Patents on Cardiovascular Drug Discovery (Discontinued)*, 2008. **3**(2): p. 84–91.

[49] Zugazagoitia, J., et al., Current challenges in cancer treatment. *Clinical Therapeutics*, 2016. **38**(7): p. 1551–1566.

[50] Atala, A., et al., *Principles of regenerative medicine*. 2018: Academic Press.

[51] Mason, C. and P. Dunnill, *A brief definition of regenerative medicine*. 2008, Taylor & Francis. p. 1–5.

[52] Nalwa, H.S., A special issue on reviews in nanomedicine, drug delivery and vaccine development. *Journal of Biomedical Nanotechnology*, 2014. **10**(9): p. 1635–1640.

[53] Rae, A., Real life applications of nanotechnology in electronics. *OnBoard Technology*, 2006. **2006**: p. 28.

[54] Štremfelj, J. and F. Smole, Nanotechnology and nanoscience – from past breakthroughs to future prospects. *Informacije MIDEM*, 2021. **51**(1): p. 25–48.

[55] Anjum, A., M. Das, and R. Garg, Introduction to nanotechnology: Transformative frontier, in *Smart and sustainable applications of nanocomposites*. 2024, IGI Global Scientific Publishing. p. 1–35.

[56] Charitidis, C.A., et al., Manufacturing nanomaterials: From research to industry. *Manufacturing Review*, 2014. **1**: p. 11.

[57] Serrano, E., G. Rus, and J. Garcia-Martinez, Nanotechnology for sustainable energy. *Renewable and Sustainable Energy Reviews*, 2009. **13**(9): p. 2373–2384.

[58] Sun, H., et al., Novel graphene/carbon nanotube composite fibers for efficient wire-shaped miniature energy devices. *Advanced Materials*, 2014. **26**(18): p. 2868–2873.

[59] Muzammil, W., et al., Nanotechnology in renewable energy: Critical reviews for wind energy, in *Nanotechnology: Applications in energy, drug and food*, 2019: p. 49–71.

[60] Mathew, J., J. Joy, and S.C. George, Potential applications of nanotechnology in transportation: A review. *Journal of King Saud University-Science*, 2019. **31**(4): p. 586–594.

[61] Malik, S., K. Muhammad, and Y. Waheed, Nanotechnology: A revolution in modern industry. *Molecules*, 2023. **28**(2): p. 661.

[62] Liu, W., et al., Applications of machine learning in computational nanotechnology. *Nanotechnology*, 2022. **33**(16): p. 162501.

[63] Aithal, P. and S. Aithal, Nanotechnology innovations & business environment of Indian automobile sector: A futuristic approach. *International Journal of Scientific Research and Modern Education (IJSRME)*, 2016. **1**: p. 296–307.

[64] Thakur, S., S. Thakur, and R. Kumar, Bio-nanotechnology and its role in agriculture and food industry. *Journal of Molecular and Genetic Medicine*, 2018. **12**(324): p. 1747–0862.

[65] Confraria, H., et al., The impact of the EU Industrial R&D Investment Scoreboard in science and policy. *2024, JRC Working Papers on Corporate R&D and Innovation (CoRDI)*.

[66] Bogusz, C., *The budget and economic outlook: Fiscal years 2013 to 2023*. 2013, Congressional Budget Office.

[67] Abdelrahman, H. and L.E. Oliveira, The rise and fall of pandemic excess savings. *FRBSF Economic Letter*, 2023. **11**: p. 2023.

[68] Bogue, R., Nanosensors: A review of recent progress. *Sensor Review*, 2008. **28**(1): p. 12–17.

[69] Javaid, M., et al., Exploring the potential of nanosensors: A brief overview. *Sensors International*, 2021. **2**: p. 100130.

[70] Nemet, G.F. and D.M. Kammen, US energy research and development: Declining investment, increasing need, and the feasibility of expansion. *Energy Policy*, 2007. **35**(1): p. 746–755.

[71] Gruber, W., D. Mehta, and R. Vernon, The R & D factor in international trade and international investment of United States industries. *Journal of Political Economy*, 1967. **75**(1): p. 20–37.

[72] Yuan, B. and Y. Zhang, Flexible environmental policy, technological innovation and sustainable development of China's industry: The moderating effect of environment regulatory enforcement. *Journal of Cleaner Production*, 2020. **243**: p. 118543.

[73] Kozhukharov, V. and M. Machkova, Nanomaterials and nanotechnology: European initiatives, status and strategy. *Journal of Chemical Technology and Metallurgy*, 2013. **48**(1): p. 3–11.

[74] Liu, L., *Emerging nanotechnology power: Nanotechnology R&D and business trends in the Asia Pacific Rim*. 2009, World Scientific.

[75] Azoulay, D., R. Senjen, and G. Foladori, *Social and environmental implications of nanotechnology development in Asia-Pacific*. 2013. Gothenburg, Sweden: IPEN.

[76] Kay, L. and P. Shapira, Developing nanotechnology in Latin America. *Journal of Nanoparticle Research*, 2009. **11**: p. 259–278.

[77] Stirling, D.A., *The nanotechnology revolution: A global bibliographic perspective*. 2018, Jenny Stanford Publishing.

[78] Nikraftar, T., E. Hosseini, and E. Mohammadi, The factors influencing technological entrepreneurship in nanotechnology businesses. *Revista de Gestão*, 2022. **29**(1): p. 76–99.

[79] Hobson, D.W., Commercialization of nanotechnology. *Wiley Interdisciplinary Reviews: Nanomedicine and Nanobiotechnology*, 2009. **1**(2): p. 189–202.

[80] Teizer, J., et al., Nanotechnology and its impact on construction: Bridging the gap between researchers and industry professionals. *Journal of Construction Engineering and Management*, 2012. **138**(5): p. 594–604.

[81] Roco, M.C., C.A. Mirkin, and M.C. Hersam, *Nanotechnology research directions for societal needs in 2020: Retrospective and outlook*. 2011, Springer.

[82] Faraji, S.N., S. Masjoodi, and S. Azmand, Biotechnology, nanotechnology, and AI: Transformative convergence in healthcare (A mini review). *Advances in Applied NanoBio-Technologies*, 2024. **5**: p. 23–28.

[83] Wu, W., Managing and incentivizing research commercialization in Chinese Universities. *The Journal of Technology Transfer*, 2010. **35**: p. 203–224.

[84] Markman, G.D., D.S. Siegel, and M. Wright, Research and technology commercialization. *Journal of Management Studies*, 2008. **45**(8): p. 1401–1423.

[85] Lazonick, W. and M. O'Sullivan, Investment in innovation. *Public Policy Brief*, 1997. **37**.

[86] Kuczmarski, T.D., What is innovation? The art of welcoming risk. *Journal of Consumer Marketing*, 1996. **13**(5): p. 7–11.

[87] Miyazaki, K. and N. Islam, Nanotechnology systems of innovation – An analysis of industry and academia research activities. *Technovation*, 2007. **27**(11): p. 661–675.

[88] Roco, M.C., *The long view of nanotechnology development: The National Nanotechnology Initiative at 10 years*. 2011, Springer. p. 427–445.

[89] Kalisz, D.E. and M. Aluchna, Research and innovation redefined. Perspectives on the European Union initiatives on Horizon 2020. *European Integration Studies*, 2012. (6).

[90] Tenhunen-Lunkka, A. and R. Honkanen, Project coordination success factors in European Union-funded research, development and innovation projects under the Horizon 2020 and Horizon Europe programmes. *Journal of Innovation and Entrepreneurship*, 2024. **13**(1): p. 7.

[91] Cao, C., R.P. Appelbaum, and R. Parker, "Research is high and the market is far away": Commercialization of nanotechnology in China. *Technology in Society*, 2013. **35**(1): p. 55–64.

[92] National Academies of Sciences, Engineering, and Medicine, et al., *A quadrennial review of the national nanotechnology initiative: Nanoscience, applications, and commercialization*. 2020, National Academies Press.

[93] Murray, G.C., Venture capital and government policy, in *Handbook of research on venture capital*. 2007, Edward Elgar Publishing.

[94] Irfan, M., et al., Nanoscale advances. *Nanoscale*, 2018. **1**(1): p. 1–200.

[95] Balachandran, S., The inside track: Entrepreneurs' corporate experience and start-ups' access to incumbent partners' resources. *Strategic Management Journal*, 2024. **45**(6): p. 1117–1150.

[96] Vanhaverbeke, W. and H. Chesbrough, A classification of open innovation and open business models. *New Frontiers in Open Innovation*, 2014. **6**: p. 50–68.

[97] Koshovets, O. and I. Frolov, Impact investing as a 'basic innovation' for the global economy and finance system post-crisis transformation. *Economy & Business*, 2015. **9**.

[98] Qiu Zhao, Q., A. Boxman, and U. Chowdhry, Nanotechnology in the chemical industry – opportunities and challenges. *Journal of Nanoparticle Research*, 2003. **5**: p. 567–572.

[99] Nordås, H.K., E. Pinali, and M.G. *Grosso, Logistics and time as a trade barrier*. 2006, OECD Publishing.

[100] Chryssochoidis, G.M. and V. Wong, Rolling out new products across country markets: An empirical study of causes of delays. *Journal of Product Innovation Management: An International Publication of the Product Development & Management Association*, 1998. **15**(1): p. 16–41.

[101] Manzoor, U., et al., Limitations and concerns of nanotechnology in obtaining the desirable products, in *Nanotechnology based microbicides and immune stimulators*. 2025, Springer. p. 217–236.

[102] Foster, L.E., *Nanotechnology: Science, innovation, and opportunity*. 2005, Prentice Hall PTR.

[103] Romig Jr, A., et al., An introduction to nanotechnology policy: Opportunities and constraints for emerging and established economies. *Technological Forecasting and Social Change*, 2007. **74**(9): p. 1634–1642.

[104] Allan, J., et al., Regulatory landscape of nanotechnology and nanoplastics from a global perspective. *Regulatory Toxicology and Pharmacology*, 2021. **122**: p. 104885.

[105] García-Quintero, A. and M. Palencia, A critical analysis of environmental sustainability metrics applied to green synthesis of nanomaterials and the assessment of environmental risks associated with the nanotechnology. *Science of the Total Environment*, 2021. **793**: p. 148524.

[106] Sahu, M.K., R. Yadav, and S.P. Tiwari, Recent advances in nanotechnology. *International Journal of Nanomaterials, Nanotechnology and Nanomedicine*, 2023. **9**(1): p. 015–023.

[107] Binns, C., *Introduction to nanoscience and nanotechnology*. 2021, John Wiley & Sons.

[108] Kumari, R., et al., Regulation and safety measures for nanotechnology-based agri-products. *Frontiers in Genome Editing*, 2023. **5**: p. 1200987.

[109] Ikumapayi, O., et al., Microfabrication and nanotechnology in manufacturing system – an overview. *Materials Today: Proceedings*, 2021. **44**: p. 1154–1162.

[110] Markandan, K. and W.S. Chai, Perspectives on nanomaterials and nanotechnology for sustainable bioenergy generation. *Materials*, 2022. **15**(21): p. 7769.

[111] Yarahmadi, A. and H. Afkhami, Nanotechnology applications in nuclear waste management: Challenges and limitations. *Journal of Radioanalytical and Nuclear Chemistry*, 2025: p. 1–23.

[112] Vaidya, S., et al., Nanotechnology in agriculture: A solution to global food insecurity in a changing climate? *NanoImpact*, 2024. **34**: p. 100502.

[113] Costello-Caulkins, M., Nanotechnology patent law: A case study of United States and European patent applications. *Santa Clara High Technology Law Journal*, 2021. **37**: p. 337.

[114] Salamanca-Buentello, F. and A.S. Daar, Nanotechnology, equity and global health. *Nature Nanotechnology*, 2021. **16**(4): p. 358–361.

[115] Wasti, S., et al., Ethical and legal challenges in nanomedical innovations: A scoping review. *Frontiers in Genetics*, 2023. **14**: p. 1163392.

[116] Caputo, F., et al., Asymmetric-flow field-flow fractionation for measuring particle size, drug loading and (in)stability of nanopharmaceuticals. The joint view of European Union Nanomedicine Characterization Laboratory and National Cancer Institute-Nanotechnology Characterization Laboratory. *Journal of Chromatography A*, 2021. **1635**: p. 461767.

[117] Rodine-Hardy, K., Nanotechnology and global environmental politics: Transatlantic divergence. *Global Environmental Politics*, 2016. **16**(3): p. 89–105.

[118] Amutha, C., et al., Nanotechnology and governance: Regulatory framework for responsible innovation, in *Nanotechnology in societal development*. 2024, Springer. p. 481–503.

[119] Kileo, A., et al., *Challenges of development of regulatory control for uranium mining in developing countries to achieve regulatory compliance – Tanzanian experience*. 2014, International Atomic Energy Agency.

[120] Roco, M.C., National nanotechnology initiative at 20 years: Enabling new horizons. *Journal of Nanoparticle Research*, 2023. **25**(10): p. 197.

[121] Mortensen, H.M., et al., NNI nanoinformatics conference 2023: Movement toward a common infrastructure for federal nanoEHS data computational toxicology. *Computational Toxicology*, 2024. **30**: p. 100316.

[122] Kawamura, K., et al., Exploring the contexts of ELSI and RRI in Japan: Case studies in dual-use, regenerative medicine, and nanotechnology. *Risks and Regulation of New Technologies*, 2021: p. 271–290.

[123] Mirza, M.A., Z. Iqbal, and H. Mishra, FDC in nanotechnology: Regulatory landscape, in *Nanocarriers for the delivery of combination drugs*. 2021, Elsevier. p. 473–496.

[124] Giakoumettis, D. and S. Sgouros, Nanotechnology in neurosurgery: A systematic review. *Child's Nervous System*, 2021. **37**: p. 1045–1054.

[125] Malik, S., K. Muhammad, and Y. Waheed, Emerging applications of nanotechnology in healthcare and medicine. *Molecules*, 2023. **28**(18): p. 6624.

[126] Talebian, S., et al., Facts and figures on materials science and nanotechnology progress and investment. *ACS Nano*, 2021. **15**(10): p. 15940–15952.

[127] Munir, M., et al., Aerosolised micro and nanoparticle: Formulation and delivery method for lung imaging. *Clinical and Translational Imaging*, 2023. **11**(1): p. 33–50.

[128] Yadav, S.K., et al., The toxic side of nanotechnology: An insight into hazards to health and the ecosystem. *Micro and Nanosystems*, 2022. **14**(1): p. 21–33.

[129] Chen, R. and C. Chen, Environment, health and safety issues in nanotechnology, in *Springer handbook of nanotechnology*, B. Bhushan, Editor. 2017, Berlin, Heidelberg: Springer Berlin Heidelberg. p. 1559–1586.

[130] Sedaghat, M.H., M. Behnia, and O. Abouali, Nanoparticle diffusion in respiratory mucus influenced by mucociliary clearance: A review of mathematical modeling. *Journal of Aerosol Medicine and Pulmonary Drug Delivery*, 2023. **36**(3): p. 127–143.

[131] Schulte, P.A., et al., Occupational safety and health criteria for responsible development of nanotechnology. *Journal of Nanoparticle Research*, 2014. **16**: p. 1–17.

[132] Shoukat, R. and M.I. Khan, Carbon nanotubes: A review on properties, synthesis methods and applications in micro and nanotechnology. *Microsystem Technologies*, 2021: p. 1–10.

[133] Gehrke, I., A. Geiser, and A. Somborn-Schulz, Innovations in nanotechnology for water treatment. *Nanotechnology, Science and Applications*, 2015: p. 1–17.

[134] Halwani, A.A., Development of pharmaceutical nanomedicines: From the bench to the market. *Pharmaceutics*, 2022. **14**(1): p. 106.

[135] Hayaliev, R., et al., Interdisciplinary approach of biomedical engineering in the development of technical devices for medical research. *Journal of Biomimetics, Biomaterials and Biomedical Engineering*, 2021. **53**: p. 85–92.

[136] Tawiah, B., E.A. Ofori, and S.C. George, Nanotechnology in societal development, in *Nanotechnology in societal development*. 2024, Springer. p. 1–64.

[137] Thakur, M., et al., Modern applications and current status of green nanotechnology in environmental industry, in *Green functionalized nanomaterials for environmental applications*. 2022, Elsevier. p. 259–281.

[138] Rambaran, T. and R. Schirhagl, Nanotechnology from lab to industry – a look at current trends. *Nanoscale Advances*, 2022. **4**(18): p. 3664–3675.

[139] Kostoff, R.N., R.G. Koytcheff, and C.G.Y. Lau, Global nanotechnology research literature overview. *Technological Forecasting and Social Change*, 2007. **74**(9): p. 1733–1747.

[140] Cooper, K.P. and R.F. Wachter, Challenges and opportunities in nanomanufacturing. *Instrumentation, Metrology, and Standards for Nanomanufacturing, Optics, and Semiconductors V*, 2011. **8105**: p. 7–12.

[141] Krajnik, P., et al., Transitioning to sustainable production – part III: Developments and possibilities for integration of nanotechnology into material processing technologies. *Journal of Cleaner Production*, 2016. **112**: p. 1156–1164.

[142] Grohn, K., et al., Lean start-up: A case study in the establishment of affordable laboratory infrastructure and emerging biotechnology business models. *Journal of Commercial Biotechnology*, 2015. **21**(2).

[143] Kaur, I.P., et al., Issues and concerns in nanotech product development and its commercialization. *Journal of Controlled Release*, 2014. **193**: p. 51–62.

[144] Aithal, S. and S. Aithal, Nanotechnology innovations and commercialization – opportunities, challenges & reasons for delay. *International Journal of Engineering and Manufacturing (IJEM)*, 2016. **6**(6): p. 15–25.

[145] Zhou, Z.J., Electron beam lithography, in *Handbook of microscopy for nanotechnology*. 2005, Springer. p. 287–321.

[146] Baumers, M., et al., The cost of additive manufacturing: Machine productivity, economies of scale and technology-push. *Technological Forecasting and Social Change*, 2016. **102**: p. 193–201.

[147] Gawin-Mikołajewicz, A., et al., Ophthalmic nanoemulsions: From composition to technological processes and quality control. *Molecular Pharmaceutics*, 2021. **18**(10): p. 3719–3740.

[148] Bramsiepe, C., et al., Low-cost small scale processing technologies for production applications in various environments – mass produced factories. *Chemical Engineering and Processing: Process Intensification*, 2012. **51**: p. 32–52.

[149] Schlenkrich, M. and S.N. Parragh, Solving large scale industrial production scheduling problems with complex constraints: An overview of the state-of-the-art. *Procedia Computer Science*, 2023. **217**: p. 1028–1037.

[150] Aithal, P. and S. Aithal, Ideal technology concept & its realization opportunity using nanotechnology. *International Journal of Application or Innovation in Engineering & Management (IJAIEM)*, 2015. **4**(2): p. 153–164.

[151] Emerich, D.F. and C.G. Thanos, Nanotechnology and medicine. *Expert Opinion on Biological Therapy*, 2003. **3**(4): p. 655–663.

[152] Daneshvar, F., et al., Critical challenges and advances in the carbon nanotube – metal interface for next-generation electronics. *Nanoscale Advances*, 2021. **3**(4): p. 942–962.

[153] Agrahari, V., et al., The role of artificial intelligence and machine learning in accelerating the discovery and development of nanomedicine. *Pharmaceutical Research*, 2024. **41**(12): p. 2289–2297.

[154] Kippers, B., *Time-to-market forecast accuracy in Nanotechnology: Do start-up and industry experience matter?* 2019, University of Twente.

[155] Hullmann, A., *The economic development of nanotechnology – an indicators based analysis*. EU report, 2006.

[156] Wang, J. and P. Shapira, Partnering with universities: A good choice for nanotechnology start-up firms? *Small Business Economics*, 2012. **38**: p. 197–215.

[157] Dimov, D., *Policy options and instruments for financing innovations: A practical guide to early-stage financing*. United Nations, 2009.

[158] Mamasoliev, S., International business law and us market expansion, in *Interdiscipline innovation and scientific research conference*. 2024.

[159] Odeyemi, O., et al., Sustainable entrepreneurship: A review of green business practices and environmental impact. *World Journal of Advanced Research and Reviews*, 2024. **21**(2): p. 346–358.

[160] Pokrajac, L., et al., *Nanotechnology for a sustainable future: Addressing global challenges with the international network4sustainable nanotechnology*. 2021, ACS Publications.

[161] Shrivastava, S. and D. Dash, Applying nanotechnology to human health: Revolution in biomedical sciences. *Journal of Nanotechnology*, 2009. **2009**(1): p. 184702.

[162] Pramanik, P.K.D., et al., Advancing modern healthcare with nanotechnology, nano-biosensors, and internet of nano things: Taxonomies, applications, architecture, and challenges. *IEEE Access*, 2020. **8**: p. 65230–65266.

[163] Ahmadi, A., Quantum computing and artificial intelligence: The synergy of two revolutionary technologies. *Asian Journal of Electrical Sciences*, 2023. **12**(2): p. 15–27.

[164] Falsini, S., et al., Sustainable strategies for large-scale nanotechnology manufacturing in the biomedical field. *Green Chemistry*, 2018. **20**(17): p. 3897–3907.

[165] Abbott, K.W. and D. Snidal, The governance triangle: Regulatory standards institutions and the shadow of the state, in *The spectrum of international institutions*. 2021, Routledge. p. 52–91.

[166] Coccia, M., Evolutionary trajectories of the nanotechnology research across worldwide economic players. *Technology Analysis & Strategic Management*, 2012. **24**(10): p. 1029–1050.

2

Nanomaterials for Green and Clean Energy Production and Storage

Shailesh S. Ankalgi, Aviraj M. Teli, Dattakumar S. Mhamane, and Mayur S. Ankalgi

2.1 Introduction

Soil, water, and energy are some of humanity's most valued and most essential resources. But how we use and exhaust these limited resources is a primary cause of speeding up climate change (Turner, 2015). As Turner (2015) reports, heightened concern regarding climate change issues is now directly linked to the continuously increasing use of fossil fuels, whose stocks are being depleted at a rapid rate and whose distribution is uneven across the globe. Ironically, although Mother Nature generously provides these resources, it is also highly susceptible to environmental disruptions and degradations created by their abuse.

The imperative for sustainable and effective management of these resources has never been greater. This has created a burning interest in renewable energy systems worldwide by both the need for clean energy and the growing evidence of climate change. Nonetheless, the absence of an integrated framework linking resource estimation to policy formulation has the consequence of leading to contradictory choices and inefficient use of accessible reserves (Yip et al., 2016). The comprehensive management of climate, land use, energy, and water is hence critical in tackling future environmental and energy crises.

The approaching depletion of fossil fuel reserves, along with rising global energy needs, has triggered extensive research in the past few decades in alternative and renewable energy technologies. Global societies have come to realize the destructive nature of human activities on the environment and the diversity of living things. This increased awareness has led to a global recognition of the importance of environmental conservation and sustainability (Yan et al., 2010). As highlighted by Yan et al. (2010), a worldwide shift toward sustainable sources of energy is crucial to counteract the environmental impact of urbanization and industrialization.

The increasing demand for low-cost, efficient, and eco-friendly sources of energy has driven spectacular development in the area of materials science. Specifically, the creation of functional materials designed for sustainable energy applications has become a potential route of scientific and technological progress. These materials are critical for use in solar cells, fuel cells, batteries, supercapacitors, biofuel cells, and redox flow batteries. They have the potential not only to optimize energy harvesting

DOI: 10.1201/9781003632498-2

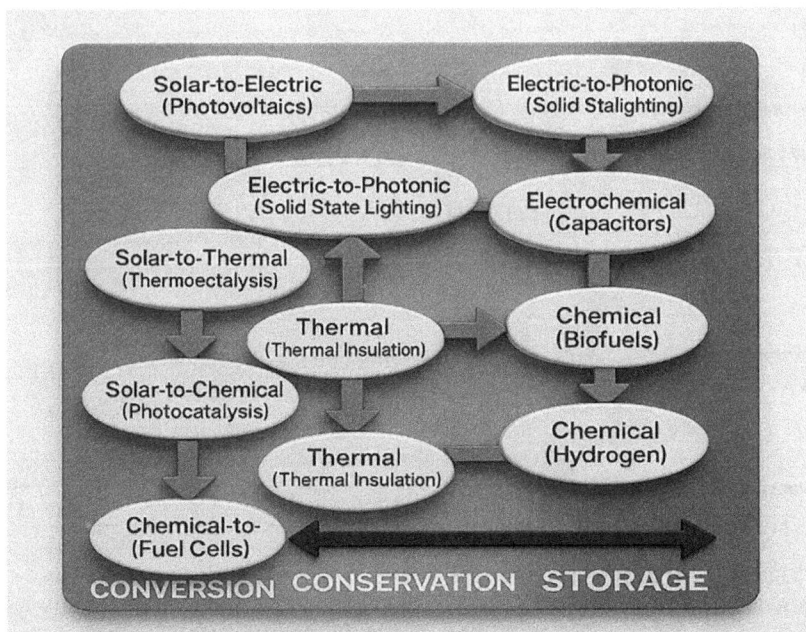

FIGURE 2.1 Energy flow from conversion to storage systems.

but also to facilitate long-term sustainability objectives. However, amid this advancement, the efficiency of many of these functional materials is still suboptimal. Performance improvement is needed to allow these technologies to dominate conventional fossil fuel-based systems.

The applications of functional nanomaterials for sustainable energy harvesting delve into the design and uses of nanomaterials in different energy technologies, address their potential for different energy storage and conversion systems (as shown in Figure 2.1), and present recent developments in the area. Nanomaterials for sustainable energy purposes are analyzed based on addressing global sustainability issues.

2.1.1 Nanotechnology: A Pillar for Sustainable Development

Energy production and usage constitute the cornerstone of modern civilization. Nevertheless, fossil fuel usage since the industrial era has generated numerous environmental challenges, such as climate change, air pollution, and water pollution, as well as the degradation of ecosystems. The way forward lies in creating clever, adaptable, and effective technologies for sustainable harvesting, transmission, and use of energy. Current interdisciplinary studies have also been increasingly concentrated on nanomaterials as a facilitator of a clean and renewable energy-based future (Liu et al., 2010).

In order to ensure maximum utilization of available limited resources, principles of dematerialization and transmaterialization should be embraced (Huang & Jiang, 2019). Transmaterialization entails moving toward the utilization of recyclable or

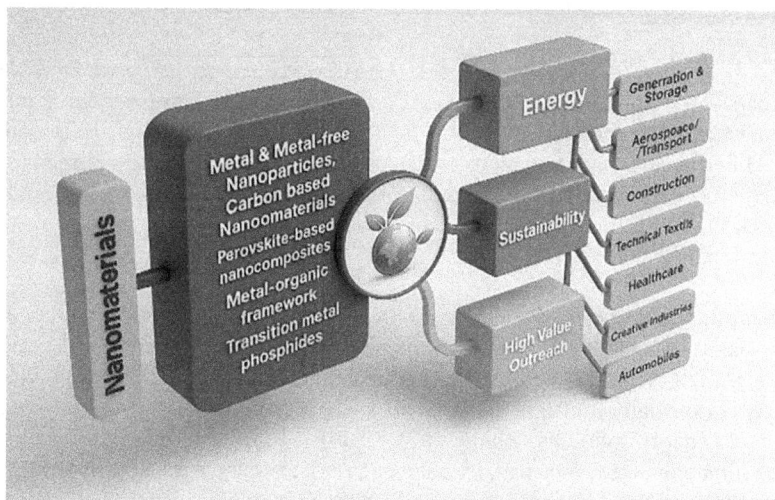

FIGURE 2.2 The nanomaterials for sustainable energy areas.

reusable materials, while dematerialization focuses on the reduction in resource utilization without loss of quality of life. The transition in energy has to be done with great regard to human and environmental well-being to ensure that technological progress is not at the expense of sustainability. While nanotechnology is not exclusively related to sustainable energy, it plays a leading role in the development of this sector. It provides new solutions for renewable and non-renewable sources of energy. Nanoemulsions, for example, are employed as anti-corrosive materials, while nanocatalysts and nanostructured membranes are being applied in the extraction of fossil fuels and nuclear power generation.

Nanotechnology has the potential to develop dematerialization through the use of minimized and highly efficient systems. Functional nanomaterials are at the forefront of the development of sustainable energy technologies. Their technologies include, but are not restricted to, solar cells, fuel cells, rechargeable batteries, supercapacitors, and hybrid energy storage devices. These technologies are not merely being developed but are also starting to be implemented in commercial and industrial applications.

Additionally, nanotechnology has enabled the production of green technologies like solar films, environment-friendly coatings, and intelligent sensor systems, as shown in Figure 2.2. Ongoing innovations in nanomaterials will greatly contribute to economic growth, conservation of the environment, and quality of life in general. Accordingly, the plan for improving functional nanomaterials should place topmost consideration on material performance challenges, scalability, environmental safety, and life cycle sustainability (Zhou et al., 2021; Arico et al., 2005).

In short, the potential of nanotechnology lies in its ability to revolutionize energy production, conversion, and storage through enhanced material properties and system efficiencies. Through interdisciplinary innovation and collaboration, nanomaterials can be a springboard for a cleaner, greener, and more sustainable future.

2.2 Types of Nanomaterials

Nanomaterials exhibit unique properties that emerge due to their behavior at atomic, molecular, and macromolecular scales, which are distinct from their bulk counterparts. Materials are classified as nanomaterials if at least one of their dimensions is less than 100 nanometers (nm). By controlling the size, shape, and composition of these materials, it is possible to tailor their physical and chemical properties, resulting in exceptional functionalities that are not seen in larger-scale materials (Kumar et al., 2020).

The primary categories of functional nanomaterials include metals, metal oxides, carbon-based nanomaterials, and quantum dots. A key characteristic of these materials is their high surface-to-volume ratio, which enhances their surface reactivity. This property is especially useful in applications such as catalysis, energy conversion, and sustainable energy solutions (Zhang et al., 2019). At the nanoscale, these materials often exhibit enhanced electrical, mechanical, optical, and chemical properties, making them suitable for a wide range of applications in medicine, electronics, renewable energy, and environmental management (Bera et al., 2019).

Nanotechnology focuses on optimizing and controlling the properties of these materials at the atomic and molecular levels, which opens up new avenues for technological breakthroughs. Understanding the unique characteristics of functional nanomaterials allows for the development of devices that can address global challenges such as clean energy production and healthcare (Ramakrishnan et al., 2021).

2.2.1 Classification

Materials can generally be divided into two categories: natural materials (such as organic, mineral, and living matter) and artificial materials, which are synthetically created. In the case of nanomaterials, much of the progress comes from manipulating materials from bulk to the nanoscale. The dimensionality of nanomaterials plays a crucial role in their properties and applications. Nanomaterials can be classified based on their dimensionality as zero-dimensional (0D), one-dimensional (1D), two-dimensional (2D), and three-dimensional (3D) materials, shown in Figure 2.3, each possessing unique characteristics and properties (Poh et al., 2018).

- **Zero-Dimensional (0D):** In 0D nanomaterials, all three dimensions are confined to the nanoscale. These materials are usually spherical nanoparticles with diameters ranging from 1 to 50 nm. Examples include gold, silver, and platinum nanoparticles, which exhibit unique optical and electronic properties due to quantum effects that arise at the nanoscale (Wang et al., 2018).
- **One-Dimensional (1D):** One-dimensional nanomaterials have one dimension at the nanoscale, while the other two dimensions are macroscopic. These materials include structures like nanotubes, nanowires, and nanorods. The confinement of charge carriers in one direction gives rise to unique electrical and mechanical properties (Deng et al., 2019).
- **Two-Dimensional (2D):** Two-dimensional nanomaterials have two dimensions in the nanoscale range and one in the macroscale. This category includes

Classification based on dimensionality of the nanomaterials			
0 D Nanospheres & Clusters	**1 D** Nanowires & Nanorods	**2 D** Layered Structures	**3 D** Bulk Nanomaterials
Quantum dots	Nanorods	Gold Nanoplates	Liposome
Ag nanosphere	Carbon Nanotubes	Graphene Sheet	Polycrystalline
Ag nanosphere	Nanowires	Layered Structures	Dendrimer

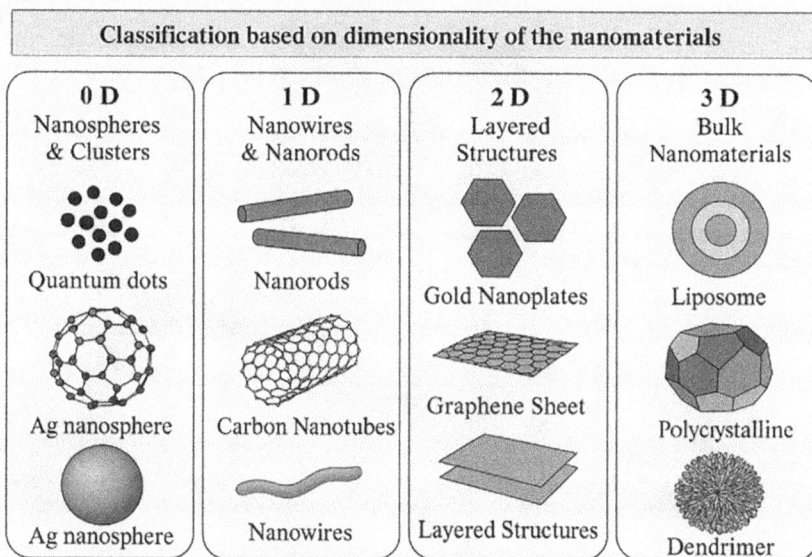

FIGURE 2.3 Schematic of nanostructured materials based on their structure of each category: zero- (OD), one- (1D), two- (2D), and three-dimensional (3D).

materials such as graphene, nanolayers, and thin films. Despite having a nanoscale thickness, 2D materials can cover large areas and exhibit properties like high electrical conductivity and large surface area, which are ideal for applications in flexible electronics and energy storage (Novoselov et al., 2016).

- **Three-Dimensional (3D):** Three-dimensional nanomaterials are those where all three dimensions extend beyond the nanoscale. While the materials themselves may exceed 100 nm in size, they may still exhibit nanoscale behavior in specific applications. Examples include bulk nanomaterials, nanocolloids, and assemblies of nanowires or nanoparticles. In these materials, electrons are free to move in all directions, contributing to their unique electrical and catalytic properties (Zhang et al., 2020).

2.2.2 Types of Nanomaterials Used in Energy Systems

Nanomaterials have become the key elements to improve the performance, efficiency, and scalability of energy systems. Their distinct properties, including high surface area, remarkable electrical conductivity, and versatility, render them suitable for different uses in energy generation, storage, and conversion. The main types of nanomaterials utilized in energy systems are carbon-based, metal-based, and hybrid nanomaterials, as shown in Figure 2.4.

- **Carbon-Based Nanomaterials:** Carbon nanomaterials, including graphene and carbon nanotubes (CNTs), fullerene, graphene sheets, and carbon hybrids (shown in Figure 2.5), are being extensively employed in energy systems based

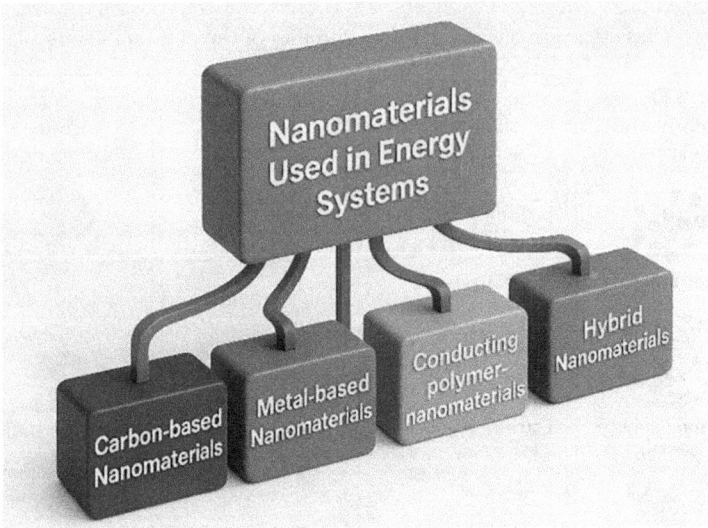

FIGURE 2.4 Classification of nanomaterials used in energy systems.

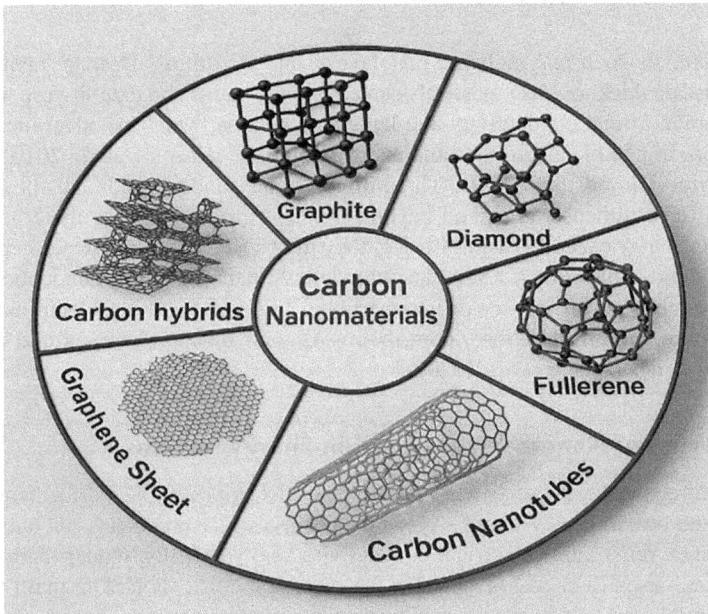

FIGURE 2.5 Carbon-based nanomaterials: graphite, diamond, fullerene, carbon nanotubes, graphene sheet, carbon hybrid materials.

on their superior electrical conductivity, toughness, and mechanical stability.

Graphene, a monolayer of carbon atoms in a 2D lattice, is characterized by high electrical conductivity and large surface area, making it a suitable material for energy storage devices such as supercapacitors and lithium-ion batteries (Novoselov et al., 2004). Graphene enhances the performance of energy storage systems by improving charge transport and increasing storage capacity (Stoller et al., 2008). CNTs, which are rolled sheets of graphene in nanocylindrical form, facilitate enhanced conductivity and structural strength for electrodes in energy storage devices, leading to increased charge/discharge rate and battery or supercapacitor lifespan (Iijima, 1991). Lightweight and long-lived materials in alternative energy technologies, such as wind turbines, benefit from CNTs.

- **Metal-Based Nanomaterials:** Metal-based nanomaterials provide enhanced catalytic activity and electronic conductivity and are hence essential for energy conversion processes. The different structures of metal-based nanomaterials are shown in Figure 2.6.

Metal nanoparticles, including platinum, gold, and silver, find extensive application as catalysts in fuel cells, facilitating efficient hydrogen conversion to electricity. The high surface area of metal nanoparticles facilitates high rates of catalytic reactions, enhancing the overall energy system efficiency. Metal nanowires are employed in batteries and solar cells to increase electrical conductivity and energy transfer, resulting in higher energy conversion and storage efficiency. Its special structure provides high performance in energy generation and storage systems (Huang et al., 2007).

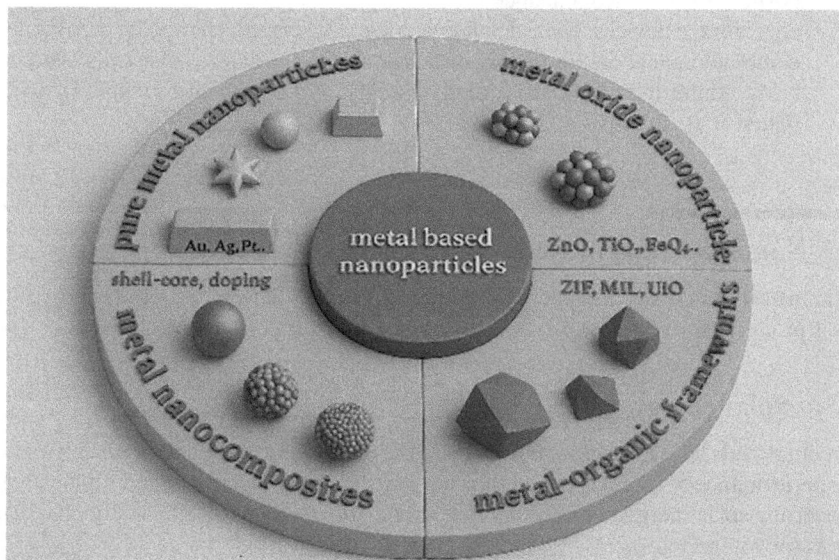

FIGURE 2.6 Different structures of metal-based nanomaterials.

- **Conducting Polymer-Based Nanomaterials:** Conducting polymer-based nanomaterials have garnered significant attention due to their unique combination of electrical conductivity, flexibility, and the enhanced properties derived from their nanostructured components. Conjugated polymers such as polyaniline (PANI), polypyrrole (PPy), and polythiophene (PT) are the foundation of these materials, as they allow for the delocalization of electrons along their backbone, enabling electrical conductivity (Bhadra et al., 2020). When incorporated with nanoparticles or nanostructures like CNTs, graphene, or metal nanoparticles, the resulting nanocomposites exhibit enhanced mechanical, optical, and electrochemical properties (Zhang et al., 2019). These conducting polymer-based nanomaterials have diverse applications and are essential in supercapacitors and batteries due to their excellent charge storage and conductivity properties (Wang et al., 2019b). They are utilized in developing highly sensitive chemical and biosensors, capitalizing on their conductivity change in response to stimuli (Gao et al., 2018). Their ability to undergo significant deformation under electrical stimuli makes them ideal for use in soft robotics and actuators (Lee et al., 2017). The integration of conducting polymers with nanomaterials opens up opportunities for advanced applications in flexible electronics, energy devices, and smart systems, driving innovation in various fields (Liu et al., 2020).
- **Hybrid Nanomaterials:** Hybrid nanomaterials are composed of distinct kinds of nanomaterials that are used to improve the performance of energy systems by benefiting from the complementary characteristics of each material. Composites integrate carbon-based materials such as graphene or CNTs with metal oxides or polymers to develop multifunctional materials to improve the performance of batteries and supercapacitors (Xie et al., 2010). Graphene-based composites, for instance, enhance the stability and charge storage capacity of energy storage devices. Hybrid nanomaterials, like organic–inorganic perovskites, combine organic and inorganic materials to form efficient solar cells. The hybrids exhibit high light absorption, stability, and energy conversion efficiency (Chen et al., 2015).

2.3 Nanomaterials in Renewable Energy Production

Nanomaterials in renewable energy production include solar energy, wind energy, biomass production, and fuel cells.

2.3.1 Solar Energy

Nanomaterials have found themselves to be the most important elements in enhancing the performance of solar cells, and they have significantly increased their efficiency in converting solar energy to electricity. The schematic diagram for solar energy harvesting is shown in Figure 2.7.

Their specific properties, including high surface area, tunable electronic properties, and capabilities to absorb light, make them exceptionally effective in solar energy usage. Solar cells nanostructured in the form of perovskite, quantum dots, and organic

FIGURE 2.7 The schematic diagram of solar energy harvesting.

photovoltaics (OPVs) are the greatest strides in the technology, leading to increased performance and efficiency.

- **Perovskite Solar Cells:** Perovskites, a group of materials with a special crystal structure, have shown outstanding potential in the field of solar energy. They possess high light absorption and charge carrier mobility, which are essential for efficient solar cells. Nanostructured perovskite solar cells involve perovskite nanocrystals or thin films, which enhance the power conversion efficiency (PCE). Their simplicity of production and capability for mass manufacturing have rendered them a viable competitor to conventional silicon solar cells, with efficiencies above 25% in recent research (Kojima et al., 2009).
- **Quantum Dot Solar Cells:** Quantum dots (QDs) are semiconductor nanomaterials with distinctive optical and electronic properties based on their quantum confinement effects. QDs can be tailored in size to maximize the bandgap to improve light absorption in a wide range of the sun's spectrum. In solar cells based on quantum dots, the materials are capable of increasing light harvesting by capturing photons in a wider range of the solar spectrum. Furthermore, QDs can enhance the charge separation process, thus leading to an improvement in the efficiency of solar cells (Mitzi, 2009).
- **Organic Photovoltaics (OPVs):** Organic solar cells employ organic materials (polymers or small molecules) as the active component. Nanostructured OPVs

are now a promising solution owing to their flexibility, low cost of production, and light weight. The addition of nanomaterials such as nanotubes and nanowires to OPVs facilitates charge transport and light absorption with enhanced efficiency. OPVs, although still under development, hold high promise for large-area, low-cost energy production (Yang et al., 2005).

Nanomaterials like perovskites, quantum dots, and OPVs are a significant step ahead in solar energy technology. Through increased light absorption, charge transport, and efficiency, nanostructured solar cells have the potential to make solar power more efficient and affordable, enabling the world's energy needs to be met sustainably.

2.3.2 Wind Energy

Wind energy is a sustainable resource that has attracted much attention as a long-term solution to world energy requirements. The durability and efficiency of wind turbines, to a great extent, are determined by the materials used to construct them, particularly the turbine blades, which are exposed to severe stresses and environmental conditions. The use of nanomaterials in turbine blade construction has been highly promising in enhancing the lightweight, high-strength nature of the blades, making them more efficient and long-lasting. Schematic diagram for wind energy production is shown in Figure 2.8.

Turbine blades are conventionally constructed using fiber-reinforced composites, commonly materials such as glass or carbon fibers in a polymer matrix. Although these materials possess good strength-to-weight ratios, nanomaterials can improve the performance of these composites further by enhancing their mechanical properties and overall durability. The incorporation of nanofillers like CNTs, graphene, and

FIGURE 2.8   Schematic diagram for wind energy production.

nanoclays into the polymer matrix results in nanocomposites, which provide improved mechanical strength, stiffness, and fatigue and wear resistance (Bonnar et al., 2006).

- **Carbon Nanotubes (CNTs):** CNTs are extensively researched for their superior mechanical strength and electrical conductivity. When added to the polymer matrix, CNTs enhance the tensile strength, flexural modulus, and impact resistance of turbine blades (Li et al., 2007). The superior mechanical properties of CNTs can be attributed to their peculiar nanoscale structure, which enables efficient load distribution and enhanced stress transfer in the composite material. This gives them lighter weights but higher strength, enabling them to better survive the high aerodynamic loads imposed upon them by the operation process.
- **Graphene:** Graphene is another very promising material for advancing turbine blade performance. Being an atomic layer of carbon atoms aligned in a 2D lattice structure, graphene contains exceptional strength, thermal conductivity, and electrical features (Novoselov et al., 2004). When incorporated into the composite matrix, graphene enhances the tensile strength and fracture toughness of the blade material. It also helps with increased durability, which allows blades to withstand the degradation caused by environmental conditions like UV radiation, humidity, and temperature variations better.
- **Nanoclays:** Nanoclays like montmorillonite are used extensively in nanocomposite formulation to improve the thermal stability and mechanical behavior of polymer-based materials (Zhou et al., 2010). The use of nanoclays adds to the molecular interaction between the composite matrix, which leads to increased strength and toughness. This improves the ability of turbine blades to withstand delamination and wear due to abrasion, enabling them to last longer.

Nanocomposite materials, reinforced with CNTs, graphene, and nanoclays, have exhibited vast potential in enhancing the performance and longevity of wind turbine blades. Nanomaterials improve the strength, stiffness, and fatigue life of turbine blades, enabling more efficient energy conversion and extended operational lifetimes. With increasing demand for renewable energy, the incorporation of nanomaterials into wind energy technologies will be pivotal in maximizing turbine performance and minimizing maintenance costs.

2.3.3 Biomass Energy

Biomass energy, which is generated from organic wastes like agricultural waste, wood, and algae, is key to the shift to renewable energy. One of the greatest challenges facing biomass and biofuel production is the effective conversion of raw biomass into energy. The use of nanomaterials, especially in catalysis, has demonstrated significant potential to enhance the efficiency and sustainability of biomass conversion processes. The schematic diagram of bioenergy production is shown in Figure 2.9. Nanocatalysts can have the ability to make the production of biofuels and algal fuels more efficient by enhancing reaction rates, selectivity, and the yield for biofuel production.

The process of converting biomass into biofuels, e.g., bioethanol, biodiesel, and biogas, typically encompasses operations such as fermentation, transesterification, and hydrolysis. The conventional catalysts adopted in these operations, e.g., metals and

FIGURE 2.9   Schematic diagram of bioenergy production.

acids, tend to be limited in terms of efficiency, selectivity, and sustainability. Nano-catalysts, on account of their high surface area, high reactivity, and ability to tune their properties at the nanoscale, possess several benefits over traditional catalysts (Figueiredo et al., 2010).

- **Biofuel Nanocatalysts:** For the manufacturing of biofuels from biomass feed-stocks, nanocatalysts, e.g., metal nanoparticles and metal oxide nanomaterials, are used to increase catalytic activity in processes like transesterification (used for the conversion of triglycerides into biodiesel) and hydroprocessing. For ex-ample, platinum (Pt), palladium (Pd), and nickel (Ni) nanoparticles have been used to significantly enhance rates of conversion and yield during such pro-cesses (Pérez et al., 2011). These nanocatalysts offer a greater active surface area for reactions, which leads to a more effective biomass conversion process, eliminating the necessity for high pressure and temperature, and thus reducing the energy input.
- **Nanocatalysts for Algae-Based Fuels:** Algae is yet another promising feed-stock for biofuel production with high lipid content and fast growth rates. Nano-catalysts are instrumental in the conversion of algae-derived oils to biodiesel. Nanocatalysts like zeolites, mesoporous silica, and graphene oxide have been investigated for the transesterification of algae oils, leading to increased yields of biodiesel with improved fuel properties (Chaudhary et al., 2017). The nano-catalysts assist in enhancing the reaction kinetics and catalytic stability, hence augmenting the overall efficiency of the biofuel process from algae. The ef-ficiency of biomass conversion is not just enhanced by nanocatalysts but also gains benefits such as reusability, environmental compliance, and performance of catalytic reactions under mild conditions. More accurate control of reaction mechanisms, along with increasing selectivity and yield of biofuels, can be ac-complished by the increased surface area, as well as variable properties of nano-materials (Figueiredo et al., 2010).

Nanomaterials, especially nanocatalysts, are critical to enhancing the efficiency and sustainability of biomass conversion processes. Through the enhancement of catalytic activity and reaction selectivity, nanocatalysts facilitate more efficient production of biofuel and algae-derived fuel, driving the shift toward cleaner, renewable energy sources. As nanocatalysis research continues, it is anticipated that these materials will increasingly optimize biomass and biofuel production, paving the way for a more sustainable energy future.

2.3.4 Fuel Cells: Use of Nanomaterials in Fuel Cell Technology

Fuel cell technology is one of the most promising substitutes for traditional energy systems, which can supply a clean and efficient means of transforming chemical energy into electrical energy. Fuel cells work by employing an electrochemical process between a fuel (hydrogen or methanol, for example) and an oxidizer (often oxygen), which generates electricity, water, and heat as byproducts. Fuel cells possess a number of benefits, such as high efficiency, zero emissions, and minimal environmental footprint, which make them suitable for use in transportation, power generation, and portable applications.

Nevertheless, there are still obstacles to be overcome, especially concerning catalysts for fuel cells. The efficiency of the catalysts determines how well fuel cells function since they determine how the electrochemical processes occur at the anode and cathode. Nanomaterials have changed fuel cell technology, providing improved properties that enhance catalyst functionality and minimize costs, thereby enabling fuel cells to become more commercially feasible. This section highlights the significance of nanocatalysts in proton exchange membrane fuel cells (PEMFCs) and direct methanol fuel cells (DMFCs), with emphasis on how nanomaterials have led to improvements in fuel cell performance, durability, and efficiency.

- **Proton Exchange Membrane Fuel Cells (PEMFCs):** PEMFCs are among the most commonly researched and economically promising fuel cell technologies. PEMFCs utilize a solid polymer electrolyte (the proton exchange membrane) to transfer protons from the anode to the cathode, and the electrons flow through an external circuit to produce electricity. The central reaction includes the oxidation of hydrogen on the anode and the reduction of oxygen on the cathode to produce water as the product. The performance of PEMFCs is highly dependent on the electrocatalysts employed in the anode and cathode. Conventional platinum (Pt)-based catalysts are not only efficient but also expensive, rare, and subject to degradation over a period of time. Nanomaterials, or nanocatalysts, can greatly enhance the efficiency, cost-effectiveness, and durability of these fuel cells. Platinum is the most common catalyst for hydrogen oxidation and oxygen reduction reactions (ORR) in PEMFCs. Nevertheless, its high price and limited supply have spurred extensive research into other nanocatalysts. Nanostructured platinum catalysts, including Pt nanoparticles, Pt nanowires, and Pt nanodendrites, have several benefits over bulk platinum. In addition to lowering the price of PEMFCs, researchers are working on non-platinum catalysts such as materials of nickel (Ni), iron (Fe), and cobalt (Co) materials, typically nanostructured materials. Carbonaceous materials like graphene, CNTs, and graphene oxide are also under study as prospective substitute catalysts or support material for non-platinum catalysts.

- **Direct Methanol Fuel Cells (DMFCs):** DMFCs are another fuel cell technology that converts methanol directly into electricity. DMFCs are useful for portable use due to the high energy density of methanol and its room-temperature liquid state, making it convenient to store and transport. Nonetheless, the efficiency of DMFCs is affected by the sluggish ORR at the cathode and the crossover of methanol from the anode to the cathode, decreasing overall efficiency. Similar to PEMFCs, DMFCs need efficient electrocatalysts to improve the hydrogen oxidation reaction (HOR) and ORR. Nanocatalysts play a significant role in enhancing the performance of DMFCs by improving the catalytic activity and stability of the materials.
- **Nanocatalysts for Methanol Oxidation Reaction (MOR):** The MOR at the anode is a vital process in DMFCs. Platinum, particularly PtRu alloys (platinum–ruthenium), is widely employed for this reaction due to its capacity to oxidize methanol with good efficiency. Platinum still suffers from poisoning from CO intermediates produced during oxidation, however. Nanocatalysts have been found to possess potential in improving the MOR by enhancing catalyst durability and efficiency of methanol oxidation. PtRu nanoparticles have been intensively investigated for their potential in enhancing the MOR due to the capability of Ru to rupture the CO bonding and avoid poisoning of the catalyst. The employment of nanostructured PtRu (such as PtRu nanodendrites or PtRu core–shell nanostructures) also supports the efficiency and stability of the catalyst by expanding the accessible active sites and enhancing the electrocatalytic activity (Ma et al., 2017).
- **Nanocatalysts for Oxygen Reduction Reaction (ORR):** The ORR at the cathode is often a rate-limiting step in DMFCs. Nanocatalysts, particularly platinum-based catalysts and carbon-based catalysts, have been shown to enhance the efficiency of this process by increasing the electronic structure and surface area. Similar to PEMFCs, nanostructured platinum materials like Pt nanoparticles, Pt nanorods, and Pt nanowires have been employed to enhance ORR performance in DMFCs. Their unique electronic properties and high surface area contribute to the enhancement of catalyst activity and stability. Besides platinum, scientists are also exploring carbon-based substances like graphene and CNTs as substitute ORR catalysts in DMFCs. Doping these substances with nitrogen or phosphorus, scientists can effectively increase their electrocatalytic activity toward oxygen reduction, optimizing the whole efficiency of the DMFC system (Wang et al., 2019a).

Nanomaterials take a transformative role in advancing fuel cell technologies, especially the improvement of the catalytic efficiency, stability, and cost-effectiveness of PEMFCs and DMFCs. Nanocatalysts, such as platinum-based and non-platinum nanomaterials, have greater electrocatalytic activity, higher surface area, and improved degradation resistance, enabling fuel cells to become more practical for commercial applications. While ongoing research is directed toward the development of non-precious metal-based catalysts and hybrid nanocatalysts, fuel cells have the potential to be an important part of moving toward clean, sustainable energy systems

2.3.5 Thermoelectric Materials

Thermoelectric materials, which directly convert heat into electricity using the See-beck effect, have been a focus area in recent times as a method to tap into waste heat energy and generate usable power. The process promises much for both energy harvesting technologies, where waste heat is readily available (industrial processes, automobile engines, and power generation plants), and in creating solid-state cooling technology for electronic devices. In spite of their promise, the effectiveness of ther-moelectric materials has hitherto been constrained by comparatively low thermoelec-tric performance. This performance is measured by the figure of merit (ZT), which is a function of the Seebeck coefficient, electrical conductivity, and thermal conduc-tivity of the material. Nanomaterials have been found to be a potential solution to increase this efficiency. The nanostructuring brings in a number of significant effects that can be exploited to increase the Seebeck coefficient and overall thermoelectric performance. These are the decrease in thermal conductivity through phonon scatter-ing (heat carriers) and the increase in the Seebeck coefficient through the quantum confinement effects.

Nanomaterials for thermoelectrics:

- **Nanostructured Bulk Materials:** Nanostructuring bulk materials is one of the most prevalent strategies to improve the thermoelectric performance. The nanograins or nanoparticles in the bulk matrix are phonon scattering centers, which decrease the thermal conductivity but will not significantly decrease the electrical conductivity. This method can lead to a dramatic enhancement of ZT. Perhaps one of the best-studied materials for thermoelectricity, bismuth telluride (Bi_2Te_3), is a high-temperature thermoelectric semiconductor at room temperature. Nanostructures like nanowires, nanoparticles, or quantum dots introduced into the matrix of Bi_2Te_3 have been used effectively to decrease thermal conductivity and increase the Seebeck coefficient. These nanostructures scatter phonons efficiently, which decreases thermal conductivity without hindering electron transport, lead-ing to an increased figure of merit (Chen et al., 2013). Lead telluride is one of the better materials for use in thermoelectric applications at higher temperatures. The incorporation of nanostructures like PbTe quantum dots and nanoinclusions helps to lower thermal conductivity substantially, as well as increase the Seebeck coefficient. Current research has proven that PbTe-based nanocomposites, accom-panied by nanoparticles of silver (Ag), can offer remarkable ZT values, especially at higher temperatures (Zhao et al., 2014).
- **Nanowires and Nanotubes:** Nanowires and nanotubes are of interest because of their 1D structures, which provide unique possibilities for the increase in thermoelectric performance. Silicon is the most ubiquitous and cheap material, and Si nanowires have been a subject of study for thermoelectric devices for a long time. The 1D structure of silicon nanowires greatly reduces thermal con-ductivity through phonon scattering along the axis of the wire. Additionally, the doping level can be controlled to tune the electrical conductivity. The high Seebeck coefficient of silicon nanowires has been attributed to the increased density of states near the Fermi level caused by quantum confinement (Wang

et al., 2016). CNTs are also good materials for thermoelectric applications as they have high electrical conductivity and low thermal conductivity along the length. Nanostructuring CNTs into arrays or composites has been found to enhance the ZT values of thermoelectric materials. CNTs have high Seebeck coefficients because of their metallic conductivity and high surface area, which are unique electronic properties. When incorporated into materials such as graphene or polymer composites, they can function as effective electron transporters, enhancing thermoelectric performance (Gao et al., 2017).

* **Quantum Dots and Nanoparticles:** QDs and nanoparticles are a second group of nanomaterials that have proven highly effective at enhancing the thermoelectric performance of materials. These nanomaterials can be introduced into bulk thermoelectric matrices in order to improve their Seebeck coefficient due to quantum effects. Incorporating PbSe quantum dots into the PbTe matrix has been proven to exhibit promising improvement in the thermoelectric properties. The quantum dots introduce novel quantum confinement effects, by which the density of states within the vicinity of the Fermi level can increase, thus also increasing the Seebeck coefficient. The incorporation of these quantum dots also decreases thermal conductivity by inducing phonon scattering, thereby showing a remarkable ZT value increase in the composite material (Li et al., 2015). Embedding nanoparticles like gold (Au) or silver (Ag) in polymer matrices has also been another means of improving thermoelectric characteristics. Nanoparticles in the polymer matrices increase the Seebeck coefficient and the electrical conductivity based on the interplay between nanoparticles and polymer chains, resulting in improved thermoelectric performance (Liu et al., 2017).

2.4 Nanomaterials in Energy Storage

2.4.1 Batteries

Batteries are now an indispensable part of contemporary life, driving everything from handheld electronics to electric vehicles (EVs) and renewable energy storage systems. Conventional battery technologies like lithium-ion (Li-ion) batteries, however, suffer from limitations regarding capacity, longevity, charging times, and safety. The use of nanomaterials in battery components, especially electrodes and electrolytes, brings about remarkable advances in performance, capacity, and longevity, solving many of these issues. This section explores the role of nanomaterials in advancing lithium-ion, sodium-ion, and solid-state batteries.

2.4.1.1 Lithium-Ion Batteries: Enhancing Performance with Nanomaterials

Lithium-ion batteries (LIBs) are the most popular rechargeable batteries at present because of their high energy density, long cycle life, and relatively low self-discharge. Nevertheless, there are a number of areas for improvement in LIBs, such as capacity, rate capability, thermal stability, and cycle life. Nanomaterials have been a main research focus to improve these characteristics.

- **Nanostructured Electrodes:** One of the major bottlenecks in LIBs is the electrode material, especially the anode and cathode, where the active materials (like graphite in the anode and lithium cobalt oxide ($LiCoO_2$) in the cathode) suffer from problems like structural damage and capacity degradation during charge/discharge cycles. The use of nanostructured materials provides a probable solution by enhancing the surface area, electrochemical activity, and efficiency of ion diffusion.
- **Nanostructured Anodes:** Silicon anodes have shown potential as a superior substitute for graphite because of their enormous theoretical capacity (approximately 10 times higher than for graphite). Silicon, however, experiences serious volume growth upon cycling, which causes mechanical deterioration and loss of capacity. The incorporation of silicon nanowires or silicon nanospheres alleviates the stress-induced degradation owing to the flexibility and greater structural integrity of the nanomaterials (Liu et al., 2012). Likewise, CNTs and graphene have been utilized as conductive additives in the anode to enhance electron transport and stabilize the electrode structure (Wang et al., 2013).
- **Nanostructured Cathodes:** Nanomaterials also enhance the performance of cathodes. For example, the use of nanostructured lithium manganese oxide ($LiMn_2O_4$), lithium iron phosphate ($LiFePO_4$), and nickel–cobalt–manganese (NCM) composites enhances the rate capability and cycling stability by increasing the surface area and providing shorter diffusion paths for lithium ions. Furthermore, nanostructured cathodes improve ion diffusion during charging and discharging, resulting in better overall performance and capacity retention over extended cycles (Cheng et al., 2015).
- **Nanostructured Electrolytes:** Besides electrodes, the electrolyte is also important in deciding the performance of LIBs. The electrolyte should be able to conduct lithium ions between the cathode and anode efficiently during cycling and have high stability and safety at a variety of temperatures. Nanomaterials have been utilized to enhance the ionic conductivity, thermal stability, and electrochemical stability of liquid- and solid-state electrolytes.
- **Nanocomposite Electrolytes:** Solid polymer electrolytes (SPEs) and liquid electrolytes may be improved by the addition of nanofillers, including nanoclays, silica, and polymer nanocomposites. The materials improve the ionic conductivity and mechanical strength of the electrolyte without sacrificing its flexibility. For example, silica nanoparticles added to polymer electrolytes facilitate the ion transport properties and thermal stability of the electrolyte, which in turn improves the overall performance of LIBs (Yuan et al., 2012).

2.4.1.2 Sodium-Ion Batteries: A Promising Replacement for Lithium-Ion

Sodium-ion batteries (SIBs) are emerging as a promising substitute for LIBs, especially for large-scale energy storage systems. Sodium is more available and less expensive than lithium, and thus SIBs are a promising candidate for applications where cost is a major concern. SIBs, however, suffer from disadvantages like lower energy density and poor cycling stability because of the larger ionic radius of sodium over lithium.

Employing nanostructured electrodes in SIBs can substantially enhance performance by solving problems with volumetric expansion and low ion diffusion rates.

- **Nanostructured Anodes:** Carbon-based anodes, including hard carbon or graphene, are widely utilized in SIBs. Through the regulation of nanostructure in carbon-based materials, sodium storage capacity and rate capability can be enhanced. For example, hard carbon nanomaterials offer a high surface area and flexible structure, allowing for better sodium ion diffusion and reversible capacity (Ji et al., 2009).
- **Nanostructured Cathodes:** Nanostructured cathodes like sodium manganese oxide ($NaMnO_2$) or sodium iron phosphate ($NaFePO_4$) have been found to promote the cycling stability and capacity retention of SIBs. Nanostructuring these cathodes enhances the structural stability and ionic conductivity upon cycling, resulting in improved overall performance in SIBs (Fang et al., 2015).
- **Nanocomposite Electrolytes in Sodium-Ion Batteries:** SIBs electrolytes are also improved through nanomaterials. Various nanofillers, such as silica and polymer-based nanocomposites, have been exploited to enhance the ion conductivity and heat stability in the solid-state electrolytes of SIBs. These all improve the restrictions of conventional liquid electrolytes for enhancing the security and overall performance of SIBs (Zhou et al., 2013).

2.4.1.3 Solid-State Batteries: The Future of Energy Storage

Solid-state batteries (SSBs) are a new generation of energy storage devices that incorporate a solid electrolyte in place of the usual liquid or gel electrolytes. Solid-state batteries can potentially deliver higher energy densities, better safety, and increased cycle lives over traditional LIBs. However, issues such as low ionic conductivity and interfacial problems between electrodes and electrolyte need to be overcome.

- **Nanomaterials in Solid-State Electrolytes:** Nanomaterials also facilitate enhanced ionic conductivity and surface stability of the solid-state electrolytes. An example of them includes nanostructured ceramics, garnet-type solid electrolytes, and perovskite-type solid electrolytes. They exhibit good conductivities and the capability to work under room temperatures. Addition of nanoparticles into such substances enhances the population density of grains within these substances, lessens the flow impedance of ions, and makes battery efficiency more satisfactory overall (Wang et al., 2015).
- **Nanostructured Electrodes for Solid-State Batteries:** In the same way, nanostructured electrodes play a vital role in the functioning of solid-state batteries. By including nanoparticles or nanowires in the anode and cathode materials, the cycling stability and rate capability can be greatly enhanced. The decreased diffusion distance of ions in nanostructured electrodes leads to accelerated charge/discharge cycles and better capacity retention over a period (Chen et al., 2014).

Nanomaterials have transformed battery technology, greatly enhancing the performance, capacity, and lifespan of lithium-ion, sodium-ion, and solid-state batteries.

Through nanostructuring of electrode materials and electrolytes, these batteries can attain greater energy densities, faster charging rates, improved cycling stability, and enhanced safety profiles. As scientific investigation advances further, the incorporation of nanomaterials will take center stage in the creation of future batteries that will address the increasing energy storage needs in EVs, renewable power systems, and portable electronics.

2.4.2 Supercapacitors

Energy storage devices play a key role in solving the growing energy demands of the world, especially in the case of renewable energy integration and EVs. Of the many energy storage devices, supercapacitors (or ultracapacitors) have attracted a lot of interest because they can offer high power density, high charge/discharge rates, and long cycle life. But conventional supercapacitors are limited by energy density, which is significantly less than batteries. To bypass these limitations, nanomaterials, such as graphene and hybrid systems, are being considered to improve the performance of supercapacitors. This part explores the application of graphene-based supercapacitors and hybrid supercapacitors, bringing together the benefits of batteries and supercapacitors, in enhancing energy storage.

2.4.2.1 Graphene-Based Supercapacitor

Graphene supercapacitors generally use graphene oxide (GO) or reduced graphene oxide (rGO) as the primary electrode material. Utilization of GO or rGO improves ionic interaction with the electrolyte, further increasing the capacitive performance. Graphene may also be mixed with CNTs, metal oxides, or conducting polymers to create nanocomposites, leading to improved energy and power density.

- **Graphene–CNT Composites:** The combination of graphene with CNTs produces a hybrid material that capitalizes on the strengths of each constituent. CNTs bring about enhanced conductivity and mechanical strength, while graphene brings surface area and charge storage capacity. Such hybrid materials have been shown to enhance the electrochemical performance of supercapacitors, specifically energy density and cycle stability (Liu et al., 2014).
- **Graphene–Metal Oxide Hybrid Composites:** Incorporation of metal oxides like manganese oxide (MnO_2) or cobalt oxide (Co_3O_4) into graphene leads to hybrid composite electrodes, displaying both electrostatic and pseudocapacitive charge storage capabilities. Metal oxides introduce faradaic charge storage mechanisms beyond electrostatic behavior, which improves the overall supercapacitance of the supercapacitor. The high energy density of these hybrid graphene–metal oxide supercapacitors is prominent, especially in the case of asymmetric supercapacitors using different materials as the anode and cathode (Zhang et al., 2017).
- **Graphene Conducting Polymers:** Graphene may also be blended with conducting polymers like polypyrrole (PPy) or polyaniline (PANI). These polymers provide high pseudo-capacitance and add to the overall device capacitance. When blended with graphene, conducting polymers improve charge storage, electrochemical stability, and cycle life (Cheng et al., 2013).

2.4.2.2 Metal Oxide-Based Supercapacitors

Supercapacitors have become key players in the field of energy storage due to their ability to charge quickly and deliver power efficiently. Among various types, metal oxide-based supercapacitors stand out for their high energy density and remarkable redox activity, offering a strong alternative to batteries and traditional capacitors in many applications. Unlike electric double-layer capacitors (EDLCs), which store energy through the electrostatic accumulation of charges at the electrode interface, metal oxide-based supercapacitors fall under the category of pseudo-capacitors. These store energy through fast, reversible faradaic redox reactions, allowing them to achieve higher capacitance and energy density.

One of the earliest and most efficient materials used is ruthenium oxide (RuO_2). It offers excellent conductivity, a broad operating voltage window, and high specific capacitance – up to 700 F/g (Conway, 1999). However, due to its high cost and limited supply, research has shifted to more affordable options like manganese oxide (MnO_2), nickel oxide (NiO), and cobalt oxide (Co_3O_4). MnO_2 is particularly attractive due to its abundance, low toxicity, and theoretical capacitance of up to 1370 F/g. However, it suffers from poor electrical conductivity, limiting its real-world application. To overcome this, scientists have created hybrid electrodes that combine MnO_2 with conductive materials such as graphene or CNTs, enhancing performance and structural integrity (Wang et al., 2012).

NiO and Co_3O_4, meanwhile, offer better electronic conductivity and mechanical strength. These materials show promising performance, especially when integrated into asymmetric supercapacitor systems, where one electrode is faradaic (like NiO) and the other is non-faradaic (like activated carbon). NiO also benefits from relatively simple synthesis and tunable nanostructures, while Co_3O_4 brings high theoretical capacity and strong redox properties (Li et al., 2013).

Despite these advantages, there are challenges to overcome. Many metal oxides suffer from low conductivity and structural degradation over time. This has led to research focusing on nanostructuring (such as creating porous nanosheets or nanowires) and composite materials, combining the high capacitance of oxides with the conductivity of carbon-based supports. Metal oxide-based supercapacitors offer a powerful and sustainable option for future energy storage technologies. With continued innovation in material design and fabrication techniques, these systems are well-positioned to meet the growing demand for high-performance, eco-friendly energy solutions.

2.4.2.3 Hybrid Supercapacitors

Hybrid supercapacitors are a promising answer to filling the gap between batteries and supercapacitors. Supercapacitors have high power density and can be charged quickly, but they have low energy density relative to batteries. Batteries have high energy density but have lower charge/discharge speeds. Hybrid supercapacitors seek to put together the best aspects of both devices by using materials that offer high energy storage capacity and rapid charge/discharge times. Hybrid supercapacitors generally consist of two different types of electrodes, each providing distinct charge storage mechanisms. One electrode is typically designed to store energy through electrostatic charge accumulation, like a traditional supercapacitor, while the other stores energy using a faradaic reaction, as in a battery. The combination of these mechanisms results in a device that offers both high energy density and power density.

- **Asymmetric Hybrid Supercapacitors:** These are designed with various materials for the cathode and anode, such that there is a mix of capacitor and battery-like qualities. For instance, the anode may consist of graphene or carbon-based materials, whereas the cathode may consist of a pseudocapacitive material such as MnO_2 or a lithium-based material. The outcome is a device that has higher energy density than conventional supercapacitors without sacrificing high power density and rapid charge/discharge (Zhou et al., 2013).
- **Lithium-Ion Hybrid Supercapacitors:** Other hybrid devices include lithium-ion hybrid supercapacitors, which incorporate Li-ion battery materials (e.g., graphite or $LiFePO_4$) with supercapacitor electrodes (usually activated carbon- or graphene-based materials). The devices possess greater energy density than conventional supercapacitors, in addition to the fast charge/discharge nature of supercapacitors (Chong et al., 2015).
- **Supercapacitors with Metal–Air Electrodes:** Hybrid systems in some cases employ metal–air electrodes, like zinc–air or lithium–air electrodes, which introduce a secondary source of chemical energy to supplement the electrostatic energy storage. These electrodes are integrated with conventional supercapacitor materials to yield hybrid supercapacitors with still greater energy densities and longer lifetimes.

The use of nanomaterials in energy storage devices, specifically graphene-based supercapacitors and hybrid supercapacitors, has greatly improved the performance and potential of supercapacitors as good substitutes or complements to conventional battery technologies. Graphene supercapacitors increase surface area, electrical conductivity, and mechanical stability, which improves power density, energy density, and cycle life. Hybrid supercapacitors, however, by integrating the strengths of both batteries and supercapacitors, offer a balanced solution for applications with high energy and fast charge/discharge requirements. With the ongoing development of nanomaterials and hybrid systems, the potential for more efficient and long-lasting energy storage devices will keep expanding, supporting wider use of renewable energy technologies and electric mobility.

2.5 Conclusion

Nanomaterials are playing a game-changing role in the journey toward a greener and cleaner energy future. Thanks to their tiny size and powerful properties, they are helping to reshape the way we generate, store, and conserve energy. While they might be invisible to the naked eye, their impact on modern energy solutions is anything but small. One of the most promising areas where nanomaterials are making a real difference is in solar energy. By using materials like quantum dots and perovskite nanocrystals, scientists have found ways to capture sunlight more efficiently and convert it into electricity. These materials can absorb more light and help solar panels work better, even when the sun is not at its brightest, which means more clean energy from the same amount of sunlight – something we can all benefit from.

When it comes to energy storage, nanotechnology is opening new doors. Batteries made with nanomaterials can charge faster, last longer, and hold more power. Imagine a phone that only needs to charge for a few minutes, or an electric car that can drive farther without needing a recharge – these kinds of improvements are becoming possible thanks to nanotech. Supercapacitors, another kind of energy storage device, are also getting a boost from nanomaterials, making them more reliable and powerful. Hydrogen, often called the fuel of the future, is another area where nanomaterials shine. They are helping make hydrogen production cleaner and more efficient and are also being used in better fuel cells and storage systems, which are key to making hydrogen a real alternative to fossil fuels.

Beyond performance and innovation, nanomaterials also support cleaner energy from an environmental point of view. They help reduce waste, make processes more efficient, and often allow us to replace more harmful materials. In many ways, they support the idea of doing more with less, which is at the heart of sustainability. Of course, this is not to say everything is perfect. There are still hurdles to overcome, like making these materials affordable and ensuring that they are safe for people and the environment over the long term. But the potential they bring is too important to ignore. With smart planning, careful research, and responsible use, nanomaterials can be a key part of our energy solutions moving forward.

In short, nanomaterials give us hope. They remind us that sometimes, the smallest things can spark the biggest changes. And as we look for ways to power our lives without harming the planet, their role will only continue to grow.

REFERENCES

Arico, A. S., Bruce, P., Scrosati, B., Tarascon, J.-M., & van Schalkwijk, W. (2005). Nanostructured Materials for Advanced Energy Conversion and Storage Devices. Nature Materials, 4(5), 366–377.

Bera, S., Roy, D., & Das, P. (2019). Nanomaterials for Energy and Environmental Applications. Wiley-VCH.

Bhadra, S., Khastgir, D., Singha, N. K., & Lee, J. H. (2020). Conducting Polymers in Nanocomposites: Synthesis, Properties, and Applications. *Journal of Materials Chemistry C*, 8(12), 4152–4163.

Bonnar, J. L., Smith, R., & Johnson, P. (2006). Nanocomposites for Wind Turbine Blade Materials. Composites Science and Technology, 66(9), 1220–1228.

Chaudhary, V., Sharma, A., & Gupta, R. (2017). Nanocatalysts in Algae-Based Biofuel Production: A Review. Renewable and Sustainable Energy Reviews, 77, 1–11.

Chen, L., Li, H., Zhang, X., & Wang, H. (2015). Organic-Inorganic Hybrid Materials for Solar Cells. Nature Materials, 14(2), 118–126.

Chen, L., Wang, H., Zhang, X., & Li, X. (2014). Nanostructured Electrodes for Solid-State Batteries. Advanced Energy Materials, 4(8), 1300840.

Chen, Z., Li, Y., Wang, H., & Zhang, X. (2013). Enhanced Thermoelectric Performance of Bismuth Telluride Nanostructures. Nature Materials, 12, 891–897.

Cheng, H., Li, X., Wang, H., & Chen, L. (2013). Conducting Polymer/Graphene Nanocomposites for Supercapacitors: A Review. Journal of Materials Chemistry A, 1(4), 1238–1253.

Cheng, J., Li, X., Wang, H., & Chen, L. (2015). Nanostructured Lithium Manganese Oxide Cathodes for Lithium-Ion Batteries. Journal of Power Sources, 299, 328–334.

Chong, L. S., Tan, B., & Lee, C. (2015). Lithium-Ion Hybrid Supercapacitors: A Review. Energy Storage Materials, 1(1), 1–17.

Conway, B. E. (1999). *Electrochemical Supercapacitors: Scientific Fundamentals and Technological Applications.* Springer. https://doi.org/10.1007/978-1-4615-4702-0

Deng, Z., Li, J., Zhang, X., Wang, H., & Chen, L. (2019). One-Dimensional Nanomaterials: Synthesis, Properties, and Applications. *Journal of Nanoscience and Nanotechnology*, 19(3), 1575–1590.

Fang, Z., Li, X., Zhang, X., & Chen, L. (2015). Nanostructured Sodium-Ion Battery Electrode Materials: Opportunities and Challenges. Energy & Environmental Science, 8(5), 1683–1701.

Figueiredo, J. L., Silva, A. M., & Almeida, P. (2010). Nanocatalysts for Biomass Conversion into Biofuels. Catalysis Science & Technology, 1(1), 1–15.

Gao, L., Wang, H., Zhang, X., & Chen, L. (2017). Thermoelectric Performance of Carbon Nanotube-Based Composites. Advanced Materials, 29(1), 1605615.

Gao, Y., Li, H., Zhang, X., Wang, H., & Chen, L. (2018). Polymer-Based Nanocomposite Sensors for Chemical and Biological Detection. *Sensors and Actuators B: Chemical*, 260, 23–38.

Huang, Z., & Jiang, L. (2019). Sustainable Materials and Dematerialization. Nature Sustainability, 2(3), 196–197.

Huang, Z., Fang, H., & Lu, A.-H. (2007). Nanowires in Energy Conversion and Storage. Nano Letters, 7(7), 2099–2106.

Iijima, S. (1991). Helical Microtubules of Graphitic Carbon. Nature, 354(6348), 56–58.

Ji, X., Li, X., Zhang, X., & Chen, L. (2009). Hard Carbon Nanomaterials for Sodium-Ion Batteries. Nature Materials, 8(9), 738–743.

Kojima, A., Teshima, K., Shirai, Y., & Miyasaka, T. (2009). Organometal Halide Perovskites as Visible-Light Sensitizers for Photovoltaic Cells. Journal of the American Chemical Society, 131(17), 6050–6051.

Kumar, A., Singh, R., Sharma, P., & Verma, K. (2020). Functional Nanomaterials: Advances and Applications in Energy, Electronics, and Medicine. Springer.

Lee, S., Kim, J., Park, S., & Lee, C. (2017). Flexible and Stretchable Conducting Polymer-Based Actuators for Soft Robotics. *Advanced Materials*, 29(50), 1701325.

Li, H., Zhang, X., Wang, H., & Chen, L. (2015). Quantum Dot-Embedded Thermoelectric Materials for Efficient Heat Conversion. Energy & Environmental Science, 8(4), 1102–1110.

Li, L., Wu, Z., Yuan, S., & Zhang, X. (2013). Advances and Challenges for Flexible Energy Storage and Conversion Devices and Systems. Energy & Environmental Science, 7, 2101–2122.

Li, X., Wang, H., Zhang, X., & Chen, L. (2007). Carbon Nanotubes Reinforced Polymer Nanocomposites for High-Performance Wind Turbine Blades. Materials Science and Engineering: A, 444(1–2), 236–242.

Liu, J., Burghaus, U., Besenbacher, F., & Wang, Z. L. (2010). Nanotechnology for Energy and Environmental Science. Nanotechnology, 21(28), 280301.

Liu, N., Zhang, X., Wang, H., & Chen, L. (2012). Silicon Nanowire Anodes for Lithium-Ion Batteries. Advanced Energy Materials, 2(6), 689–696.

Liu, X., Zhang, X., Wang, H., & Chen, L. (2017). Nanoparticle-Embedded Polymers for Enhanced Thermoelectric Properties. Nano Energy, 36, 170–178.

Liu, Y., Wang, H., Zhang, X., & Chen, L. (2014). Carbon Nanotubes and Graphene-Based Composites for Supercapacitors. Journal of Power Sources, 262, 91–106.

Liu, Y., Wang, H., Zhang, X., & Chen, L. (2020). Conducting Polymer Nanomaterials: Synthesis, Properties, and Applications in Energy Storage Devices. *Nano Energy*, 71, 104675.

Ma, Y., Li, J., Zhang, W., & Chen, L. (2017). Platinum-Ruthenium Nanocatalysts for Methanol Oxidation in Direct Methanol Fuel Cells. Electrochimica Acta, 225, 201–211.

Mitzi, D. B. (2009). Quantum Dots and Their Applications in Photovoltaics. Nature Materials, 8(3), 166–174.

Novoselov, K. S., Geim, A. K., Morozov, S. V., Jiang, D., Zhang, Y., Dubonos, S. V., Grigorieva, I. V., & Firsov, A. A. (2004). Electric Field Effect in Atomically Thin Carbon Films. Science, 306(5696), 666–669.

Novoselov, K. S., Mishchenko, A., Carvalho, A., & Castro Neto, A. H. (2016). 2D Materials and Van der Waals Heterostructures. *Science*, 353(6298), 628–633.

Pérez, J. L., Pérez, J. L., González, M., & Torres, A. (2011). Nanocatalysts in Biodiesel Production. Energy & Environmental Science, 4(6), 1820–1831.

Poh, H. L., Tan, S. J., Lee, C. K., & Lim, K. H. (2018). Functional Nanomaterials: Design, Synthesis, and Applications. *Nanomaterials*, 8(12), 1025.

Ramakrishnan, V., Gupta, R., & Sinha, P. (2021). Nanomaterials for Clean Energy Applications. Elsevier.

Stoller, M. D., Park, S., Zhu, Y., An, J., & Ruoff, R. S. (2008). Graphene-Based Ultracapacitors. Nano Letters, 8(10), 3466–3470.

Turner, J. A. (2015). Sustainable Hydrogen Production. Science, 305(5686), 972–974.

Wang, H., Casalongue, H. S., Liang, Y., & Dai, H. (2012). Ni(OH)2 Nanoplates Grown on Graphene as Advanced Electrochemical Pseudocapacitor Materials. Journal of the American Chemical Society, 132(21), 7472–7477.

Wang, H., Li, X., Zhang, X., & Chen, L. (2013). Carbon Nanotube-Graphene Hybrid Materials for Lithium-Ion Battery Applications. Journal of Materials Chemistry A, 1(5), 1908–1915.

Wang, H., Li, X., Zhang, Y., Chen, L., & Liu, Y. (2018). Zero-Dimensional Nanomaterials and Their Applications in Sensing, Catalysis, and Environmental Remediation. *Nano Today*, 19, 39–56.

Wang, X., Li, H., Zhang, X., & Chen, L. (2019a). Carbon-Based Non-Precious Metal Catalysts for Oxygen Reduction Reaction. Energy & Environmental Science, 12(5), 1675–1691.

Wang, X., Li, J., Zhang, X., & Chen, L. (2016). Silicon Nanowires for Thermoelectric Applications. Nano Letters, 16(5), 3461–3467.

Wang, Y., Li, X., Zhang, X., & Chen, L. (2019b). Conducting Polymers for Supercapacitors and Batteries: A Review. *Electrochimica Acta*, 318, 128–138.

Wang, Y., Zhang, X., Liu, Y., & Chen, L. (2015). Nanostructured Solid Electrolytes for Lithium-Ion Batteries. Journal of Solid State Electrochemistry, 19(6), 1859–1870.

Xie, X., Li, H., Zhang, X., & Chen, L. (2010). Graphene and its Hybrid Nanocomposites for Energy Storage. Nanotechnology, 21(25), 254018.

Yan, Z.-F., Hao, G.-P., & Lu, A.-H. (2010). Materials for Renewable and Sustainable Energy. Advanced Materials, 22(28), 3026–3033.

Yang, X., Li, H., Wang, H., & Chen, L. (2005). Organic Photovoltaics: Materials, Devices, and Applications. Materials Science and Engineering: R: Reports, 47(1–2), 1–38.

Yip, N. Y., Brogioli, D., Hamelers, H. V. M., & Nijmeijer, K. (2016). Salinity Gradients for Sustainable Energy: Primer, Progress, and Prospects. Environmental Science & Technology, 50(24), 12072–12094.

Yuan, X., Li, H., Zhang, X., & Chen, L. (2012). Silica Nanoparticles as Additives in Solid Polymer Electrolytes for Lithium-Ion Batteries. Solid State Ionics, 217, 47–53.

Zhang, J., Li, X., Wang, H., & Chen, L. (2019). Nanostructured Conducting Polymers: Synthesis, Characterization, and Applications. *Journal of Nanomaterials*, 2019, 5210421.

Zhang, L., Wang, H., Liu, Y., & Chen, L. (2017). Graphene/Metal Oxide Supercapacitors: Materials, Performance, and Applications. Energy & Environmental Science, 10(1), 24–56.

Zhang, X., Li, H., Wang, Y., Chen, L., & Liu, Y. (2020). Three-Dimensional Nanomaterials: Fundamentals, Applications, and Advances. *Journal of Materials Chemistry A*, 8(4), 2200–2221.

Zhao, L. D., Tan, G., Hao, Q., He, J., Pei, Y., Chi, H., Wang, H., & Kanatzidis, M. G. (2014). Enhanced Thermoelectric Performance of PbTe Nanocomposites. Science, 351(6275), 141–146.

Zhou, H., Li, J., Wang, H., & Chen, L. (2013). Polymer Nanocomposite Electrolytes for Sodium-Ion Batteries. Journal of Materials Chemistry A, 1(18), 5852–5862.

Zhou, Y., Li, C., Wang, J., Zhang, X., & Chen, Y. (2010). The Influence of Nanoclays on the Mechanical Properties of Polymer Nanocomposites for Wind Turbine Blades. Materials Science and Engineering: A, 527(10–11), 2454–2460.

Zhou, Y., Li, X., Wang, H., Zhang, X., & Chen, L. (2021). Advanced Functional Materials for Sustainable Energy and Environmental Applications. Nature Reviews Materials, 6(4), 1–22.

3

Life Cycle Thinking of Nanotechnology

Himanshi Goel and Kumar Rakesh Ranjan

3.1 Introduction

Nanotechnology has rapidly established itself as a groundbreaking field, opening new possibilities across diverse areas like healthcare, energy, electronics, and environmental solutions. By working at the nanoscale, scientists exploit unique material properties such as increased surface area, quantum effects, and enhanced reactivity traits that are often not observed in bulk materials [1]. These remarkable characteristics have fueled the development of novel applications, yet they also bring to light important questions about the broader impacts of these technologies over time. This is where life cycle thinking (LCT) becomes essential [2]. Rather than focusing solely on the immediate benefits or isolated risks of nanomaterials, LCT encourages a comprehensive view that spans from raw material extraction and nanoparticle synthesis to product manufacturing, usage, and end-of-life management, including recycling or safe disposal [3]. It provides a framework to evaluate the full spectrum of environmental, health, and social implications associated with nanotechnology, allowing for a more balanced assessment of both its promises and potential hazards. Nanomaterials pose distinct challenges within this context. Their small size and unique behaviors can lead to unexpected interactions with biological systems and the environment, often in ways that traditional materials do not [4]. For example, nanoparticles may transform once released into natural environments or during product use, potentially leading to new exposure pathways that are still not fully understood. Moreover, there are still significant knowledge gaps in toxicity data and environmental fate, which complicate accurate life cycle assessments (LCA). To address these complexities, researchers are turning to advanced tools and approaches, such as dynamic life cycle models, sophisticated analytical methods, and safer-by-design strategies [5]. These efforts aim to anticipate and mitigate risks early in the development process, making it possible to harness the advantages of nanotechnology while safeguarding human health and the environment [6]. Incorporating LCT from the outset not only supports more responsible research and innovation but also helps industries and policymakers make informed choices (Figure 3.1). Ultimately, it paves the way to design safer, more sustainable nanomaterials and products, ensuring that the future of nanotechnology aligns with global sustainability goals [7].

DOI: 10.1201/9781003632498-3

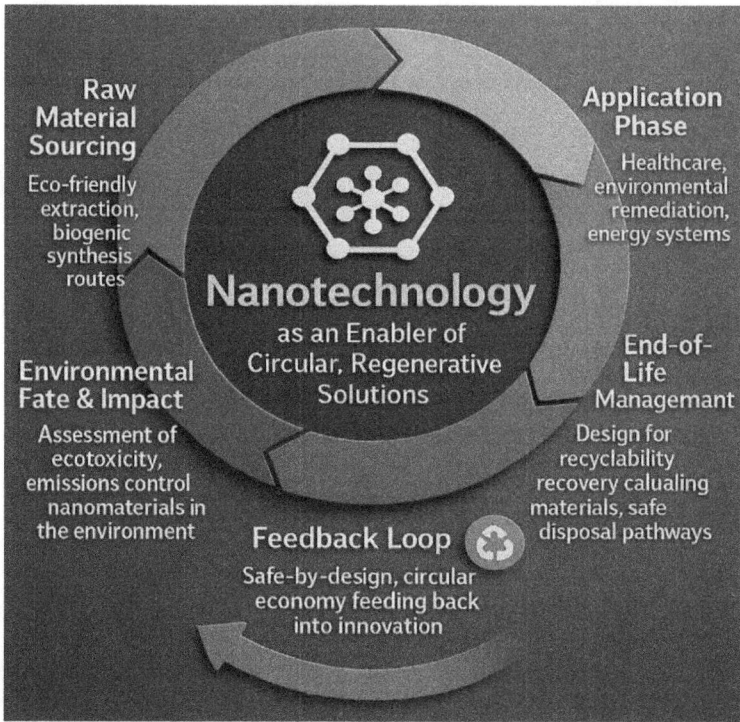

FIGURE 3.1 Schematic depiction of life cycle of nanotechnology

3.2 Nanomaterials in Focus: A Life Cycle Approach to Balancing Innovation and Ecological Impact

LCT provides a comprehensive lens to evaluate the performance and impacts of nanomaterials across their entire life span, from raw material extraction to production, application, and disposal. In the construction sector, nanomaterials have been engineered to enhance durability, functionality, and resource efficiency [8]. For example, nano-silica improves concrete strength, while titanium dioxide (TiO_2) facilitates self-cleaning surfaces through photocatalytic reactions, thereby extending material life and reducing maintenance needs [9]. During the *operational phase*, nanomaterials contribute to energy efficiency and environmental health. Zinc oxide (ZnO) and TiO_2 coatings reflect solar radiation, lowering building cooling demands, whereas silver nanoparticles offer antimicrobial protection in indoor environments. However, the gradual wear of these materials may release nanoparticles into the environment, posing risks that LCT helps to identify and mitigate. *End-of-life* considerations are equally [10]. Nanoparticles in construction debris may resist degradation, complicating recycling or disposal. Nonetheless, advanced recovery technologies offer pathways for reusing these materials, aligning with circular economy strategies [11]. Safer-by-design approaches, involving surface modifications and controlled particle engineering, aim to preserve functional benefits while minimizing ecological and health hazards. While LCA tools support the evaluation of environmental impacts, data gaps and methodological challenges persist, particularly concerning nanoparticle release and toxicity profiles [12]. Despite these hurdles, adopting LCT ensures that the benefits of nanomaterials are balanced with environmental stewardship and human safety, guiding the construction industry toward more sustainable practices [13].

3.3 Nanotechnology: A Catalyst for Circular Economy in Sustainable Industries

Building a truly sustainable economy, one that balances economic progress with environmental responsibility and social well-being, is no longer a choice, but a necessity. Achieving this vision requires a decisive shift away from the traditional "take, make, dispose" model, which depletes natural resources and strains ecosystems. In its place, regenerative systems must take root, prioritizing resource longevity, recovery, and waste minimization. Central to this transformation is the circular economy, a model that reimagines the entire life cycle of materials and products [14]. Unlike the linear approach that ends in disposal, the circular economy is designed to keep materials circulating within the system for as long as possible, eliminating waste and regenerating natural resources. By doing so, it not only reduces our dependence on finite resources but also unlocks new economic opportunities through material recovery and innovative industrial solutions [15]. Within this evolving framework, nanotechnology has emerged as a transformative force, accelerating the practical application of circular economic principles. The ability to engineer materials at the nanoscale enables the creation of products with enhanced physical, chemical, and functional properties. By carefully adjusting material structures at this level, nanotechnology extends the useful life of products, boosts their performance, and increases efficiency in production processes [16]. These improvements lead to lower material consumption and significantly reduce waste generation, making nanotechnology a natural ally in the pursuit of sustainable resource management. When combined with circular economic strategies, nanotechnology offers a compelling pathway toward achieving sustainability targets [17]. Together, they drive innovative approaches to resource efficiency, extend the working lifespan of materials, and foster industrial practices that support environmental goals while enhancing economic resilience. This synergy not only advances technological progress but also strengthens the foundation for sustainable development [18].

To ensure the effectiveness of circular economic initiatives, it is crucial to establish reliable methods for measuring progress. Carefully tracking indicators such as the proportion of products that are recycled, remanufactured, or reused at the end of their life cycle provides valuable insights into the efficiency of resource circulation. These quantitative assessments help identify areas for improvement, guide decision-making, and confirm that circular strategies are delivering meaningful outcomes on the path to a more sustainable future [18].

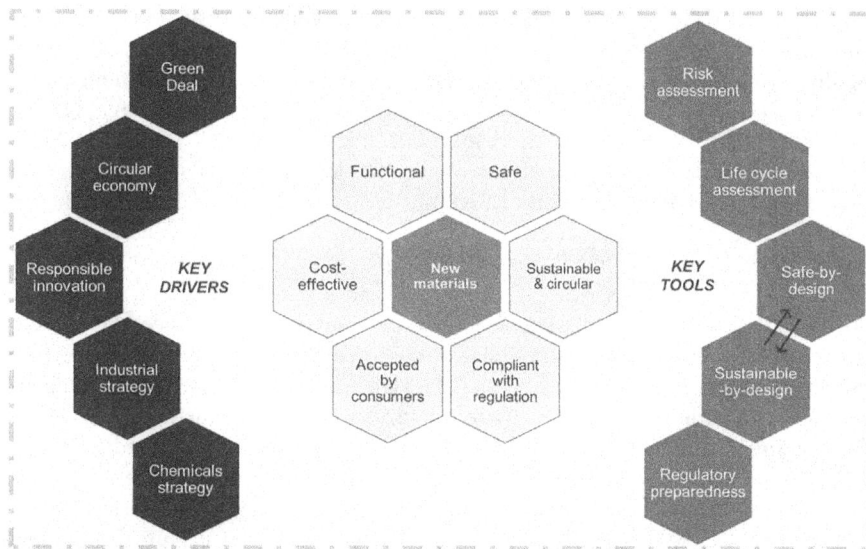

FIGURE 3.2 Schematic representation of the journey of nanomaterials from laboratory to industrial fate.
[14]

Nanotechnology plays a transformative role in building smarter, more sustainable materials and systems. One of its remarkable applications is the development of intelligent, self-healing materials that can significantly extend the lifespan of products, from protective coatings for infrastructure to advanced materials in electronics [19]. In the realm of waste management, nanotechnology enables innovative recycling techniques by breaking down complex materials into their basic components, making recycling and upcycling far more efficient (Figure 3.2). Furthermore, nanoscale catalysts offer the ability to convert waste substances into valuable resources, such as turning carbon emissions into useful chemical compounds, contributing to waste valorization [20]. The advancement of nanoelectronics brings additional benefits by producing devices that are smaller, more durable, and energy-efficient, effectively reducing the growing burden of electronic waste. In agriculture, nanotechnology enhances precision farming using nano-sensors and advanced nanomaterials, optimizing resource use and lowering environmental impact [21].

Bringing all these perspectives together, it becomes clear that integrating nanotechnology with circular economic strategies provides a powerful path toward a sustainable future. By promoting resource efficiency, extending the life cycle of materials, enabling advanced waste management solutions, and fostering sustainable innovation across industries, nanotechnology stands as a cornerstone in the transition toward a circular, regenerative, and environmentally conscious economy [22, 23].

3.4 Catalyzing Net-Zero Goals with Nanotechnology Solutions

Nanocatalysts, including metals, metal oxides, low-dimensional materials, and structured supports like zeolites and MOFs, offer superior surface area-to-volume ratios, providing more active sites for reactant interactions compared to traditional catalysts. This design enhances catalytic efficiency, selectivity, durability, and recyclability, making them vital for cleaner industrial processes [24]. For instance, rare-earth metals are efficient for water splitting to produce green hydrogen, but their scarcity and high cost hinder scalability (Figure 3.3). Alternatively, atomically dispersed iridium on MnO_2 achieves equivalent hydrogen production with 95% less iridium. Triple junction catalysts, uniquely formed at the nanoscale, combine three materials to deliver heightened activity and selectivity by precisely controlling active site distribution and minimizing side reactions [25]. Furthermore, nano-catalytic pyrolysis of captured methane presents a dual benefit: sustainable hydrogen production and formation of carbon nanomaterials, potentially offsetting emissions from hard-to-abate sectors like steel, aluminium, and cement [26]. Nanocatalysts also play a role in converting captured CO_2 into fuels and chemicals using photocatalysts such as TiO_2 and metal sulfides, supporting the transition toward net-zero synthetic fuels for sectors like aviation and shipping. Waste valorization is another promising avenue, with CaO nanocatalysts derived from eggshell waste improving biodiesel production via transesterification of free fatty acids. In carbon management, nanoporous polymers, low-dimensional nanomaterials, and advanced MOFs like CALF-20 and aluminum formate selectively capture CO_2 through physisorption and chemisorption. Notably, aluminium formate offers a cost-effective point-source capture solution at around $2/kg [27]. By coupling CO_2 capture with catalytic conversion to formate, a near-circular process for MOF production is achievable, further reducing raw material demand. While scalability challenges persist, particularly regarding material stability in humid environments and synthesis costs, nano-enabled technologies are poised to transform energy, industrial, and environmental sectors, driving progress toward global carbon neutrality targets [28].

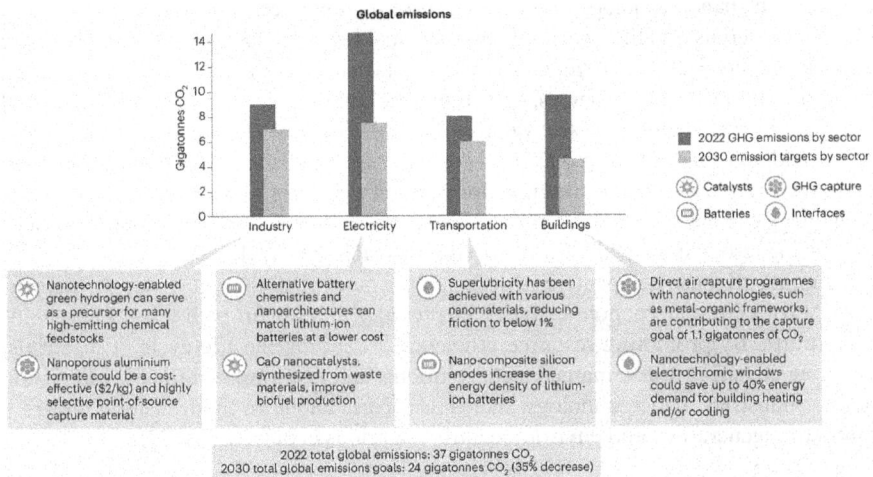

FIGURE 3.3 Graphical depiction of global CO_2 emission.

[29]

3.5 Sustainable Valorization of Agri-Food Waste through Nanotechnology

Nanotechnology has emerged as a transformative tool in addressing the challenges associated with agricultural and food waste management. By improving enzymatic reactions, nanomaterials facilitate the more efficient degradation of complex organic substrates, enhancing processes such as composting and anaerobic digestion. This results in the accelerated generation of valuable byproducts, including biofertilizers and biopesticides derived from agricultural residues [30]. Furthermore, nano-enabled delivery systems ensure the precise and controlled release of agrochemicals, thereby reducing environmental contamination and improving the efficacy of fertilizers and pesticides. In food packaging and preservation, nanotechnology is driving advancements through the incorporation of antimicrobial nanocomposites, which extend product shelf life by preventing microbial growth [31]. Concurrently, the integration of nano-sensors within packaging materials allows real-time monitoring of food freshness, supporting effective supply chain management and reducing food spoilage. Precision agriculture also benefits from nanoscale sensing technologies, which provide accurate, real-time data on soil health, crop status, and environmental conditions. This empowers farmers to make data-driven decisions, optimizing resource allocation and maximizing yields [32]. Agricultural and food wastes are rich in bioactive molecules such as polyphenols, carotenoids, vitamins, and tocopherols, many of which possess potent health-promoting properties. However, these compounds are prone to degradation under environmental stresses. Advanced encapsulation techniques, particularly nanoencapsulation, have been developed to protect these sensitive compounds. Nanoencapsulation enhances solubility, bioavailability, and ensures controlled release, while also improving sensory attributes by masking undesirable flavors and odors. Furthermore, agricultural residues are increasingly utilized as feedstocks to produce biopolymers, enzymes, organic acids, and single-cell proteins. These value-added products have wide-ranging applications across pharmaceuticals, cosmetics, food processing, and environmental sectors. For example, microbial fermentation of food waste produces organic acids and protein-rich biomass, offering sustainable and cost-effective alternatives to conventional resources. Insect-based bioconversion, notably using black soldier fly larvae, is another innovative method transforming organic waste into protein-rich animal feed and biofertilizers, contributing to circular bioeconomy models. The development of biodegradable bioplastics such as polyhydroxyalkanoates (PHA) and polyhydroxybutyrate (PHB) from food waste is also gaining momentum, providing eco-friendly alternatives to petrochemical plastics and contributing to waste reduction goals. In conclusion, while nanotechnology and bioconversion strategies present promising avenues for waste valorization, further research is necessary to overcome challenges related to scalability, cost-effectiveness, and environmental impacts. Continued advancements in these fields will play a pivotal role in promoting sustainable agricultural practices and fostering a circular economy.

3.6 Nano-Enabled Cosmeceuticals: Transforming Skin Delivery and Bioactivity of Actives

The integration of nanotechnology into cosmetic science has unlocked transformative potential, especially in overcoming the persistent challenges of bioactive skin delivery. Nanocarriers, particularly lipid-based systems like liposomes and nanoliposomes, present remarkable physicochemical advantages. By encapsulating active compounds within lipid bilayers or polymeric matrices, these carriers enhance solubility, stability, and controlled release of unstable molecules such as curcumin, resveratrol, and various plant-derived antioxidants. These improvements are critical, as natural bioactive compounds often suffer from poor water solubility, photodegradation, and enzymatic breakdown, limiting their efficacy in conventional formulations [33].

Lipid-based nanocarriers, such as solid lipid nanoparticles (SLNs) and nanostructured lipid carriers (NLCs), possess amphiphilic character, allowing them to interact effectively with the stratum corneum. Their nanoscale dimensions (typically < 200 nm) increase surface area-to-volume ratio, thus improving dermal absorption by enhancing passive diffusion and bypassing cutaneous enzymatic activity [34]. These delivery vehicles promote sustained release kinetics, maintaining therapeutic levels of actives at the target site and minimizing the need for multiple applications. The encapsulation of agro-industrial bio-wastes like pectin (from fruit peels) and chitosan (from crustacean shells) into nanocarriers has emerged as a sustainable innovation. These polysaccharides offer inherent antioxidant properties and biocompatibility, reinforcing skin barrier function while promoting environmental responsibility in product development [35]. Furthermore, recent advancements in surface modification of nanoparticles – such as hydrophobic–hydrophilic balance adjustment, surface charge tuning, and morphological optimization – enable precise control over skin permeation behavior. Positively charged nanocarriers, for instance, exhibit enhanced interaction with the negatively charged epidermal surface, facilitating deeper penetration and prolonged retention time [36].

Global cosmetic leaders like Dior, Lancôme, and Boticário have capitalized on these advances, employing nano-encapsulated retinoids, ceramides, and vitamin complexes to amplify anti-aging and skin-brightening effects. Similarly, gold nanoparticles (AuNPs), leveraged by brands such as Chantecaille and Tony Moly, function not only as luxurious actives but also as photothermal agents that stimulate microcirculation and aid in skin rejuvenation. Emerging systems such as cyclodextrin-based complexes further augment formulation stability by forming host–guest inclusion complexes. These cyclic oligosaccharides possess a hydrophobic internal cavity capable of entrapping lipophilic molecules, thereby increasing aqueous solubility and reducing volatility. Cyclodextrins have found extensive use in treatments targeting dermatological concerns like psoriasis, acne, and photodamage, providing controlled release and improved dermal bioavailability with minimal irritation. Moreover, pioneering studies have demonstrated the efficacy of elastic nanoliposomes in transporting collagen peptides derived from marine sources like *Asterias pectinifera*. These flexible vehicles promote dermal penetration and exhibit anti-photoaging properties by downregulating matrix metalloproteinase-1 (MMP-1), a key enzyme implicated in

UV-induced skin aging. Similarly, creams fortified with donkey milk-derived nanoliposomes have demonstrated superior moisturizing effects, attributable to deeper skin permeation and faster hydration kinetics compared to traditional formulations [37].

Haircare formulations have equally benefited from nanotechnological advances. Nano-encapsulated botanical oils, like batana oil and sericin protein nanoparticles, integrate seamlessly with hair fibers, delivering actives more uniformly and deeply into the hair shaft. These systems extend the longevity of hair color and softness while reducing reliance on aggressive surfactants or preservatives. Despite these promising developments, certain limitations persist. Challenges such as low drug loading capacity, batch-to-batch reproducibility, and nanoparticle instability under varying environmental conditions remain areas of active research. Techniques like high-pressure homogenization, ultrasonication, and supercritical fluid processing are being refined to address these issues and scale production effectively.

3.7 Incorporating Nanotechnology into Construction: Advancing Sustainability While Addressing Emerging Risks

Nanotechnology has introduced significant advancements in the construction sector, where engineered nanomaterials (ENMs) have been employed to improve structural performance, provide multifunctionality, and enable innovations such as self-cleaning surfaces and smart sensing capabilities [38]. These nanomaterials exhibit enhanced physicochemical properties such as durability, thermal, mechanical, and electrical properties, fire resistance (in modified mortar composition), nucleating effect, and high surface-to-volume ratio that surpass those of their bulk counterparts. However, it requires more water for hydration as a greater number of particles are participating in the reaction (see Figure 3.4) [39].

FIGURE 3.4 SEM image of the hydration process of nano-cement particles.

[40]

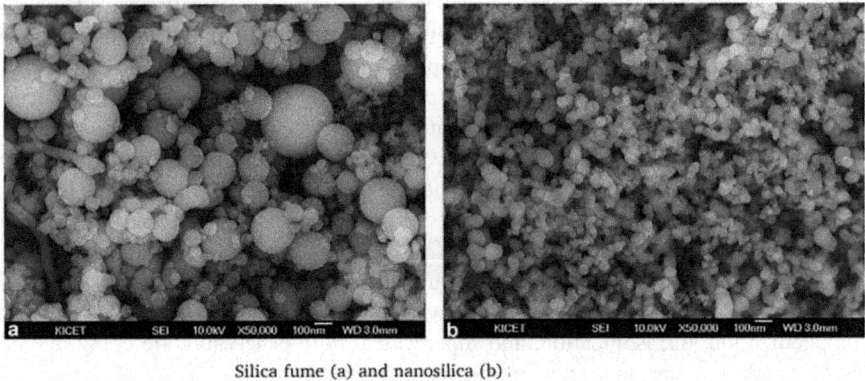

Silica fume (a) and nanosilica (b)

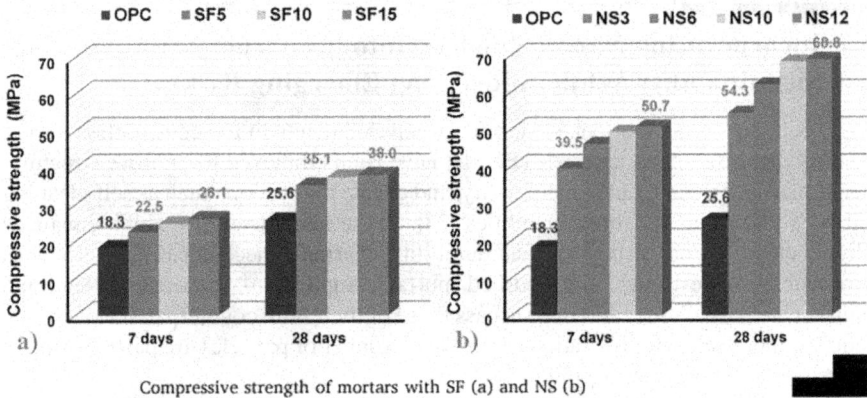

Compressive strength of mortars with SF (a) and NS (b)

FIGURE 3.5 Comparison in compressive strength of silica fumes and nano-silica. [42]

- **Nano-silica (SiO$_2$):** It consists of amorphous silica particles typically within the 1–500 nm range. It derives primarily from natural silica sources and contributes to enhanced mechanical strength, flexibility, and abrasion resistance in building materials. The performance of nano-silica is strongly influenced by particle size, surface area, and synthesis parameters (see Figure 3.5) [41].
- **Titanium dioxide (TiO$_2$):** these nanoparticles, which have been obtained from mineral sources such as ilmenite and rutile, are fabricated in various morphologies, including spheres, nanotubes, and fibers. These nanoparticles are renowned for their photocatalytic efficiency, which is utilized in the development of self-cleaning surfaces due to light-induced superhydrophilicity of thin crystalline TiO$_2$ films [43]. The surface of titania consists of five coordinated Ti atoms, with the sixth position occupied by water (OH- ion). Oxygen vacancies at the two co-ordinated oxygen bridging sites at the surface are formed under UV irradiation. Simultaneously, Ti^{4+} ions convert to Ti^{3+}. The presence of such structural defects is responsible for the increased affinity for hydroxyl ions created by dissociation

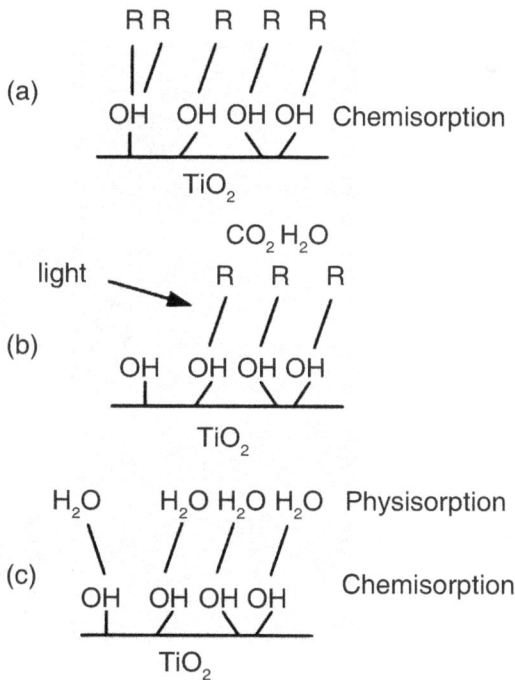

FIGURE 3.6 Schematic depiction of self-cleansing activity of hydrophilic crystalline TiO_2 films. [46]

of chemisorbed water molecules, resulting in the formation of hydrophilic domains (see Figure 3.6) and the degradation of environmental pollutants. TiO_2 is among the most widely utilized nanomaterials in construction, alongside silica [44, 45].

- **Carbon Nanotubes (CNTs):** CNTs, 2D carbon allotropes, offer excellent optical, electrochemical properties with high tensile strength, electrical transmission, and light weight. They act as additives – either single-walled (SWCNT) or multi-walled (MWCNT) cylindrical carbon structures – to enhance the properties of other building materials (such as glass and cement) (see Figure 3.7). They are produced at a large scale predominantly via catalytic chemical vapor deposition. CNTs improve the flexibility and durability of construction composites and act as sorbents for pollutant removal [47, 48].
- **Iron oxide (Fe₂O₃):** These nanoparticles exhibit better catalytic efficiency and crack healing ability via hyperthermia. These magnetically active nanoparticles generate heat when dispersed homogenously in the matrix of bituminous (thermoplastic polymeric material). They act as the best alternative for crack healing based upon the concentration ratio of matrix and nanoparticles (see Figure 3.8). Additionally, these nanoparticles exhibit high chemical stability, corrosion resistance, magnetic properties, and photocatalytic potential [50].

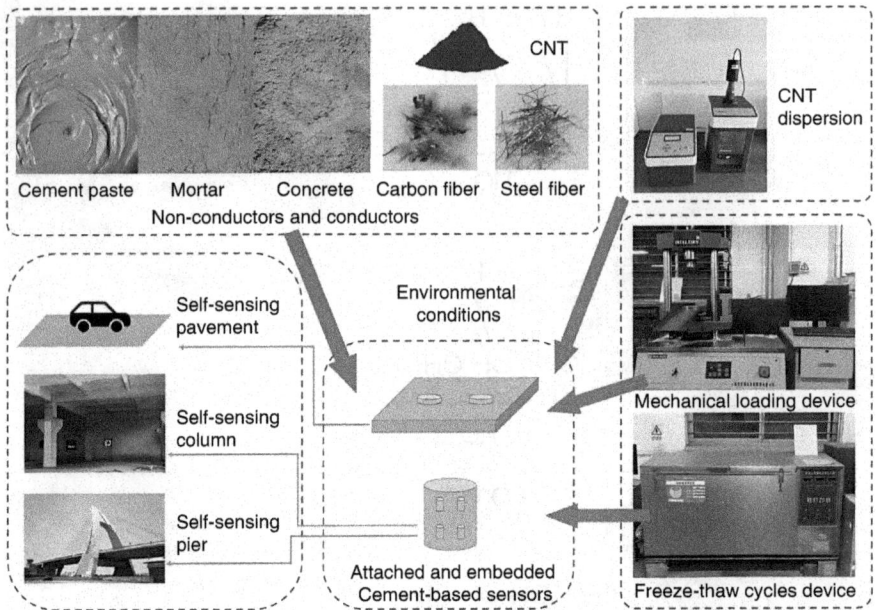

FIGURE 3.7 Application of CNTs-based cement materials for construction.

[49]

- **Silver (Ag):** Silver nanoparticles, ranging from 1 to 100 nm, possess potent antimicrobial activity, making them suitable for incorporation into construction materials used in hygiene-critical environments (Figure 3.9). Their properties, including shape and oxidation state, are dictated by synthesis conditions [52].
- **Aluminum Oxide (Al_2O_3):** Nano-alumina is typically synthesized via sol–gel processes, pyrolysis, sputtering, or laser ablation. Compared to its bulk form, nano-alumina demonstrates superior catalytic activity and adsorption potential, finding applications in industrial and construction contexts [53–55].
- **Zinc Oxide (ZnO):** ZnO nanoparticles exhibit excellent optical, electrical, and piezoelectric properties. Derived from natural zincite, ZnO enhances light retention, reduces resistivity, and is considered non-toxic, making it suitable for optoelectronic applications and protective coatings in construction [56].

3.8 Conclusion and Perspective

The integration of these nanomaterials has notably improved the mechanical robustness, longevity, and functional versatility of materials. However, the increasing use of nanotechnology raises significant concerns about environmental persistence and potential health hazards. Their unique surface properties and high reactivity lead to distinct environmental behaviors compared to bulk materials. Factors such as particle modification, environmental conditions, and interactions with biological systems (e.g., protein corona

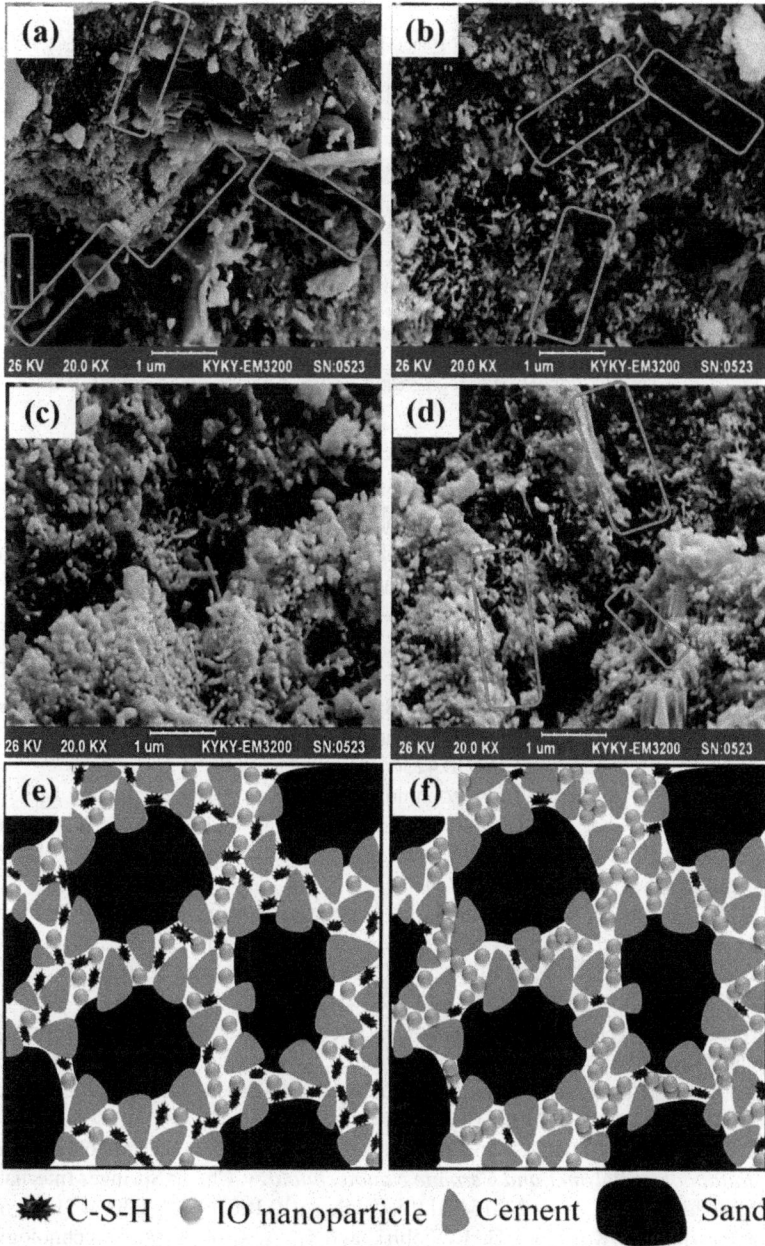

FIGURE 3.8 SEM images depicting different concentrations of Fe_2O_3 in cement mortars. [51]

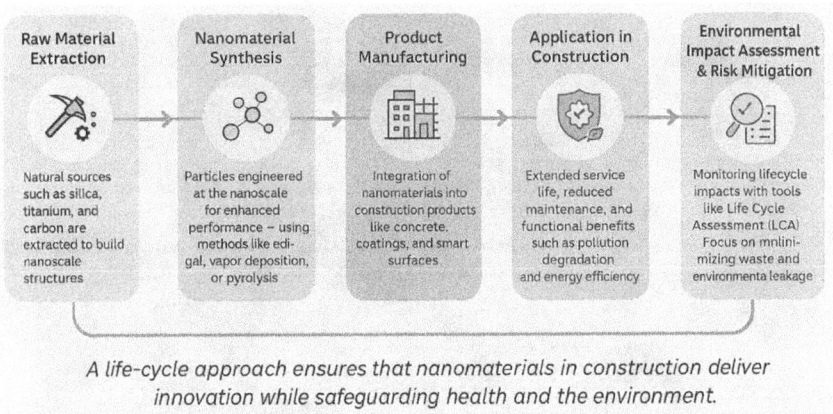

Raw Material Extraction	Nanomaterial Synthesis	Product Manufacturing	Application in Construction	Environmental Impact Assessment & Risk Mitigation
Natural sources such as silica, titanium, and carbon are extracted to build nanoscale structures	Particles engineered at the nanoscale for enhanced performance – using methods like edigal, vapor deposition, or pyrolysis	Integration of nanomaterials into construction products like concrete, coatings, and smart surfaces	Extended service life, reduced maintenance, and functional benefits such as pollution degradation and energy efficiency	Monitoring lifecycle impacts with tools like Life Cycle Assessment (LCA) Focus on mnlinimizing waste and environmenta leakage

A life-cycle approach ensures that nanomaterials in construction deliver innovation while safeguarding health and the environment.

FIGURE 3.9 Schematic representation of the life cycle of nanomaterials in the construction industry.

formation) influence their degradation, mobility, bioavailability, and toxicity profiles. Furthermore, concerns about bioaccumulation necessitate a detailed investigation into their absorption, distribution, metabolism, and excretion (ADME) characteristics.

REFERENCES

[1] N. M. Rehan, "Nanotechnology as a Sustainable Approach for Achieving Sustainable Future," *World Journal of Engineering and Technology*, vol. 09, no. 04, pp. 877–890, 2021, https://doi.org/10.4236/wjet.2021.94060.

[2] S. Stoycheva, A. Zabeo, L. Pizzol, and D. Hristozov, "Socio-Economic Life Cycle-Based Framework for Safe and Sustainable Design of Engineered Nanomaterials and Nano-Enabled Products," *Sustainability*, vol. 14, no. 9, p. 5734, May 2022, https://doi.org/10.3390/su14095734.

[3] M. K. Murthy, P. Khandayataray, D. Samal, R. Pattanayak, and C. S. Mohanty, "Green Nanotechnology: A Roadmap to Long-Term Applications in Biomedicine, Agriculture, Food, Green Buildings, Coatings, and Textile Sectors," pp. 231–261, 2023, https://doi.org/10.1007/978-981-99-4149-0_12.

[4] M. Menegaldo *et al.*, "Environmental and Economic Sustainability in Cultural Heritage Preventive Conservation: LCA and LCC of Innovative Nanotechnology-Based Products," *Cleaner Environmental Systems*, vol. 9, p. 100124, Jun. 2023, https://doi.org/10.1016/j.cesys.2023.100124.

[5] S. Shah and M. Shah, "Nanotechnology: A Scope for a Sustainable Future," in *Handbook of Polymer and Ceramic Nanotechnology*, Cham: Springer International Publishing, 2021, pp. 1627–1649, https://doi.org/10.1007/978-3-030-40513-7_62.

[6] B. Tawiah, R. K. Seidu, G. K. Fobiri, and C. A. Chichi, "Nanotechnology and Education: Preparing the Next Generation of Future Leaders," pp. 247–270, 2024, https://doi.org/10.1007/978-981-97-6184-5_7.

[7] A. García-Quintero and M. Palencia, "A Critical Analysis of Environmental Sustainability Metrics Applied to Green Synthesis of Nanomaterials and the Assessment of Environmental Risks Associated with the Nanotechnology," *Science of the Total Environment*, vol. 793, p. 148524, Nov. 2021, https://doi.org/10.1016/j.scitotenv.2021.148524.

[8] R. A. Soni, Mohd. A. Rizwan, and S. Singh, "Opportunities and Potential of Green Chemistry in Nanotechnology," *Nanotechnology for Environmental Engineering*, vol. 7, no. 3, pp. 661–673, Sep. 2022, https://doi.org/10.1007/s41204-022-00233-5.

[9] C. Bauer, J. Buchgeister, R. Hischier, W. R. Poganietz, L. Schebek, and J. Warsen, "Towards a Framework for Life Cycle Thinking in the Assessment of Nanotechnology," *Journal of Cleaner Production*, vol. 16, nos. 8–9, pp. 910–926, May 2008, https://doi.org/10.1016/j.jclepro.2007.04.022.

[10] R. Dhingra, S. Naidu, G. Upreti, and R. Sawhney, "Sustainable Nanotechnology: Through Green Methods and Life-Cycle Thinking," *Sustainability*, vol. 2, no. 10, pp. 3323–3338, Oct. 2010, https://doi.org/10.3390/su2103323.

[11] M. Cossutta and J. McKechnie, "Environmental Impacts and Safety Concerns of Carbon Nanomaterials," in *Carbon Related Materials*, Singapore: Springer Singapore, 2021, pp. 249–278, https://doi.org/10.1007/978-981-15-7610-2_11.

[12] R. Gaur, "Environmental Impact and Life Cycle Analysis of Green Nanomaterials," in *Green Functionalized Nanomaterials for Environmental Applications*, Elsevier, 2022, pp. 513–539, https://doi.org/10.1016/B978-0-12-823137-1.00018-X.

[13] J. B. Guinée, R. Heijungs, M. G. Vijver, W. J. G. M. Peijnenburg, and G. Villalba Mendez, "The Meaning of Life . . . Cycles: Lessons from and for Safe by Design Studies," *Green Chemistry*, vol. 24, no. 20, pp. 7787–7800, 2022, https://doi.org/10.1039/D2GC02761E.

[14] S. Gottardo *et al.*, "Towards Safe and Sustainable Innovation in Nanotechnology: State-of-Play for Smart Nanomaterials," *NanoImpact*, vol. 21, p. 100297, Jan. 2021, https://doi.org/10.1016/j.impact.2021.100297.

[15] A. Kumar, P. K. Tyagi, S. Tyagi, and M. Ghorbanpour, "Integrating Green Nanotechnology with Sustainable Development Goals: A Pathway to Sustainable Innovation," *Discover Sustainability*, vol. 5, no. 1, p. 364, Oct. 2024, https://doi.org/10.1007/s43621-024-00610-x.

[16] L. Pokrajac *et al.*, "Nanotechnology for a Sustainable Future: Addressing Global Challenges with the International Network4Sustainable Nanotechnology," *ACS Nano*, vol. 15, no. 12, pp. 18608–18623, Dec. 2021, https://doi.org/10.1021/acsnano.1c10919.

[17] C. Mesa Alvarez and T. Ligthart, "A Social Panorama Within the Life Cycle Thinking and the Circular Economy: A Literature Review," *The International Journal of Life Cycle Assessment*, vol. 26, no. 11, pp. 2278–2291, Nov. 2021, https://doi.org/10.1007/s11367-021-01979-x.

[18] C. Visentin, A. W. da S. Trentin, A. B. Braun, and A. Thomé, "Life Cycle Sustainability Assessment of the Nanoscale Zero-Valent Iron Synthesis Process for Application in Contaminated Site Remediation," *Environmental Pollution*, vol. 268, p. 115915, Jan. 2021, https://doi.org/10.1016/j.envpol.2020.115915.

[19] P. Thakur and A. Thakur, "Introduction to Nanotechnology," in *Synthesis and Applications of Nanoparticles*, Singapore: Springer Nature Singapore, 2022, pp. 1–17, https://doi.org/10.1007/978-981-16-6819-7_1.

[20] Y. Khan *et al.*, "Classification, Synthetic, and Characterization Approaches to Nanoparticles, and Their Applications in Various Fields of Nanotechnology: A Review," *Catalysts*, vol. 12, no. 11, p. 1386, Nov. 2022, https://doi.org/10.3390/catal12111386.

[21] L. G. Soeteman-Hernández, C. F. Blanco, M. Koese, A. J. A. M. Sips, C. W. Noorlander, and W. J. G. M. Peijnenburg, "Life Cycle Thinking and Safe-and-Sustainable-by-Design Approaches for the Battery Innovation Landscape," *iScience*, vol. 26, no. 3, p. 106060, Mar. 2023, https://doi.org/10.1016/j.isci.2023.106060.

[22] M. Angrisano and F. Fabbrocino, "The Relation Between Environmental Risk Analysis and the Use of Nanomaterials in the Built Environment Sector: A Circular Economy Perspective," *Recent Progress in Materials*, vol. 05, no. 01, pp. 1–21, Jan. 2023, https://doi.org/10.21926/rpm.2301005.

[23] S. F. Hansen *et al.*, "Nanotechnology Meets Circular Economy," *Nature Nanotechnology*, vol. 17, no. 7, pp. 682–685, Jul. 2022, https://doi.org/10.1038/s41565-022-01157-6.

[24] A. Tiwari, "Emerging Global Trends in the Potential of Nanotechnology for Achieving the Net Zero Goals," pp. 31–41, 2024, https://doi.org/10.1007/978-3-031-44603-0_3.

[25] A. Li *et al.*, "Atomically Dispersed Hexavalent Iridium Oxide from MnO_2 Reduction for Oxygen Evolution Catalysis," *Science (1979)*, vol. 384, no. 6696, pp. 666–670, May 2024, https://doi.org/10.1126/science.adg5193.

[26] Y. Chu *et al.*, "Tuning Proton Transfer and Catalytic Properties in Triple Junction Nanostructured Catalyts," *Nano Energy*, vol. 86, p. 106046, Aug. 2021, https://doi.org/10.1016/j.nanoen.2021.106046.

[27] H. A. Evans *et al.*, "Aluminum Formate, $Al(HCOO)_3$: An Earth-Abundant, Scalable, and Highly Selective Material for CO_2 Capture," *Science Advances*, vol. 8, no. 44, Nov. 2022, https://doi.org/10.1126/sciadv.ade1473.

[28] Sarika, A. Anand, R. Meena, U. Mina, A. Shukla, and A. Sharma, "Adoption of the Green Energy Technology for the Mitigation of Greenhouse Gas Emission: Embracing the Goals of the Paris Agreement," pp. 47–72, 2023, https://doi.org/10.1007/978-981-99-6924-1_4.

[29] M. F. Campa *et al.*, "Nanotechnology Solutions for the Climate Crisis," *Nature Nanotechnology*, vol. 19, no. 10, pp. 1422–1426, Oct. 2024, https://doi.org/10.1038/s41565-024-01772-5.

[30] J. N. Akinniyi, "Transforming Agricultural Food Waste in Nigeria into Sustainable Nanoparticles: A Revolution in Green Nanotechnology: A Mini Review," *Open Journal of Agricultural Science (ISSN: 2734-214X)*, vol. 4, no. 2, pp. 29–53, Nov. 2023, https://doi.org/10.52417/ojas.v4i2.750.

[31] R. Sharma, S. Lata, and R. Garg, "Valorisation of Agricultural Waste and Their Role in Green Synthesis of Value-Added Nanoparticles," *Environmental Technology Reviews*, vol. 13, no. 1, pp. 40–59, Dec. 2024, https://doi.org/10.1080/21622515.2023.2283412.

[32] E. Capanoglu, E. Nemli, and F. Tomas-Barberan, "Novel Approaches in the Valorization of Agricultural Wastes and Their Applications," *Journal of Agricultural and Food Chemistry*, vol. 70, no. 23, pp. 6787–6804, Jun. 2022, https://doi.org/10.1021/acs.jafc.1c07104.

[33] V. Gupta *et al.*, "Nanotechnology in Cosmetics and Cosmeceuticals – a Review of Latest Advancements," *Gels*, vol. 8, no. 3, p. 173, Mar. 2022, https://doi.org/10.3390/gels8030173.

[34] X. Xu, A. P. Costa, M. A. Khan, and D. J. Burgess, "Application of Quality by Design to Formulation and Processing of Protein Liposomes," *International Journal of Pharmaceutics*, vol. 434, no. 1–2, pp. 349–359, Sep. 2012, https://doi.org/10.1016/j.ijpharm.2012.06.002.

[35] J. Joseph, B. N. Vedha Hari, and D. Ramya Devi, "Experimental Optimization of Lornoxicam Liposomes for Sustained Topical Delivery," *European Journal of Pharmaceutical Sciences*, vol. 112, pp. 38–51, Jan. 2018, https://doi.org/10.1016/j.ejps.2017.10.032.

[36] S. K. Dubey, A. Dey, G. Singhvi, M. M. Pandey, V. Singh, and P. Kesharwani, "Emerging Trends of Nanotechnology in Advanced Cosmetics," *Colloids and Surfaces B: Biointerfaces*, vol. 214, p. 112440, Jun. 2022, https://doi.org/10.1016/j.colsurfb.2022.112440.

[37] P. Angelopoulou, E. Giaouris, and K. Gardikis, "Applications and Prospects of Nanotechnology in Food and Cosmetics Preservation," *Nanomaterials*, vol. 12, no. 7, p. 1196, Apr. 2022, https://doi.org/10.3390/nano12071196.

[38] P. Mercader-Moyano and P. Porras-Pereira, Eds., *Life Cycle Analysis Based on Nanoparticles Applied to the Construction Industry*, Cham: Springer Nature Switzerland, 2025, https://doi.org/10.1007/978-3-031-79115-4.

[39] S. C. Paul, A. S. van Rooyen, G. P. A. G. van Zijl, and L. F. Petrik, "Properties of Cement-Based Composites Using Nanoparticles: A Comprehensive Review," *Construction and Building Materials*, vol. 189, pp. 1019–1034, Nov. 2018, https://doi.org/10.1016/j.conbuildmat.2018.09.062.

[40] Y. Reches, "Nanoparticles as Concrete Additives: Review and Perspectives," *Construction and Building Materials*, vol. 175, pp. 483–495, Jun. 2018, https://doi.org/10.1016/j.conbuildmat.2018.04.214.

[41] L. Laím, H. Caetano, and A. Santiago, "Review: Effects of Nanoparticles in Cementitious Construction Materials at Ambient and High Temperatures," *Journal of Building Engineering*, vol. 35, p. 102008, Mar. 2021, https://doi.org/10.1016/j.jobe.2020.102008.

[42] B.-W. Jo, C.-H. Kim, G. Tae, and J.-B. Park, "Characteristics of Cement Mortar with Nano-SiO2 Particles," *Construction and Building Materials*, vol. 21, no. 6, pp. 1351–1355, Jun. 2007, https://doi.org/10.1016/j.conbuildmat.2005.12.020.

[43] M. Janczarek, Ł. Klapiszewski, P. Jędrzejczak, I. Klapiszewska, A. Ślosarczyk, and T. Jesionowski, "Progress of Functionalized TiO2-Based Nanomaterials in the Construction Industry: A Comprehensive Review," *Chemical Engineering Journal*, vol. 430, p. 132062, Feb. 2022, https://doi.org/10.1016/j.cej.2021.132062.

[44] D. Ihnatiuk *et al.*, "Photoelectrochemical, Photocatalytic and Electrocatalytic Behavior of Titania Films Modified by Nitrogen and Platinum Species," *Applied Nanoscience*, vol. 12, no. 3, pp. 565–577, Mar. 2022, https://doi.org/10.1007/s13204-021-01690-1.

[45] S. Banerjee, D. D. Dionysiou, and S. C. Pillai, "Self-Cleaning Applications of TiO2 by Photo-Induced Hydrophilicity and Photocatalysis," *Applied Catalysis B: Environment and Energy*, vol. 176–177, pp. 396–428, Oct. 2015, https://doi.org/10.1016/j.apcatb.2015.03.058.

[46] K. Guan, "Relationship Between Photocatalytic Activity, Hydrophilicity and Self-Cleaning Effect of TiO2/SiO2 Films," *Surface and Coatings Technology*, vol. 191, nos. 2–3, pp. 155–160, Feb. 2005, https://doi.org/10.1016/j.surfcoat.2004.02.022.

[47] S. Sharma and N. C. Kothiyal, "Facile Growth of Carbon Nanotubes Coated with Carbon Nanoparticles: A Potential Low-Cost Hybrid Nanoadditive for Improved Mechanical, Electrical, Microstructural and Crystalline Properties of Cement Mortar Matrix," *Construction and Building Materials*, vol. 123, pp. 829–846, Oct. 2016, https://doi.org/10.1016/j.conbuildmat.2016.07.045.

[48] S. Yang, "Properties, Applications, and Prospects of Carbon Nanotubes in the Construction Industry," *Architecture, Structures and Construction*, vol. 3, no. 3, pp. 289–298, Sep. 2023, https://doi.org/10.1007/s44150-023-00090-z.

[49] K. Cui, J. Chang, L. Feo, C. L. Chow, and D. Lau, "Developments and Applications of Carbon Nanotube Reinforced Cement-Based Composites as Functional Building Materials," *Frontiers in Materials*, vol. 9, Mar. 2022, https://doi.org/10.3389/fmats.2022.861646.

[50] M. Maiti, M. Sarkar, M. A. Malik, S. Xu, Q. Li, and S. Mandal, "Iron Oxide NPs Facilitated a Smart Building Composite for Heavy-Metal Removal and Dye Degradation," *ACS Omega*, vol. 3, no. 1, pp. 1081–1089, Jan. 2018, https://doi.org/10.1021/acsomega.7b01545.

[51] M. Valizadeh Kiamahalleh, A. Alishah, F. Yousefi, S. Hojjati Astani, A. Gholampour, and M. Valizadeh Kiamahalleh, "Iron Oxide Nanoparticle Incorporated Cement Mortar Composite: Correlation Between Physico-Chemical and Physico-Mechanical Properties," *Materials Advances*, vol. 1, no. 6, pp. 1835–1840, 2020, https://doi.org/10.1039/D0MA00295J.

[52] G. D. da Silva, E. J. Guidelli, G. M. de Queiroz-Fernandes, M. R. M. Chaves, O. Baffa, and A. Kinoshita, "Silver Nanoparticles in Building Materials for Environment Protection Against Microorganisms," *International Journal of Environmental Science and Technology*, vol. 16, no. 3, pp. 1239–1248, Mar. 2019, https://doi.org/10.1007/s13762-018-1773-0.

[53] R. C. Congreve, C. P. Quezada, and V. Kokkarachedu, "Aluminum Oxide Nanoparticles: Properties and Applications Overview," pp. 265–288, 2024, https://doi.org/10.1007/978-3-031-50093-0_12.

[54] A. M. Baghdadi, A. A. Saddiq, A. Aissa, Y. Algamal, and N. M. Khalil, "Structural Refinement and Antimicrobial Activity of Aluminum Oxide Nanoparticles," *Journal of the Ceramic Society of Japan*, vol. 130, no. 3, p. 21140, Mar. 2022, https://doi.org/10.2109/jcersj2.21140.

[55] J. Gou, G. Wang, H. M. Al-Tamimi, T. Alkhalifah, F. Alturise, and H. E. Ali, "Application of Aluminum Oxide Nanoparticles in Asphalt Cement Toward Non-Polluted Green Environment Using Linear Regression," *Chemosphere*, vol. 321, p. 137925, Apr. 2023, https://doi.org/10.1016/j.chemosphere.2023.137925.

[56] V. P. Klienchen de Maria *et al.*, "Advances in ZnO Nanoparticles in Building Material: Antimicrobial and Photocatalytic Applications – Systematic Literature Review," *Construction and Building Materials*, vol. 417, p. 135337, Feb. 2024, https://doi.org/10.1016/j.conbuildmat.2024.135337.

4

Current Status and Future Prospects of Nanotechnology in the Healthcare Industry

Vishakha Rathi, Umesh Kumar, Garima Rathi, and Pritipriyambada Baisakh

4.1 Introduction

One of the key factors promoting economic growth across all levels is technology. The smallest objects in the universe always fascinate people, and that is why magnifying lenses were created following the first microscopes. The discovery of electron microscopy in the mid-1900s enabled researchers to witness an unknown side of the universe's biogenesis and biodiversity. The ongoing advancement in this area made it possible for scientists and researchers to create pictures of molecules and their bonds, which eventually gave rise to the field of study known as nanotechnology, such as managing, modifying, and developing systems according to their atomic or molecular properties. The ability to alter matter at the atomic, molecular, and supramolecular levels in order to create newer structures and gadgets is the essence of nanotechnology, according to the US National Science and Technology Council. This science often works with structures that are at least one dimension in size between 1 and 100 nanometres (nm), and it involves the manufacture and modification of nanomaterials and nanodevices. It has persisted as a subject of considerable scientific investigation in a number of domains, including biological, optical, and electrical. More than 100-nm-sized bacterial, plant, and mammalian cells may readily absorb or internalize nanoparticles such as viruses (75–100 nm), proteins (5–50 nm), nucleic acids (2 nm wide), and atoms (0.1 nm). Human hair is 50,000 times larger than a 1 nm nanofibre when compared to a single human hair diameter of 50 μm. It uses interdisciplinary fields of research like chemistry, biology, physics, electrical sciences, and materials sciences, and has the ability to offer technical solutions in a wide range of application areas. Nonetheless, the most significant influence on human welfare is most likely to be found in the biological and medical applications of nanotechnology (Patel & Nanda, 2015; Pramanik et al., 2020; Shah et al., 2021).

4.1.1 The Creative Founders in Nanotechnology

In 1959, Nobel Prize–winning American physicist Richard Feynman introduced the idea of nanotechnology. Feynman delivered a talk titled *"There's Plenty of Room at the Bottom"* at the California Institute of Technology (Caltech) during the American

Physical Society's annual meeting. Feynman proposed the question, "Why can't we write the entire 24 volumes of the Encyclopaedia Britannica on the head of a pin?" at this lecture. He also outlined a plan to use machines to build smaller machines, and even at the molecular level. Feynman is regarded as the founder of contemporary nanotechnology because of this novel concept, which showed that his theories were validated. Fifteen years later, in 1974, Japanese scientist Norio Taniguchi was the first to utilize and define the term "nanotechnology," stating that it primarily refers to the processing of material separation, consolidation, and deformation by a single atom or molecule (Bayda et al., 2019).

4.1.2 Current Developments in Healthcare Nanotechnology

The development of nanotechnology has been an innovator in healthcare in the past few years, delivering novel possibilities for improving the performance and uses of medical supplies, as well as ongoing advances in medication research. Nanoscale matter modification has opened the door for novel findings that could lead to breakthroughs in patient care, therapies, and diagnostics. Healthcare diagnostics is one field that highlights the possible applications of nanotechnology in medical care. Thus, nanoparticles are applied to increase the efficiency and accuracy of scanning and biological sensing procedures, making it possible for quicker and more precise diagnosis of conditions. Nanomaterials in tissue engineering improve tissue regeneration and cell interactions, providing prospective breakthroughs in fields like skin, bone, dental, and neural repair. Nanostructures are being studied for gene and medicine delivery methods, with their potential to focus on cells that may reduce negative effects and substantially boost therapy effects. The curative effects of typical cancer therapies like radiotherapy, chemotherapy, and surgery may be strengthened by nanotechnology while unlocking the door for the advancement of novel treatments, including photodynamic therapy, photothermal therapy, and biotherapy. Also, nanomaterials are currently utilized in antibacterial and antiviral functions, such as the development of materials and coverings that can combat resistant pathogens without causing infections (Kumar et al., 2024).

New nanotechnologies for the detection, visualization, and therapeutic management of a variety of conditions, including cancer, as well as disorders relating to the cardiovascular, ophthalmic, and central nervous systems, have surfaced primarily as a consequence of the increasing variety of nanotechnology components in medical applications. Since most biological systems are nanoscale, nanomaterials fit in well with biomedical equipment. Compared to traditional methods, nanosystems in the drug delivery industry provide precise drug administration to the intended tissues or organs with a regulated release and improved retention time (Sahu et al., 2024). One of the best examples of nanosystems now being developed for targeted medication delivery to treat cardiovascular illnesses and other cancers is nano-liposomes. The main justifications for using nano-liposomes are medication delivery to the intended tissue, high biocompatibility, and bloodstream drug flow regulation (Ma et al., 2024) (Figure 4.1).

4.1.2.1 Advancing Healthcare with Nanotechnology and Nanostructures

Targeted Drug Delivery: Nanotechnology in Modern Pharmaceuticals – Therapeutic chemicals can now be delivered to diseased or damaged parts of the body in

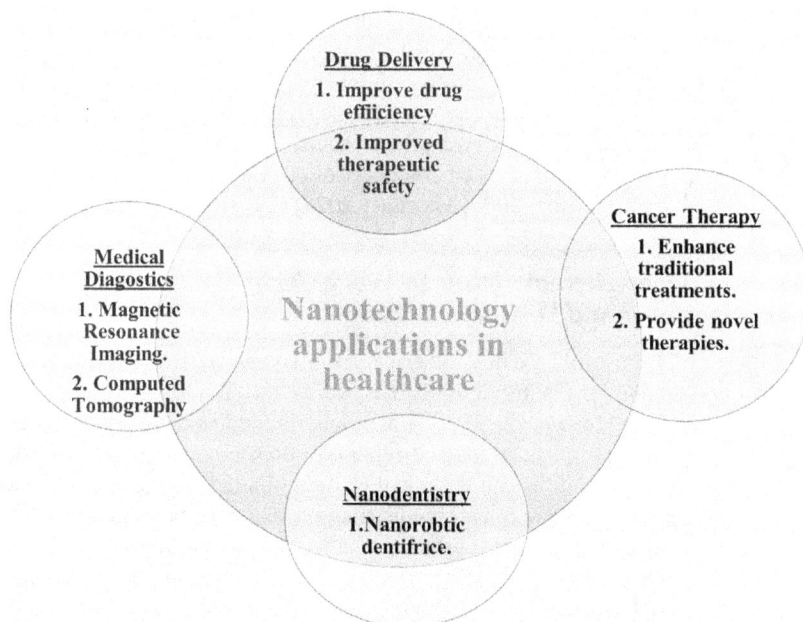

FIGURE 4.1 Venn diagram showing various applications of nanotechnology in healthcare.

a regulated and precisely targeted manner by means of nanotechnology. Like after being prepared, very insoluble but active chemicals can be used in pharmaceuticals by entering into nanoscale particles with specific delivery abilities. For example, peptides and polypeptides that are prone to bacterial or enzymatic breakdown in infected tissues or during transit through the intestinal tract might be incorporated into nanoparticles to prevent these conditions. NPs can improve cancer treatments because of their small size, making it possible for them to enter tumour cells deeply. Targeted drug distribution is essential when hydrophobic medicines contain harmful chemicals. Such solvents have the potential to damage bodily fluids, including blood circulation, if they are released somewhere away from the target cell. Drugs in desired dosages can be continuously released under control because of nanostructures. Medication dosages are also decreased via targeted and localized medication delivery. The size, shape, and other fundamental biophysical or chemical properties of NPs affect how effective they are as drug carriers. A particle core, an outer biocompatible protective layer, and a linking molecule for enhanced bioactivity are essential parts of NPs used for drug delivery (Singh et al., 2024). The binding molecule binds the NPs' core to bioactive molecules due to the reactive compounds at both ends. Before being delivered, nanovectors undergo modification, which involves covering them with ligands such as peptides, folic acid, and antibodies. To further improve selectivity, ligands are affixed to NPs so that they can bind to specific places. Multiple ligands must be attached because if only one is attached, it could bind to receptors that are not at the intended site. For example, molecularly imprinted polymeric nanoparticles are new possibilities for drug delivery techniques that combine both passive and active targeting techniques. In their polymer matrix, MIP NPs have been engineered to have distinct and complementary binding sites for medicinal molecules known as "templates." Through covalent or non-covalent binding, these regions allow MIP NPs to recognize and load medications or other templates with antibody-like specificity and selectivity. MIP NPs can increase both the accuracy of medication release and delivery. Personalized medication development requires immediate attention because, as we have seen in several instances, emergent crises can only be appropriately treated when this goal is met (Hobson, 2009; Ma et al., 2024).

4.1.2.2 In Cancer Therapy: Drug Delivery Using Liposomal and Polymeric NPs

- **Liposome:** Liposomes are tiny, spherical artificial vesicles composed of cholesterol and amphiphilic phospholipids that self-assemble into one or more bilayers that surround an aqueous core. Water-soluble medications can be encapsulated in liposomal formulations and released into the aqueous space, whereas hydrophobic medications can be linked to lipid bilayers. The size, lipid makeup, medication release rate, and biodistribution of liposomes can all differ. To enhance their qualities, their surface can also be altered. There are only four liposome-based formulations that have received clinical approval: Myocet®, DaunoXome®, DepoCyt®, and Doxil® (Caelyx® in Europe).

  Doxil® is doxorubicin that has been PEGylated liposomal to treat multiple myeloma, ovarian cancer, metastatic breast cancer, and Kaposi's sarcoma associated with AIDS. Myocet® is a non-PEGylated liposomal doxorubicin that is prescribed in conjunction with cyclophosphamide to treat breast cancer. DaunoXome®, a PEGylated liposomal daunorubicin, has currently proven to be an efficient and secure remedy for severe AIDS-related KS. DepoCyt®, the generic name of a non-PEGylated liposomal cytarabine, is a hydrophilic product. Localized intrathecal therapy of lymphomatous meningitis, a serious adverse reaction linked to brain cancer, is accepted for DepoCyt® (Egusquiaguirre et al., 2012).

- **Polymeric nanoparticles:** Drugs can be dissolved, entrapped, adsorbed, attached, or encapsulated within polymeric nanoparticles, and these are solid, biodegradable, colloidal systems with submicron diameters. Synthetic or natural biodegradable polymers are used to create polymeric nanoparticles. Heparin, chitosan, dextran, albumin, and collagen are a few examples of natural polymers. Among the synthetic polymers, extensive research has also been done on polyethylene glycol (PEG), polyglutamic acid, polyglycolic acid (PGA), polylactic acid (PLA), polycaprolactone (PCL), and poly(D,L lactide-co-glycolic) acid (PLGA) (Egusquiaguirre et al., 2012).

- **Treating brain disorders using nanotechnology:** The purpose of the blood–brain barrier (BBB) outer layer, composed of endothelial cells (ECs), is to preserve and control the passage of nutrients and other essential compounds into the brain, thereby maintaining its integrity. The regulation of substance movement into and out of the brain, ionic equilibrium, and defence against the movement of circulating agents, neurotransmitters, xenobiotics, and other molecules that can harm the integrity of the brain are the primary roles of the BBB. If we can solve the problem of the BBB, we are capable of curing brain illnesses. The BBB shields the brain's neuronal tissues from the bloodstream. The primary challenge to treating brain disorders is the presence of the BBB, which keeps the brain in a state of homeostasis and prevents medications from entering the central nervous system (CNS). A broken BBB prevents medications from invading the brain, while any disruption to the BBB can lead to neuro-inflammatory and neurodegenerative conditions like Parkinson's disease, Alzheimer's disease, etc. Although different kinds of NPs may penetrate the BBB and effectively transport medications to brain regions that have been injured and pass through the BBB, NPs use both organic and inorganic components as a core. Silica, molybdenum, cerium, iron, and gold are examples of inorganic materials; PLA, PLGA, and trehalose are examples of organic materials that can be utilized. Small size, high drug loading capacity, and effective imaging capability (especially for inorganic NPs) are the unique characteristics that enable NPs to treat neurodegenerative disorders. Several NPs themselves exhibit some therapeutic efficacy, such as lowering ROS levels, expressing antioxidant qualities, and limiting aggregation. The best results are obtained when NPs are coupled with ligands and interact with BBB receptors at low densities (Ngowi et al., 2021).

- **Dentistry in the nanoindustry:** Another branch of nanomedicine is nanodentistry, which deals with extensive uses of nanotechnology in dental care, including diagnosis, prevention, cure, prognosis, and treatment choices. Dentition denaturalization, hypersensitivity cure, orthodontic realignment issues, and updated enamelling alternatives for oral health maintenance are a few significant uses of oral nanotechnology. Following the same way, mechanical dentifrobots sense nerve impulse traffic at the tooth's core in real time, allowing them to control tooth tissue penetration and provide care for regular operation. The procedure is combined with nanocomputers that have been configured to respond to stimuli from outside by connecting to specific internal nerve stimuli. Similarly, nanotechnology has a wide range of uses in denaturalization, tooth repositioning, hypersensitivity treatment, and tooth healing (Malik et al., 2023).

- **Nanotechnology in ophthalmology:** Applications of nanotechnology in ophthalmology include oxidative stress management, intraocular pressure measurement, agnostics, the use of nanoparticles for choroidal new vessel treatment, scar prevention following glaucoma surgery, gene therapy for retinal degenerative disease, prosthetics, and regenerative nanomedicine. In addition to helping with many unresolved issues, including sight-restoring therapy for individuals with retinal degenerative disease, nanotechnology will revolutionize the current therapeutic hurdles in drug administration and postoperative scarring. It is believed that this developing field will provide treatments for eye conditions. An innovative nanoscale-dispersed eye ointment (NDEO) has been effectively created to treat severe evaporative dry eye. Petrolatum and lanolin, which are semi-solid lipids found in traditional eye ointment, were combined with medium-chain triglycerides (MCT), a liquid lipid, and both phases were subsequently dispersed in a polyvinyl pyrrolidone solution to create a nanodispersion. In the field of ocular drug delivery, recent studies have demonstrated the use of a variety of nanoparticulate systems, including microemulsions, nanosuspensions, nanoparticles, liposomes, niosomes, dendrimers, and cyclodextrins. They also illustrate how the frontiers of ocular drug delivery and therapy can be explored through the use of emerging nanotechnology, such as nanodiagnostics, nanoimaging, and nanomedicine (Yezdani et al., 2018).

- **Biosensor-based diagnostics advancement through nanotechnology:** Biosensors are chemical sensors whose recognition processes rely on the use of biochemical mechanisms and comprise a biological element (as the charge of sampling) and a physical element (referred to as a transducer that transmits testing results for subsequent processing). The biological part of a biosensor has a biosensitive section that may comprise bioreceptors or bioreceptors covalently linked to the transducer. Bioreceptors are in charge of binding a compound of interest to the sensor for the measurement; they can be biological molecular species (e.g., antibodies, enzymes, or proteins) or living biological mechanisms (e.g., cells, tissues, or organisms). The type of biological component determines the process used to grant biological specificity Information from the biological element is translated by the physical element into a chemical or physical output signal with a defined sensitivity. Thus, the chemical data, from the concentration of a particular sample component to an analysis of the sample's overall composition, is converted into an analytically meaningful signal in the biosensor. The type of biological element, the signal transduction mechanism, or a combination of the two can be used to categorize biosensors. Six major types of biosensors can be identified based on the biological specificity-granting mechanism that is applied: 1) antibody/antigen-based; 2) enzymes-based (mono- or multi-enzyme systems); 3) nucleic acids-based (DNA, cDNA, and RNA); 4) cellular interactions-based (cellular structures/cells); 5) biomimetic materials-based (e.g., synthetic bioreceptors); and 6) whole cells-based (microorganisms, such as bacteria, fungi, eukaryotic cells, or yeast; and cell organelles or particles such as mitochondria, cell walls, and tissue slices) (Kubik et al., 2005).

The binding of an antigen (Ag) to a particular antibody (Ab) is the basis for biosensors that use antibody or antigen as bioreceptors. The formation of these Ab-Ag complexes must be monitored in a way that minimizes non-specific interactions. Enzymes make up bioreceptors in enzyme-based biosensors. The ability of these bioreceptors to bind and/or catalyse is necessary for the detection process. Biocatalysts are macromolecules that catalyse a reaction that amplifies detection in biocatalytic recognition systems. All enzymes are proteins, with the exception of a tiny subset of catalytic ribonucleic acid molecules. More than their amino acid residues, certain enzymes can function without the presence of any additional chemical groups. Some require a cofactor, which might be a more complex organic or metalloorganic molecule known as a coenzyme, or one or more inorganic ions. The strength of an enzyme's initial protein structure affects its catalytic activity. The catalytic activity of an enzyme is eliminated if it is weakened, broken into its constituent amino acids, or fragmented into its subunits. The detection processes can also be impacted by enzyme-coupled receptors. For example, when a ligand attaches to the receptor, the functioning of an enzyme can be changed. DNA biosensors, also known as genosensors or biodetectors, are biosensors that rely on the association of biological components' nucleic acids. Sensors are used to identify small quantities of DNA (microorganisms such as bacteria or viruses) in an extensive sample. This depends on matching the DNA of the sample to that of a known organism (DNA probe). Several copies of the sample DNA must be made for accurate analysis because the sample solution might only contain a few molecules. Micro-electro-mechanical systems (MEMS) devices that can carry

TABLE 4.1

Nanotechnology's Significance in the Healthcare Area

Emerging Nanotechnology	Potential Uses	Advantages	Drawbacks/Things to Think About
Nanorobots	Targeted drug delivery mechanism, cellular-level surgery	Precision treatment, minimal invasion	Complex design, ethical concerns, high cost
Nano-vaccines	mRNA and DNA vaccine delivery	Enhanced immune response, reduced side effects	Safety validation, regulatory approval
Nano-biosensors and wearable nanosensors	Real-time disease monitoring, glucose sensors	Early disease detection, personalized health monitoring	Data privacy, long-term durability
Artificial nanomaterials for tissue regeneration	Bone, nerve, skin regeneration	Mimics natural tissues, enhances wound healing	Immune response, integration with biological tissues

out the polymerase chain reaction are used for this purpose. A lab-on-a-chip system like this has chambers, valves, and channels. A particular kind of biosensor, which employs a whole cell as a biological component, is one based on bacteria with genetic modification. Worldwide variables like stress, toxicity, or agents that damage DNA (e.g., heat shock or mutagenic agents) can be identified by them, as well as particular organic (e.g., m-xylene and benzene derivatives, naphthalene, and hydrocarbons) and inorganic (e.g., cadmium, mercury, aluminium, zinc, and iron) compounds. Drugs and treatments can be tested and their efficacy recorded using cellular biosensors. Also, the microbe-based biosensor can track corrosion of metallic materials triggered by microbiological factors (Kubik et al., 2005) (Table 4.1).

4.1.3 Challenges for Nanotechnology

Even though the field of nanotechnology is expanding quite quickly, there are still several obstacles at different phases of development that prevent the product from becoming widely available. If the growth obstacles listed below are removed, the medical and healthcare industries may undergo a radical transformation.

Toxicology is an important challenge to the use of nano-based products in biological systems for medical purposes. Unwanted allergic reactions and other possibly dangerous bodily reactions have been brought on by a variety of nanomaterials. As it relies on multiple variables, like the morphology, size, dosage, surface area, method, and period of administration, toxicity is an extremely complicated notion in and of itself. Additionally, there is still work to be done on the repeatability and dependability of research employing NPs. It is particularly challenging to regulate their activity in delicate conditions because these are minuscule organisms. Their high price, impurity levels, effects on the ecology, etc., are some other drawbacks. These risks could have extremely deadly consequences if they are not handled correctly (Patel & Nanda, 2015).

4.1.3.1 Lack of Understanding of the Properties of NP Components

Nanostructures come in a wide variety, each with unique properties and functions. It is unclear how these NPs behave physicochemically in vitro and in vivo. Therefore, it is essential to choose the appropriate nanomaterial for the specified indication. PEI is increasingly recognized as a great cargo for targeting intracellular nucleic acids. However, it is also thought to be a powerful cytotoxic agent. Because of its greater ability to deliver drugs, strategies to lessen its toxicity have been developed, such as connecting low-molecular-weight PEI to dithiodipropionic acid di(N-succinimidyl ester).

4.1.3.2 Awareness About the Implications of Our Biological System

Regarding toxicity to organs or cells, genotoxicity, and carcinogenicity, it remains uncertain how it will affect safety and health. These substances are small enough to be inhaled, and when they deposit in the lung alveoli, they can cause inflammation or cancer. Due to the possibility of occupational hazards to workers, this is of utmost importance.

4.1.3.3 Variability in the Level of Toxicity

Different types of cells may be harmed by nanomaterials with varying compositions, sizes, or shapes under various exposure scenarios. The composition, size, shape, charge, aggregation, coating, and solubility of the nanoparticles all affect the target cell and the target moieties for toxicity. Human T-cells and alveolar macrophages are cytotoxically affected by CNTs at 400 µg/ml and 3.06 µg/cm^2, respectively, but cell cultures exposed to 3.8 µg/ml show no cytotoxicity.

4.1.4 Environmental Risks and Toxicity Associated with Nanotechnology

4.1.4.1 Nanotoxicity and Safety Risks

Environmental risks: Nanotechnology is capable of preventing and treating pollution from hazardous substances, and it may also constitute challenges to the ecosystem if not managed accordingly; however, it is likely to be a useful tool in environmental remediation efforts. Another cause for concern is that unexpected ecological effects may result from the special qualities of nanomaterials, such as their small size and strong reactivity. As per research, nanoparticles like Ag and TiO_2 NPs may result in toxicity in both aquatic and terrestrial plants and animals, possibly harming biological diversity and the surrounding ecosystems. Nanomaterials could have a direct impact on the productivity of crops, the metabolism of metabolites, and the development and growth of rare medicinal herbs after they reach ecosystems through industrial discharges or farming procedures. The effect of TiO_2 NPs used in daily life, which is a sensitive anticancer flavonoid luteolin, was studied. The results demonstrated a direct correlation between the impact and the degree of nanoparticle contamination. Depending on variables including pH, temperature, and the presence of other chemical species, flavonoids can change chemically in a variety of ways when exposed to metallic oxide nanoparticles (Ma et al., 2024).

Risk management: It covers all five phases of the nanomaterials life cycle: resource acquisition, production, utilization, elimination, and waste management. The development of reliable environmental monitoring methods that can track the concentration, transformation, and bioavailability of nanomaterials in various environmental compartments (such as air, water, and soil) remains a challenge in this evaluation. When nanomaterials enter natural environments, they may undergo various forms of degradation due to environmental processes, leading to the formation of new nanomaterial species with transformed reactivity and toxicity, as well as changes like surface transformation and dissolution. Because aged nanomaterials may pose diverse dangers, risk assessments should not solely concentrate on pristine nanomaterials.

4.1.4.2 Safety Risk Evaluation

Researchers indicate the following suggestions for medical product producers to conduct risk analysis of medical products generated from nanotechnology, based on the principles mentioned earlier: Nanoparticles should receive more attention regarding their greater mobility, and all materials' surface qualities should be attentively taken into consideration. For instance, when in direct reach, a nanostructured covering may not release NPs but still be harmful due to physical interaction mechanisms. No universal rule has been established when examining the characteristics of a medical device derived from nanotechnology; specifically, the nanoscale materials it may contain or produce throughout its life cycle (such as size, size distribution, surface modification, etc.). Since both documentation and result reliability require a characterization protocol grounded in scientific rationale, consult relevant technical documents for assistance on creating a possible characterization protocol. These mentors, however, are just meant to serve as a beginning point for further evaluation by distinguishing between "normal bulk materials" and "nano materials."

4.1.4.3 Toxicity of Nanoparticles

In addition to the environmental consequences of commercialization and market entry, the question of biocompatibility must be taken into consideration whenever any foreign material is injected into the body for therapeutic purposes. For both experts and companies, conducting a thorough and trustworthy assessment of their toxicity throughout their life cycle is difficult. This is due to the fact that shrinking in size or dimensions, especially to the nanoscale in one or two dimensions, can result in both unique features and possibly unusual toxicity behaviour. Usually, possible concerns regarding safety can be divided into four major groups: 1) local toxicity, 2) systemic toxicity, 3) genotoxicity, and 4) carcinogenicity, which is very important for long-term exposure in the case of long-lasting implants that vary depending on the device's application and include blood contact or interactions with the CNS (Ma et al., 2024).

- **Systemic toxicity:** nanomaterials, whether synthetically developed or present in nature, their biokinetics impact damage to the body. This phenomenon is studied using the ADME (T) framework, in which "A" denotes administration, "D" distribution, "M" metabolism, "E" elimination, and "T" toxicity. The technical complexity and difficulties of understanding the biodistribution and clearance rates of nanomaterials have significantly impeded the broader clinical application of nanomedicines and nanotechnology-based technologies (Ma et al., 2024).

- Location of implantation or the method of therapy, like topical application, is linked to the local toxicity of nanomaterials. Based on research, Ag NPs have the ability to cause local toxicity at the site of entrance, triggering an immunological response. They cause tissue injury and inflammation when they come into contact with nearby tissues. For example, fibrosis occurs through the development of extra fibrous connective tissue, which can affect organ function and cause illness; it may be caused by repeated exposure to long asbestos fibres. Breast implant-associated anaplastic large cell lymphoma (BIA-ALCL) is a prime example of a local toxicity resulting from the substances' inherent characteristics or from their interactions with biological components, causing persistent inflammation (Ma et al., 2024).

- **Carcinogenicity:** When the risks may be sufficiently assessed or controlled without further research, the evaluation of carcinogenicity risk for nanomaterials usually depends on theoretical analysis and available data rather than new carcinogenicity testing. Manufacturers have serious financial worries because long-term carcinogenicity studies are expensive and time-consuming. There are two types of primary genotoxicity: direct and indirect mechanisms. When nanomaterials physically interact with DNA inside the nucleus, they can cause chromosomal changes, breakage, or lesions, known as direct genotoxicity. Indirect genotoxicity, on the other hand, is frequently linked to the production of ROS brought on by nanomaterials or the discharge of hazardous ions as a result of their disintegration. The main mechanism of genotoxicity caused by nanomaterials is secondary genotoxicity, which is mediated by ROS generated by inflammatory cells (Ma et al., 2024).

4.1.5 Future Trends for Healthcare Applications

Nanotechnology has an everyday impact on human life today. However, prolonged population exposure to nanoparticles presents significant questions about future health and environmental risks. Other scientific fields like nanotoxicology and nanomedicine emerged as a result of these worries. The study of possible harmful health effects of nanoparticles is known as nanotoxicology. The field of nanomedicine was created to investigate the advantages and disadvantages of using nanomaterials in medicine and medical equipment. It includes subsectors such as tissue engineering, biomaterials, biosensors, and bioimaging. Improved drug delivery, decreased inflammation, enhanced surgical tissue repair, and monitoring of circulating cancer cells are several advantages of medicinal nanostructures. However, the lack of reliable toxicity data is a serious concern, as it raises the possibility of significant risks to human health (Hulla et al., 2015).

Theranostics and novel methods for delivering drugs are ways that continue to transform healthcare through novel approaches to increase the efficacy and personalization of medical services. Scientists are creating nanoformulations to link to each person's unique genetic traits to improve their therapeutic results. Individualized treatment and nanoformulation that blends artificial intelligence and nanotechnology require precise and error-free delivery of therapeutic chemicals. These breakthroughs have future consequences, allowing more precise and effective medical operations. For example, optical imaging-guided nanotheranostics and advanced imaging methods like MRI and CT enhance the reliability of cancer therapies, which is an important aspect in this field. Theranostics has significant potential since real-time monitoring assures accurate medication delivery to tumour areas, minimizing harm to healthy tissues and lowering systemic toxicity. Translating these developments from the lab to clinical settings is important to utilize the developments presented in this area to maximum effect, and establishing scalable manufacturing techniques, completing stringent clinical trials, and gaining regulatory permission are all steps in this process. Smart drug delivery instruments have been designed to release medicinal substances in a controlled way. Polymeric nanodroplets, for example, are flexible systems that function as both fuel and drug delivery vehicles. They have demonstrated potential in preventing and treating conditions like cancer and hypoxia by responding to stimuli such as variations in pH, temperature, or magnetism (Ma et al., 2024).

4.2 Conclusion

Nanotechnology is believed to have transformational potential, from offering creative approaches in dentistry to modifying drug delivery and enhancing therapies for brain conditions. Despite these developments, a number of difficulties still exist. Its broad clinical adoption continues to be challenged by the absence of thorough knowledge of nanoparticle behaviour and questions about biological interactions, toxicity, and environmental hazards. The clinical application of nanotechnology in healthcare is likely to have a promising future by providing more specific and effective therapies. Nanostructures play a significant role in healthcare systems with persistent innovation, enhanced regulatory structures, and ongoing multidisciplinary studies.

REFERENCES

Application of nanotechnology in diagnosis and treatment of various diseases and its future advances in medicine. (n.d.). https://doi.org/10.20959/wjpps201818-12703

Bayda, S., Adeel, M., Tuccinardi, T., Cordani, M., & Rizzolio, F. (2019). The history of nanoscience and nanotechnology: From chemical – physical applications to nanomedicine. *Molecules, 25*(1), 112. https://doi.org/10.3390/molecules25010112

Egusquiaguirre, S. P., Igartua, M., Hernández, R. M., & Pedraz, J. L. (2012). Nanoparticle delivery systems for cancer therapy: Advances in clinical and preclinical research. *Clinical and Translational Oncology, 14*(2), 83–93. https://doi.org/10.1007/s12094-012-0766-6

Hobson, D. W. (2009). Commercialization of nanotechnology. *WIREs Nanomedicine and Nanobiotechnology, 1*(2), 189–202. https://doi.org/10.1002/wnan.28

Hulla, J., Sahu, S., & Hayes, A. (2015). Nanotechnology: History and future. *Human & Experimental Toxicology, 34*(12), 1318–1321. https://doi.org/10.1177/0960327115603588

Kubik, T., Bogunia-Kubik, K., & Sugisaka, M. (2005). Nanotechnology on duty in medical applications. *Current Pharmaceutical Biotechnology, 6*(1), 17–33. https://doi.org/10.2174/1389201053167248

Kumar, U., Kumari, M., Sahu, B., Vij, D., Virmani, L., Thakur, N., & Chatterji, T. (2024). Nanobiotechnology in personalized oncology. In *Nanotherapeutics for inflammatory arthritis: Design, diagnosis, and treatment* (pp. 231–242). https://doi.org/10.1201/9781003348672-12

Ma, X., Tian, Y., Yang, R., Wang, H., Allahou, L. W., Chang, J., Williams, G., Knowles, J. C., & Poma, A. (2024). Nanotechnology in healthcare, and its safety and environmental risks. *Journal of Nanobiotechnology*, *22*(1), 715. https://doi.org/10.1186/s12951-024-02901-x

Malik, S., Muhammad, K., & Waheed, Y. (2023). Nanotechnology: A revolution in modern industry. *Molecules*, *28*(2), 661. https://doi.org/10.3390/molecules28020661

Ngowi, E. E., Wang, Y.-Z., Qian, L., Helmy, Y. A. S. H., Anyomi, B., Li, T., Zheng, M., Jiang, E.-S., Duan, S.-F., Wei, J.-S., Wu, D.-D., & Ji, X.-Y. (2021). The application of nanotechnology for the diagnosis and treatment of brain diseases and disorders. *Frontiers in Bioengineering and Biotechnology*, *9*, 629832. https://doi.org/10.3389/fbioe.2021.629832

Patel, S., & Nanda, R. (2015). Nanotechnology in healthcare: Applications and challenges. *Medicinal Chemistry*, *05*(12). https://doi.org/10.4172/2161-0444.1000312

Pramanik, P. K. D., Solanki, A., Debnath, A., Nayyar, A., El-Sappagh, S., & Kwak, K.-S. (2020). Advancing modern healthcare with nanotechnology, nanobiosensors, and internet of nano things: Taxonomies, applications, architecture, and challenges. *IEEE Access*, *8*, 65230–65266. https://doi.org/10.1109/ACCESS.2020.2984269

Sahu, B., Sharma, J., Behera, B., & Kumar, U. (2024). Nanotechnology-based radiation therapy to cure cancer. In *Nanoparticles in cancer therapy: Innovations and clinical applications* (pp. 118–141). https://doi.org/10.1201/9781003515630-9

Shah, S. S., Shaikh, M. N., Khan, M. Y., Alfasane, Md. A., Rahman, M. M., & Aziz, Md. A. (2021). Present status and future prospects of jute in nanotechnology: A review. *The Chemical Record*, *21*(7), 1631–1665. https://doi.org/10.1002/tcr.202100135

Singh, G., Himanshu, Kumar, U., & Thakur, N. (2024). Theranostic nanoparticles: Revolutionizing cancer diagnosis and treatment. In *Nanoparticles in cancer therapy: Innovations and clinical applications* (pp. 187–2021). CRC Press. https://doi.org/10.1201/9781003515630-13

Yezdani, U., Khan, M. G., Kushwah, N., Verma, A., & Khan, F. (2018). Application of nanotechnology in diagnosis and treatment of various diseases and its future advances in medicine. *World Journal of Pharmacy and Pharmaceutical Sciences*, *7*, 1611–1633.

5

Nanotechnology in Environmental Remediation

Monika Verma, Ruchi Bharti, and Renu Sharma

5.1 Introduction to Nanotechnology

The rapid pace of industrialization, urbanization, and agricultural expansion has led to an alarming rise in environmental pollution worldwide. Our air, water, and soil eco-systems are now burdened with a complex array of contaminants – including heavy metals, persistent organic pollutants, dyes, pesticides, pharmaceutical residues, and pathogenic microorganisms. According to the World Health Organization, air pollu-tion alone causes approximately 4 million premature deaths annually, making it one of the leading environmental health risks globally (1). Simultaneously, over 1.1 billion people still lack access to safely managed drinking water, and an estimated 80% of wastewater globally is discharged untreated into the environment (2). Soil degradation also poses serious concerns, with nearly one-third of global arable land already classi-fied as degraded, affecting agricultural productivity and food security (3)

Traditional remediation techniques – such as chemical precipitation, incineration, and adsorption using activated carbon, electrocoagulation, and biological treatments – are widely employed for tackling pollution (4). However, these methods often have limitations, including poor efficiency at low pollutant concentrations, non-specificity, sludge generation, high energy requirements, and limited adaptability to in situ condi-tions (5). This has catalyzed the search for innovative, efficient, and sustainable alter-natives that can overcome the challenges of existing technologies.

Nanotechnology, which involves the design, synthesis, and application of materials at the nanometer scale (1–100 nm), has emerged as a revolutionary tool in environ-mental remediation. At this scale, materials exhibit unique physicochemical proper-ties such as increased surface area-to-volume ratio, high surface energy, quantum confinement, and enhanced reactivity. These attributes make nanomaterials excep-tionally well-suited to interact with and transform environmental pollutants in ways not achievable by conventional bulk materials (6–8).

Various classes of engineered nanomaterials (ENMs) – including metal and metal oxide nanoparticles (e.g., nano-Fe^0, TiO_2, ZnO, MnO_2), carbon-based nanomaterials (e.g., carbon nanotubes, graphene oxide, fullerenes), dendritic polymers (dendrimers), and hybrid nanocomposites – have been developed for environmental applications (9–11). Their functionalities can be tailored through surface modifications or doping, enhancing selectivity toward specific contaminants such as arsenic, lead, chromium,

DOI: 10.1201/9781003632498-5

organic dyes, or volatile organic compounds (VOCs). Among these, nanoscale zero-valent iron (nZVI) stands out as a flagship nanomaterial due to its dual capability to adsorb and reductively degrade a variety of pollutants (12–14). Field studies have shown that nZVI can reduce concentrations of chlorinated solvents like trichloroethylene (TCE) and perchloroethylene (PCE) by over 70–90%, making it a promising tool for in situ groundwater remediation (15, 16).

Nanomaterials exhibit diverse mechanisms of pollutant removal: (i) adsorption, where pollutants are immobilized on the surface of nanomaterials through electrostatic, covalent, or π–π interactions (17); (ii) catalytic degradation, including advanced oxidation processes (AOPs), photocatalysis, or redox transformations (18–20); and (iii) filtration and sieving, through nanostructured membranes and nanofiber mats that physically trap or reject contaminants based on size and charge (21, 22). These mechanisms often work synergistically – for example, a single nanoparticle may simultaneously adsorb heavy metals, reduce toxic ions, and degrade organic molecules, providing multi-functional remediation performance.

The versatility of nanotechnology is not limited to laboratory settings. Several real-world demonstrations and pilot projects have validated its potential for field applications. For example, iron nanoparticles stabilized with biopolymers have been successfully injected into aquifers to degrade TCE and immobilize hexavalent chromium (Cr^{6+}), resulting in notable improvements in groundwater quality (23). Similarly, TiO_2-based photocatalytic coatings on building surfaces have been deployed in urban areas to reduce ambient levels of nitrogen oxides (NO_x) (24, 25), a key contributor to smog and respiratory problems (26). Nanotechnology is also being used in water purification systems, where nano-enabled membranes and nanocomposite filters remove heavy metals, pathogens, and organic micropollutants with high efficiency and low energy consumption.

A critical advantage of nanotechnology lies in its potential for in situ remediation, which eliminates the need for costly excavation, transportation, or off-site treatment (27). Nanoparticles can be injected directly into contaminated groundwater plumes or soil matrices, allowing them to travel with the fluid and interact with dispersed pollutants (28). Furthermore, nano-enabled materials can be integrated into filters, membranes, sensors, and coatings for continuous, passive pollutant removal in both point-of-use and municipal-scale systems (29, 30).

Despite the promising performance, the widespread use of nanomaterials for environmental applications raises important concerns about nanotoxicity, environmental persistence, bioaccumulation, and unintended ecological impacts (31, 32). Nanoparticles – due to their small size – can cross biological barriers, interact with cellular components, and potentially accumulate in aquatic or terrestrial food chains (33). Therefore, understanding the fate, transport, and toxicity of ENMs in the environment is critical. Regulatory agencies like the US EPA and EU REACH have begun formulating frameworks for nano-specific risk assessment, but standardization and long-term monitoring remain key challenges (34–37).

Recent advancements in green synthesis, nano-bio hybrid systems, and smart, stimuli-responsive nanomaterials are addressing some of these concerns. For instance, nanoparticles synthesized using plant extracts, microbes, or biopolymers offer a more eco-friendly alternative to traditional chemical methods (38). Additionally, the integration of artificial intelligence (AI) and machine learning in nanomaterial design is enabling predictive optimization of nano-remediation systems, further enhancing efficiency and safety (39, 40).

In this chapter, we will explore the full spectrum of nanotechnology's role in environmental remediation. Starting with the fundamental principles and synthesis of nanomaterials, we delve into their mechanisms of action, followed by application case studies in water, soil, and air treatment. We also discuss toxicity concerns, regulatory issues, and sustainability frameworks, concluding with a look into emerging trends and future directions of this evolving field.

5.2 Fundamentals of Nanotechnology

Nanotechnology operates at the scale of atoms and molecules, typically between 1 and 100 nanometers, where materials exhibit unique physicochemical behaviors not seen in their bulk forms. This section explores the core types of nanomaterials, their size-based advantages, synthesis approaches, and the functionalization techniques that make them applicable to environmental remediation.

5.2.1 Types of Nanomaterials

Nanomaterials used in environmental applications can be broadly categorized by their composition and structure. The major classes include:

(a) **Carbon-based nanomaterials** – such as fullerenes (spherical carbon cages), carbon nanotubes (cylindrical tubes of graphene layers), and graphene sheets. These are prized for their high surface areas, thermal and chemical stability, and strong adsorption affinity (especially the π-electron systems in carbon nanotubes or graphene that bind organic molecules) (41, 42).

(b) **Metal and metal oxide nanoparticles** include zero-valent metals (e.g., nZVI, nano-Ag, nano-Zn), metal oxides (TiO_2, Fe_3O_4, Al_2O_3, etc.), and quantum dots (semiconductor nanocrystals). These inorganic nanomaterials often exhibit size-dependent optical and catalytic properties; for example, nano-TiO_2 is a more effective photocatalyst than bulk TiO_2 due to higher surface area and altered bandgap physics (43).

(c) **Polymeric nanomaterials and dendrimers** – dendrimers are monodisperse, tree-like polymer molecules with nanoscale dimensions and multiple functional end groups. Poly(amidoamine) (PAMAM) dendrimers, for instance, have abundant amine terminals that can chelate metal ions, making them useful "nanosponges" for heavy metals (44).

(d) **Nanocomposites** – hybrid materials where nanoparticles are embedded in or attached to another matrix (polymer, ceramic, etc.), combining properties of both. For example, iron oxide nanoparticles embedded in porous carbon or polymer beads can create a nanocomposite with high adsorption capacity and easy magnetic separability (45).

To quantify the significance of nanomaterials based on their category, their environmental applications are listed in Table 5.1.

TABLE 5.1

Summary of Nanomaterial Types and Their Environmental Applications

Nanomaterial Type	Examples	Key Features	Environmental Applications
Carbon-based	Graphene oxide, CNTs, fullerenes	High surface area, π–π interactions, conductivity	Adsorption of dyes (46), VOCs (47), heavy metals (48)
Metal/metal oxide NPs	nZVI, Ag, TiO_2, ZnO, Fe_3O_4	High redox potential, photocatalysis, magnetic recovery	Degradation of organic compounds (49, 50), disinfection (51), Cr(VI) reduction (52)
Polymeric/ dendrimers	PAMAM, PEG-functionalized polymers	Chelation, high solubility, tailored surface chemistry	Metal ion capture, sensing [23]
Nanocomposites	Fe_3O_4@GO, TiO_2@ zeolite	Synergistic adsorption and catalysis	Removal of heavy metals and dyes (53, 54)

5.2.2 Nanoscale Dimension Advantages

The shift from bulk to nanoscale imparts several key advantages. The most cited benefit is the tremendous increase in surface area per unit mass. As particle size decreases to the nanometer range, a larger fraction of atoms resides at the surface relative to the interior, providing more active sites for contaminant interaction (55).

This translates to faster reaction rates and higher capacity for pollutant removal. Quantum effects also emerge at the nanoscale, altering optical and catalytic behavior – especially in semiconductors like nano-TiO_2, which show enhanced UV absorption and photocatalysis (56).

High surface energy renders many nanomaterials highly reactive and often necessitates stabilization (e.g., by polymer coatings) (57). Their small size facilitates penetration into soil pores or aquifers, making them suitable for in situ remediation (58). Colloidal stability also enables prolonged suspension and migration in aqueous systems, particularly when surface coatings like carboxymethyl cellulose (CMC) or polyethylene glycol (PEG) are applied (59).

5.2.3 Synthesis Methods

Nanomaterials can be produced by two general strategies, as shown in Figure 5.1, which discusses the comparative steps in top-down versus bottom-up synthesis pathways.

Top-down methods are those in which bulk materials are mechanically or physically broken down to the nanoscale. Examples include high-energy ball milling of iron to synthesize nZVI and lithography (for patterned nanostructures).

The bottom-up methods are those where nanomaterials are built from atomic or molecular precursors. Examples include chemical reduction (e.g., metal salts reduced

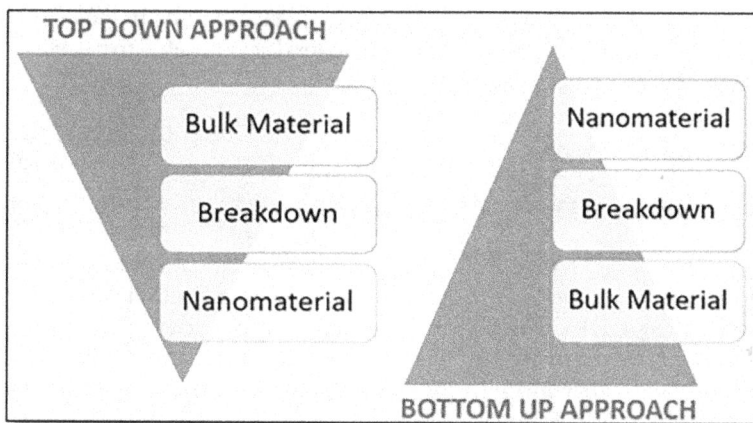

FIGURE 5.1   A flowchart visualizing the comparative steps in top-down versus bottom-up synthesis pathways.

by NaBH₄ to form nanoparticles), co-precipitation, sol–gel, hydrothermal, and microbial or green synthesis using plant extracts as eco-friendly reducing and capping agents (60).

5.2.4 Surface Functionalization

Post-synthesis surface modification enhances nanomaterial stability, dispersibility, selectivity, and reusability. Functionalization techniques include (61):

- Silane coupling agents (e.g., APTES) for amine group attachment on silica or TiO₂ surfaces (62).
- PEGylation to prevent aggregation and increase aqueous stability (63).
- Oxidation of carbon nanotubes or graphene oxide to introduce carboxyl (-COOH) or hydroxyl (-OH) groups for metal chelation and dispersion (64).
- Polymer coatings like polydopamine or CMC to enhance nZVI reactivity and prevent premature oxidation (65).

These modifications improve compatibility with environmental systems, enable targeted pollutant removal (e.g., Cr(VI), Pb²⁺, dyes), and extend material reusability.

5.3 Mechanisms of Nanomaterial-Based Remediation

Nanomaterials can remediate environmental pollutants through a variety of physico-chemical mechanisms (66). The dominant mechanisms include adsorption, catalysis (including redox reactions), photocatalytic degradation, and size-selective filtration.

Often, a single nanomaterial may employ multiple mechanisms in tandem. A thorough understanding of these processes is crucial for designing effective nanoremediation systems and predicting their performance.

5.3.1 Adsorption

Many nanomaterials act as exceptionally effective adsorbents, scavenging contaminants from soil or water by binding them onto their surfaces. The high surface area and often high-affinity functional groups on nanomaterials drive this process. For example, carbon nanotubes and graphene have extensive π-electron surfaces and can adsorb organic molecules (like polyaromatic hydrocarbons or dyes) via π–π stacking and hydrophobic interactions. They also adsorb heavy metal ions from water, especially when functionalized with oxygen- or nitrogen-containing groups that complex with metal cations (67). Nanoscale metal oxides, such as nano-Fe_3O_4 or nano-TiO_2, have surfaces that can electrostatically attract and bind heavy metals or phosphates. In soils, nanoclays (nanometer-thick clay mineral layers) present charged sites and interlayer spacing that can trap pollutants by ion exchange or intercalation (68). Adsorption by nanomaterials is often fast and can achieve high removal efficiencies because the contaminants are literally "mopped up" by the abundant binding sites on the nanoparticle surfaces. One key aspect is that adsorption is typically reversible, which is beneficial for regenerating and reusing nanoadsorbents (desorb the pollutant by changing pH, ionic strength, etc.), but also means that without proper disposal, the pollutant-laden nanoparticles might release contaminants back into the environment if conditions change (69, 70).

5.3.2 Redox Reactions and Catalytic Degradation

Certain nanomaterials serve as potent reductants or catalysts that chemically transform pollutants into less harmful forms (71). nZVI is a prime example: it is essentially nano-sized elemental iron (Fe^0), which is a strong reducing agent. nZVI particles can donate electrons to oxidized contaminants, for instance:

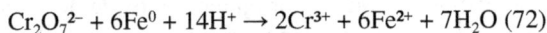

$$Cr_2O_7^{2-} + 6Fe^0 + 14H^+ \rightarrow 2Cr^{3+} + 6Fe^{2+} + 7H_2O \ (72)$$

This reduces toxic hexavalent chromium (Cr^{6+}) to the less mobile and less toxic trivalent chromium (Cr^{3+}), while iron is oxidized. Similarly, nZVI can reductively dechlorinate organic solvents like TCE into ethene or ethane (72, 73). This process occurs on the nanoparticle surface (often an Fe^0 core with an Fe-oxide shell) and yields precipitates (like insoluble chromium hydroxide or chloride salts) that immobilize the pollutants. Bimetallic nanoparticles such as Fe/Pd or Fe/Ni combine iron's reducing power with a secondary metal catalyst (Pd or Ni) to accelerate dechlorination of solvents and PCBs (polychlorinated biphenyls) via catalytic hydrogenolysis (74). On the oxidative side, nanoscale metal oxides (e.g., MnO_2, CeO_2) can catalyze the oxidation of pollutants like arsenic (III) to arsenic (V) (which then adsorbs to surfaces) or convert CO to CO_2 in air cleanup (75). Many nanomaterials also exhibit Fenton-like catalytic activity, where iron or other transition metal NPs generate hydroxyl radicals from hydrogen peroxide to oxidize

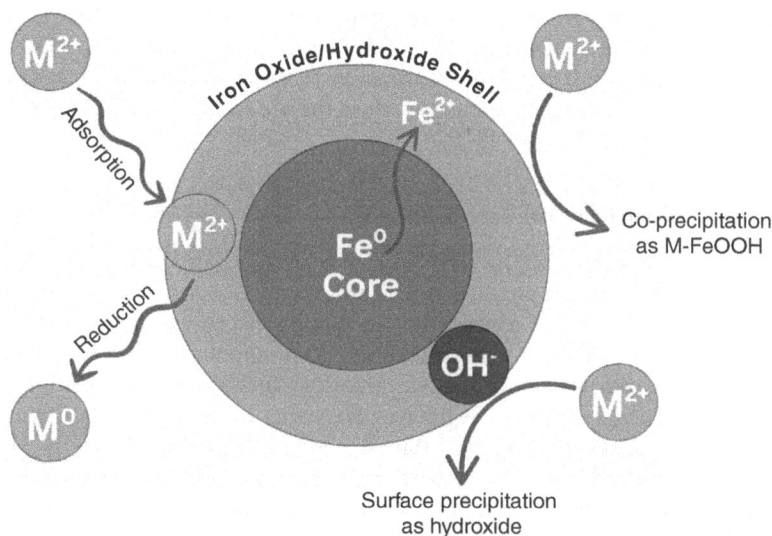

FIGURE 5.2 Core–shell structure of nanoscale zero-valent iron (nZVI) and its remediation mechanisms.

organic contaminants (an AOP). Because nanoparticles have high reactivity, these redox processes tend to be faster and more complete than with bulk materials (76, 77). Here, Figure 5.2 illustrates the core–shell structure of an nZVI particle and how it reduces or adsorbs pollutants.

The Fe^0 core (gray) serves as an electron source for reducing pollutants, while the outer shell consists of iron oxide/hydroxide (orange), which adsorbs metal ions. For example, a chlorinated hydrocarbon (R–Cl) is reduced to R–H via electron transfer at the Fe^0 surface. Similarly, a heavy metal cation (Me^{n+}) is either reduced to a lower oxidation state ($Me^{(n-x)+}$) or adsorbed onto the oxide shell. This dual action (reduction + adsorption) enables nZVI to detoxify both organic and inorganic contaminants (78).

5.3.3 Photocatalysis

Photocatalytic nanomaterials harness light energy to drive chemical reactions that degrade pollutants. The most extensively studied photocatalyst is titanium dioxide (TiO_2) in nano form. When TiO_2 nanoparticles are illuminated with ultraviolet (UV) light (energy above its bandgap), they generate electron–hole pairs that can initiate redox reactions: typically, the hole oxidizes water or hydroxide to produce hydroxyl radicals (•OH), while the electron can reduce O_2 to superoxide radicals. These reactive oxygen species non-selectively oxidize organic pollutants, breaking them down into carbon dioxide, water, or other mineralized products. For instance, nano-TiO_2 can photocatalytically degrade pesticides, dyes, and VOCs in water or air. Because of the nanoscale size, TiO_2 particles provide a large reactive surface and can be suspended in water to treat pollutants uniformly. Modified photocatalysts (such as metal-doped TiO_2 or composite nanostructures) are being developed to extend activity into visible light and enhance efficiency. Other semiconductor nanomaterials like ZnO, CdS, and newer graphene-based composites also show photocatalytic remediation capabilities (79, 80). Photocatalysis is particularly attractive for organic contaminants, as it can completely mineralize toxic compounds without the need for chemical reagents – sunlight or UV provides the energy (81).

5.3.4 Nanofiltration and Size Exclusion

The term "nanofiltration" usually refers to membrane filtration with pore sizes in the nanometer range, which is a pressure-driven process to separate contaminants by size and charge. In the context of nanotechnology-based remediation, this can mean two things: (1) the development of nanostructured membranes that have nanoscale pores or incorporate nanomaterials to improve filtration performance; and (2) the direct use of nanoparticle suspensions to filter or flocculate contaminants. Nanofiltration membranes are a class of semipermeable membranes with pore sizes typically <1 nm, capable of rejecting multivalent ions, small organic molecules, and even viruses. These membranes often have a thin-film composite structure and can be enhanced by embedding nanoparticles (creating thin-film nanocomposite membranes) to add functionalities. For example, adding silver or graphene oxide nanoparticles into a membrane can confer antibacterial properties and reduce biofouling, while titanium oxide or zeolite nanoparticles can increase water flux and selectivity. In air filtration, electrospun nanofiber filters act on a similar principle of size exclusion and interception: a mat of polymer nanofibers creates nanoscale pores that can physically trap fine particulate matter (PM2.5, PM1.0) more efficiently than conventional microfiber filters, with relatively low airflow resistance due to the high porosity (82, 83).

5.3.5 Pollutant-Specific Interactions

Nanomaterials can be tailored to target specific contaminants through surface functionalization. For heavy metals, many nanomaterials rely on chemisorption – binding metals via surface hydroxyl, carboxylate, or sulfhydryl groups and often reducing them to insoluble forms (84). For example, thiol-coated silica nanoparticles show high affinity for soft metal ions like lead and mercury, forming strong mercaptide bonds (85). In the case of organic pollutants, hydrophobic carbon-based nanomaterials partition organic toxins out of water and onto their surfaces, whereas catalytic nanoparticles break specific bonds (like carbon–halogen bonds in chlorinated organics) (86). Molecular sieving is another interaction mode. Certain nanomaterials – such as metal–organic frameworks at the nanoscale, or carbon nanotubes with defined pore structure – can preferentially allow some molecules to enter and exclude others based on size or shape, thus separating contaminants (87). An emerging strategy involves nanoscale sensors coupled with reactive elements, where a nanomaterial first detects or binds a target pollutant and then triggers a reaction to neutralize it (88).

It is evident that nanomaterials provide a versatile platform for remediation – they can adsorb pollutants like mini-sponges, destroy pollutants like catalysts, or separate pollutants like ultra-fine filters. These mechanisms can also complement each other. For instance, an nZVI particle can adsorb a chlorinated compound on its surface and then immediately reduce it to a benign product, effectively combining adsorption and reaction. The next sections will look at concrete implementations of these mechanisms in water, soil, and air cleanup applications, highlighting real-world examples and case studies.

5.4 Nanotechnology in Water Treatment

Nanotechnology offers promising solutions for water and wastewater treatment, leveraging the high surface area and tunable reactivity of nanomaterials to address various pollutants. Nanomaterials can remove contaminants via adsorption, redox transformation, membrane filtration, and photocatalytic degradation (Figure 5.3). This section elaborates on their roles and incorporates tabulated data, emerging issues, and nanomaterial recyclability.

5.4.1 Nanomaterials as Adsorbent

Carbon nanotubes, graphene-based materials, nanometal oxides, and silsesquioxane-based materials predominantly behave as adsorbents and help in the treatment of water (see Table 5.2). Nanoadsorbents not only function rapidly but also possess significant pollutant-binding capacities (90). They also undergo chemical regeneration when they become exhausted. To obtain improved outcomes for the elimination of contaminants from wastewater, nanoparticles are emerging as new options for the treatment of wastewater (91).

5.4.2 Nanomaterials as Photocatalyst

Different mechanisms have been suggested for the degradation of dyes by the photocatalyst materials. One mechanism indicates that the free radicals first initiate oxidation of the organic compounds primarily due to the electron–hole (e^-/h^+) pairs at the surface of the photocatalyst (121). The other mechanism says that the organic

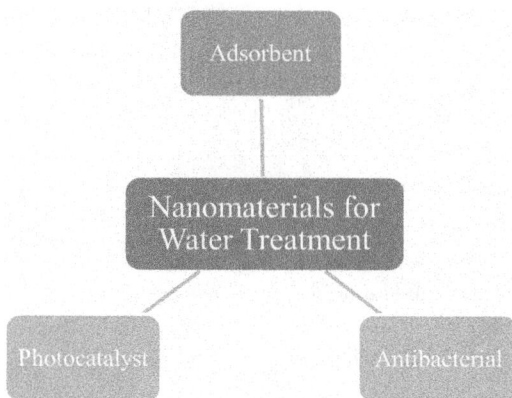

FIGURE 5.3 Nanomaterials for water treatment.

(89)

TABLE 5.2

Removal of Organic Pollutants Using Nanoadsorbents

S. No.	Adsorbent	Adsorbate
Carbon nanotubes		
1.	Multiwalled carbon nanotubes (MWCNTs)	Ni^{2+} (92)
		Pb^{2+}, Cu^{2+}, Cd^{2+} (93)
		Methyl orange and methylene blue (94)
		4-Chloro-2-nitrophenol (95)
		Tetracycline (96)
		Ciprofloxacin (97)
		Sulpfamethoxazole (98)
		Norfloxacin (99)
2.	Single walled carbon nanotubes (SWCNTs)	Ni^{2+} (92)
		Reactive red 120 dye(RR-120) (100)
		4-Chloro-2-nitrophenol (95)
		Ciprofloxacin (97)
Graphene based		
3.	Graphene oxide (GO)	Pb^{2+}(101), tetracycline (102)
4.	GO/Fe_3O_4	Cu^{2+}, fulvic acid (103)
5.	$GO/silica/Fe_3O_4$	Pb^{2+}, Cd^{2+}, humic acid (104)
6.	Graphene nanosheets (GNs)	Pb^{2+} (105)
7.	MnO_2/GNs	Hg^{2+} (106)
8.	Porous graphene hydrogels	Ciprofloxacin (107)
Nano metal oxides		
9.	Goethite (α-FeOOH)	Cu^{2+} (108)
10.	ZnO	Pb^{2+}(109)
11.	CeO_2	Pb^{2+}(110)
12.	TiO_2	Pb^{2+}(111)
13.	Modified Al_2O_3	Pb^{2+}, Cd^{2+}(112)
14.	Fe3O4-CTAB	Acid red 27 (113)
15.	γ-Fe_2O_3	Acridine orange (114)
Supported nano metal oxide		
16.	Magnesium oxide-coated bentonite (MCB)	Cu^{2+} (115)
17.	Goethite-coated sand	Pb^{2+}, Cd^{2+} (116)
Miscellaneous nanoadsorbents		
18.	Chitosan nanoparticles	Eosin Y (117)
19.	Fe-Al-Ce nanoadsorbent suspension	Fluoride (118)
20.	Nano-$ZnTiO_3$ ceramic	Malachite green (119)
Silsesquioxane based		
21.	HP-TPPs and HP-TPPOs	Rhodamine B (120)

TABLE 5.3

Photocatalytic Nanomaterial for Pollutant Removal

S. No.	Photocatalytic nanomaterial	Pollutant
1.	RGO-ZnCdS nanocomposite	Methyl orange, rhodamine B (124)
2.	Nitrogen-doped graphene/ZnSe nanocomposite	Methyl orange (125)
3.	Graphene–gold nanocomposite	Rhodamine B (126)
4.	Graphene–metal–oxide composites	Rhodamine B (127)
5.	ZnO–graphene nano hybrid	Methylene blue (128)

TABLE 5.4

Antibacterial Nanomaterials

S. No.	Nanomaterial	Bacteria
1.	Graphite, graphite oxide, graphene oxide, reduced graphene oxide	*E. coli* K12 (132)
2.	GO-Ag	*E. coli* ATCC 25922 and *S. aureus* ATCC 6538 (133)
3.	GO-Ag	Pseudomonas *aeruginosa* (134)

compound is first adsorbed onto the photocatalyst surface and then reacts with the excited superficial e^-/h^+ pairs or OH radical to produce the end products (122). Multiple reaction mechanisms rely on surface adsorbed and solution phase species, which lead to various kinetics of photodegradation. Adsorption of the organic pollutants is usually taken to be a significant parameter in the determination of photo-catalytic oxidation degradation rates (123). Some examples of photocatalytic nanomaterials along with targeted pollutants are given in Table 5.3.

5.4.3 Nanomaterials as Antibacterial Agents

One of the severe issues the world is currently encountering is infectious diseases and the rising resistance of microorganisms against antibiotics (129). The majority of the infection-inducing bacteria are strongly resistant to minimum one of the antibiotics that are commonly utilized to eliminate the infection (130). To prevent such kind of infectious microorganisms, nano-antimicrobials have been established as an effective treatment option (131) (Table 5.4).

5.5 Nanotechnology in Soil Remediation

Soil pollution is often complex, involving mixtures of heavy metals, hydrocarbons, pesticides, and other chemicals bound in the soil matrix (135). Traditional soil remediation (like excavation, washing) can be costly and disruptive. Although bioremediation can help in saving the quality of soil to an extent, it is a very slow

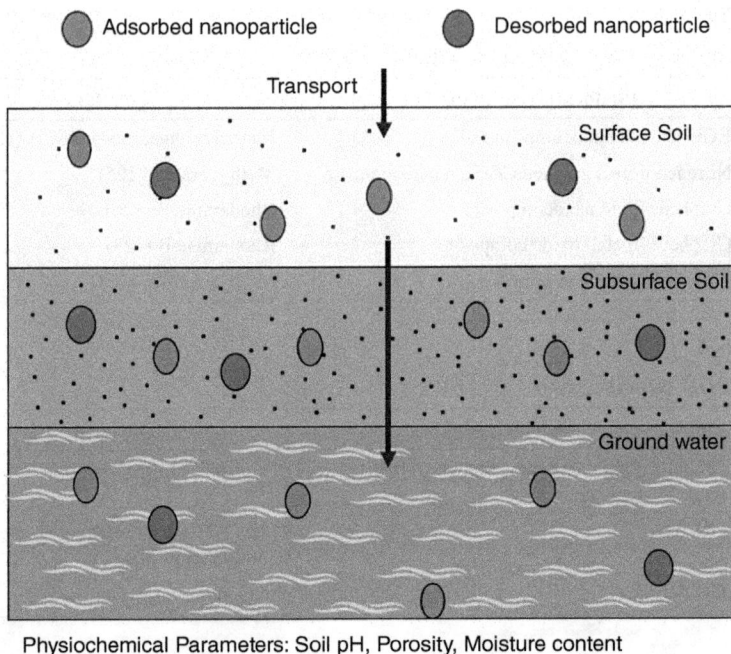

FIGURE 5.4   Schematic representation of nanoparticle transport in soil layers.

process, suitable only for biodegradable compounds, and is affected by environmental conditions such as temperature, pH, humidity, and oxygen level (136). For a quick remediation, nanotechnology offers tools for in situ soil cleanup that minimize disturbance: nanoparticles can be delivered into soils to immobilize or degrade pollutants right where they are (27). Key applications in soil include the use of nZVI and other reactive nanoparticles for in situ treatment, nanoclays for stabilization of contaminants, and nanoparticles that enhance breakdown of organic pollutants (137, 138). Figure 5.4 (schematic of nanoparticle movement in soil layers) visually explains transport.

5.5.1 Physicochemical Parameters Affecting Nanoparticle Transport in Soil

Nanoparticle transport and remediation efficacy are influenced by key soil parameters (139–141):

- **pH:** Determines surface charge of particles and solubility of metals. Lower pH often leads to enhanced metal solubilization (142).
- **Porosity and Texture:** Sandy soils allow deeper penetration, while clayey soils restrict nanoparticle mobility due to small pore size (142).
- **Ionic Strength:** Elevated ionic concentrations cause nanoparticle aggregation, reducing transport distance (143).

- **Organic Matter:** Can stabilize or hinder particle mobility depending on inter- actions (144).
- **Moisture Content:** Affects hydraulic connectivity and nanoparticle distribution (140).

5.5.2 Soil Remediation Using nZVI

As discussed for groundwater, nZVI is also applied to soil (vadose zone and saturated zone) remediation. In soil, the goal is often to immobilize metals or degrade haloge- nated organics that are present in soil pore water or loosely sorbed to soil particles. When nZVI is injected or mixed into soil, it can reduce toxic metal ions:

- **Removal of Lead:** The nZVI promotes reduction of soluble Pb^{2+} to metallic Pb or PbS (if sulfide is present), hence greatly reduces its mobility and bioavail- ability (145).
- **Removal of Mercury and Arsenic:** Iron nanoparticles have achieved great suc- cess in the removal of mercury and arsenic (146).
- **Removal of Chlorinated Organic Pollutants:** In soil containing residual DNAPL or pesticide residues, nZVI promotes dechlorination reactions similar to those in groundwater (147). The advantage in soil is that nZVI can navigate the small pore spaces and deliver treatment agents uniformly in ways that larger iron particles or additives cannot.

5.5.3 Remediation of Soil Using Nanoclays

Nanoclays are another set of nanomaterials used for soil remediation (148). Clays are naturally occurring minerals, but "nanoclay" typically refers to clay particles that have been processed to nanoscale dimensions (very thin platelets) or to highly exfoli- ated clays with enormous surface area.

The exfoliated nanosheet or nanotube clay adsorbents exhibit significantly enhanced performance due to the highly exposed reactive sites and extremely high specific sur- face area. Clay nanomaterials, like nano-zeolite, nano-halloysite, nano-attapulgite, and nano-bentonite, are very effective owing to their high cation exchange capac- ity and surface area, which facilitates the adsorption of heavy metals and organic pollutants (Table 5.5). Likewise, nano-cloisite and nano-montmorillonite are layered silicate materials that exhibit excellent capabilities for removing contaminants from soil environments (149).

5.5.4 Nanoparticles for Degradation of Pesticides and Hydrocarbons

Soil contamination by organic compounds like pesticides, petroleum hydrocarbons, and polycyclic aromatic hydrocarbons (PAHs) is widespread, especially in agricul- tural lands and former industrial sites (161–163). Nano-sulphonated graphene, multi- walled carbon nanotubes, carbon nanotubes, and amphiphilic nanoscale polyurethane are a few nanomaterials used against PAHs (Table 5.6) (164). Nanoparticles can aid in the breakdown of these organics, either by direct catalytic action or by enhancing bioremediation processes. For example, certain photoactive nanoparticles (e.g., nano- TiO_2 or ZnO) can be sprayed onto or mixed into surface soils, whereupon sunlight

TABLE 5.5

Nanoclay for Soil Remediation

S. No.	Nanoclay	Pollutant
1.	Natural-based nano-zeolite	Lead and cadmium (150)
2.	Nano-zeolite	Cadmium (151)
3.	Nano-zeolite	Lead (152)
4.	Nanosized halloysite/biochar	Arsenic and copper (153)
5.	Halloysite nanotubes	Cadmium, chromium, copper, nickel, lead, zinc (154)
6.	Nano zero-valent iron modified attapulgite clay	Cadmium, chromium, lead (155)
7.	Combined compost–attapulgite	Cadmium (156)
8.	Attapulgite	Vanadium (157)
9.	Nano-bentonite	Nickel, copper, lead (158)
10.	Nano-closite	Lead and copper (159)
11.	Nano-montmorillonite	Cadmium (160)

TABLE 5.6

Nanomaterial for Sensing and Destroying Pesticides (167)

S. No.	Nanomaterial	Pesticide
1.	Quantum dots (QDs)	Pyrethroid
2.	Polystyrene-Eu(III) chelate NP	Atrazine
3.	AuNPs	Paraoxon, parathion
4.	AgNPs	β-Endosulfan
5.	Chitosan-AuNps membrane	Picloram
6.	MWCNT-AChE/PB/MWCNT	Carbofuran
7.	TiO_2 NPs SPE	Dichlofenthion
8.	Nanomagnetite	DNOC
9.	MHNTs composites	TCP

exposure, they generate radicals that oxidize pesticide residues (165). This approach has been studied for persistent pesticides on agricultural soil; nano-TiO_2 under UV has degraded organochlorine pesticides and even some emerging contaminants like antibiotics in manure-amended soils. However, delivering light into deeper soil layers is challenging, so this is mainly for surface treatments (166).

5.5.5 Nanobioremediation

Nanoparticles are used to assist microbial degradation. It involves using nanomaterials as additives to stimulate soil microorganisms or enzymes (168). For example, adding nano-sized nutrients or nano-oxides can provide microbes with better access to essential elements, thus boosting their metabolism of pollutants (169). Enzyme-mimetic nanoparticles, known as nanozymes, are also being investigated. These nanoparticles (such as CeO_2 or Fe_3O_4) can mimic peroxidase enzymes and may help degrade pollutants like phenols in soil by catalyzing oxidation in the presence of peroxides (170). While much of this is at the research stage, the idea is to have nanomaterials augment the natural attenuation processes.

5.6 Nanotechnology for Air Pollution Treatment

Air pollution includes gaseous pollutants (e.g., NO_x, SO_x, VOCs) and particulate matter (PM) (171). Nanotechnology-based solutions for air purification leverage nanomaterials as catalysts, filters, or sensors to remove or detect airborne contaminants. The nanomaterials can be deployed in active air treatment systems (e.g., photocatalytic reactors, air filters) or as passive coatings on surfaces that clean the air. Key contributions of nanotech in air quality management include nanosensors, nanocatalysts, nanoadsorbents, nanofilters, and nanocoating (see Figure 5.5) (172).

5.6.1 Nanosensors for Air Remediation

Nanosensors are important for air remediation as they provide the accurate detection and monitoring of pollutants, a requirement for effective air quality control. The sensors, based on nanoscale materials, are highly sensitive, portable, and low-cost in detection, a feature that suits them for multiple environmental applications. Nanosensors are nanoscale sensors that gather data at the nanoscale, detect, and convert physical values into analyzable and detectable signals. Nanosensors are highly sensitive, portable, inexpensive, and easy-to-use sensing devices to detect chemical and

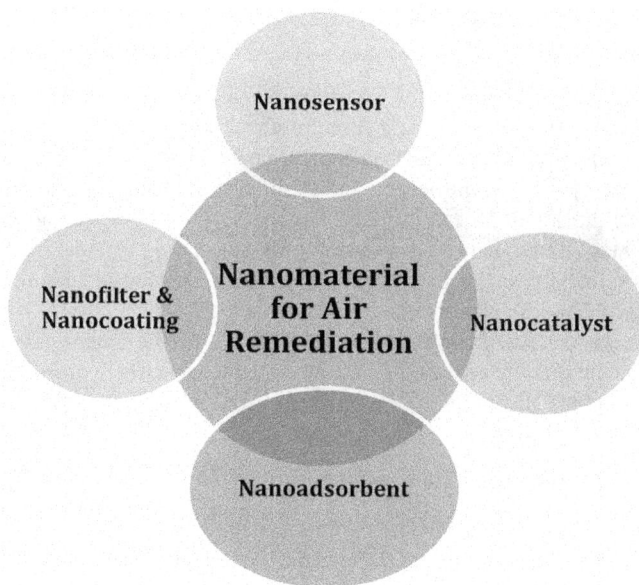

FIGURE 5.5 Nanomaterial for air remediation.

TABLE 5.7

Nanosensors Used for Air Pollutant Detection (176)

S.no.	Nanosensor		Air Pollutant
1.	**Carbon-based nanosensors**	$Ca_{12}O_{12}$ nanocage	CO_2, SO_2, NO_2 (177)
2.		Laser-induced graphene	Nox (178)
3.		B24N24 fullerene	COs, H_2S, SO_2, CS_2 (179)
4.	**Metal oxide-based nanosensors**	rGO/Pd-coated SnO_2 film	NO_2 (180)
5.		Single SnO2 nanowire	C_3H_6O, NH_3, CO, C_2H_6O, H_2, NO_2, C_7H_8 (181)
6.		Comb-like ZnO	H_2S (182)
7.	**Electrospun nanofiber-based nanosensor**	PAN	Chloroform (183)
8.		WO3	NO_2 (184)
9.	**Quantum dot-based nanosensor**	ZnO	H_2S (185)
10.		SnS	NO_2 (186)
11.	**Polymer-based nanosensors**	Poly(3-aminophenylboronic acid) (PAPBA)	CO, NO, NO_2, SO_2, SO_3 (187)
12.		PANI	NH_3 (188)
13.	**Nanosensor for detection of heavy metals in water**	Carbon dots from microalgae biochar	Cr^{6+} (189)
14.		Rice husk carbon quantum dots	Fe^{3+} (190)
15.	**Nanosensor for drug detection**	Carbon dots embedded in hydrogel spheres	Rifampicin (191)
16.		Chitosan–molybdenum vanadate nanocomposite	Hydroxychloroquine sulfate (192)

biological pollutants. A nanosensor is composed of three basic elements, namely, a receptor probe, a transducer element, and an amplifier. A receptor responds to the air pollutants and produces a reaction that is changed into an electric signal by a transducer, subsequently amplified by an amplifier and changed to an output that is quantifiable by the signal processing unit. For the purpose of detecting chemical and biological contaminants, nanosensors are highly sensitive, portable, affordable, and user-friendly sensing devices. The sensing element in nanosensors is employed using nanomaterials (173–175). Nanomaterials possess a high surface area-to-volume ratio, and their surface sensitivity can be improved through surface functionalization. They are more efficient and stronger as a result of their higher surface area. The nanosensors broadly used for air pollution detection are listed in Table 5.7.

5.6.2 Nanocatalysts for Air Purification

Just as in water, photocatalysis using nano-sized semiconductors is a potent method for air purification (193). Titanium dioxide (TiO_2) nanoparticles coated on surfaces can oxidize airborne pollutants when exposed to UV light (either sunlight or UV lamps) (194). In urban environments, TiO_2-based coatings on building facades or road surfaces have been tested to reduce nitrogen oxides (NO_x) from vehicle emissions. When NO_x gases contact a UV-illuminated TiO_2 surface, they get oxidized to nitrate, which can be washed away by rain. Field studies in some cities found measurable NO_x reduction on streets treated with photocatalytic materials (although results vary with environmental conditions) (24). Similarly, VOCs like benzene, formaldehyde, or toluene can be broken down by photocatalytic reactors installed in air ducts or near emission sources. Nano-TiO_2 is preferred because it provides a high surface area and can be applied as a thin film on substrates (glass, metal meshes, and concrete). It has been used in indoor air purifiers to destroy VOCs and kill airborne microbes using near-UV or even visible light (for modified TiO_2) (195).

Apart from TiO_2, other nano-photocatalysts include zinc oxide (ZnO) (193) and emerging composites like titanium dioxide with graphene (to enhance charge separation and visible light absorption) (196). There are also nanostructured photoreactive coatings, for instance, nano-sized anatase TiO_2 embedded in paints for indoor walls; these paints claim to continuously break down low-level pollutants (like formaldehyde emitted from furniture) under room light (197). The advantage of using nanomaterials here is the creation of transparent, high-surface area coatings that do not visibly alter the surface but provide functionality. In summary, photocatalytic nanocoatings represent a self-cleaning and passive way to remove gaseous pollutants from air, leveraging the power of light-driven reactions at the nanoscale.

5.6.3 Nanoadsorbents for Air Treatment

Nanoadsorbents are nanomaterials with large adsorption capacity, employed in the removal of pollutants from air. These can be employed in numerous applications such as air filters, catalytic converters, and sensors for detecting contaminants (198). One of the most common uses of nanoadsorbents is for VOC removal, which are toxic organic chemicals usually released during industrial activities, paints, and consumer products. Nanomaterials, including carbon nanotubes (CNTs), graphene oxide, and metal–organic frameworks (MOFs), have proven to possess very good VOC adsorption ability based on their porous structures and strong organic molecule affinity (199). Particularly, CNT composites represent a cutting-edge material class for air purification owing to their exceptional surface area, porosity, and tunable surface chemistry. These composites can effectively adsorb various airborne pollutants, including VOCs like benzene and toluene, PM, and hazardous gases, with high selectivity and efficiency. Functionalization of CNTs enhances their adsorption capacity, selectivity, and antimicrobial properties, making them especially promising for indoor air purification systems (200).

The other significant application is the capture and sequestration of greenhouse gases, especially carbon dioxide (CO_2) (201). Amine-modified silica nanoparticles and MOFs have been studied widely for CO_2 capture owing to their high selectivity and adsorption capacities (202, 203). Likewise, nanoadsorbents are efficient in the removal of acidic gases like sulfur dioxide (SO_2) and nitrogen oxides (NO_x), which are significant contributors to acid rain and respiratory diseases (204). Metal oxide nanoparticles such as titanium dioxide (TiO_2) and zinc oxide (ZnO) not only adsorb these gases but also catalyze their degradation under UV or visible light by photocatalytic reactions (205). This double role increases the overall efficiency of air purification systems.

5.6.4 Nanofilters and Nanocoatings for Air Purification

The removal of particulate pollutants (dust, smoke, soot, and microorganisms) from air is crucial for respiratory health (206). Conventional HEPA filters are effective but typically use micro-scale fibers (207). Nanofiber filters, made of polymer fibers only a few hundred nanometers thick, offer improved performance (208). Due to the small fiber diameter, the inter-fiber pores are also very small, which increases the capture efficiency for ultrafine particles (<0.1 µm) by mechanisms like Brownian diffusion and interception (209). At the same time, nanofiber mats can be made highly porous (because so little material is needed to form the filter), resulting in lower pressure drop (210). Research has demonstrated that air filters composed of electrospun nanofibers can achieve high filtration efficiencies for PM2.5 (fine PM) while maintaining good airflow, translating to energy savings in HVAC systems (211). Some commercial face masks and air purifiers have started integrating nanofiber layers to improve performance against viruses and smoke particles (212).

Nanofibers can also incorporate active functionalities. For example, a nanofiber filter can be loaded with adsorbent nanoparticles (like activated carbon nanospheres or alumina) to capture gaseous pollutants like VOCs and odors in addition to particles (213). Or they can contain catalytic nanoparticles (like MnO_2 or noble metals) that can catalyze the breakdown of certain pollutants (some filters for cabin air include Pt/Pd nanoparticles on fibers to help oxidize carbon monoxide) (75). Antimicrobial nanofiber filters have silver or copper nanoparticles embedded to kill trapped bacteria and prevent growth on the filter. This is particularly useful for air handling units to maintain hygiene (214).

Additionally, nanocoatings on conventional filters can give them new properties. A thin coating of nano-TiO_2 on a filter, when illuminated by UV, can prevent organic matter buildup (a self-cleaning filter) and inactivate pathogens (215). Hydrophobic nanocoatings (using silica or fluorinated nanoparticles) on filters can make them resist moisture and maintain performance in humid conditions (216).

Another air-related application is nanocoatings for surfaces in indoor environments that continually clean the air. For instance, a coating containing TiO_2 and silver nanoparticles on tiles or HVAC ducts can reduce both biological and chemical contaminants in the passing air (217, 218). While not a filter present, it is a complementary technique ensuring any pollutant that touches those surfaces is eliminated.

5.7 Toxicity and Environmental Risk Assessment of Nanomaterials

While the benefits of nanotechnology for remediation are promising, it is equally important to consider the potential risks and environmental health implications of introducing ENMs into the environment. Engineered nanoparticles have sizes comparable to biomolecules and can interact with living organisms in unintended ways (219). Therefore, understanding nanomaterial toxicity (nanotoxicology), environmental fate, transport, and developing appropriate risk assessment frameworks is crucial for the responsible deployment of nanoremediation technologies.

ENMs can be seen as a double-edged sword: they can remove pollutants, but once their job is done, those nanomaterials themselves become a new potential contaminant in the environment (220). Key questions include: Do nanoparticles persist in the environment? Do they bioaccumulate in organisms? Are they toxic to aquatic or soil life? Research indicates that the nanoparticle impacts vary greatly depending on their composition, size, surface coating, dosage, and the environmental matrix (pmc.ncbi.nlm. nih.gov). For example, silver nanoparticles (widely known for antimicrobial properties) can be toxic to beneficial bacteria in soil or to aquatic organisms if they leach into waterways, primarily through the release of Ag+ ions, which are highly toxic to many forms of life (221). Nanoparticles of titanium dioxide or zinc oxide can induce oxidative stress in fish under illumination (since they generate reactive oxygen species similar to how they kill microbes) (222, 223). Carbon nanotubes, if inhaled by animals, have been shown in some studies to cause lung inflammation or fibrosis reminiscent of ultrafine dust effects (224).

The fate of nanoparticles refers to how they transform or move in the environment. Many metal-based NPs will oxidize or dissolve over time. For instance, nZVI quickly becomes iron oxides (rust) after reacting – these oxides are generally less toxic and may even naturally occur in soils (pmc.ncbi.nlm.nih.gov). However, the byproducts (like the coating materials or any impurities) should be considered (225). Nanoparticles also tend to agglomerate (stick together to form larger clusters) in environmental media due to ionic strength or natural organic matter; aggregation usually reduces their mobility and biological uptake, which can be positive for reducing bioavailability (226). However, under certain conditions, stabilized nanoparticles might remain dispersed and travel with groundwater or run-off, potentially reaching non-target sites. Transport studies suggest that factors like pH, ionic composition, and the presence of natural colloids greatly influence nanoparticle mobility. For example, in sandy aquifer media, bare nanoparticles may get filtered out quickly, but if coated with a polymer (for stabilization in remediation), they might travel further (227).

Another concern is bioaccumulation. Some organisms can take up nanoparticles (228). Plants have been shown to absorb metal nanoparticles through roots (although often they dissolve and the plant takes up the metal ions) (229). Aquatic filter feeders might ingest nanoparticles present in water, and those could pass up the food chain (230). The long-term effects of chronic low-level nanoparticle exposure are still under investigation. Studies on freshwater zooplankton and fish have reported subtle effects on growth and reproduction at certain nanoparticle concentrations, whereas other studies find no significant harm at environmentally relevant levels (230).

5.8 Conclusion

Nanoparticles are strong agents for environmental remediation with their peculiar physicochemical attributes, which include high surface area, reactivity, and tailoring. Nanoparticles present novel and efficient means to remediate contaminants in air, water, and soil, i.e., heavy metals, organic pollutants, and microbes. As development advances, more eco-friendly, cost-efficient, and upscalable nanomaterials help improve the performance of nanoparticles for practical uses. Yet it is also necessary to evaluate the long-term health and environmental effects of nanoparticles in order to have safe and sustainable applications. With ongoing progress, nanoparticles are full of promise to lead to a cleaner and greener environment.

REFERENCES

[1] Organización Mundial de la Salud (OMS). WHO global air quality guidelines. Particulate matter (PM25 and PM10), ozone, nitrogen dioxide, sulfur dioxide and carbon monoxide. 2021;1–360.

[2] Mishra RK. Fresh water availability and its global challenge. *Br J Multidiscip Adv Stud*. 2023;4(3):1–78.

[3] Smith P, Poch RM, Lobb DA, Bhattacharyya R, Alloush G, Eudoxie GD, . . . Hallett P. Status of the world's soils. *Annu Rev Environ Resour*. 2024;49(2024):73–104.

[4] Rajasulochana P, Preethy V. Comparison on efficiency of various techniques in treatment of waste and sewage water – A comprehensive review. *Resour Eff Technol*. 2016 Dec;2(4):175–84.

[5] Mishra S, Chowdhary P, Bharagava RN. Conventional methods for the removal of industrial pollutants, their merits and demerits. In *Emerging and Eco-Friendly Approaches for Waste Management*. Springer Singapore; 2018. p. 1–31.

[6] Banerjee S, Gautam RK, Gautam PK, Jaiswal A, Chattopadhyaya MC. Recent trends and advancement in nanotechnology for water and wastewater treatment: Nanotechnological approach for water purification. In *Advanced Research on Nanotechnology for Civil Engineering Applications*. IGI Global; 2016. p. 208–252.

[7] Asghar N, Hussain A, Nguyen DA, Ali S, Hussain I, Junejo A, et al. Advancement in nanomaterials for environmental pollutants remediation: A systematic review on bibliometrics analysis, material types, synthesis pathways, and related mechanisms. *J Nanobiotechnol*. 2024 Jan 10;22(1):26.

[8] Mehndiratta P, Jain A, Srivastava S, Gupta N. Environmental pollution and nanotechnology. *Environ Pollut*. 2013 Mar 13;2(2).

[9] Sohail MI, Waris AA, Ayub MA, Usman M, ur Rehman MZ, Sabir M, Faiz T. Environmental application of nanomaterials: A promise to sustainable future. In *Comprehensive Analytical Chemistry* (Vol. 87). Elsevier; 2019. p. 1–54.

[10] Dhanapal AR, Thiruvengadam M, Vairavanathan J, Venkidasamy B, Easwaran M, Ghorbanpour M. Nanotechnology approaches for the remediation of agricultural polluted soils. *ACS Omega*. 2024 Mar 26;9(12):13522–33.

[11] Ravi S, Vadukumpully S. Sustainable carbon nanomaterials: Recent advances and its applications in energy and environmental remediation. *J Environ Chem Eng*. 2016 Mar;4(1):835–56.

[12] Garg R, Mittal M, Tripathi S, Eddy NO. Core to concept: Synthesis, structure, and reactivity of nanoscale zero-valent iron (NZVI) for wastewater remediation. *Environ Sci Pollut Res Int*. 2024;31(60):67496–67520.

[13] Li L, Hu J, Shi X, Fan M, Luo J, Wei X. Nanoscale zero-valent metals: A review of synthesis, characterization, and applications to environmental remediation. *Environ Sci Pollut Res Int*. 2016;23(18):17880–17900.

[14] Wang C, Feng C, Gao Y, Ma X, Wu Q, Wang Z. Preparation of a graphene-based magnetic nanocomposite for the removal of an organic dye from aqueous solution. *Chem Eng J*. 2011 Sep;173(1):92–7.

[15] Qian L, Chen Y, Ouyang D, Zhang W, Han L, Yan J, et al. Field demonstration of enhanced removal of chlorinated solvents in groundwater using biochar-supported nanoscale zero-valent iron. *Sci Total Environ*. 2020 Jan;698:134215.

[16] Chen H, Qian L. Performance of field demonstration nanoscale zero-valent iron in groundwater remediation: A review. *Sci Total Environ*. 2024;912:169268.

[17] Singh S, Kapoor D, Khasnabis S, Singh J, Ramamurthy PC. Mechanism and kinetics of adsorption and removal of heavy metals from wastewater using nanomaterials. *Environ Chem Lett*. 2021 Jun 11;19(3):2351–81.

[18] Giwa A, Yusuf A, Balogun HA, Sambudi NS, Bilad MR, Adeyemi I, et al. Recent advances in advanced oxidation processes for removal of contaminants from water: A comprehensive review. *Process Saf Environ Prot.* 2021 Feb;146:220–56.

[19] Bethi B, Sonawane SH, Bhanvase BA, Gumfekar SP. Nanomaterials-based advanced oxidation processes for wastewater treatment: A review. *Chem Eng Process – Process Intensif.* 2016 Nov;109:178–89.

[20] Liu H, Wang C, Wang G. Photocatalytic advanced oxidation processes for water treatment: Recent advances and perspective. *Chem Asian J.* 2020 Oct 16;15(20):3239–53.

[21] Moattari RM, Mohammadi T. Nanostructured membranes for water treatments. In *Nanotechnology in the Beverage Industry.* Elsevier; 2020. p. 129–150

[22] Gohil JM, Choudhury RR. Introduction to nanostructured and nano-enhanced polymeric membranes: Preparation, function, and application for water purification. In *Nanoscale Materials in Water Purification.* Elsevier; 2019. p. 25–57

[23] Pandey K, Sharma S, Saha S. Advances in design and synthesis of stabilized zerovalent iron nanoparticles for groundwater remediation. *J Environ Chem Eng.* 2022 Jun;10(3):107993.

[24] Perera R, Prasad R, Seneviratne I, Perera B. Adaptability of Tio2-based photocatalytic coating for air pollution mitigation in Sri Lanka. Available from: *SSRN 5079411.*

[25] Ballari MM, Hunger M, Hüsken G, Brouwers HJH. NOx photocatalytic degradation employing concrete pavement containing titanium dioxide. *Appl Catal B.* 2010 Apr;95(3–4):245–54.

[26] César ACG, Carvalho Jr. JA, Nascimento LFC. Association between NOx exposure and deaths caused by respiratory diseases in a medium-sized Brazilian city. *Braz J Med Biol Res.* 2015 Dec;48(12):1130–5.

[27] Kuppusamy S, Palanisami T, Megharaj M, Venkateswarlu K, Naidu R. In-situ remediation approaches for the management of contaminated sites: A comprehensive overview. *Rev Environ Contam Toxicol.* 2016;236:1–115.

[28] Liu W, Tian S, Zhao X, Xie W, Gong Y, Zhao D. Application of stabilized nanoparticles for in situ remediation of metal-contaminated soil and groundwater: A critical review. *Curr Pollut Rep.* 2015 Dec 23;1(4):280–91.

[29] Al Harby NF, El-Batouti M, Elewa MM. Prospects of polymeric nanocomposite membranes for water purification and scalability and their health and environmental impacts: A review. *Nanomaterials.* 2022 Oct 17;12(20):3637.

[30] Elzein B. Nano revolution: "Tiny tech, big impact: How nanotechnology is driving SDGs progress". *Heliyon.* 2024 May;10(10):e31393.

[31] El-Kady MM, Ansari I, Arora C, Rai N, Soni S, Verma DK, et al. Nanomaterials: A comprehensive review of applications, toxicity, impact, and fate to environment. *J Mol Liq.* 2023 Jan;370:121046.

[32] Kumar CV, Karthick V, Kumar VG, Inbakandan D, Rene ER, Suganya KU, . . . Sowmiya P. The impact of engineered nanomaterials on the environment: Release mechanism, toxicity, transformation, and remediation. *Environ Res.* 2022;212:113202.

[33] Nel A, Xia T, Mädler L, Li N. Toxic potential of materials at the nanolevel. *Science* (1979). 2006 Feb 3;311(5761):622–7.

[34] Porter RD, Breggin L, Falkner R, Pendergrass J, & Jaspers N. Regulatory responses to nanotechnology uncertainties. *The nanotechnology challenge.* Cambridge University Press, New York. 2012.

[35] Amenta V, Aschberger K, Arena M, Bouwmeester H, Botelho Moniz F, Brandhoff P, et al. Regulatory aspects of nanotechnology in the agri/feed/food sector in EU and non-EU countries. *Regul Toxicol Pharmacol.* 2015 Oct;73(1):463–76.

[36] Allan J, Belz S, Hoeveler A, Hugas M, Okuda H, Patri A, et al. Regulatory landscape of nanotechnology and nanoplastics from a global perspective. *Regul Toxicol Pharmacol.* 2021 Jun;122:104885.

[37] Chávez-Hernández JA, Velarde-Salcedo AJ, Navarro-Tovar G, Gonzalez C. Safe nanomaterials: From their use, application, and disposal to regulations. *Nanoscale Adv.* 2024;6(6):1583–610.

[38] Pandit C, Roy A, Ghotekar S, Khusro A, Islam MN, Emran T Bin, et al. Biological agents for synthesis of nanoparticles and their applications. *J King Saud Univ Sci.* 2022 Apr;34(3):101869.

[39] Chen H, Zheng Y, Li J, Li L, Wang X. AI for nanomaterials development in clean energy and carbon capture, utilization and storage (CCUS). *ACS Nano.* 2023 Jun 13;17(11):9763–92.

[40] Jia Y, Hou X, Wang Z, Hu X. Machine learning boosts the design and discovery of nanomaterials. *ACS Sustain Chem Eng.* 2021 May 10;9(18):6130–47.

[41] Mauter MS, Elimelech M. Environmental applications of carbon-based nanomaterials. *Environ Sci Technol.* 2008 Aug 1;42(16):5843–59.

[42] Díez-Pascual AM. Carbon-based nanomaterials. *Int J Mol Sci.* 2021 Jul 20;22(14):7726.

[43] Nair GM, Sajini T, Mathew B. Advanced green approaches for metal and metal oxide nanoparticles synthesis and their environmental applications. *Talanta Open.* 2022 Aug;5:100080.

[44] Viltres H, López YC, Leyva C, Gupta NK, Naranjo AG, Acevedo–Peña P, . . . Kim KS. Polyamidoamine dendrimer-based materials for environmental applications: A review. *J Mol Liq.* 2021;334:116017.

[45] Zhu J, Wei S, Chen M, Gu H, Rapole SB, Pallavkar S, et al. Magnetic nanocomposites for environmental remediation. *Adv Powder Technol.* 2013 Mar;24(2):459–67.

[46] Rezania S, Darajeh N, Rupani PF, Mojiri A, Kamyab H, Taghavijeloudar M. Recent advances in the adsorption of different pollutants from wastewater using carbon-based and metal-oxide nanoparticles. *Appl Sci.* 2024 Dec 10;14(24):11492.

[47] Yin F, Yue W, Li Y, Gao S, Zhang C, Kan H, et al. Carbon-based nanomaterials for the detection of volatile organic compounds: A review. *Carbon N Y.* 2021 Aug;180:274–97.

[48] Gargiulo V, Alfè M, Lisi L, Manfredi C, Volino S, Di Natale F. Colloidal carbon-based nanoparticles as heavy metal adsorbent in aqueous solution: Cadmium removal as a case study. *Water Air Soil Pollut.* 2017 May 28;228(5):192.

[49] Nagajyothi PC, Prabhakar Vattikuti SV, Devarayapalli KC, Yoo K, Shim J, Sreekanth TVM. Green synthesis: Photocatalytic degradation of textile dyes using metal and metal oxide nanoparticles-latest trends and advancements. *Crit Rev Environ Sci Technol.* 2020 Dec 16;50(24):2617–723.

[50] Shanaah HH, Alzaimoor EFH, Rashdan S, Abdalhafith AA, Kamel AH. Photocatalytic degradation and adsorptive removal of emerging organic pesticides using metal oxide and their composites: Recent trends and future perspectives. *Sustainability.* 2023 Apr 28;15(9):7336.

[51] Chen W, Tsai P, Chen Y. Functional Fe3O4/TiO2 core/shell magnetic nanoparticles as photokilling agents for pathogenic bacteria. *Small.* 2008 Apr 23;4(4):485–91.

[52] Besharat F, Ahmadpoor F, Nasrollahzadeh M. Graphene-based (nano)catalysts for the reduction of Cr(VI): A review. *J Mol Liq.* 2021 Jul;334:116123.

[53] Pang AL, Arsad A, Ahmad Zaini MA, Garg R, Saqlain Iqbal M, Pal U, et al. A comprehensive review on photocatalytic removal of heavy metal ions by polyaniline-based nanocomposites. *Chem Eng Commun.* 2024 Feb 28;211(2):275–99.

[54] Sakib A, Masum S, Hoinkis J, Islam R, Molla Md. Synthesis of CuO/ZnO nanocomposites and their application in photodegradation of toxic textile dye. *J Compos Sci.* 2019 Sep 17;3(3):91.

[55] Paras, Yadav K, Kumar P, Teja DR, Chakraborty S, Chakraborty M, et al. A review on low-dimensional nanomaterials: Nanofabrication, characterization and applications. *Nanomaterials.* 2022 Dec 29;13(1):160.

[56] Gatou MA, Syrrakou A, Lagopati N, Pavlatou EA. Photocatalytic TiO2-based nanostructures as a promising material for diverse environmental applications: A review. *Reactions.* 2024 Feb 1;5(1):135–94.

[57] Xu L, Liang HW, Yang Y, Yu SH. Stability and reactivity: Positive and negative aspects for nanoparticle processing. *Chem Rev.* 2018 Apr 11;118(7):3209–50.

[58] Mukhopadhyay R, Sarkar B, Khan E, Alessi DS, Biswas JK, Manjaiah KM, et al. Nanomaterials for sustainable remediation of chemical contaminants in water and soil. *Crit Rev Environ Sci Technol.* 2022 Aug 3;52(15):2611–60.

[59] Schubert J, Chanana M. Coating matters: Review on colloidal stability of nanoparticles with biocompatible coatings in biological media, living cells and organisms. *Curr Med Chem.* 2018 Dec 3;25(35):4553–86.

[60] Abid N, Khan AM, Shujait S, Chaudhary K, Ikram M, Imran M, . . . Maqbool M. Synthesis of nanomaterials using various top-down and bottom-up approaches, influencing factors, advantages, and disadvantages: A review. *Adv Colloid Interface Sci.* 2022;300:102597.

[61] Ahmad F, Salem-Bekhit MM, Khan F, Alshehri S, Khan A, Ghoneim MM, et al. Unique properties of surface-functionalized nanoparticles for bio-application: Functionalization mechanisms and importance in application. *Nanomaterials.* 2022 Apr 13;12(8):1333.

[62] Zhao J, Milanova M, Warmoeskerken MMCG, Dutschk V. Surface modification of TiO2 nanoparticles with silane coupling agents. *Colloids Surf A Physicochem Eng Asp.* 2012 Nov;413:273–9.

[63] Reznickova A, Slavikova N, Kolska Z, Kolarova K, Belinova T, Hubalek Kalbacova M, et al. PEGylated gold nanoparticles: Stability, cytotoxicity and antibacterial activity. *Colloids Surf A Physicochem Eng Asp.* 2019 Jan;560:26–34.

[64] Liu H, Wang X, Ding C, Dai Y, Sun Y, Lin Y, et al. Carboxylated carbon nanotubes-graphene oxide aerogels as ultralight and renewable high performance adsorbents for efficient adsorption of glyphosate. *Environ Chem.* 2020;17(1):6.

[65] Wang X, Tong J, Ma J. Design of sisal fibre biochar/poly(dopamine)/NZVI@PAN membrane for efficient degradation of tetracycline. *React Funct Polym.* 2023 Nov;192:105704.

[66] Roy A, Sharma A, Yadav S, Jule LT, Krishnaraj R. Nanomaterials for remediation of environmental pollutants. *Bioinorg Chem Appl.* 2021 Dec 28;2021:1–16.

[67] Sajid M, Asif M, Baig N, Kabeer M, Ihsanullah I, Mohammad AW. Carbon nanotubes-based adsorbents: Properties, functionalization, interaction mechanisms, and applications in water purification. *J Water Process Eng.* 2022 Jun;47:102815.

[68] Zewdu D, Krishnan C M, Nikhil Raj PP. Nanoclays for environmental remediation: A review. In: *Nanoclay-Based Sustainable Materials.* Elsevier; 2024. p. 167–200.

[69] Bundschuh M, Filser J, Lüderwald S, McKee MS, Metreveli G, Schaumann GE, et al. Nanoparticles in the environment: Where do we come from, where do we go to? *Environ Sci Eur.* 2018 Dec 8;30(1):6.

[70] Martínez G, Merinero M, Pérez-Aranda M, Pérez-Soriano E, Ortiz T, Villamor E, et al. Environmental impact of nanoparticles' application as an emerging technology: A review. *Materials.* 2020 Dec 31;14(1):166.

[71] Saxena M, Jain K, Saxena R. Role of nanomaterials in catalytic reduction of organic pollutants. *IJBB*. 2022;59(4):415–430.

[72] Li X qin, Elliott DW, Zhang W xian. Zero-valent iron nanoparticles for abatement of environmental pollutants: Materials and engineering aspects. *Crit Rev Solid State Mater Sci*. 2006 Dec 21;31(4):111–22.

[73] Wang CB, Zhang W xian. Synthesizing nanoscale iron particles for rapid and complete dechlorination of TCE and PCBs. *Environ Sci Technol*. 1997 Jul 1;31(7):2154–6.

[74] Liu WJ, Qian TT, Jiang H. Bimetallic Fe nanoparticles: Recent advances in synthesis and application in catalytic elimination of environmental pollutants. *Chem Eng J*. 2014 Jan;236:448–63.

[75] Li Q, Yan X, Shi M, Wang Q, Liu H, Lin Z, et al. Recent advances in metal/ceria catalysts for air pollution control: Mechanism insight and application. *Environ Sci Nano*. 2021;8(10):2760–79.

[76] Bhaskar S, Apoorva KV, Ashraf S, Athul Devan T. Synthesis and application of iron nanoparticles from scrap metal for triclosan degradation in water via Fenton and Sono-Fenton oxidation. *Waste Manag Bull*. 2025 Apr;3(1):293–300.

[77] Raj SI, Jaiswal A. Nanoscale transformation in CuS Fenton-like catalyst for highly selective and enhanced dye degradation. *J Photochem Photobiol A Chem*. 2021 Apr;410:113158.

[78] Yang J, Hou B, Wang J, Tian B, Bi J, Wang N, et al. Nanomaterials for the removal of heavy metals from wastewater. *Nanomaterials*. 2019 Mar 12;9(3):424.

[79] Van Thuan D, Nguyen TBH, Pham TH, Kim J, Hien Chu TT, Nguyen MV, et al. Photodegradation of ciprofloxacin antibiotic in water by using ZnO-doped g-C3N4 photocatalyst. *Chemosphere*. 2022 Dec;308:136408.

[80] Chen F, Jin X, Jia D, Cao Y, Duan H, Long M. Efficient treament of organic pollutants over CdS/graphene composites photocatalysts. *Appl Surf Sci*. 2020 Feb;504:144422.

[81] Rajaitha PM, Hajra S, Sahu M, Mistewicz K, Toroń B, Abolhassani R, et al. Unraveling highly efficient nanomaterial photocatalyst for pollutant removal: A comprehensive review and future progress. *Mater Today Chem*. 2022 Mar;23:100692.

[82] Van der Bruggen B, Vandecasteele C. Removal of pollutants from surface water and groundwater by nanofiltration: Overview of possible applications in the drinking water industry. *Environ Pollut*. 2003 Apr;122(3):435–45.

[83] Canalli Bortolassi AC, Guerra VG, Aguiar ML, Soussan L, Cornu D, Miele P, Bechelany M. Composites based on nanoparticle and pan electrospun nanofiber membranes for air filtration and bacterial removal. *Nanomaterials*. 2019;9(12):1740.

[84] Jawed A, Saxena V, Pandey LM. Engineered nanomaterials and their surface functionalization for the removal of heavy metals: A review. *J Water Process Eng*. 2020 Feb;33:101009.

[85] Rostamian R, Najafi M, Rafati AA. Synthesis and characterization of thiol-functionalized silica nano hollow sphere as a novel adsorbent for removal of poisonous heavy metal ions from water: Kinetics, isotherms and error analysis. *Chem Eng J*. 2011 Jul;171(3):1004–11.

[86] Boulkhessaim S, Gacem A, Khan SH, Amari A, Yadav VK, Harharah HN, et al. Emerging trends in the remediation of persistent organic pollutants using nanomaterials and related processes: A review. *Nanomaterials*. 2022 Jun 22;12(13):2148.

[87] Xu Z, Fan Z, Shen C, Meng Q, Zhang G, Gao C. Porous composite membrane based on organic substrate for molecular sieving: Current status, opportunities and challenges. *Adv Membr*. 2022;2:100027.

[88] Soni A. Nanotechnology and its implications for controlling air pollution: A mini review. *Nanomed Nanotechnol Open Access*. 2024;9(4):1–6.

[89] Santhosh C, Velmurugan V, Jacob G, Jeong SK, Grace AN, Bhatnagar A. Role of nanomaterials in water treatment applications: A review. *Chem Eng J*. 2016 Dec;306:1116–37.

[90] Yang K, Xing B. Desorption of polycyclic aromatic hydrocarbons from carbon nanomaterials in water. *Environ Pollut*. 2007 Jan;145(2):529–37.

[91] Tratnyek PG, Johnson RL. Nanotechnologies for environmental cleanup. *Nano Today*. 2006 May;1(2):44–8.

[92] Lu C, Liu C. Removal of nickel(II) from aqueous solution by carbon nanotubes. *J Chem Technol Biotechnol*. 2006 Dec 6;81(12):1932–40.

[93] Li YH, Ding J, Luan Z, Di Z, Zhu Y, Xu C, et al. Competitive adsorption of Pb2+, Cu2+ and Cd2+ ions from aqueous solutions by multiwalled carbon nanotubes. *Carbon N Y*. 2003;41(14):2787–92.

[94] Ma J, Yu F, Zhou L, Jin L, Yang M, Luan J, et al. Enhanced adsorptive removal of methyl orange and methylene blue from aqueous solution by alkali-activated multi-walled carbon nanotubes. *ACS Appl Mater Interfaces*. 2012 Nov 28;4(11):5749–60.

[95] Mehrizad A, Aghaie M, Gharbani P, Dastmalchi S, Monajjemi M, Zare K. Comparison of 4-chloro-2-nitrophenol adsorption on single-walled and multi-walled carbon nanotubes. *Iran J Environ Health Sci Eng*. 2012 Dec 3;9(1):5.

[96] Álvarez-Torrellas S, Rodríguez A, Ovejero G, García J. Comparative adsorption performance of ibuprofen and tetracycline from aqueous solution by carbonaceous materials. *Chem Eng J*. 2016 Jan;283:936–47.

[97] Ncibi MC, Sillanpää M. Optimized removal of antibiotic drugs from aqueous solutions using single, double and multi-walled carbon nanotubes. *J Hazard Mater*. 2015 Nov;298:102–10.

[98] Wang F, Sun W, Pan W, Xu N. Adsorption of sulfamethoxazole and 17β-estradiol by carbon nanotubes/CoFe2O4 composites. *Chem Eng J*. 2015 Aug;274:17–29.

[99] Yang W, Lu Y, Zheng F, Xue X, Li N, Liu D. Adsorption behavior and mechanisms of norfloxacin onto porous resins and carbon nanotube. *Chem Eng J*. 2012 Jan;179:112–8.

[100] Bazrafshan E, Mostafapour FK, Hosseini AR, Raksh Khorshid A, Mahvi AH. Decolorisation of reactive red 120 dye by using single-walled carbon nanotubes in aqueous solutions. *J Chem*. 2013 Jan 16;2013(1).

[101] Lee YC, Yang JW. Self-assembled flower-like TiO2 on exfoliated graphite oxide for heavy metal removal. *J Ind Eng Chem*. 2012 May;18(3):1178–85.

[102] Gao Y, Li Y, Zhang L, Huang H, Hu J, Shah SM, et al. Adsorption and removal of tetracycline antibiotics from aqueous solution by graphene oxide. *J Colloid Interface Sci*. 2012 Feb;368(1):540–6.

[103] Li J, Zhang S, Chen C, Zhao G, Yang X, Li J, et al. Removal of Cu(II) and fulvic acid by graphene oxide nanosheets decorated with Fe3O4 nanoparticles. *ACS Appl Mater Interfaces*. 2012 Sep 26;4(9):4991–5000.

[104] Wang Y, Liang S, Chen B, Guo F, Yu S, Tang Y. Synergistic removal of Pb(II), Cd(II) and humic acid by Fe3O4@mesoporous silica-graphene oxide composites. *PLoS One*. 2013 Jun 11;8(6):e65634.

[105] Huang ZH, Zheng X, Lv W, Wang M, Yang QH, Kang F. Adsorption of lead(II) ions from aqueous solution on low-temperature exfoliated graphene nanosheets. *Langmuir*. 2011 Jun 21;27(12):7558–62.

[106] Sreeprasad TS, Maliyekkal SM, Lisha KP, Pradeep T. Reduced graphene oxide – metal/metal oxide composites: Facile synthesis and application in water purification. *J Hazard Mater.* 2011 Feb;186(1):921–31.

[107] Ma J, Yang M, Yu F, Zheng J. Water-enhanced removal of ciprofloxacin from water by porous graphene hydrogel. *Sci Rep.* 2015 Sep 4;5(1):13578.

[108] Grossl PR, Sparks DL, Ainsworth CC. Rapid kinetics of Cu(II) adsorption/desorption on goethite. *Environ Sci Technol.* 1994 Aug 1;28(8):1422–9.

[109] Ma X, Wang Y, Gao M, Xu H, Li G. A novel strategy to prepare ZnO/PbS heterostructured functional nanocomposite utilizing the surface adsorption property of ZnO nanosheets. *Catal Today.* 2010 Dec;158(3–4):459–63.

[110] Cao CY, Cui ZM, Chen CQ, Song WG, Cai W. Ceria hollow nanospheres produced by a template-free microwave-assisted hydrothermal method for heavy metal ion removal and catalysis. *J Phys Chem C.* 2010 Jun 3;114(21):9865–70.

[111] Engates KE, Shipley HJ. Adsorption of Pb, Cd, Cu, Zn, and Ni to titanium dioxide nanoparticles: Effect of particle size, solid concentration, and exhaustion. *Environ Sci Pollut Res.* 2011 Mar 9;18(3):386–95.

[112] Afkhami A, Saber-Tehrani M, Bagheri H. Simultaneous removal of heavy-metal ions in wastewater samples using nano-alumina modified with 2,4-dinitrophenylhydrazine. *J Hazard Mater.* 2010 Sep;181(1–3):836–44.

[113] Zargar B, Parham H, Hatamie A. Fast removal and recovery of amaranth by modified iron oxide magnetic nanoparticles. *Chemosphere.* 2009 Jul;76(4):554–7.

[114] Qadri S, Ganoe A, Haik Y. Removal and recovery of acridine orange from solutions by use of magnetic nanoparticles. *J Hazard Mater.* 2009 Sep;169(1–3):318–23.

[115] Eren E, Tabak A, Eren B. Performance of magnesium oxide-coated bentonite in removal process of copper ions from aqueous solution. *Desalination.* 2010 Jul;257(1–3):163–9.

[116] Lai CH, Chen CY, Wei BL, Lee CW. Adsorptive characteristics of cadmium and lead on the goethite-coated sand surface. *J Environ Sci Health A.* 2001 May 31;36(5):747–63.

[117] Du WL, Xu ZR, Han XY, Xu YL, Miao ZG. Preparation, characterization and adsorption properties of chitosan nanoparticles for eosin Y as a model anionic dye. *J Hazard Mater.* 2008 May;153(1–2):152–6.

[118] Chen L, Wang TJ, Wu HX, Jin Y, Zhang Y, Dou XM. Optimization of a Fe – Al – Ce nano-adsorbent granulation process that used spray coating in a fluidized bed for fluoride removal from drinking water. *Powder Technol.* 2011 Jan;206(3):291–6.

[119] Raveendra RS, Prashanth PA, Hari Krishna R, Bhagya NP, Nagabhushana BM, Raja Naika H, et al. Synthesis, structural characterization of nano ZnTiO3 ceramic: An effective azo dye adsorbent and antibacterial agent. *J Asian Ceram Soc.* 2014 Dec 20;2(4):357–65.

[120] Shen R, Liu H. Construction of bimodal silsesquioxane-based porous materials from triphenylphosphine or triphenylphosphine oxide and their size-selective absorption for dye molecules. *RSC Adv.* 2016;6(44):37731–9.

[121] Kormann C, Bahnemann DW, Hoffmann MR. Photolysis of chloroform and other organic molecules in aqueous titanium dioxide suspensions. *Environ Sci Technol.* 1991 Mar 1;25(3):494–500.

[122] OLLIS D. Heterogeneous photoassisted catalysis: Conversions of perchloroethylene, dichloroethane, chloroacetic acids, and chlorobenzenes. *J Catal.* 1984 Jul;88(1):89–96.

[123] Yu Y, Yu JC, Chan CY, Che YK, Zhao JC, Ding L, et al. Enhancement of adsorption and photocatalytic activity of TiO2 by using carbon nanotubes for the treatment of azo dye. *Appl Catal B.* 2005 Oct;61(1–2):1–11.

[124] Shen J, Huang W, Li N, Ye M. Highly efficient degradation of dyes by reduced graphene oxide – ZnCdS supramolecular photocatalyst under visible light. *Ceram Int.* 2015 Jan;41(1):761–7.

[125] Chen P, Xiao TY, Li HH, Yang JJ, Wang Z, Yao HB, et al. Nitrogen-doped graphene/ ZnSe nanocomposites: Hydrothermal synthesis and their enhanced electrochemical and photocatalytic activities. *ACS Nano.* 2012 Jan 24;6(1):712–9.

[126] Xiong Z, Zhang LL, Ma J, Zhao XS. Photocatalytic degradation of dyes over graphene – gold nanocomposites under visible light irradiation. *Chem Commun.* 2010;46(33):6099.

[127] Zhang J, Xiong Z, Zhao XS. Graphene – metal – oxide composites for the degradation of dyes under visible light irradiation. *J Mater Chem.* 2011;21(11):3634.

[128] Fu D, Han G, Chang Y, Dong J. The synthesis and properties of ZnO – graphene nano hybrid for photodegradation of organic pollutant in water. *Mater Chem Phys.* 2012 Feb;132(2–3):673–81.

[129] www.who.int/news-room/fact-sheets/detail/antimicrobial-resistance

[130] Allahverdiyev AM, Abamor ES, Bagirova M, Rafailovich M. Antimicrobial effects of TiO_2 and Ag_2O nanoparticles against drug-resistant bacteria and *Leishmania* parasites. *Future Microbiol.* 2011 Aug 23;6(8):933–40.

[131] Díez-Pascual AM. Antibacterial activity of nanomaterials. *Nanomaterials.* 2018;(6):359.

[132] Liu S, Zeng TH, Hofmann M, Burcombe E, Wei J, Jiang R, et al. Antibacterial activity of graphite, graphite oxide, graphene oxide, and reduced graphene oxide: Membrane and oxidative stress. *ACS Nano.* 2011 Sep 27;5(9):6971–80.

[133] Shao W, Liu X, Min H, Dong G, Feng Q, Zuo S. Preparation, characterization, and antibacterial activity of silver nanoparticle-decorated graphene oxide nanocomposite. *ACS Appl Mater Interfaces.* 2015 Apr 1;7(12):6966–73.

[134] de Faria AF, Martinez DST, Meira SMM, de Moraes ACM, Brandelli A, Filho AGS, et al. Anti-adhesion and antibacterial activity of silver nanoparticles supported on graphene oxide sheets. *Colloids Surf B Biointerfaces.* 2014 Jan;113:115–24.

[135] Osman KT. *Soil Degradation, Conservation and Remediation* (Vol. 820). Springer Netherlands; 2014. https://doi.org/10.1007/978-94-007-7590-9

[136] Vogel TM. Bioaugmentation as a soil bioremediation approach. *Curr Opin Biotechnol.* 1996 Jun;7(3):311–6.

[137] Zhao X, Liu W, Cai Z, Han B, Qian T, Zhao D. An overview of preparation and applications of stabilized zero-valent iron nanoparticles for soil and groundwater remediation. *Water Res.* 2016 Sep;100:245–66.

[138] Hussain A, Rehman F, Rafeeq H, Waqas M, Asghar A, Afsheen N, et al. In-situ, Ex-situ, and nano-remediation strategies to treat polluted soil, water, and air – a review. *Chemosphere.* 2022 Feb;289:133252.

[139] Darlington TK, Neigh AM, Spencer MT, Guyen OTN, Oldenburg SJ. Nanoparticle characteristics affecting environmental fate and transport through soil. *Environ Toxicol Chem.* 2009 Jun 1;28(6):1191–9.

[140] Wang M, Gao B, Tang D. Review of key factors controlling engineered nanoparticle transport in porous media. *J Hazard Mater.* 2016 Nov;318:233–46.

[141] Fazeli Sangani M, Owens G, Fotovat A. Transport of engineered nanoparticles in soils and aquifers. *Environ Rev.* 2019 Mar;27(1):43–70.

[142] Chen H. Metal based nanoparticles in agricultural system: Behavior, transport, and interaction with plants. *Chem Speciat Bioavailab.* 2018 Jan 1;30(1):123–34.

[143] Praetorius A, Labille J, Scheringer M, Thill A, Hungerbühler K, Bottero JY. Hetero-aggregation of titanium dioxide nanoparticles with model natural colloids under environmentally relevant conditions. *Environ Sci Technol.* 2014 Sep 16;48(18):10690–8.

[144] Miao AJ, Zhang XY, Luo Z, Chen CS, Chin WC, Santschi PH, et al. Zinc oxide – engineered nanoparticles: Dissolution and toxicity to marine phytoplankton. *Environ Toxicol Chem.* 2010 Sep 1;29(12):2814–22.

[145] Vasarevičius S, Danila V, Paliulis D. Application of stabilized nano zero valent iron particles for immobilization of available Cd2+, Cu2+, Ni2+, and Pb2+ ions in soil. *Int J Environ Res.* 2019 Jun 10;13(3):465–74.

[146] Gil-Díaz M, Alonso J, Rodríguez-Valdés E, Gallego JR, Lobo MC. Comparing different commercial zero valent iron nanoparticles to immobilize As and Hg in brownfield soil. *Sci Total Environ.* 2017 Apr;584–585:1324–32.

[147] Li Y, Zhao HP, Zhu L. Remediation of soil contaminated with organic compounds by nanoscale zero-valent iron: A review. *Sci Total Environ.* 2021;760:143413.

[148] Soleimani M, & Amini N. Remediation of environmental pollutants using nano-clays. In *Nanoscience and Plant–Soil Systems* 2017 (pp. 279–289). Cham: Springer International Publishing.

[149] Mohanapragash AG, Kaleeswari RK, Madhupriyaa D, S Salma Santhosh, Meena S, Baskar M, et al. Nanotechnology in sustainable soil remediation: Exploring a review on carbon-based, clay-based, dendrimer, polymer, and nanocomposite based nano-materials for environmental cleanup. *Int J Environ Anal Chem.* 2025 Mar 24;1–26.

[150] Hasanabadi T, Lack S, Modhej A, Ghafourian H, Alavifazel M, Ardakani MR. Feasibility study on reducing lead and cadmium absorption by alfalfa (*Medicago scutellata* L.) in a contaminated soil using nano-activated carbon and natural based nano-zeolite. *Not Bot Horti Agrobot Cluj Napoca.* 2019 Nov 28;47(4):1185–93.

[151] Hu Q, Zeng W ai, Li F, Huang Y, Gu S, Cai H, et al. Effect of nano zeolite on the transformation of cadmium speciation and its uptake by tobacco in cadmium-con-taminated soil. *Open Chem.* 2018 Jul 18;16(1):667–73.

[152] Abdel-Salam M. Remediation of a Pb-contaminated soil cultivated with rose gera-nium (*Pelargonium graveolens*) using nano-zeolite. *J Soil Sci Agric Eng.* 2018 Oct 1;9(10):473–9.

[153] Radziemska M, Gusiatin ZM, Kumar V, Brtnicky M. Co-application of nanosized hallo-ysite and biochar as soil amendments in aided phytostabilization of metal(-oid)s-contam-inated soil under different temperature conditions. *Chemosphere.* 2022 Feb;288:132452.

[154] Radziemska M, Gusiatin MZ, Cydzik-Kwiatkowska A, Blazejczyk A, Majewski G, Jaskulska I, et al. Effect of freeze – thaw manipulation on phytostabilization of industrially contaminated soil with halloysite nanotubes. *Sci Rep.* 2023 Dec 13;13(1):22175.

[155] Xu C, Qi J, Yang W, Chen Y, Yang C, He Y, et al. Immobilization of heavy metals in vegetable-growing soils using nano zero-valent iron modified attapulgite clay. *Sci Total Environ.* 2019 Oct;686:476–83.

[156] Yang Z, Guo W, Cheng Z, Wang G, Xian J, Yang Y, et al. Possibility of using com-bined compost – attapulgite for remediation of Cd contaminated soil. *J Clean Prod.* 2022 Sep;368:133216.

[157] Yang J, Gao X, Li J, Zuo R, Wang J, Song L, et al. The stabilization process in the remediation of vanadium-contaminated soil by attapulgite, zeolite and hydroxyapa-tite. *Ecol Eng.* 2020 Sep;156:105975.

[158] El-Nagar DA, Abdel-Halim KY. Remediation of heavy metals in contaminated soil by using nano-bentonite, nano-hydroxyapatite, and nano-composite. *Land Degrad Dev*. 2021 Oct;32(16):4562–73.

[159] Mohajeri P, Smith CMS, Chau HW, Lehto N, Azimi A, Farraji H. Adsorption behavior of Na-bentonite and nano cloisite Na+ in interaction with Pb(NO3)2 and Cu(NO3)2·3H2O contamination in landfill liners: Optimization by response surface methodology. *J Environ Chem Eng*. 2019 Dec;7(6):103449.

[160] Zhao C, Wang S, He X, Sun H, Yan H, Zhao S, et al. The joint action of nano-montmorillonite and plant roots on the remediation of cadmium-contaminated soil and the improvement of rhizosphere bacterial characterization. *J Environ Chem Eng*. 2025 Apr;13(2):115462.

[161] Wołejko E, Jabłońska-Trypuć A, Wydro U, Butarewicz A, Łozowicka B. Soil biological activity as an indicator of soil pollution with pesticides – A review. *Appl Soil Ecol*. 2020 Mar;147:103356.

[162] Pinedo J, Ibáñez R, Lijzen JPA, Irabien Á. Assessment of soil pollution based on total petroleum hydrocarbons and individual oil substances. *J Environ Manage*. 2013 Nov;130:72–9.

[163] Sakshi, Singh SK, Haritash AK. Polycyclic aromatic hydrocarbons: Soil pollution and remediation. *Int J Env Sci Technol*. 2019 Oct 25;16(10):6489–512.

[164] Mazarji M, Minkina T, Sushkova S, Mandzhieva S, Bidhendi GN, Barakhov A, et al. Effect of nanomaterials on remediation of polycyclic aromatic hydrocarbons-contaminated soils: A review. *J Environ Manage*. 2021 Apr;284:112023.

[165] Šebesta M, Kolenčík M, Sunil BR, Illa R, Mosnáček J, Ingle AP, et al. Field application of ZnO and TiO2 nanoparticles on agricultural plants. *Agronomy*. 2021 Nov 11;11(11):2281.

[166] Anucha CB, Altin I, Bacaksiz E, Stathopoulos VN. Titanium dioxide (TiO₂)-based photocatalyst materials activity enhancement for contaminants of emerging concern (CECs) degradation: In the light of modification strategies. *Chem Eng J Adv*. 2022 May;10:100262.

[167] Aragay G, Pino F, Merkoçi A. Nanomaterials for sensing and destroying pesticides. *Chem Rev*. 2012 Oct 10;112(10):5317–38.

[168] Patra Shahi M, Kumari P, Mahobiya D, Kumar Shahi S. Nano-bioremediation of environmental contaminants: Applications, challenges, and future prospects. In: *Bioremediation for Environmental Sustainability*. Elsevier; 2021. p. 83–98.

[169] Bairwa P, Kumar N, Devra V, Abd-Elsalam K. Nano-biofertilizers synthesis and applications in agroecosystems. *Agrochemicals*. 2023 Feb 22;2(1):118–34.

[170] Korschelt K, Tahir MN, Tremel W. A step into the future: Applications of nanoparticle enzyme mimics. *Chem Eur J*. 2018 Jul 11;24(39):9703–13.

[171] Jyethi DS. Air quality: Global and regional emissions of particulate matter, SOx, and NOx. In *Plant Responses to Air Pollution*. Springer Singapore; 2016. p. 5–19.

[172] Saleem H, Zaidi SJ, Ismail AF, Goh PS. Advances of nanomaterials for air pollution remediation and their impacts on the environment. *Chemosphere*. 2022 Jan;287:132083.

[173] Abdel-Karim R, Reda Y, Abdel-Fattah A. Review – nanostructured materials-based nanosensors. *J Electrochem Soc*. 2020 Jan 2;167(3):037554.

[174] Chern M, Kays JC, Bhuckory S, Dennis AM. Sensing with photoluminescent semiconductor quantum dots. *Methods Appl Fluoresc*. 2019 Jan 24;7(1):012005.

[175] Roy A, Ray A, Sadhukhan P, Naskar K, Lal G, Bhar R, et al. Polyaniline-multiwalled carbon nanotube (PANI-MWCNT): Room temperature resistive carbon monoxide (CO) sensor. *Synth Met*. 2018 Nov;245:182–9.

[176] Grozdanov A, Dimitrievska I, Paunovic P. Recent advancements in nano sensors for air and water pollution control. *Mater Sci Eng Int J*. 2023 May 26;7(2):113–28.

[177] Louis H, Egemonye TC, Unimuke TO, Inah BE, Edet HO, Eno EA, et al. Detection of carbon, sulfur, and nitrogen dioxide pollutants with a 2D Ca12O12 nanostructured material. *ACS Omega*. 2022 Oct 4;7(39):34929–43.

[178] Yang L, Zheng G, Cao Y, Meng C, Li Y, Ji H, et al. Moisture-resistant, stretchable NOx gas sensors based on laser-induced graphene for environmental monitoring and breath analysis. *Microsyst Nanoeng*. 2022 Jul 8;8(1):78.

[179] Ding S, Gu W. Evaluate the potential utilization of B24N24 fullerene in the recognition of COS, H2S, SO2, and CS2 gases (environmental pollution). *J Mol Liq*. 2022 Jan;345:117041.

[180] Akshya S, Juliet AV. A computational study of a chemical gas sensor utilizing Pd – rGO composite on SnO2 thin film for the detection of NOx. *Sci Rep*. 2021 Jan 13;11(1):970.

[181] Tonezzer M. Selective gas sensor based on one single SnO_2 nanowire. *Sens Actuators B Chem*. 2019 Jun;288:53–9.

[182] Faisal AD. Synthesis of ZnO comb-like nanostructures for high sensitivity $$\hbox{H}_{2}$$ H 2 S gas sensor fabrication at room temperature. *Bull Mater Sci*. 2017 Oct 15;40(6):1061–8.

[183] Ince Yardimci A, Yagmurcukardes N, Yagmurcukardes M, Capan I, Erdogan M, Capan R, et al. Electrospun polyacrylonitrile (PAN) nanofiber: Preparation, experimental characterization, organic vapor sensing ability and theoretical simulations of binding energies. *Appl Phys A*. 2022 Mar 3;128(3):173.

[184] Qiu Y, Wang Y. Synthesis, growth kinetics and ultra-sensitive performance of electrospun WO3 nanofibers for NO2 detection. *Appl Surf Sci*. 2023 Jan;608:155112.

[185] Zhang B, Li M, Song Z, Kan H, Yu H, Liu Q, et al. Sensitive H2S gas sensors employing colloidal zinc oxide quantum dots. *Sens Actuators B Chem*. 2017 Oct;249:558–63.

[186] Li H, Li M, Kan H, Li C, Quan A, Fu C, et al. Surface acoustic wave NO2 sensors utilizing colloidal SnS quantum dot thin films. *Surf Coat Technol*. 2019 Mar;362:78–83.

[187] Taremi S, Rouhani M, & Mirjafary Z. Poly (3-aminophenylboronic acid) as a sensitive electrical and optical sensor material for detection of some air pollutants: A computational study. *Computational and Theoretical Chemistry*. 2012;1214:113801.

[188] Zhu C, Dong X, Guo C, Huo L, Gao S, Zheng Z, et al. Template-free synthesis of a wafer-sized polyaniline nanoscale film with high electrical conductivity for trace ammonia gas sensing. *J Mater Chem A Mater*. 2022;10(22):12150–6.

[189] Pena ACC, Raymundo LM, Trierweiler LF, Gutterres M. Green carbon dots synthesized from Chlorella Sorokiniana microalgae biochar for chrome detection. *J Ind Eng Chem*. 2023 Jan;117:130–9.

[190] Kundu A, Maity B, Basu S. Rice husk-derived carbon quantum dots-based dual-mode nanoprobe for selective and sensitive detection of Fe3+ and fluoroquinolones. *ACS Biomater Sci Eng*. 2022 Nov 14;8(11):4764–76.

[191] Li Y, Mou C, Xie Z, Zheng M. Carbon dots embedded hydrogel spheres for sensing and removing rifampicin. *Dyes Pigm*. 2022 Feb;198:110023.

[192] Monsef R, Salavati-Niasari M. Electrochemical sensor based on a chitosan-molybdenum vanadate nanocomposite for detection of hydroxychloroquine in biological samples. *J Colloid Interface Sci*. 2022 May;613:1–14.

[193] Linley S, Thomson NR. Environmental applications of nanotechnology: Nanoenabled remediation processes in water, soil and air treatment. *Water Air Soil Poll*. 2021;232(2):59.

[194] Mahlambi MM, Ngila CJ, Mamba BB. Recent developments in environmental photocatalytic degradation of organic pollutants: The case of titanium dioxide nanoparticles – a review. *J Nanomater.* 2015 Jan;2015(1).

[195] Zhao W, Adeel M, Zhang P, Zhou P, Huang L, Zhao Y, et al. A critical review on surface-modified nano-catalyst application for the photocatalytic degradation of volatile organic compounds. *Environ Sci Nano.* 2022;9(1):61–80.

[196] Shaik BB, Katari NK, Raghupathi JK, Jonnalagadda SB, Rana S. Titanium dioxide/graphene-based nanocomposites as photocatalyst for environmental applications: A review. *ChemistrySelect.* 2024 Dec 10;9(46).

[197] Dhawale P, Gadhave R V. Indoor formaldehyde removal techniques through paints: Review. *Green Sustain Chem.* 2024;14(01):1–15.

[198] Hashemi M, Maleky S, Rajabi S. Nanoadsorbents in air pollution control. In: *Adsorption through Advanced Nanoscale Materials.* Elsevier; 2023. p. 289–324.

[199] Ali A, Majhi SM, Siddig LA, Deshmukh AH, Wen H, Qamhieh NN, et al. Recent advancements in MXene-based biosensors for health and environmental applications – a review. *Biosensors (Basel).* 2024 Oct 12;14(10):497.

[200] Wang L, Zhu F, Liu E, Yang Y, Yu Q, He Y, . . . Chen X. Progress in advanced carbon nanotubes composites for air purification. *Adv Compos Hybrid Ma.* 2024;7(3):100.

[201] Alguacil FJ. Nanomaterials for CO_2 capture from gas streams. *Separations.* 2023 Dec 19;11(1):1.

[202] Girimonte R, Testa F, Turano M, Leone G, Gallo M, Golemme G. Amine-functionalized mesoporous silica adsorbent for CO2 capture in confined-fluidized bed: Study of the breakthrough adsorption curves as a function of several operating variables. *Processes.* 2022 Feb 21;10(2):422.

[203] Yu J, Xie LH, Li JR, Ma Y, Seminario JM, Balbuena PB. CO2 capture and separations using MOFs: Computational and experimental studies. *Chem Rev.* 2017;117(14):9674–9754.

[204] Akhter P, Arshad A, Tahir M. Innovative nanomaterials for sustainable environment for reducing the smog effects: A technical review. *Environ Sci Pollut Res.* 2025;1–22.

[205] Tamanna NJ, Sahadat Hossain Md, Tabassum S, Bahadur NM, Ahmed S. Easy and green synthesis of nano-ZnO and nano-TiO2 for efficient photocatalytic degradation of organic pollutants. *Heliyon.* 2024 Sep;10(17):e37469.

[206] Mack SM, Madl AK, Pinkerton KE. Respiratory health effects of exposure to ambient particulate matter and bioaerosols. *Compr Physiol.* 2020 Jan;10(1):1–20.

[207] Zhou M, Fang M, Quan Z, Zhang H, Qin X, Wang R, & Yu J. Large-scale preparation of micro-gradient structured sub-micro fibrous membranes with narrow diameter distributions for high-efficiency air purification. *Environmental Science: Nano.* 2019;6(12):3560–3578.

[208] Wang J, Kim SC, Pui DYH. Investigation of the figure of merit for filters with a single nanofiber layer on a substrate. *J Aerosol Sci.* 2008 Apr;39(4):323–34.

[209] Ji X, Huang J, Teng L, Li S, Li X, Cai W, et al. Advances in particulate matter filtration: Materials, performance, and application. *Green Energy Environ.* 2023 Jun;8(3):673–97.

[210] Liu R, Hou L, Yue G, Li H, Zhang J, Liu J, et al. Progress of fabrication and applications of electrospun hierarchically porous nanofibers. *Adv Fiber Mater.* 2022 Aug 4;4(4):604–30.

[211] Niu Z, Bian Y, Xia T, Zhang L, Chen C. An optimization approach for fabricating electrospun nanofiber air filters with minimized pressure drop for indoor PM2.5 control. *Build Environ.* 2021 Jan;188:107449.

[212] Akduman C, Akçakoca Kumbasar EP. Nanofibers in face masks and respirators to provide better protection. *IOP Conf Ser Mater Sci Eng.* 2018 Dec 24;460:012013.

[213] Bode-Aluko CA, Pereao O, Ndayambaje G, Petrik L. Adsorption of toxic metals on modified polyacrylonitrile nanofibres: A review. *Water Air Soil Pollut.* 2017 Jan 21;228(1):35.

[214] Pardo-Figuerez M, Chiva-Flor A, Figueroa-Lopez K, Prieto C, Lagaron JM. Antimicrobial nanofiber based filters for high filtration efficiency respirators. *Nanomaterials.* 2021 Apr 1;11(4):900.

[215] Kumaravel V, Nair KM, Mathew S, Bartlett J, Kennedy JE, Manning HG, et al. Antimicrobial TiO2 nanocomposite coatings for surfaces, dental and orthopaedic implants. *Chem Eng J.* 2021 Jul;416:129071.

[216] Zeng G, Gong B, Li Y, Wang K, & Guan Q. Nano-silica modified with silane and fluorinated chemicals to prepare a superhydrophobic coating for enhancing self-cleaning performance. *Water Science & Technology.* 2024;90(3):777–790.

[217] Chotigawin R, Kandasamy B, Asa P, Semangoen T, Ajawatanawong P, Phibanchon S, . . . & Suwannahong K. Next-generation eco-friendly hybrid air purifier: Ag/TiO2/PLA biofilm for enhanced bioaerosols removal. *International Journal of Molecular Sciences.* 2025;26(10):4584.

[218] Chirumbolo S, Gibellini D, Berto L, Cirrito C, Vella A, Bjørklund G, . . . & Tirelli U. TiO2–Ag–NP adhesive photocatalytic films able to disinfect living indoor spaces with a straightforward approach. *Scientific Reports.* 2023;13(1):4200.

[219] Saleem H, Zaidi SJ. Developments in the application of nanomaterials for water treatment and their impact on the environment. *Nanomaterials.* 2020 Sep 7;10(9):1764.

[220] Garner KL, Keller AA. Emerging patterns for engineered nanomaterials in the environment: A review of fate and toxicity studies. *J Nanopart Res.* 2014 Aug 3;16(8):2503.

[221] Gambardella C, Costa E, Piazza V, Fabbrocini A, Magi E, Faimali M, et al. Effect of silver nanoparticles on marine organisms belonging to different trophic levels. *Mar Environ Res.* 2015 Oct;111:41–9.

[222] Hou J, Wang L, Wang C, Zhang S, Liu H, Li S, et al. Toxicity and mechanisms of action of titanium dioxide nanoparticles in living organisms. *J Environ Sci.* 2019 Jan;75:40–53.

[223] Deenathayalan U, Nandita R, Kavithaa K, Kavitha VS., Govindasamy C, Al-Numair KS, . . . Brindha D. Evaluation of developmental toxicity and oxidative stress caused by zinc oxide nanoparticles in zebra fish embryos/larvae. *Appl Biochem Biotechnol.* 2024;196(8):4954–4973.

[224] Kobayashi N, Izumi H, Morimoto Y. Review of toxicity studies of carbon nanotubes. *J Occup Health.* 2017 Sep 20;59(5):394–407.

[225] Dwivedi AD, Dubey SP, Sillanpää M, Kwon YN, Lee C, Varma RS. Fate of engineered nanoparticles: Implications in the environment. *Coord Chem Rev.* 2015 Mar;287:64–78.

[226] Harrison DM, Briffa SM, Mazzonello A, Valsami-Jones E. A review of the aquatic environmental transformations of engineered nanomaterials. *Nanomaterials.* 2023 Jul 18;13(14):2098.

[227] Mlih R, Liang Y, Zhang M, Tombácz E, Bol R, Klumpp E. Transport and retention of poly(acrylic acid-co-maleic acid) coated magnetite nanoparticles in porous media: Effect of input concentration, ionic strength and grain size. *Nanomaterials.* 2022 May 2;12(9):1536.

[228] Wang H, Ho KT, Scheckel KG, Wu F, Cantwell MG, Katz DR, et al. Toxicity, bioaccumulation, and biotransformation of silver nanoparticles in marine organisms. *Environ Sci Technol.* 2014 Dec 2;48(23):13711–7.

[229] Murali M, Gowtham HG, Singh SB, Shilpa N, Aiyaz M, Alomary MN, et al. Fate, bioaccumulation and toxicity of engineered nanomaterials in plants: Current challenges and future prospects. *Sci Total Environ.* 2022 Mar;811:152249.

[230] Shi X, Li Z, Chen W, Qiang L, Xia J, Chen M, et al. Fate of TiO2 nanoparticles entering sewage treatment plants and bioaccumulation in fish in the receiving streams. *NanoImpact.* 2016 Jul;3–4:96–103.

6

Application and Prospects of Nanotechnology for Water Treatment

Bibhudendu Behera, Bhupender Sahu, Shalini Rai, Siya Sharma, and Umesh Kumar

6.1 Introduction

Nanotechnology has proved to be one of the most promising areas for water treatment research in existence. With increased progress, it presents real and practical means to surpass established and conventional water quality challenges anywhere. In contrast to most other conventional treatment technologies that are susceptible to failure in their applications in remediating contaminants, nanotechnology-based technologies have shown immeasurable potential to close such gaps (Chausali et al., 2023; Nishu & Kumar, 2023). Nanotechnology refers to the manipulation of material at extremely small scales – below 100 nanometers. At this level, materials begin to behave in a unique physicochemical way. Their increased surface area to volume, reactivity, and surface properties that can be controlled open new avenues, particularly for environmental use (Joudeh & Linke, 2022; Khan et al., 2019). Nanomaterials perform splendidly in purification and decomposition of all types of pollutants – anything from chemicals and heavy metals to bacteria and even microplastics (Anoob et al., 2024). "Emerging pollutants" such as pharmaceuticals, hormones, microplastics, flame retardants, and surfactants are appearing increasingly in waters all over the globe. They are unregulated and pose ecological and health monstrosities. They span from endocrine disruption to facilitating antibiotic resistance as a wide spectrum of potential effects (Alijagic et al., 2024; Das et al., 2024; Mariano et al., 2023; Matavos-Aramyan, 2024; Ramírez-Malule et al., 2020). The chemicals are likely to be from domestic cleaning products and sunscreens, industrial and agricultural waste like pesticides and solvents, and abandoned healthcare waste like hormones and antibiotics. Since the traditional water treatment plants cannot remove such chemicals, they will remain in the environment and drinking water, which leads to bioaccumulation and eventually to long-term environmental and human health problems (Chakraborty et al., 2023; Rasheed et al., 2019).

Availability-based conflicts are most common and arise due to scarcity of water, but the usability of water equally depends on quality, too. There are also policy and structural failures that constrain existing responses to the current crisis. Water infrastructure is not typically equipped for unconventional sources like municipal wastewater and brackish water, for instance (Gleick & Cooley, 2021; Koseoglu-Imer

DOI: 10.1201/9781003632498-6

et al., 2023). There is also a shortage of proper utilization of water reuse technology, such as reuse treatment and desalination. There are monitoring systems in which there is no approach for the surveillance of the surrounding environments for secondary contamination; for instance, keeping water in overhead tanks or tankers for transportation. All these shortcomings would demand more long-term and extensive policy planning and more coordination among the utility providers, the regulators, and the local communities (Cagno et al., 2022; Kesari et al., 2021; S. Mishra et al., 2023; Salgot & Folch, 2018; Voulvoulis, 2018). The scarcity of water will be the most adverse problem of the 21st century. Although water constitutes nearly 70% of the earth, only 3% of the fresh water, and a large portion of it is inaccessible for use (Musie & Gonfa, 2023). While 1.1 billion have been estimated not to have access to safe water, an additional 2.7 billion experience water scarcity for at least one month of the year. Other than that, around 2.4 billion individuals get infected as contaminated water is one of the contributing factors toward growing diseases such as typhoid and cholera (WHO, 2019, 2023; WWF, 2014). These diseases affect both children and other weaker sections of society. Global warming changed the rainfall pattern, and pollution of the air acidified the soil by acid rain, and such cascading effects resulted in flooding and other natural calamities. Farm runoff, industrial effluent, and raw sewerage continue to pollute enormous amounts of water. Agricultural usage generates around 70% of global freshwater consumption, further draining available supplies. Furthermore, increased populations are placing further demands with increased stress on available water resources (EPA, 2025; Moghimi Dehkordi et al., 2024; NCBI, 2022).

India is the perfect example of the shortage of water in the context that it has 17.5% of the world's population, but only about 4% of the world's freshwater in the entire globe (World Bank, 2023). India also has its own set of problems, some of which are the rapidly declining groundwater that typically has fluoride, arsenic, and nitrates. Surface water sources are highly contaminated with industrial process effluent and municipal sewage wastewater (Costa Santos et al., 2024; Esha Lohia, 2025; Shaji et al., 2021). The same is also augmented by inadequate infrastructure, thus resulting in untreated water infiltrating supply pipes and frequent pipe bursts. Most of the chlorinated drinking water disinfection byproducts are also health-hazardous and a complication for public safe drinking water (He et al., 2023; Li & Mitch, 2018; Liu et al., 2017). The traditional water treatment processes, including coagulation, sedimentation, filtration, and chlorination systems, are also vulnerable to emerging contaminants. They are unable to break down any new chemicals, such as drugs and heavy metals (R. K. Mishra et al., 2023; Samal et al., 2022; Sauvé & Desrosiers, 2014; Sharma & Bhattacharya, 2017). Their use among the poor or rural communities is questionable regarding their high maintenance and operation costs. Again, conventional treatment cannot destroy conditions such as ongoing water hardness, removal of other organic wastes, or microbial growth, making the water unconsumable (Razzak et al., 2022; Wysowska et al., 2021). Secondary contamination of distribution systems may also occur, and disinfection processes may create toxic byproducts such as carcinogenic chlorinated chemicals. Even sophisticated treatments such as reverse osmosis or ultraviolet treatment are not able to remove dissolved organic matter, and water quality is, thus, an ongoing issue (Kumar et al., 2023; Li & Mitch, 2018; Sharma & Bhattacharya, 2017).

The quantification and quality issues need to be addressed in an eco-friendly response to the water shortage, which is uncommon. This would entail installing and employing novel treatment technologies with the capability against new and emerging contaminants (Essamlali et al., 2024; Nishu & Kumar, 2023; Salehi, 2022). Point-of-use technology must be economically and scale-deployed to achieve balanced and equitable access. Coordination with drivers on the regional scale toward water quality, along with modulation in monitoring infrastructure in responding to treatment of a wide range of classes of chemical and biological contaminants, must be included in the plan (Ejiohuo et al., 2025; Krishnan et al., 2024; Quaranta et al., 2023). It should demand a global coalition of policy reform, scientific development, infrastructure development, and engaged citizenship to bring the UN Sustainable Development Goal 6 – world clean water and sanitation targets – achievable (Herrera, 2019). With more clean water demand increases, which resulted in the form of shortage or contamination, the requirement for green and cost-effective treatment of wastewater also grew. Conventional techniques remain functional under regular circumstances but cannot handle the challenges of modern-day stubborn microbes and drug contaminants (Qi & He, 2025). Nanotechnology is the answer with a multi-tool set of equipment, as it offers extra filtration, adsorption, and even catalytic annihilation of harmful contaminants. Metal oxides and metal nanomaterials like silver (Ag), titanium dioxide (TiO_2), and iron oxide (Fe_3O_4) have been found to be highly efficient in microorganism destruction and organic pollutant decomposition due to their strong antimicrobial and photocatalytic activities (Alhalili, 2023; Sahu et al., 2024). Likewise, carbon-based compounds such as carbon nanotubes (CNTs) and graphene have also been shown to potentially trap impurities and are conductors of electricity, as well as being optimally utilized for future applications both in electrochemistry and filtration (Arora & Attri, 2020).

Nanotechnology applications in water treatment have progressed stepwise and in a multi-disciplinary manner, as initially much focus was on fabricating and investigating nanomaterials (Nagar & Pradeep, 2020). By the early 2000s, these branched out and found new uses – nanomaterial-based filters, sun-activated photocatalysts, and magnetically directed cleaning nanoparticles to destroy polluting agents in lab tests (Ahire et al., 2022). Pioneering work with silver nanoparticles for microbial killing and with titanium dioxide as a photo-degrader led to portable, miniaturized treatment equipment. They were discovered to be extremely efficient where the infrastructure is weak (Chand et al., 2020; Razzaq et al., 2021).

The experiments were conducted in the direction of optimal performance material amalgams. The CNTs were incorporated with metal oxides or polymers to construct hybrid membranes of higher strength with improved removal efficiency (Abubakre et al., 2023; Rashid et al., 2024). At the sustainability end, work was focused on nanomaterial synthesis by greener methods, i.e., microbial or plant extract to keep environmental as well as costing impacts to a minimum (Osman et al., 2024). The recent innovation is the combination of nanomaterials and real-time monitoring. These "smart" devices can sense and react to ultra-low concentrations of contaminants, which is a more fundamental shift (Sharma et al., 2025) – a shift from installed treatment systems to adaptive, responsive technology that reacts to a change in the environment (Mondejar et al., 2021).

6.2 Fundamentals of Nanotechnology in Water Treatment

Nanotechnology promises revolutionary transformation of wastewater treatment through material nanomanipulation at the nanoscale (1–100 nm) with unprecedented physicochemical properties elicited by high surface-to-volume ratios and quantum effects (Nishu & Kumar, 2023; Verma et al., 2024). These properties enable nanoparticles to be more adsorptive, selective, and reactive, hence highly efficient in contaminant removal like heavy metals, dyes, organic pollutants, and microorganisms. The structure of the nanomaterial, as dictated by particle size, geometry, and surface composition, is an important determinant of maximum contaminant contact for effective treatment (Naseem & Durrani, 2021b; Naseer, 2024).

Nanomaterials such as metal oxides, CNTs, and functionalized membranes are being utilized to enhance conventional treatment systems increasingly. The nanomaterials possess favorable surface area and reactivity, due to which advanced oxidation activities, adsorption activities, and photocatalytic degradation activities are attributed to them (Abu-Dief et al., 2025). Both top-down and bottom-up synthesis methods allow us to effectively manage the nanoparticles' properties for their corresponding needs of each filtration, disinfection, and even resource recovery process. Nanocomposites from biodegradable materials are also researched further with enhanced biocompatibility and environmental sustainability (Namakka et al., 2023).

While nanotechnology enables a more efficient treatment and allows for compacted, effective system design, mass production is of concern, not to mention treating nanomaterials in treated wastewater residues and long-term safety. Despite this, it still has to fill these loopholes through multidisciplinary research and make sustainable use of actual wastewater treatment plants a possibility (Yaqoob et al., 2020).

6.3 Mechanisms of Nanomaterial-Based Water Treatment

Nanomaterials have now become the focus of water treatment technologies due to their distinctive physical and chemical properties. Nanomaterials enhance the effectiveness of conventional treatment methods such as adsorption, filtration, photocatalysis, and electrochemical treatment by exhibiting enhanced interaction with impurities, enhanced surface reactivity, and enhanced contaminant selectivity (Tripathy et al., 2024). Their nanoscale nature makes them capable of interacting with contaminants at the molecular level, making them far from ideal to be able to cope with intricate events in modern wastewater treatment. The following sections clarify the general mechanisms under which nanomaterials are utilized in water cleansing, outlining their processes in adsorption, filtration, photocatalytic degradation, and electrochemical treatment (Pérez et al., 2023).

6.3.1 Adsorption Mechanism

Adsorption is the most advanced technology that utilizes the application of nanomaterials for purifying water. Highly surface area nanostructures like porous architecture nanostructures – particularly carbon nanomaterials – are effective in removing a wide range of contaminants from trace amounts. They have high surface activity capable of interacting strongly with heavy metals, dyes, and other impurities. Certain nano-materials produced, such as functionalized CNTs and ferrite metals, possess higher adsorption capacity and kinetic capacity through enhanced interaction with particular pollutants (Satyam & Patra, 2024). The nanocomposite engineering technologies have also enhanced adsorption efficiency, as the technologies are not only enhanced to increase the adsorption of the pollutant, but also the desorption process is enhanced to increase the reusability and regeneration of the adsorbent material. Surface chem-istry optimization of nanomaterials can even improve target pollutant selectivity, too, through functionalization with target chemical groups. Optimization makes it a pos-sibility for adsorbent design for targeted ions or molecule adsorption at high levels of specificity, thereby optimizing treatment processes overall (Sadegh, Ali, Gupta, et al., 2017).

6.3.2 Filtration Techniques

Nanomaterials have also transformed the technology of filtration to a great extent in the form of high-performance membranes. Nanomaterial-loaded membranes, such as graphene oxide or CNTs, are more selective and more permeable. These are the attri-butes by which efficient suspended solid removal, microbes, and chemical impurities from water occur. In addition, such nanomaterial-based membranes would also be expected to be anti-fouling in nature, thus increasing the life of filtration systems as well as lowering the maintenance needs (Hebbar et al., 2017; Januário et al., 2021). Hybrid membranes reinforced with nanomaterials are exhibiting record improvement in liquid flux values and water permeability, as well as pollutant rejection rates. Mem-brane performance can be enhanced by the addition of two-way properties, such as chemical resistance and some selectivity to impurities. Being exceptions, the addi-tion of electrostatic interaction capabilities of nanomaterials to filtration membranes also enables selective targeting of charged impurities and immobilization, enhancing purification of intractable wastewater streams like oily or industrial effluent (Eniola et al., 2024).

6.3.3 Photocatalysis

Photocatalysis is among the highly developed and advanced technologies that make use of the photo-excited property of nanomaterials for the elimination of microbial and organic pollutants from water. Zinc oxide and titanium dioxide are common semiconductor nanoparticles. When exposed to light, nanoparticles produce reac-tive oxygen species with the ability to oxidize and break down complex pollutants through oxidative degradation. Quantum-sized dimensions and high surface area enhance photocatalytic activity and reaction kinetics (Njema & Kibet, 2024). Zinc oxide and titanium dioxide exhibit a strong catalytic activity, due to which the pol-lutants are oxidized into less toxic substances. This is also used to remove mercury metals present in wastewater (Jain et al., 2021). Zinc oxide is a semiconducting oxide that possesses several properties, such as large excitation energy, deep violet/border-line ultraviolet (UV) absorption at room temperature, and also has good photocata-lytic activity (Mclain, 2018). The presence of organic dyes such as MO, RhB, and acid orange (AO7) have significantly increased in wastewater and to remove these dyes from wastewater several researchers are investigating various methods to remove these dyes from wastewater.

Margan & Haghighi (2018) prepared cadmium oxide–zinc oxide nano-photocatalyst (CdO-ZnO) for the elimination of AO7 organic dye, which has 69% photocatalytic degradation ability in 140 min. TiO_2 NPs are inexpensive, stable, reusable, and considered a good photocatalyst. These NPs play an important role in the treatment of toxic organic and chlorinated compounds generated from paper and pulp industries (Zahra et al., 2020). The other main concern is the presence of antibiotic residues in water bodies, so the titanium dioxide photocatalyst doped with boron and cerium is investigated for breaking down ciprofloxacin and norfloxacin (fluoroquinolone-based antibiotics) using sunlight irradiation, and showed 90–93% degradation of both the antibiotics at optimal conditions (Manasa et al., 2021).

Recent progress in photocatalytic systems includes the application of diverse semiconductor materials for increased light absorption windows and charge carrier kinetics. Incorporation of plasmonic nanostructures into such systems has led to significant augmentation of light harvesting, increasing the performance of refractory contaminant degradation. These recent photocatalytic reactions can now be applied not only to treat chemical contaminants but also to treat microbial pathogens, offering integrated water treatment and disinfection options (Li et al., 2025).

6.3.4 Electrochemical Processes

Electrochemical reactions represent a significant area in which nanomaterials are used to boost the degradation of pollutants. Metal and metal oxide nanoparticles that are electrically conductive and catalytically active are used in electrochemical devices, catalyzing redox reactions for converting toxic pollutants to more or less biodegradable compounds. Nanomaterial-supported electrochemical advanced oxidation processes yield highly reactive intermediate species that aid in efficient pollutant degradation (Rathi et al., 2024). Ag NPs have distinct optical, electrical, catalytic, and antimicrobial properties, which are used for various range of applications (Abbas et al., 2024). AgNPs have an antimicrobial nature as these NPs can penetrate the cell or get attached to the cell wall due to their large surface area and cause the disruption in permeability of the membrane, creating pores and ultimately causing content to leak out of the cell (Kambale et al., 2020). The oxidized Ag ions integrate with proteins and enzymes, interfere with the respiratory system, and cause the distressing of the bacterial cell wall (Urnukhsaikhan et al., 2021). The antimicrobial activity of NPs is also beneficial for dye removal, water purification, and wastewater treatments (Tarannum & Gautam, 2019).

The study conducted by Khaldoun et al. (2024) on MOS-AgNPs, prepared from the extract of *Moringa oleifera* seed, shows antimicrobial activity toward Gram-negative bacteria, including *Salmonella enterica typhimurium*: 29 mm and *Escherichia coli*: 30.6 mm; and Gram-positive bacteria, including *Staphylococcus aureus*: 14.6 mm and *Pseudomonas aeruginosa*: 22.8 mm. These AgNPs also have extraordinary photocatalytic action against organic dyes like methylene blue (> 81%), orange-red (> 82%), and 4-nitrophenol (> 75%) under sunlight irradiation. These NPs also eradicate about 80% of toxic lead metal ions from water. The collaborative use of nanomaterials in electrochemical systems provides dual-mode functionality with contaminant removal by chemical reaction and physical separation. Magnetic nanoparticles, for instance, may be introduced to treatment systems to facilitate simple recovery after treatment and prevent secondary pollution risks. Electrode material modification by design with nanotechnology-based methods further facilitates such systems to provide selective treatment of intricate wastewater streams. Such integration not only enhances the efficiency of pollutant removal but also facilitates the construction of sustainable and resilient treatment infrastructures (Díez et al., 2022).

6.4 Nanomaterials in Water Treatment

Nanomaterials are increasingly becoming more and more versatile for use in water treatment due to their physicochemical properties, like high surface-to-volume ratio, surface chemistries that can be tuned, and increased reactivity. These characteristics render them more efficient in removing a broad spectrum of contaminants, including heavy metal ions, organic compounds, and microbially derived pathogens, from water systems. This section explores the roles of carbon-derived, metal-derived, polymer-derived, and biologically derived nanomaterials, and the technological innovations and challenges prevalent now are reflected (Ahmed et al., 2022).

Carbon nanomaterials such as graphene derivatives and CNTs possess excellent adsorption properties, particularly in removing harmful metals from polluted water. Their nanoscale sizes give them excellent contaminant capture and form new filtration modes. Similarly, metal nanostructures and metal oxides such as zero-valent iron and titanium dioxide possess excellent adsorptive and catalytic activity. They are excellent at degrading organic pollutants and possessing scalable solutions to complex remediation mechanisms (Krishna et al., 2023). Nanobiotechnology is an interdisciplinary field and is expanding day by day as researchers are developing new technologies that contribute to the new application-based aspect of nanoparticles (Sivakami et al., 2021). Various metals and metal oxides, such as Fe_3O_4, TiO_2, MnO_2, MgO, ZnO, CdO, etc., are used extensively to remove heavy metals due to unique properties like small size, a high surface-to-volume ratio, a high sensitivity and reactivity, and a high adsorption capacity (Sadegh et al., 2017). Heavy metals are a class of metallic elements that are characterized by their relatively high density, and due to their significant toxicity, they are considered harmful to the environment. These can cause damage to various organs when exposed to even low concentrations (Bhateria & Singh, 2019).

Wastewater from an oil refinery contains calcium (Ca^{2+}) and copper (Cu^{2+}), so nano-adsorbents based on Fe_3O_4/GO-COOH are prepared that remove about 78.4% and 51% of Ca^{2+} and Cu^{2+}, respectively, at 60 min and have 82.1% Ca^{2+} and 91.8% Cu^{2+} recovery rates (L. He et al., 2021). The surface modification of nanoparticles with functional groups, such as thiol, carboxylic acid, silanol, and phosphoric acid, is used to enhance the activity, like magnetite nanoparticles surface modification by ligand 1,6-hexadiamine, which is useful for the removal of 98% Cu(II) from polluted river and tap water (Bhateria & Singh, 2019). Others include the synthesis of polyacrylic acid (PAA)-conjugated magnetic ferric oxide (Fe_3O_4) NPs that are functionalized with CR azo dye that has maximal heavy metal removal efficiency at 6.5 pH for a 45-minute reaction (Sadak et al., 2020).

Polymer nanomaterials, typically enhanced by nanofillers, are of utmost significance to membrane filtration technology. Hybrid membranes, being more porous, selective, and mechanically stress-resistant, are formed by such hybrid membranes. Moreover, nature-inspired materials, especially biopolymer-based nanocomposites, also possess potential for ecologically safe water treatment with the combination of ecological safety and performance efficiency (Panigrahy et al., 2024). Notwithstanding this, there are a few challenges in the area of the ecotoxicological impact of nanomaterials. Unrestricted release in the environment has the potential to cause ecosystem and human injury. In addition, scaling innovation at the laboratory scale to field scale application involves proof of performance, economic efficiency, and longevity over extended time intervals under genuine conditions (R. K. Mishra et al., 2023). Lastly, the development of nanotechnology in water treatment depends on knowledge-led interdisciplinary research and policy-making to ensure it is sustainable, safe, and effective globally (Ma et al., 2024).

6.5 Nanomaterial-Enhanced Biosensors for Water Quality Monitoring

Biosensor coupling with nanomaterial-amplified water quality monitoring is a humongous step toward technological advancement in water quality monitoring. Since water pollution is gaining more international attention, there has been an increasing demand for real-time sensing devices that are portable in hand and have high sensitivity to detect a wide variety of pollutants. This section is dedicated to a thorough review of recent advancements in nanomaterial-based biosensor technologies and their application in water quality monitoring to detect heavy metals, organic pollutants, and microplastics in water bodies (Fdez-Sanromán et al., 2025).

6.5.1 Nanomaterial-Based Biosensors for Water Quality Monitoring

Biosensors have been demonstrated in recent studies to play a very significant role in enabling real-time, in situ monitoring of water quality parameters. Sensors have been extremely effective for the identification of chemical and biological pollutants at extremely high sensitivity and selectivity (Bhatia et al., 2024).

Pandey & Mishra (2024) emphasize growing sensor network complexities, especially when paired with sophisticated data treatment systems, since they enhance real-time environmental observation and decision-making. Biosensors are sensitive to pollutants, and bioeffects of toxins are sensed directly from water samples (Su et al., 2024). Incorporation of nanomaterial-based biosensors has also enhanced their analytical sensitivity exponentially, miniaturization, and application. Nanocomposites such as nanomaterial-based carbon and enzymatic components are more selective and sensitive for phenolic pollutant detection (Hemdan et al., 2025). Similarly, the use of nanomaterials has been successful in detecting microbial pollutants such as *Escherichia coli* at a high percentage (Takallu et al., 2024). Electrochemical biosensors have been of special interest due to their field condition variability and real-time sensitivity. Equipment is increasingly being utilized for the detection of organic and inorganic pollutants. Their ability to operate under a broad spectrum of environmental conditions has made them suitable for continuous and self-powered monitoring (Wang et al., 2024).

6.5.2 Applications in Aquaculture and Environmental Safety

Nanomaterial-based biosensors are also being used in aquaculture systems that are employed as early warning sensors for the monitoring of toxicants (Khan et al., 2024).

Badiola et al. (2012) explain that biosensing devices of recirculation aquaculture systems (RAS) are being designed so that they ensure optimal aquatic health with the lowest possible fish mortality by coming into contact with contaminants. DNA nanobiosensors are also studied for their potential use as highly sensitive sensors for environmental diagnosis with precise biological recognition properties.

Among the most notable advancements, the assimilation of machine learning algorithms on biosensing platforms has facilitated effective processing of sophisticated data sets gathered during monitoring. The platforms integrated facilitate predictive models and real-time alarms for contamination events. Such technologies propel smart water quality monitoring systems through global initiatives to sustainable water resource management (Zhou et al., 2024).

6.5.3 Detection of Heavy Metals Using Nanotechnology

Heavy metals like lead, arsenic, and cadmium are a special case of interest since they are irreversible and toxic to the environment and human health. Nanotechnology research has been followed by the emergence of extremely sensitive nanosensors that can detect minute quantities of the metal. Carbon quantum dots (CQDs), for instance, have been demonstrated to have improved detection sensitivity that overcomes the drawback of traditional sensing methods (Maghsoudi et al., 2021).

Nanozymes – artificial nanomaterials that mimic the activity of natural enzymes – have also been useful tools in the determination of heavy metals and degradation of related pollutants. Two-dimensional layer nanomaterials have also been useful as electrochemical biosensors with high sensitivity and real-time response (Nagendran et al., 2024).

Other advanced techniques, such as surface-enhanced Raman spectroscopy (SERS), also increase the efficiency of such sensors. They become even better to use, which ensures ultra-sensitivity with minimal quantities. Success hence justifies nanomaterial's application toward the development of solid and stable platforms for heavy metal detection (Bahlol et al., 2024).

6.5.4 Detection of Organic Pollutants

Organic pollutants such as pesticides, dyes, and residues from drugs contaminate the water media. Nanotechnology has effective and selective sensors for them. Nanomaterials and molecularly imprinted polymers (MIPs) have proved to be useful selective single organic compound sensors. The sensors are response-quick and very selective and therefore suitable for real-time analysis (Aoulad El Hadj Ali et al., 2024).

Nanocomposite biosensors were also effective in the analysis of aromatic and phenolic compounds, with a large diversity of applications from industrial wastewater to drinking water sources. Nanomaterial-based electrochemical biosensors are being used more and more to address specificity and sensitivity issues in real water matrices (Kanoun et al., 2021).

6.5.5 Detection of Microplastics

Microplastics are among the most critical types of pollutants to be quantified sensitively due to their very small particle size and heterogeneous physicochemical composition. Fluorescence-based assays on quantum dots and other nanomaterials have been shown to be selective for the determination and quantification of microplastics in samples of environmental samples. Sensitive detection is enhanced due to the high optical efficiency of the methods. In addition, microfluidics–nanosensor integration is a feasible method for the isolation, detection, and concentration of microplastics in nearly all bodies of water (Costa et al., 2021). Faramarzi et al. (2024) describe that multicriteria analysis of microplastic dispersion is enabled by microfluidic–nano integrated systems, to shed light on their influence on the environment.

6.5.6 Electrochemical Biosensing and Energy-Generating Devices

The new generation of biosensors based on nanomaterials and MFCs finds applications in electric power generation and monitoring environmental quality. The platforms can be employed in biochemical energy conversion to electrical signals, and therefore, self-sustaining pathogen detection and toxicity is possible. Multimodal platforms reflect the wide range of applications of nanomaterials in green environment technology (Cui et al., 2019).

6.6 Challenges and Environmental Implications

With the rapid industrialization, there is an increase in the deposition of heavy metals in the soil, air, and water as well. As we are moving toward sustainable development, there is a need to counter it. The nanoparticles are very efficient in removing heavy metals, organic pollutants, and inorganic pollutants from the wastewater due to their large surface area, and even at low concentration, they possess strong adsorption reactivity and capacity (Naseem & Durrani, 2021a). With this, there is an increase in the nanomaterial production, and there is an increasing concern related to the environmental effects of nanomaterials that needs to be addressed. It is necessary to understand the fate and behavior of manufactured nanoparticles – whether these particles retain their nominal nanoscale size or original structure, or they will change over the period of time (Chávez-Hernández et al., 2024).

6.6.1 Toxicity and Environmental Risk of Nanomaterials

Various types of nanoparticles are used in our day-to-day life, from skin products to food supplements to drug delivery systems to sewage and wastewater treatment. Nanoparticles (1–100 nm) can easily enter the cell and cell organelles (Gupta & Xie, 2018)). The various NPs, such as ZnO, are considered good photocatalysts and show antibacterial activity with 90% growth reduction of *Bacillus subtilis* at 10 ppm

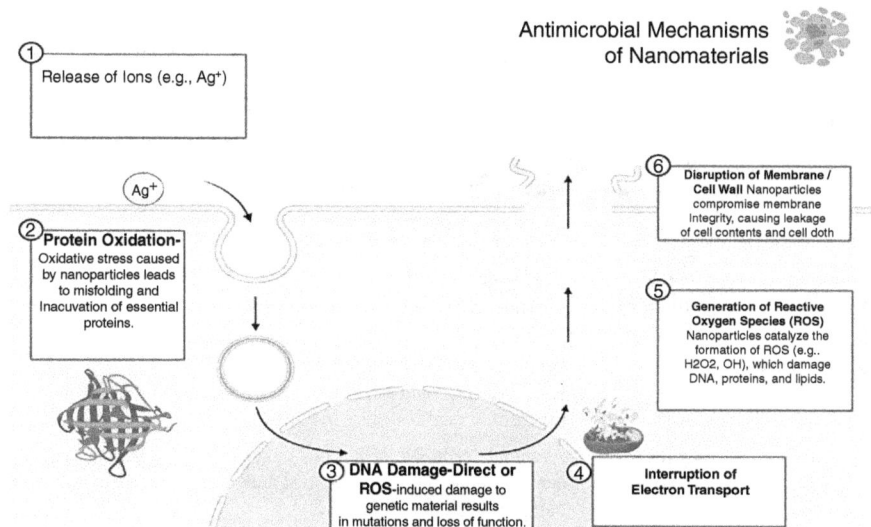

FIGURE 6.1 Antimicrobial mechanisms of nanomaterials in a microbial cell.

(Figure 6.1). The photocatalysis of titanium oxide removes various impurities from wastewater (Naseem & Durrani, 2021a). On the other side, the Ag NP, TiO2 NP also affect the beneficial microbial communities like the nitrifying bacteria. CNTs are also used in heavy metals removal, but cause a reduction in the soil microbial communities and microbial consortium, such as rhizobium and mycorrhiza (Hegde et al., 2016). Some studies on rats conclude that ingesting TiO_2 causes vital organ damage and DNA damage (Jain et al., 2021). Iron oxide (Fe_3O_4) shows excellent sorption potential and is used for the removal of heavy metals like Cu^{2+}, Zn^{2+}, Pb^{2+}, and Cd^{2+}, but causes inflammation, pulmonary fibrosis, and genotoxicity (Kornberg et al., 2017). NPs can enter cells and bind to hormone receptors, thereby activating or inhibiting downstream signaling pathways. Ag and ZnO bind to estrogenic receptors and exert estrogenic activity, which disrupts the endocrine balance in pregnant women and children (Xuan et al., 2023) (Table 6.1).

6.6.2 Cost, Scalability, and Manufacturing Challenges

The scale-up process includes extensive research and development by monitoring various parameters and conditions of the system, chemical reactions, and instruments. Desirable properties may change while scaling up. There are two bottom-up approaches for synthesizing nanoparticles, which involve building molecules from the smallest units, followed by a self-assembly process. Top-down approaches include the reduction of bulk material into the desired structure with the desired properties (Kumari et al., 2023). The characteristic of the influent is cost-dependent. The synthesis of nanoparticles also requires raw material, labor, and space, which are cost-consuming. The methods for manufacturing NPs include physical and chemical processes, which are also cost-intensive. However, cost can increase if a catalyst is

TABLE 6.1

Examples of Nanoparticle Toxicity on Various Passages of Entry of Nanoparticles into the Body

Adverse Effects	Type of NPs	Toxicity	References
Neurotoxicity	SiO2	SiO_2 decreases cell viability, increases lactate dehydrogenase levels, triggers oxidative stress, and accumulates in the striatum of the brain in humans.	(Xuan et al., 2023)
Respiratory immune toxicity	ZnO, SiO2, carbon-based NPs	Oxidative stress, respiratory epithelial damage, and inflammation.	(Guo et al., 2022)
Cardiotoxicity	SiO_2	Penetrate air–blood barrier in the lungs, enter the systemic circulation, and invade the cardiovascular system in humans.	(Liu et al., 2021)
Hepatotoxicity	AgNPs	Accumulates in the liver, alters enzyme activity, induces oxidative stress, and damages hepatocytes in mice.	(Olugbodi et al., 2023)
Reproductive toxicity	TiO_2	Disrupts hormone levels and reduces sperm motility and oocyte viability in zebrafish and rodents.	(Dutta et al., 2023; Wang et al., 2011)
Genotoxicity	ZnO	Causes DNA strand breaks and chromosomal aberrations in fish and rodents.	(Sharma et al., 2012)
Nephrotoxicity	CuO	Leads to renal oxidative stress, glomerular damage, and tubular necrosis in rats.	(Chen et al., 2006)
Immunotoxicity	Carbon nanotubes (CNTs)	Impairs cytokine expression, suppresses macrophage activity, and alters immune responses in mice.	(Mitchell et al., 2007)

needed to speed up the reaction or if some solvent is used to make the media soluble for chemical interaction. So, the manufacturing of NPs is very challenging, also there is a limited study on the techno-economic feasibility of nanoparticles with respect to wastewater treatment (Mpongwana & Rathilal, 2022).

6.6.3 Regulatory and Safety Considerations

Includes rigorous testing, proper labelling, environmental impact assessments, and training for handling and disposal. In the risk and exposure assessment, hazard identification is the first step. In 2017, the World Health Organization published the guidelines on protecting workers from potential risks of manufactured nanomaterials. There are some regulatory authorities, such as the UK Office for Product Safety and Standards (OPSS), to regulate nanomaterials applied to cosmetics (Gonzalez, 2024). In India, the Ministry of Environment, Forest, and Climate Change (MoEFCC) controls the use, import, export, and manufacture of hazardous compounds, including nanomaterials. In Europe, chemical management is governed by the Registration, Evaluation, Authorization, and Restriction of Chemicals (REACH) law, and nanoparticle disposal follows the same rules, which are outlined for the disposal of chemicals

(Dhall et al., 2024). The British national regulator, Health and Safety Executive (HSE), 2013, issued a guidance document related to the control of exposure to manufactured nanomaterials. SINANOTOX 2015 is an initiative to create a platform to evaluate the impact of NMs on various animal models in Mexico (Gonzalez, 2024). The guidelines for nanoparticles will assist decision makers to minimize the exposure and limit it to the working personnel and the environment (Herbeck-engel et al., 2021).

6.6.4 Need for Life Cycle Assessment in Nanotechnology Applications

LCA enables the analysis of environmental loads of a product throughout its entire life cycle and the potential impacts on the environment (Liu et al., 2024). LCA consists of four phases according to the provisions of ISO 14040, including Goal and Scope, Inventory Analysis, Impact Assessment, and Interpretation. Three aspects to consider about the use of LCAs include the lack of inventory information on average life cycle, characterization factors, and the need for an adequate functional unit for NPs (Olanrewaju et al., 2024). Toxicity assessments are used to evaluate the safety of nanoparticles, which include cell culture and animal toxicity tests. The computational nanotoxicity methods can bridge the gaps between cell culture, animal models, and human subjects. The development of regulatory frameworks, legislations, and guidelines can be considered as actions and strategies toward achieving the SDGs in all countries to promote prosperity while protecting the planet (Yang et al., 2021). Nanotechnology could be considered as a potential field in wastewater treatment as the nanoparticles possess excellent properties that can be exploited for the better management of the wastewater (Jain et al., 2021). Continued scientific research and knowledge are required to monitor and evaluate the toxicity of NPs (Yang et al., 2021).

6.7 Future Prospects and Emerging Trends in Green Nanotechnology

6.7.1 Advancements in Green Nanotechnology

Increasing pressure for sustainability has led nanotechnology to more eco-friendly, greener ecologically stable directions. Green nanotechnology is focused on materials and synthesis techniques that will exert minimal or zero environmental effect and maximum performance. Recent developments have focused on biodegradable and biocompatible nanomaterials with fewer aspects of pollution without compromising functionality. Biopolymers, for instance, are progressively being seen as nanocomposite matrix materials with versatility of applications and mechanical functionality from biomedical devices to packaging (Sahu et al., 2025). These composites, especially those from natural polymers and nanoparticles, are not only structurally stable but also biodegradable, as per life cycle sustainability and ecoefficiency perspectives (Jafarzadeh et al., 2025).

Green synthesis methods enable this by the substitution of harmful chemicals with plant, microbial, or enzymatic processes. Plant-mediated synthesis is one of the potential routes for the synthesis of functional nanomaterials with less energy input and fewer toxic substances. Enzyme and microbial synthesis pathways also possess higher ecological compatibility and can construct nanostructures with less toxic operations and those less detrimental to human and environmental health. These are fortified by the shift toward circular economy, whereby production and product design become infused with sustainability right from the onset (Alsaiari et al., 2023).

6.7.2 Next-Generation Nanomaterials for Water Treatment

Next-generation nanomaterials such as MXenes, metal–organic frameworks (MOFs), and quantum dots are revolutionizing water treatment technology. MXenes, a series of two-dimensional transition metal carbides and nitrides, exhibit high surface area, high electric conductivity, and tunable hydrophilicity. Such properties make them most appropriate to use in membrane filtration, electrochemical sensing, and capacitive deionization water treatment. Their conductivity also positions them in the best way to use in real-time monitoring systems for integrating functional intelligence into water treatment plants (Aldhaher et al., 2024; Gu et al., 2024).

Tunability of MOFs' chemistries and pore sizes has been found to be very effective against targeted adsorptions of impurities like heavy metals, pharmaceuticals, and dyes. Tunability is extremely helpful toward the targeted removal of various water pollutants. Thus, quantum dot native photophysics are utilized in visible-light-irradiated photocatalytic cleaning of organic contaminations to provide energy-saving and decentralized water purification technology that is usable even under off-grid or infrastructurally poor scenarios (Fard et al., 2024). Their employment not only supports the technical efficiency of existing water cleaning technologies but also enables water management resilience owing to the potential of decentralized, mobile, and flexible purification technology design (Zodrow et al., 2017).

6.7.3 Hybrid Systems Integrating Nanotechnology with Bioremediation

The most promising area of future research is hybrid nanotechnology and bioremediation process systems. Hybrid systems leverage the high surface reactivity of nanomaterials and the metabolic diversity of microbes to provide removal of contaminants from soil and water. Functionalized nanoparticles like MXenes are being planned for the adsorption of target pollutants with microbial functionality for breakdown, conversion, and degradation into harmless material. The system can be designed to deliver a synergistic effect for pollutant degradation enabled along with ecological balance and rehabilitation (Hidangmayum et al., 2023).

Latest work with genetically modified microbes in combination with nanomaterials to achieve specificity-induced pathway disruption has also been encouraging. Bio-nano systems have the potential to be engineered for identifying and hitting known contaminants, e.g., heavy metals or hydrocarbons, and therefore offer site-specific and effective mechanisms of clean-up. Furthermore, the majority of these systems have been developed to harness industrial or agricultural waste streams as substrates, thereby also promoting sustainability through recycling of low-value products in environmental remediation applications (Vázquez-Núñez et al., 2020).

6.7.4 Policy and Research Directions for Sustainable Implementation

With further developments unfolding in green nanotechnology, its eco-friendly future application will rely ever more heavily on effective policy interventions along with interdisciplinarity. Policy formation will be forced to heavily concentrate on framing guidelines for encouraging the safe application of nanomaterials with stringent life cycle consideration and transparent reporting regulations. These hold the secret to estimating long-term human and environmental effects, particularly for the diversity and complexity of nanomaterials (Pokrajac et al., 2021).

Trans-material interfaces, toxicology, environmental engineering, and regulatory science from interdisciplinary studies are needed in order to design evidence-based standards that will shape nanomaterial use and design. Compliance tool co-evolution with technology can be assisted through models for interactions among academia, industry, and government, such that innovation comes forward in ways socially and ecologically sound. Policy must involve the integration of ethical concerns, including nanoparticle waste management, occupational health, and potential for unforeseen environmental impact (Cassee et al., 2024).

6.8 Conclusion

As this chapter makes clear, nanotechnology is no longer future science fiction; it is a concrete, multidisciplinary science transforming how we treat, monitor, and clean water systems today. Nanomaterials' unusual surface properties and reactivity make it possible to effectively remove all sorts of contaminants, from heavy metals and organic chemicals to microbial pathogens and microplastics. Nanotechnology, starting from titanium dioxide and silver nanospheres up to hybrid membranes and CNTs, leads the pack evenly. On top of this comes incorporation in biosensing devices for direct-time detection of emerging pollutants, with the information required for public health and environmental safety. But greater use of nanomaterials also raises key questions like environmental hazard, cost of production, regulatory loopholes, and danger of misapplication. These need to be tackled by evidence-based life cycle analysis, environmentally friendly production processes, and open regulatory processes. The integration of nanotechnology with bioremediation, artificial intelligence, and circular economies holds a promising future. Lastly, bringing out the true potential of nanotechnology in water treatment will be as much a function of our collective capacity to use it responsibly for the individuals, environment, and future.

REFERENCES

Abbas, R., Luo, J., Qi, X., Naz, A., Khan, I. A., Liu, H., Yu, S., & Wei, J. (2024). Silver nanoparticles: Synthesis, structure, properties and applications. *Nanomaterials*, *14*(17), 1425. https://doi.org/10.3390/NANO14171425

Abubakre, O. K., Medupin, R. O., Akintunde, I. B., Jimoh, O. T., Abdulkareem, A. S., Muriana, R. A., James, J. A., Ukoba, K. O., Jen, T. C., & Yoro, K. O. (2023). Carbon nanotube-reinforced polymer nanocomposites for sustainable biomedical applications: A review. *Journal of Science: Advanced Materials and Devices*, *8*(2), 100557. https://doi.org/10.1016/J.JSAMD.2023.100557

Abu-Dief, A. M., Alsaedi, W. H., & Zikry, M. M. (2025). A collective study on the fabrication of nano-materials for water treatment. *Journal of Umm Al-Qura University for Applied Sciences*, *2025*, 1–23. https://doi.org/10.1007/S43994-025-00227-4

Ahire, S. A., Bachhav, A. A., Pawar, T. B., Jagdale, B. S., Patil, A. V., & Koli, P. B. (2022). The augmentation of nanotechnology era: A concise review on fundamental concepts of nanotechnology and applications in material science and technology. *Results in Chemistry*, *4*, 100633. https://doi.org/10.1016/J.RECHEM.2022.100633

Ahmed, S. F., Mofijur, M., Ahmed, B., Mehnaz, T., Mehejabin, F., Maliat, D., Hoang, A. T., & Shafiullah, G. M. (2022). Nanomaterials as a sustainable choice for treating wastewater. *Environmental Research*, *214*. https://doi.org/10.1016/J.ENVRES.2022.113807

Aldhaher, A., Rabiee, N., & Iravani, S. (2024). Exploring the synergistic potential of MXene-MOF hybrid composites: A perspective on synthesis, properties, and applications. *Hybrid Advances*, *5*, 100131. https://doi.org/10.1016/J.HYBADV.2023.100131

Alhalili, Z. (2023). Metal oxides nanoparticles: General structural description, chemical, physical, and biological synthesis methods, role in pesticides and heavy metal removal through wastewater treatment. *Molecules*, *28*(7), 3086. https://doi.org/10.3390/MOLECULES28073086

Alijagic, A., Suljević, D., Fočak, M., Sulejmanović, J., Šehović, E., Särndahl, E., & Engwall, M. (2024). The triple exposure nexus of microplastic particles, plastic-associated chemicals, and environmental pollutants from a human health perspective. *Environment International*, *188*. https://doi.org/10.1016/j.envint.2024.108736

Alsaiari, N. S., Alzahrani, F. M., Amari, A., Osman, H., Harharah, H. N., Elboughdiri, N., & Tahoon, M. A. (2023). Plant and microbial approaches as green methods for the synthesis of nanomaterials: Synthesis, applications, and future perspectives. *Molecules*, *28*(1), 463. https://doi.org/10.3390/MOLECULES28010463

Anoob, F., Arachchi, S., Azamathulla, H. M., Al-mahbashi, N., & Rathnayake, U. (2024). Nanoadsorbents as an effective wastewater treatment candidate for pharmaceutical contaminants; towards sustainable policy development. *Case Studies in Chemical and Environmental Engineering*, *9*. https://doi.org/10.1016/j.cscee.2024.100639

Aoulad El Hadj Ali, Y., Azzouz, A., Ahrouch, M., Lamaoui, A., Raza, N., & Ait Lahcen, A. (2024). Molecular imprinting technology for next-generation water treatment via photocatalysis and selective pollutant adsorption. *Journal of Environmental Chemical Engineering*, *12*(3), 112768. https://doi.org/10.1016/J.JECE.2024.112768

Arora, B., & Attri, P. (2020). Carbon nanotubes (CNTs): A potential nanomaterial for water purification. *Journal of Composites Science*, *4*(3), 135. https://doi.org/10.3390/JCS4030135

Badiola, M., Mendiola, D., & Bostock, J. (2012). Recirculating aquaculture systems (RAS) analysis: Main issues on management and future challenges. *Aquacultural Engineering*, *51*, 26–35. https://doi.org/10.1016/J.AQUAENG.2012.07.004

Bahlol, H. S., Li, J., Deng, J., Foda, M. F., & Han, H. (2024). Recent progress in nanomaterial-based surface-enhanced Raman spectroscopy for food safety detection. *Nanomaterials*, *14*(21), 1750. https://doi.org/10.3390/NANO14211750

Bhateria, R., & Singh, R. (2019). A review on nanotechnological application of magnetic iron oxides for heavy metal removal. *Journal of Water Process Engineering*, *31*, 100845. https://doi.org/10.1016/j.jwpe.2019.100845

Bhatia, D., Paul, S., Acharjee, T., & Ramachairy, S. S. (2024). Biosensors and their widespread impact on human health. *Sensors International*, *5*, 100257. https://doi.org/10.1016/J.SINTL.2023.100257

Cagno, E., Garrone, P., Negri, M., & Rizzuni, A. (2022). Adoption of water reuse technologies: An assessment under different regulatory and operational scenarios. *Journal of Environmental Management*, *317*. https://doi.org/10.1016/j.jenvman.2022.115389

Cassee, F. R., Bleeker, E. A. J., Durand, C., Exner, T., Falk, A., Friedrichs, S., Heunisch, E., Himly, M., Hofer, S., Hofstätter, N., Hristozov, D., Nymark, P., Pohl, A., Soeteman-Hernández, L. G., Suarez-Merino, B., Valsami-Jones, E., & Groenewold, M. (2024). Roadmap towards safe and sustainable advanced and innovative materials. (Outlook for 2024–2030). *Computational and Structural Biotechnology Journal, 25*, 105–126. https://doi.org/10.1016/J.CSBJ.2024.05.018

Chakraborty, A., Adhikary, S., Bhattacharya, S., Dutta, S., Chatterjee, S., Banerjee, D., Ganguly, A., & Rajak, P. (2023). Pharmaceuticals and personal care products as emerging environmental contaminants: Prevalence, toxicity, and remedial approaches. *ACS Chemical Health and Safety, 30*(6), 362–388. https://doi.org/10.1021/ACS.CHAS.3C00071

Chand, K., Cao, D., Fouad, D. E., Shah, A. H., Lakhan, M. N., Dayo, A. Q., Sagar, H. J., Zhu, K., & Mohamed, A. M. A. (2020). Photocatalytic and antimicrobial activity of biosynthesized silver and titanium dioxide nanoparticles: A comparative study. *Journal of Molecular Liquids, 316*. https://doi.org/10.1016/J.MOLLIQ.2020.113821

Chausali, N., Saxena, J., & Prasad, R. (2023). Nanotechnology as a sustainable approach for combating the environmental effects of climate change. *Journal of Agriculture and Food Research, 12*. https://doi.org/10.1016/j.jafr.2023.100541

Chávez-Hernández, J. A., Velarde-Salcedo, A. J., Navarro-Tovar, G., & Gonzalez, C. (2024). Safe nanomaterials: From their use, application, and disposal to regulations. *Nanoscale Advances, 6*(6), 1583–1610. https://doi.org/10.1039/D3NA01097J

Costa, C. Q. V., Cruz, J., Martins, J., Teodósio, M. A. A., Jockusch, S., Ramamurthy, V., & Da Silva, J. P. (2021). Fluorescence sensing of microplastics on surfaces. *Environmental Chemistry Letters, 19*(2), 1797–1802. https://doi.org/10.1007/S10311-020-01136-0

Costa Santos, R., Aguilar da Silva, R., Moreira dos Santos, M., Botelho Bovo, A., & Furtado da Silva, A. (2024). Assessing nitrate contamination in groundwater for public supply: A study in a small Brazilian town. *Groundwater for Sustainable Development, 25*, 101084. https://doi.org/10.1016/J.GSD.2024.101084

Cui, Y., Lai, B., & Tang, X. (2019). Microbial fuel cell-based biosensors. *Biosensors, 9*(3), 92. https://doi.org/10.3390/BIOS9030092

Das, S., Parida, V. K., Tiwary, C. S., Gupta, A. K., & Chowdhury, S. (2024). Emerging contaminants in the aquatic environment: Fate, occurrence, impacts, and toxicity. *ACS Symposium Series, 1475*, 1–32. https://doi.org/10.1021/BK-2024-1475.CH001

Dhall, S., Nigam, A., Harshavardhan, M., Mukherjee, A., & Srivastava, P. (2024). A comprehensive overview of methods involved in nanomaterial production and waste disposal from research labs and industries and existing regulatory guidelines for handling engineered nanomaterials. *Environmental Chemistry and Ecotoxicology, 6*, 269–282. https://doi.org/10.1016/j.enceco.2024.06.002

Díez, A. G., Rincón-Iglesias, M., Lanceros-Méndez, S., Reguera, J., & Lizundia, E. (2022). Multicomponent magnetic nanoparticle engineering: The role of structure-property relationship in advanced applications. *Materials Today Chemistry, 26*, 101220. https://doi.org/10.1016/J.MTCHEM.2022.101220

Dutta, S., Sengupta, P., Bagchi, S., Chhikara, B. S., Pavlík, A., Sláma, P., & Roychoudhury, S. (2023). Reproductive toxicity of combined effects of endocrine disruptors on human reproduction. *Frontiers in Cell and Developmental Biology, 11*, 1162015. https://doi.org/10.3389/FCELL.2023.1162015

Ejiohuo, O., Onyeaka, H., Akinsemolu, A., Nwabor, O. F., Siyanbola, K. F., Tamasiga, P., & Al-Sharify, Z. T. (2025). Ensuring water purity: Mitigating environmental risks and safeguarding human health. *Water Biology and Security*. https://doi.org/10.1016/j.watbs.2024.100341

Eniola, J. O., Kujawa, J., Nwokoye, A., Al-Gharabli, S., Avornyo, A. K., Giwa, A., Amusa, H. K., Yusuf, A. O., & Okolie, J. A. (2024). Advances in electrochemical membranes for water treatment: A comprehensive review. *Desalination and Water Treatment*, *319*, 100450. https://doi.org/10.1016/J.DWT.2024.100450

EPA. (2025, March 19). *Effects of acid rain | US EPA*. www.epa.gov/acidrain/effects-acid-rain

Esha Lohia. (2025). *Government report highlights groundwater contamination across India*. Mongabay. https://india.mongabay.com/2025/01/government-report-highlights-groundwater-contamination-across-india/

Essamlali, I., Nhaila, H., & El Khaili, M. (2024). Advances in machine learning and IoT for water quality monitoring: A comprehensive review. *Heliyon*, *10*(6), e27920. https://doi.org/10.1016/J.HELIYON.2024.E27920

Faramarzi, P., Jang, W., Oh, D., Kim, B., Kim, J. H., & You, J. B. (2024). Microfluidic detection and analysis of microplastics using surface nanodroplets. *ACS Sensors*, *9*(3), 1489–1498. https://doi.org/10.1021/ACSSENSORS.3C02627

Fard, N. E., Ali, N. S., Saady, N. M. C., Albayati, T. M., Salih, I. K., Zendehboudi, S., Harharah, H. N., & Harharah, R. H. (2024). A review on development and modification strategies of MOFs Z-scheme heterojunction for photocatalytic wastewater treatment, water splitting, and DFT calculations. *Heliyon*, *10*(13), e32861. https://doi.org/10.1016/J.HELIYON.2024.E32861

Fdez-Sanromán, A., Bernárdez-Rodas, N., Rosales, E., Pazos, M., González-Romero, E., & Sanromán, M. Á. (2025). Biosensor technologies for water quality: Detection of emerging contaminants and pathogens. *Biosensors*, *15*(3), 189. https://doi.org/10.3390/BIOS15030189

Gleick, P. H., & Cooley, H. (2021). Freshwater scarcity. *Annual Review of Environment and Resources*, *46*, 319–348. https://doi.org/10.1146/ANNUREV-ENVIRON-012220-101319

Gonzalez, C. (2024). *Nanoscale Advances*. 6(6). https://doi.org/10.1039/d3na01097j

Gu, P., Liu, S., Cheng, X., Zhang, S., Wu, C., Wen, T., & Wang, X. (2024). Recent strategies, progress, and prospects of two-dimensional metal carbides (MXenes) materials in wastewater purification: A review. *Science of the Total Environment*, *912*, 169533. https://doi.org/10.1016/J.SCITOTENV.2023.169533

Guo, C., Lv, S., Liu, Y., & Li, Y. (2022). Biomarkers for the adverse effects on respiratory system health associated with atmospheric particulate matter exposure. *Journal of Hazardous Materials*, *421*, 126760. https://doi.org/10.1016/J.JHAZMAT.2021.126760

Gupta, R., & Xie, H. (2018). Nanoparticles in daily life: Applications, toxicity and regulations. *Journal of Environmental Pathology, Toxicology and Oncology: Official Organ of the International Society for Environmental Toxicology and Cancer*, *37*(3), 209. https://doi.org/10.1615/JENVIRONPATHOLTOXICOLONCOL.2018026009

He, H., Li, F., Liu, K., Zhan, J., Wang, X., Lai, C., Yang, X., Huang, B., & Pan, X. (2023). The disinfectant residues promote the leaching of water contaminants from plastic pipe particles. *Environmental Pollution*, *327*. https://doi.org/10.1016/j.envpol.2023.121577

He, L., Wang, L., Zhu, H., Wang, Z., Zhang, L., Yang, L., Dai, Y., Mo, H., Zhang, J., & Shen, J. (2021). A reusable Fe3O4/GO-COOH nanoadsorbent for Ca2+ and Cu2+ removal from oilfield wastewater. *Chemical Engineering Research and Design*, *166*, 248–258. https://doi.org/10.1016/j.cherd.2020.12.019

Hebbar, R. S., Isloor, A. M., Inamuddin, & Asiri, A. M. (2017). Carbon nanotube- and graphene-based advanced membrane materials for desalination. *Environmental Chemistry Letters*, *15*(4), 643–671. https://doi.org/10.1007/S10311-017-0653-Z

Hegde, K., Kaur, S., & Mausam, B. (2016). Current understandings of toxicity, risks and regulations of engineered nanoparticles with respect to environmental microorganisms. *Nanotechnology for Environmental Engineering, 1*(1), 1–12. https://doi.org/10.1007/s41204-016-0005-4

Hemdan, M., Abuelhaded, K., Shaker, A. A. S., Ashour, M. M., Abdelaziz, M. M., Dahab, M. I., Nassar, Y. A., Sarguos, A. M. M., Zakaria, P. S., Fahmy, H. A., Abdel Mageed, S. S., Hamed, M. O. A., Mubarak, M. F., Taher, M. A., Gumaah, N. F., & Ragab, A. H. (2025). Recent advances in nano-enhanced biosensors: Innovations in design, applications in healthcare, environmental monitoring, and food safety, and emerging research challenges. *Sensing and Bio-Sensing Research, 48*, 100783. https://doi.org/10.1016/J.SBSR.2025.100783

Herbeck-engel, P., Tavernaro, I., Dekkers, S., Soeteman-hern, L. G., Noorlander, C., & Kraegeloh, A. (2021). NanoImpact safe-by-design part II : A strategy for balancing safety and functionality in the different stages of the innovation process. *24*. https://doi.org/10.1016/j.impact.2021.100354

Herrera, V. (2019). Reconciling global aspirations and local realities: Challenges facing the sustainable development goals for water and sanitation. *World Development, 118*, 106–117. https://doi.org/10.1016/j.worlddev.2019.02.009

Hidangmayum, A., Debnath, A., Guru, A., Singh, B. N., Upadhyay, S. K., & Dwivedi, P. (2023). Mechanistic and recent updates in nano-bioremediation for developing green technology to alleviate agricultural contaminants. *International Journal of Environmental Science and Technology, 20*(10), 11693–11718. https://doi.org/10.1007/S13762-022-04560-7

Jafarzadeh, S., Oladzadabbasabadi, N., Dheyab, M. A., Lalabadi, M. A., Sheibani, S., Ghasemlou, M., Esmaeili, Y., Barrow, C. J., Naebe, M., & Timms, W. (2025). Emerging trends in smart and sustainable nano-biosensing: The role of green nanomaterials. *Industrial Crops and Products, 223*, 120108. https://doi.org/10.1016/J.INDCROP.2024.120108

Jain, K., Patel, A. S., Pardhi, V. P., & Flora, S. J. S. (2021). Nanotechnology in wastewater management: A new paradigm towards wastewater treatment. *Molecules, 26*(6), 1797. https://doi.org/10.3390/MOLECULES26061797

Januário, E. F. D., Vidovix, T. B., Beluci, N. de C. L., Paixão, R. M., Silva, L. H. B. R. da, Homem, N. C., Bergamasco, R., & Vieira, A. M. S. (2021). Advanced graphene oxide-based membranes as a potential alternative for dyes removal: A review. *Science of the Total Environment, 789*, 147957. https://doi.org/10.1016/J.SCITOTENV.2021.147957

Joudeh, N., & Linke, D. (2022). Nanoparticle classification, physicochemical properties, characterization, and applications: A comprehensive review for biologists. *Journal of Nanobiotechnology, 20*(1). https://doi.org/10.1186/S12951-022-01477-8

Kambale, E. K., Nkanga, C. I., Mutonkole, B.-P. I., Bapolisi, A. M., Tassa, D. O., Liesse, J.-M. I., Krause, R. W. M., & Memvanga, P. B. (2020). Green synthesis of antimicrobial silver nanoparticles using aqueous leaf extracts from three Congolese plant species (*Brillantaisia patula, Crossopteryx febrifuga* and *Senna siamea*). *Heliyon, 6*(8), e04493. https://doi.org/10.1016/j.heliyon.2020.e04493

Kanoun, O., Lazarević-Pašti, T., Pašti, I., Nasraoui, S., Talbi, M., Brahem, A., Adiraju, A., Sheremet, E., Rodriguez, R. D., Ali, M. Ben, & Al-Hamry, A. (2021). A review of nanocomposite-modified electrochemical sensors for water quality monitoring. *Sensors, 21*(12), 4131. https://doi.org/10.3390/S21124131

Kesari, K. K., Soni, R., Jamal, Q. M. S., Tripathi, P., Lal, J. A., Jha, N. K., Siddiqui, M. H., Kumar, P., Tripathi, V., & Ruokolainen, J. (2021). Wastewater treatment and reuse: A review of its applications and health implications. *Water, Air, and Soil Pollution, 232*(5). https://doi.org/10.1007/S11270-021-05154-8

Khaldoun, K., Khizar, S., Saidi, S., Nadia, B., & Abdelhamid, Z. (2024). Synthesis of silver nanoparticles as an antimicrobial mediator. *Journal of Umm Al-Qura University for Applied Sciences*, 0123456789. https://doi.org/10.1007/s43994-024-00159-5

Khan, I., Saeed, K., & Khan, I. (2019). Nanoparticles: Properties, applications and toxicities. *Arabian Journal of Chemistry*, *12*(7), 908–931. https://doi.org/10.1016/j.arabjc.2017.05.011

Khan, S. K., Dutta, J., Ahmad, I., & Rather, M. A. (2024). Nanotechnology in aquaculture: Transforming the future of food security. *Food Chemistry: X*, *24*, 101974. https://doi.org/10.1016/J.FOCHX.2024.101974

Kornberg, T. G., Id, T. A. S., Antonini, J. M., Rojanasakul, Y., Castranova, V., Id, Y. Y., & Rojanasakul, L. W. (2017). Potential toxicity and underlying mechanisms associated with pulmonary exposure to iron oxide nanoparticles: Conflicting literature and unclear risk. 1–26. https://doi.org/10.3390/nano7100307

Koseoglu-Imer, D. Y., Oral, H. V., Coutinho Calheiros, C. S., Krzeminski, P., Güçlü, S., Pereira, S. A., Surmacz-Górska, J., Plaza, E., Samaras, P., Binder, P. M., van Hullebusch, E. D., & Devolli, A. (2023). Current challenges and future perspectives for the full circular economy of water in European countries. *Journal of Environmental Management*, *345*. https://doi.org/10.1016/j.jenvman.2023.118627

Krishna, R. H., Chandraprabha, M. N., Samrat, K., Krishna Murthy, T. P., Manjunatha, C., & Kumar, S. G. (2023). Carbon nanotubes and graphene-based materials for adsorptive removal of metal ions – a review on surface functionalization and related adsorption mechanism. *Applied Surface Science Advances*, *16*. https://doi.org/10.1016/J.APSADV.2023.100431

Krishnan, A., Sundaram, T., Nagappan, B., Devarajan, Y., & Bhumika. (2024). Integrating artificial intelligence in nanomembrane systems for advanced water desalination. *Results in Engineering*, *24*. https://doi.org/10.1016/j.rineng.2024.103321

Kumar, M., Shekhar, S., Kumar, R., Kumar, P., Govarthanan, M., & Chaminda, T. (2023). Drinking water treatment and associated toxic byproducts: Concurrence and urgence. *Environmental Pollution*, *320*. https://doi.org/10.1016/j.envpol.2023.121009

Kumari, S., Raturi, S., Kulshrestha, S., Chauhan, K., Dhingra, S., András, K., Thu, K., Khargotra, R., & Singh, T. (2023). A comprehensive review on various techniques used for synthesizing nanoparticles. *Journal of Materials Research and Technology*, *27*, 1739–1763. https://doi.org/10.1016/j.jmrt.2023.09.291

Li, B., Ren, L., Jiang, D., Jia, M., Zhang, M., Xu, G., Sun, Y., Hou, L., Yuan, C., & Yuan, Y. (2025). Optimizing charge carrier dynamics in photocatalysts for enhanced CO2 photoreduction: Fundamental principles, advanced strategies, and characterization techniques. *Next Energy*, *7*, 100222. https://doi.org/10.1016/J.NXENER.2024.100222

Li, X. F., & Mitch, W. A. (2018). Drinking water disinfection byproducts (DBPs) and human health effects: Multidisciplinary challenges and opportunities. *Environmental Science and Technology*, *52*(4), 1681–1689. https://doi.org/10.1021/ACS.EST.7B05440/ASSET/IMAGES/LARGE/ES-2017-05440A_0003.JPEG

Liu, G., Zhang, Y., Knibbe, W. J., Feng, C., Liu, W., Medema, G., & van der Meer, W. (2017). Potential impacts of changing supply-water quality on drinking water distribution: A review. *Water Research*, *116*, 135–148. https://doi.org/10.1016/j.watres.2017.03.031

Liu, M., Zhu, G., & Tian, Y. (2024). The historical evolution and research trends of life cycle assessment. *Green Carbon*, *2*(4), 425–437. https://doi.org/10.1016/J.GREENCA.2024.08.003

Liu, X., Wei, W., Liu, Z., Song, E., Lou, J., Feng, L., Huang, R., Chen, C., Ke, P. C., & Song, Y. (2021). Serum apolipoprotein A-I depletion is causative to silica nanoparticles-induced cardiovascular damage. *Proceedings of the National Academy of Sciences of the United States of America*, *118*(44). https://doi.org/10.1073/PNAS.2108131118

Ma, X., Tian, Y., Yang, R., Wang, H., Allahou, L. W., Chang, J., Williams, G., Knowles, J. C., & Poma, A. (2024). Nanotechnology in healthcare, and its safety and environmental risks. *Journal of Nanobiotechnology, 22*(1), 1–81. https://doi.org/10.1186/S12951-024-02901-X

Maghsoudi, A. S., Hassani, S., Mirnia, K., & Abdollahi, M. (2021). Recent advances in nanotechnology-based biosensors development for detection of arsenic, lead, mercury, and cadmium. *International Journal of Nanomedicine, 16*, 803. https://doi.org/10.2147/IJN.S294417

Manasa, M., Chandewar, P. R., & Mahalingam, H. (2021). Photocatalytic degradation of ciprofloxacin & norfloxacin and disinfection studies under solar light using boron & cerium doped TiO2 catalysts synthesized by green EDTA-citrate method. *Catalysis Today, 375*, 522–536. https://doi.org/10.1016/j.cattod.2020.03.018

Margan, P., & Haghighi, M. (2018). Sono-coprecipitation synthesis and physicochemical characterization of CdO-ZnO nanophotocatalyst for removal of acid orange 7 from wastewater. *Ultrasonics Sonochemistry, 40*, 323–332. https://doi.org/10.1016/J.ULTSONCH.2017.07.003

Mariano, S. M. F., Angeles, L. F., Aga, D. S., Villanoy, C. L., & Jaraula, C. M. B. (2023). Emerging pharmaceutical contaminants in key aquatic environments of the Philippines. *Frontiers in Earth Science, 11*. https://doi.org/10.3389/FEART.2023.1124313/FULL

Matavos-Aramyan, S. (2024). Addressing the microplastic crisis: A multifaceted approach to removal and regulation. *Environmental Advances, 17*. https://doi.org/10.1016/j.envadv.2024.100579

McLain, A. A. (2018). Photocatalytic properties of zinc oxide and graphene nanocomposites. *Proceedings of the Wisconsin Space Conference*. https://doi.org/10.17307/WSC.V1I1.253

Mishra, R. K., Mentha, S. S., Misra, Y., & Dwivedi, N. (2023). Emerging pollutants of severe environmental concern in water and wastewater: A comprehensive review on current developments and future research. *Water-Energy Nexus, 6*, 74–95. https://doi.org/10.1016/j.wen.2023.08.002

Mishra, S., Kumar, R., & Kumar, M. (2023). Use of treated sewage or wastewater as an irrigation water for agricultural purposes – environmental, health, and economic impacts. *Total Environment Research Themes, 6*. https://doi.org/10.1016/j.totert.2023.100051

Moghimi Dehkordi, M., Pournuroz Nodeh, Z., Soleimani Dehkordi, K., Salmanvandi, H., Rasouli Khorjestan, R., & Ghaffarzadeh, M. (2024). Soil, air, and water pollution from mining and industrial activities: Sources of pollution, environmental impacts, and prevention and control methods. *Results in Engineering, 23*, 102729. https://doi.org/10.1016/J.RINENG.2024.102729

Mondejar, M. E., Avtar, R., Diaz, H. L. B., Dubey, R. K., Esteban, J., Gómez-Morales, A., Hallam, B., Mbungu, N. T., Okolo, C. C., Prasad, K. A., She, Q., & Garcia-Segura, S. (2021). Digitalization to achieve sustainable development goals: Steps towards a smart green planet. *Science of the Total Environment, 794*, 148539. https://doi.org/10.1016/J.SCITOTENV.2021.148539

Mitchell, J. F., Sundberg, K. A., & Reynolds, J. H. (2007). Differential attention-dependent response modulation across cell classes in macaque visual area V4. *Neuron, 55*(1), 131–141. https://doi.org/10.1016/j.neuron.2007.06.018

Mpongwana, N., & Rathilal, S. (2022). A review of the techno-economic feasibility of nanoparticle application for wastewater treatment. *Water, 14*(10), 1550. https://doi.org/10.3390/W14101550

Musie, W., & Gonfa, G. (2023). Fresh water resource, scarcity, water salinity challenges and possible remedies: A review. *Heliyon*, *9*(8), e18685. https://doi.org/10.1016/J.HELIYON.2023.E18685

Nagar, A., & Pradeep, T. (2020). Clean water through nanotechnology: Needs, gaps, and fulfillment. *ACS Nano*, *14*(6), 6420–6435. https://doi.org/10.1021/ACSNANO.9B01730/SUPPL_FILE/NN9B01730_LIVESLIDES.MP4

Nagendran, V., Goveas, L. C., Vinayagam, R., Varadavenkatesan, T., & Selvaraj, R. (2024). Nanozymes in environmental remediation: A bibliometric and comprehensive review of their oxidoreductase-mimicking capabilities. *Microchemical Journal*, *207*, 111748. https://doi.org/10.1016/J.MICROC.2024.111748

Namakka, M., Rahman, M. R., Said, K. A. M. Bin, Abdul Mannan, M., & Patwary, A. M. (2023). A review of nanoparticle synthesis methods, classifications, applications, and characterization. *Environmental Nanotechnology, Monitoring & Management*, *20*, 100900. https://doi.org/10.1016/J.ENMM.2023.100900

Naseem, T., & Durrani, T. (2021a). Environmental chemistry and ecotoxicology the role of some important metal oxide nanoparticles for wastewater and antibacterial applications : A review. *Environmental Chemistry and Ecotoxicology*, *3*, 59–75. https://doi.org/10.1016/j.enceco.2020.12.001

Naseem, T., & Durrani, T. (2021b). The role of some important metal oxide nanoparticles for wastewater and antibacterial applications: A review. *Environmental Chemistry and Ecotoxicology*, *3*, 59–75. https://doi.org/10.1016/J.ENCECO.2020.12.001

Naseer, A. (2024). Role of nanocomposites and nano adsorbents for heavy metals removal and dyes. An overview. *Desalination and Water Treatment*, *320*, 100662. https://doi.org/10.1016/J.DWT.2024.100662

NCBI. (2022). *Microbes and climate change – science, people & impacts*. https://doi.org/10.1128/AAMCOL.NOV.2021

Nishu, & Kumar, S. (2023). Smart and innovative nanotechnology applications for water purification. *Hybrid Advances*, *3*, 100044. https://doi.org/10.1016/j.hybadv.2023.100044

Njema, G. G., & Kibet, J. K. (2024). A review of novel materials for nano-photocatalytic and optoelectronic applications: Recent perspectives, water splitting and environmental remediation. *Progress in Engineering Science*, *1*(4), 100018. https://doi.org/10.1016/J.PES.2024.100018

Olanrewaju, O. I., Enegbuma, W. I., & Donn, M. (2024). Challenges in life cycle assessment implementation for construction environmental product declaration development: A mixed approach and global perspective. *Sustainable Production and Consumption*, *49*, 502–528. https://doi.org/10.1016/J.SPC.2024.06.02 1

Olugbodi, J. O., Lawal, B., Bako, G., Onikanni, A. S., Abolenin, S. M., Mohammud, S. S., Ataya, F. S., & Batiha, G. E. S. (2023). Effect of sub-dermal exposure of silver nanoparticles on hepatic, renal and cardiac functions accompanying oxidative damage in male Wistar rats. *Scientific Reports*, *13*(1), 10539. https://doi.org/10.1038/S41598-023-37178-X

Osman, A. I., Zhang, Y., Farghali, M., Rashwan, A. K., Eltaweil, A. S., Abd El-Monaem, E. M., Mohamed, I. M. A., Badr, M. M., Ihara, I., Rooney, D. W., & Yap, P. S. (2024). Synthesis of green nanoparticles for energy, biomedical, environmental, agricultural, and food applications: A review. *Environmental Chemistry Letters*, *22*(2), 841–887. https://doi.org/10.1007/S10311-023-01682-3

Pandey, D. K., & Mishra, R. (2024). Towards sustainable agriculture: Harnessing AI for global food security. *Artificial Intelligence in Agriculture*, *12*, 72–84. https://doi.org/10.1016/J.AIIA.2024.04.003

Panigrahy, S. K., Nandha, A., Chaturvedi, M., & Mishra, P. K. (2024). Novel nanocomposites with advanced materials and their role in waste water treatment. *Next Sustainability*, *4*, 100042. https://doi.org/10.1016/J.NXSUST.2024.100042

Pérez, H., García, O. J. Q., Amezcua-Allieri, M. A., & Vázquez, R. R. (2023). Nanotechnology as an efficient and effective alternative for wastewater treatment: An overview. *Water Science and Technology*, *87*(12), 2971–3001. https://doi.org/10.2166/WST.2023.179

Pokrajac, L., Abbas, A., Chrzanowski, W., Dias, G. M., Eggleton, B. J., Maguire, S., Maine, E., Malloy, T., Nathwani, J., Nazar, L., Sips, A., Sone, J., Van Den Berg, A., Weiss, P. S., & Mitra, S. (2021). Nanotechnology for a sustainable future: Addressing global challenges with the international network4Sustainable nanotechnology. *ACS Nano*, *15*(12), 18608–18623. https://doi.org/10.1021/ACSNANO.1C10919/ASSET/IMAGES/MEDIUM/NN1C10919_0018.GIF

Qi, Y., & He, K. (2025). Science and technology for water purification: Achievements and strategies. *Water*, *17*(1), 91. https://doi.org/10.3390/W17010091

Quaranta, E., Bejarano, M. D., Comoglio, C., Fuentes-Pérez, J. F., Pérez-Díaz, J. I., Sanz-Ronda, F. J., Schletterer, M., Szabo-Meszaros, M., & Tuhtan, J. A. (2023). Digitalization and real-time control to mitigate environmental impacts along rivers: Focus on artificial barriers, hydropower systems and European priorities. *Science of the Total Environment*, *875*. https://doi.org/10.1016/j.scitotenv.2023.162489

Ramírez-Malule, H., Quiñones-Murillo, D. H., & Manotas-Duque, D. (2020). Emerging contaminants as global environmental hazards. A bibliometric analysis. *Emerging Contaminants*, *6*, 179–193. https://doi.org/10.1016/j.emcon.2020.05.001

Rasheed, T., Bilal, M., Nabeel, F., Adeel, M., & Iqbal, H. M. N. (2019). Environmentally-related contaminants of high concern: Potential sources and analytical modalities for detection, quantification, and treatment. *Environment International*, *122*, 52–66. https://doi.org/10.1016/j.envint.2018.11.038

Rashid, A. Bin, Haque, M., Islam, S. M. M., & Uddin Labib, K. M. R. (2024). Nanotechnology-enhanced fiber-reinforced polymer composites: Recent advancements on processing techniques and applications. *Heliyon*, *10*(2), e24692. https://doi.org/10.1016/J.HELIYON.2024.E24692

Rathi, B. S., Ewe, L. S., Sanjay, S., Sujatha, S., Yew, W. K., Baskaran, R., & Tiong, S. K. (2024). Recent trends and advancement in metal oxide nanoparticles for the degradation of dyes: Synthesis, mechanism, types and its application. *Nanotoxicology*. https://doi.org/10.1080/17435390.2024.2349304

Razzak, S. A., Faruque, M. O., Alsheikh, Z., Alsheikhmohamad, L., Alkuroud, D., Alfayez, A., Hossain, S. M. Z., & Hossain, M. M. (2022). A comprehensive review on conventional and biological-driven heavy metals removal from industrial wastewater. *Environmental Advances*, *7*. https://doi.org/10.1016/j.envadv.2022.100168

Razzaq, Z., Khalid, A., Ahmad, P., Farooq, M., Khandaker, M. U., Sulieman, A. A. M., Rehman, I. U., Shakeel, S., & Khan, A. (2021). Photocatalytic and antibacterial potency of titanium dioxide nanoparticles: A cost-effective and environmentally friendly media for treatment of air and wastewater. *Catalysts*, *11*(6), 709. https://doi.org/10.3390/CATAL11060709

Sadak, O., Hackney, R., Sundramoorthy, A., Yilmaz, G., & Gunasekaran, S. (2020). Azo dye-functionalized magnetic Fe3O4/polyacrylic acid nanoadsorbent for removal of lead (II) ions. *Environmental Nanotechnology Monitoring & Management*, *14*. https://doi.org/10.1016/j.enmm.2020.100380

Sadegh, H., Ali, G. A. M., & Gupta, V. K. (2017). The role of nanomaterials as effective adsorbents and their applications in wastewater treatment. 1–14. https://doi.org/10.1007/s40097-017-0219-4

Sadegh, H., Ali, G. A. M., Gupta, V. K., Makhlouf, A. S. H., Shahryari-ghoshekandi, R., Nadagouda, M. N., Sillanpää, M., & Megiel, E. (2017). The role of nanomaterials as effective adsorbents and their applications in wastewater treatment. *Journal of Nanostructure in Chemistry, 7*(1), 1–14. https://doi.org/10.1007/S40097-017-0219-4

Sahu, B., Behera, B., & Kumar, U. (2025). Transgene delivery system: Viral, nonviral, and other methods for central nervous system. *Genome Editing for Neurodegenerative Diseases*, 135–155. https://doi.org/10.1016/B978-0-443-23826-0.00011-8

Sahu, B., Sharma, J., Behera, B., & Kumar, U. (2024). Nanotechnology-based radiation therapy to cure cancer. *Nanoparticles in Cancer Therapy: Innovations and Clinical Applications*, 118–141. https://doi.org/10.1201/9781003515630-9/NANOTECH-NOLOGY-BASED-RADIATION-THERAPY-CURE-CANCER-BHUPEND-ER-SAHU-JANVI-SHARMA-BIBHUDENDU-BEHERA-UMESH-KUMAR

Salehi, M. (2022). Global water shortage and potable water safety; Today's concern and tomorrow's crisis. *Environment International, 158*. https://doi.org/10.1016/j.envint.2021.106936

Salgot, M., & Folch, M. (2018). Wastewater treatment and water reuse. *Current Opinion in Environmental Science and Health, 2*, 64–74. https://doi.org/10.1016/j.coesh.2018.03.005

Samal, K., Mahapatra, S., & Hibzur Ali, M. (2022). Pharmaceutical wastewater as emerging contaminants (EC): Treatment technologies, impact on environment and human health. *Energy Nexus, 6*. https://doi.org/10.1016/j.nexus.2022.100076

Satyam, S., & Patra, S. (2024). Innovations and challenges in adsorption-based wastewater remediation: A comprehensive review. *Heliyon, 10*(9), e29573. https://doi.org/10.1016/J.HELIYON.2024.E29573

Sauvé, S., & Desrosiers, M. (2014). A review of what is an emerging contaminant. *Chemistry Central Journal, 8*(1). https://doi.org/10.1186/1752-153X-8-15

Shaji, E., Santosh, M., Sarath, K. V., Prakash, P., Deepchand, V., & Divya, B. V. (2021). Arsenic contamination of groundwater: A global synopsis with focus on the Indian Peninsula. *Geoscience Frontiers, 12*(3), 101079. https://doi.org/10.1016/J.GSF.2020.08.015

Sharma, M., Mahajan, P., Alsubaie, A. S., Khanna, V., Chahal, S., Thakur, A., Yadav, A., Arya, A., Singh, A., & Singh, G. (2025). Next-generation nanomaterials-based biosensors: Real-time biosensing devices for detecting emerging environmental pollutants. *Materials Today Sustainability, 29*, 101068. https://doi.org/10.1016/J.MTSUST.2024.101068

Sharma, S., & Bhattacharya, A. (2017). Drinking water contamination and treatment techniques. *Applied Water Science, 7*(3), 1043–1067. https://doi.org/10.1007/S13201-016-0455-7

Sharma, V., Singh, P., Pandey, A. K., & Dhawan, A. (2012). Induction of oxidative stress, DNA damage and apoptosis in mouse liver after sub-acute oral exposure to zinc oxide nanoparticles. *Mutation Research, 745*(1–2), 84–91. https://doi.org/10.1016/J.MRGENTOX.2011.12.009

Sivakami, A., Sarankumar, R., & Vinodha, S. (2021). Introduction to nanobiotechnology: Novel and smart applications. In K. Pal (Ed.), *Bio-manufactured nanomaterials: Perspectives and promotion* (pp. 1–22). Springer International Publishing. https://doi.org/10.1007/978-3-030-67223-2_1

Su, H., Yan, J., Yan, X., Zhao, Q., Liao, C., Li, N., & Wang, X. (2024). Highly sensitive standardized toxicity biosensors for rapid water quality warning. *Bioresource Technology, 406*, 130985. https://doi.org/10.1016/J.BIORTECH.2024.130985

Takallu, S., Aiyelabegan, H. T., Zomorodi, A. R., Alexandrovna, K. V., Aflakian, F., Asvar, Z., Moradi, F., Behbahani, M. R., Mirzaei, E., Sarhadi, F., & Vakili-Ghartavol, R. (2024). Nanotechnology improves the detection of bacteria: Recent advances and future perspectives. *Heliyon*, *10*(11), e32020. https://doi.org/10.1016/J.HELI-YON.2024.E32020

Tarannum, N., & Gautam, Y. K. (2019). Nanoparticles : A state-of-the-art review. 34926–34948. https://doi.org/10.1039/c9ra04164h

Tripathy, J., Mishra, A., Pandey, M., Thakur, R. R., Chand, S., Rout, P. R., & Shahid, M. K. (2024). Advances in nanoparticles and nanocomposites for water and wastewater treatment: A review. *Water*, *16*(11), 1481. https://doi.org/10.3390/W16111481

Urnukhsaikhan, E., Bold, B.-E., Gunbileg, A., Sukhbaatar, N., & Mishig-Ochir, T. (2021). Antibacterial activity and characteristics of silver nanoparticles biosynthesized from Carduus crispus. *Scientific Reports*, *11*(1), 21047. https://doi.org/10.1038/s41598-021-00520-2

Vázquez-Núñez, E., Molina-Guerrero, C. E., Peña-Castro, J. M., Fernández-Luqueño, F., & de la Rosa-Álvarez, M. G. (2020). Use of nanotechnology for the bioremediation of contaminants: A review. *Processes*, *8*(7), 826. https://doi.org/10.3390/PR8070826

Verma, G., Mondal, K., Islam, M., & Gupta, A. (2024). Recent advances in advanced micro and nanomanufacturing for wastewater purification. *ACS Applied Engineering Materials*, *2*(2), 262–285. https://doi.org/10.1021/ACSAENM.3C00711

Voulvoulis, N. (2018). Water reuse from a circular economy perspective and potential risks from an unregulated approach. *Current Opinion in Environmental Science and Health*, *2*, 32–45. https://doi.org/10.1016/j.coesh.2018.01.005

Wang, J., Zhu, X., Zhang, X., Zhao, Z., Liu, H., George, R., Wilson-Rawls, J., Chang, Y., & Chen, Y. (2011). Disruption of zebrafish (*Danio rerio*) reproduction upon chronic exposure to TiO_2 nanoparticles. *Chemosphere*, *83*(4), 461–467. https://doi.org/10.1016/J.CHEMOSPHERE.2010.12.069

Wang, X., Zhou, J., & Wang, H. (2024). Bioreceptors as the key components for electrochemical biosensing in medicine. *Cell Reports Physical Science*, *5*(2), 101801. https://doi.org/10.1016/J.XCRP.2024.101801

WHO. (2019, June 18). *1 in 3 people globally do not have access to safe drinking water – UNICEF, WHO*. www.who.int/news/item/18-06-2019-1-in-3-people-globally-do-not-have-access-to-safe-drinking-water-unicef-who

WHO. (2023, September 13). *Drinking-water*. www.who.int/news-room/fact-sheets/detail/drinking-water

World Bank. (2023). *How is India addressing its water needs?* www.worldbank.org/en/country/india/brief/world-water-day-2022-how-india-is-addressing-its-water-needs

WWF. (2014, November). *Water for Our Future: Americas Regional Process Event | Publications | WWF*. https://www.worldwildlife.org/publications/water-for-our-future-americas-regional-process-event

Wysowska, E., Wiewiórska, I., & Kicińska, A. (2021). The impact of different stages of water treatment process on the number of selected bacteria. *Water Resources and Industry*, *26*. https://doi.org/10.1016/j.wri.2021.100167

Xuan, L., Ju, Z., Skonieczna, M., Zhou, P.-K., & Huang, R. (2023). Nanoparticles-induced potential toxicity on human health: Applications, toxicity mechanisms, and evaluation models. *MedComm*, *4*(4), e327. https://doi.org/10.1002/mco2.327

Yang, W., Wang, L., Mettenbrink, E. M., Deangelis, P. L., & Wilhelm, S. (2021). Nanoparticle toxicology. *Annual Review of Pharmacology and Toxicology*, *61*, 269–289. https://doi.org/10.1146/ANNUREV-PHARMTOX-032320-110338

Yaqoob, A. A., Parveen, T., Umar, K., & Ibrahim, M. N. M. (2020). Role of nanomaterials in the treatment of wastewater: A review. *Water, 12*(2), 495. https://doi.org/10.3390/W12020495

Zahra, Z., Habib, Z., Chung, S., & Badshah, M. A. (2020). Exposure route of TiO2 NPs from industrial applications to wastewater treatment and their impacts on the agro-environment. *Nanomaterials (Basel, Switzerland), 10*(8). https://doi.org/10.3390/nano10081469

Zhou, Z., Xu, T., & Zhang, X. (2024). Empowerment of AI algorithms in biochemical sensors. *TrAC Trends in Analytical Chemistry, 173*, 117613. https://doi.org/10.1016/J.TRAC.2024.117613

Zodrow, K. R., Li, Q., Buono, R. M., Chen, W., Daigger, G., Dueñas-Osorio, L., Elimelech, M., Huang, X., Jiang, G., Kim, J. H., Logan, B. E., Sedlak, D. L., Westerhoff, P., & Alvarez, P. J. J. (2017). Advanced materials, technologies, and complex systems analyses: Emerging opportunities to enhance urban water security. *Environmental Science and Technology, 51*(18), 10274–10281. https://doi.org/10.1021/ACS.EST.7B01679

7

Eco-Friendly Alchemy: Harnessing Nature's Power for the Green Synthesis of Nanomaterials

Priya Kaushik, Ruchi Bharti, Renu Sharma, and Monika Verma

7.1 Introduction

In recent years, nanotechnology has gained global attention due to its applications in diverse fields. Nanomaterials are typically classified as substances ranging from 1 to 100 nm. The morphology and size of compounds play significant roles in defining their characteristics, even if their size is what defines them as a nanomaterial. Nano-sized materials are extensively utilized in many industries, including electronics, agriculture, and medicine. The technological progress in nanotechnology has enhanced the performance of materials, leading to notable improvements in their thermal conductivity, mechanical strength, and insulating properties [1, 2].

The synthesis of nano-sized materials can be generally categorized into two main approaches: top-down and bottom-up synthesis. The top-down approach focuses on breaking down larger substances into nanoscale particles, while the bottom-up approach constructs nanomaterials from individual atoms or molecules, as illustrated in Figure 7.1.

Nanomaterial synthesis falls under two categories: traditional and green approaches. Traditional synthesis methods provide various advantages, including synthesizing an extensive range of nanoparticles for various applications. Specific methods allow excellent scalability [3] and fine control over nanoparticle shape [4–6]. These advantages enable traditional methods suitable for employment in applications that include electrical systems, enhanced battery conductivity [7–11], target disease medication [12, 13], and storage of energy [14, 15].

Organic solvents are widely used to fabricate these nanomaterials, posing considerable risks to neurobehavioral function and reproductive health [16–18]. Moreover, the need for intense pressure and elevated temperatures can lead to potentially dangerous workplace environments [19–21].

Developing improved methodologies that align with the 12 principles of green chemistry is significant for addressing the climate catastrophe. Green synthesis is a sustainable, cost-effective, and dependable method for manufacturing nanomaterials. The increasing demand for sustainable and eco-friendly methods has enabled the field of nanoscience to focus on the study of biology and offer an innovative environment for environmentally friendly synthesis of nanomaterials using microbes, fungi, and

DOI: 10.1201/9781003632498-7

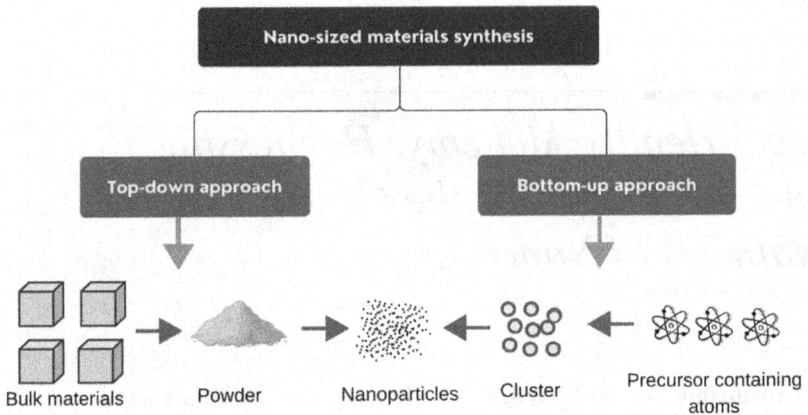

FIGURE 7.1 Top-down and bottom-up approach for synthesis of nano-sized materials.

plant extracts. Green synthesis produces nanomaterials using fungi, microorganisms, proteins, plant extract, enzymes, and sugars [22, 23].

This chapter presents an overview of green synthesis techniques for nanoparticles, focusing on the role of active molecules from microorganisms in controlling nanoparticle shape, size, and function. Understanding these biomolecules provides a sustainable approach to environmental challenges, which helps to create tailored nanomaterials for applications like wastewater treatment [24].

7.2 Resources for Green Synthesis

Various biological sources, including bacteria, yeast, fungi, algae, and plants, can be used for the synthesis of nanoparticles. Each of these systems needs methods to obtain effective nanoparticle production. These biological methods offer a green alternative to standard approaches, as shown in the following section:

7.2.1 Bacteria

Bacteria used in green nanomaterials fabrication belong to unicellular organisms having a cell wall structure but lacking organelles and nuclei. Although specific bacterial variants can be harmful, others exist naturally in the body and pose minimal risks to humans. Many of these variants, including *Escherichia coli* and *Bacillus subtilis*, are easily cultured and have adaptable genetics. Bacteria can synthesize nanoparticles due to their unique features. Bacteria are cultivated under aerobic conditions to an ideal optical density and mixed with nanoparticle precursors to synthesize nanomaterials. After incubation and a noticeable color shift, the solution is spun up at tremendous speeds (≥ 10000 rpm). The final product includes a suspension of nanomaterials [25]. The size of nanoparticles and its structure are influenced by bacteria strains and precursors, as illustrated in Table 7.1.

TABLE 7.1

Nanoparticles Synthesized by Bacteria

Nanoparticles	Species/Source	Active Molecules	Size	References
Ag	*Bacillus licheniformis*	NADPH-dependent nitrate reductase	50 nm	[26]
CdS	*Escherichia coli*	Glutathione	2–5 nm	[27]
CdS	*Rhodobacter sphaeroides*	Cysteine desulfhydrase	10.5 nm	[28]
ZnS	*Desulfobacteraceae*	Cysteine desulfhydrase	2–5 nm	[29]
MnO	*Bacillus* sp.	Cardiolipins	4.62 nm	[30]

7.2.2 Fungi

Nanoparticles employing fungi are widely known due to the affordable price and ease of preparing biomass, making it a practical nanoparticle production method [31]. The proteins and enzymes found within fungi serve as reducing agents, enabling simpler hydrolysis of metals. Nitrate-dependent reductants, hydrogenase, and other enzymes in fungi facilitate the bio-reduction process. Fungi produce proteins, organic acids, enzymes, and polysaccharides, which influence the size and structure of nanocrystals [32, 33].

Mukherjee et al. reported that the fungus *Verticillium*, which is the cause of Verticillium Wilt in crops, can produce silver nanoparticles on its cell walls by reducing silver nitrate in a solution [34]. Ahmad et al. and Gericke et al. found that Trichothecium's internal enzymes and *Verticillium luteoalbum* produce gold nanospheres and nanorods [35, 36]. Nanoparticles manufactured by fungi have a variety of applications, including medicine and optoelectronics, as illustrated in Table 7.2 [37, 38].

The amino acids in the cell wall of fungi act as capping and stabilizing agents. Furthermore, according to Phillip et al., nanoparticles manufactured using this approach are non-toxic when employed in medications, whereas those synthesized using traditional methods are not.

7.2.3 Yeast

Yeasts are single-celled fungus family members, like bacteria. Saccharomyces cerevisiae is a conventional and widely utilized yeast that converts carbohydrates into carbon dioxide and alcohol [42]. Various nanosystems can be synthesized using yeast cells as an alternative to bacteria. Yeast species have been successfully used to synthesize a variety of nanoparticles, including silver, gold, cadmium sulfide, lead sulfide, ferrous oxide, selenium, and antimony. Several research studies have successfully synthesized nanoparticles and nanomaterials employing yeast. Kowshik et al. demonstrated that MKY-3 yeast cells that are silver-tolerant can synthesize nanoparticles. The morphology and size of these nanoparticles varied according to the synthesis factors (pH, concentration of silver chloride, time, and temperature, etc.) [42]. According

TABLE 7.2

Nanoparticles Synthesized by Fungi

Nanoparticles	Species/Source	Active Molecules	Size	References
Au	*Trichothecium* Sp.	Enzymes	5–180 nm	[35]
Si	*Fusarium oxysporum*	Proteins	5–16 nm	[39]
Ag	*Coriolus versicolor*	Amine and cysteine amino residue	25–75 nm	[40]
ZrO$_2$	*Fusarium oxysporum*	Amide amino acids	4–11 nm	[41]

to this study, eliminating biochemical reducing agents resulted in an extracellular reduction of silver chloride.

7.2.4 Algae

Algae are a broad category of photosynthetic eukaryotes distinct from true plants. These chlorophyll-containing organisms may exist in unicellular or multicellular forms. However, they predominantly inhabit aquatic environments; they lack the characteristic morphological features of plants, such as genuine stems, leaves, and vascular tissues. Furthermore, they can impact humans, from medicinal species like Spirulina, which contains significant concentrations of natural nutrients [43]. However, Anabaena can be lethal if taken due to its toxins [44]. Singaravelu et al. proved that *Sargassum wightii* can synthesize stable gold nanoparticles. However, the process yields only a limited number of nanoparticles with a controlled size, which may restrict their applicability [45].

7.2.5 Plants

The green production of nanoparticles has emerged as a significant advancement over traditional techniques as it is easier, less expensive, and more reproducible as illustrated in Figure 7.2. This method also often produces nanoparticles with enhanced stability, making it an effective and sustainable option for nanomaterials. With the green synthesis technique, there is no reason for hazardous components, high pressure, energy consumption, or high temperatures. Researchers are, therefore, increasingly moving away from traditional synthetic methods. Particularly, plant-mediated synthesis provides an accessible approach for large-scale production and produces nanoparticles with enhanced stability compared to other techniques. Furthermore, there is a significantly reduced chance of contamination, increasing compatibility for various applications [46].

Plant extracts have naturally occurring compounds that act as capping, stabilizing, or reducing agents, making them a perfect substrate for nanoparticle production. The cost of production is reduced by employing this single-step procedure. Plant extracts

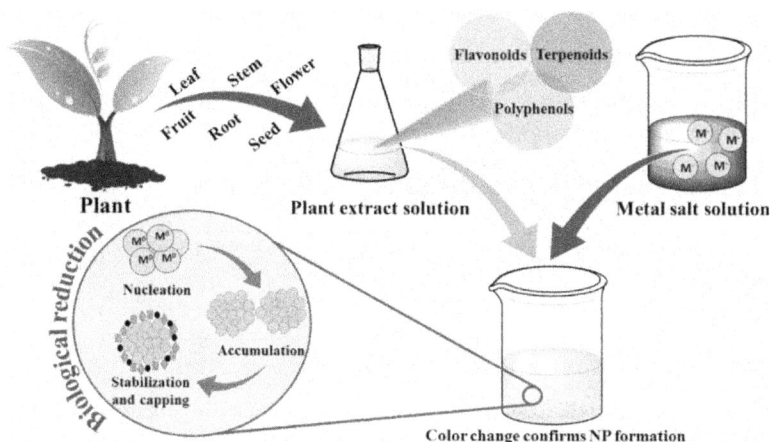

FIGURE 7.2 A schematic illustration of nanoparticle biosynthesis facilitated by plant-based synthesis.

[47]

TABLE 7.3

Nanoparticles Synthesized by Plants

Nanoparticles	Species/Source	Active Molecules	Size	References
Ag	*Allium cepa* L.	Flavonoids, glucosides	12 nm	[1]
Cu	*Tecoma castanifolia* leaf extract	ND (Not Determined)	100 nm	[54]
ZnO	α-amylase	ND	10–15 nm	[55]
Zn	*Aloe barbadensis*	ND	34 nm	[56]

can be derived from various parts of the plant, including leaves, stems, bark, and naturally occurring fruits [48]. This makes the entire synthesis process inexpensive. Phytomolecules such as phenolic chemicals, alkaloids, and flavonoids are present in plant extracts. Such phytomolecules derived from polyphenols effectively adsorb to the surface of nanoparticles and can operate as effective reducing agents. These phenolic chemicals have antioxidant properties and are found in the fleshy portion of plants [49–51]. The stabilizing and reducing agents in plant extracts enable the eco-friendly reduction of metal compounds into metallic nanoparticles. This process removes the need for poisonous chemical-reducing and capping agents. Metal nanoparticles, such as gold, silver, zinc, or iron, have been produced from plant extracts from leaves, fruits, and bark [52, 53]. To synthesize the silver and gold nanoparticles, a range of plants have been employed, including aloe vera (*Aloe barbadensis* Miller), tulsi (*Ocimum sanctum*), and mustard (*Brassica juncea*), as listed in Table 7.3.

7.3 Factors Affecting Synthesis

Although the exact method of producing nanoparticles is still unreported, many studies have been conducted worldwide. However, many study findings demonstrate that during the production of nanoparticles, biomolecules serve as capping or reducing agents, either separately or in combined form. A broad range of factors may affect the synthesis process, including pH, temperature, concentration of extract, and size of the particles. Some key factors that influence the biogenesis of nanoparticles are described below:

7.3.1 Specific Method Approach

Nanoparticles can be created using various processes, including mechanical processes and chemical or biological approaches. These approaches employ a variety of organic and inorganic compounds, as well as living organisms. Every approach offers its own advantages and disadvantages. Biological synthesis uses non-toxic, environmentally friendly, suitable materials combined with green technology, making it a more sustainable and preferred alternative to traditional methods [57, 58].

7.3.2 Temperature

It is an important factor and plays a vital role in nanoparticle production using all techniques. Physical approaches require the highest temperatures (above 350°C), while chemical methods operate at temperatures lower than this. Green synthesis, on the other hand, typically occurs at temperatures below 100°C, if not at ambient conditions. Temperature directly affects the properties and features of nanoparticles [59].

7.3.3 pH

pH has a significant impact on green technology nanoparticle synthesis. Researchers discovered that the pH of the solution significantly influences the size, shape, and surface texture of the synthesized nanoparticles [60, 61]. Modifying the pH of the solution is an effective way to control the size and morphology of nanoparticles during synthesis. Soni and Prakash demonstrated the effect pH has over the shape and dimension of the synthesized silver nanoparticle [62].

7.3.4 Time

The duration of time in the solution used for the reaction significantly impacts the amount and properties of nanoparticles manufactured with sustainable methods [63]. Similarly, their characteristics vary with time and are impacted by factors such as the manufacturing process involving light exposure and specific parameters [64]. Variations in storage time cause particle aggregation, shrinking, or growth, thereby compromising stability and efficacy. Furthermore, nanoparticles may have a short life, affecting their effectiveness [65].

7.3.5 Concentration of Metal Ions

The concentration of the metal ions in the reaction mixture influences the size of nanoparticles. Due to high metal ion concentrations and specific functional groups, nanoparticles agglomerate [66].

7.4 Characterization of Green-Synthesized Nanoparticles

Nanoparticle size, morphology, porous structure, and particle size distribution are all determined through characterization. UV–visible spectrophotometry, X-ray diffraction (XRD), scanning electron microscopy (SEM), transmission electron microscopy (TEM), energy dispersive X-ray spectroscopy, dynamic light scattering (DLS), Fourier transform infrared (FTIR) spectroscopy, and zeta potential analysis are the analytical techniques employed for characterization, which are illustrated as follows:

7.4.1 UV–Vis Spectroscopy

UV–vis is a straightforward but efficient technique for quantitative and qualitative classification of biochemical substances [67]. It is an approach to analyze the physicochemical properties of nanomaterials, including size, morphology, concentration, and agglomeration of particles [68]. Figure 7.3 illustrates the various parts of the UV–vis spectrophotometer, which include the light source and optical surface, monochromator, sample and reference interfaces, and detector. UV–vis spectroscopy characterizes materials using their optical characteristics, such as color, absorbance, transmittance, and reflection [69].

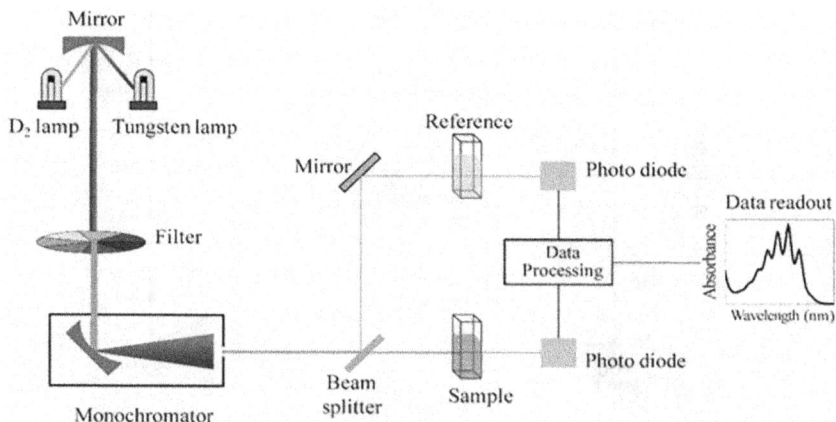

FIGURE 7.3   UV–vis spectrophotometer.

Source: [70]

The 200–600 nm wavelength range is generally accepted for confirmed nanoparticle fabrication. Metal nanoparticles such as silver (Ag), gold (Au), and copper (Cu) exhibit surface plasmon resonance (SPR), which is highly sensitive to the size, shape, and surrounding environment of the nanoparticles. The characteristics of surface plasmon resonance have been employed in biosensors, surface-enhanced Raman scattering, catalysis, and other fields. The optical and catalytic features of nanoparticles that are impacted by their composition and shape measure their performance characteristics [71]. Kouvaris et al. employed leaf extract from *Arbutus unedo* for their research on biosynthesis and characterization of silver nanoparticles (Ag NPs). The investigation demonstrated that the free electrons in metal nanoparticles produced plasmon resonance, which changed their color from colorless to reddish-yellow. Silver colloids exhibited a clear adsorption peak at 436 nm, and this shift was attributed to excitation effects [72]. A previous investigation by Singhal and Bhavesh demonstrated that adding *O. sanctum* leaf extract to a silver nitrate solution resulted in a shade shift from colorless to darker yellow. The conversion of $AgNO_3$ to silver nanoparticles can be observed by a peak at 413 nm in the UV spectrum [73]. A study by Parida et al. evaluated the UV–visible absorption characteristics of gold nanoparticles synthesized using *Allium cepa* (onion) extract. Gold nanoparticle stimulation exhibits surface plasmon resonance, which results in a strong resonance that can be detected at 540 nm, as illustrated in Figure 7.4 [74].

Silveira et al. reported another investigation into detecting and analyzing silver nanoparticles' antibacterial properties using an extract of *Ilex paraguariensis*. This investigation demonstrated that Ag formed in the extract with a maximal absorption of 460 nm and a blue shift to 440 nm [75].

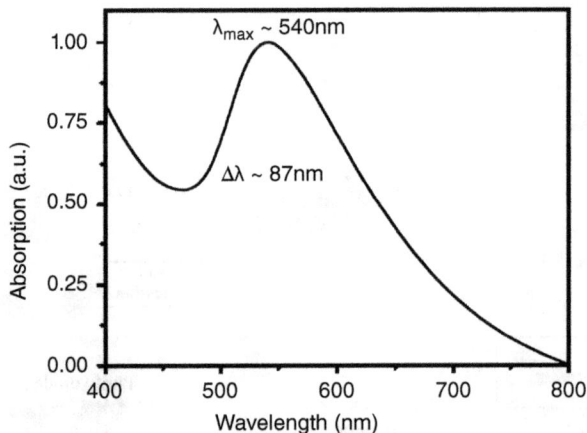

FIGURE 7.4   UV–visible absorption spectrum of gold nanoparticles synthesized using *Allium cepa* extract. A strong absorption peak is observed at 540 nm, corresponding to the excitation of surface plasmon resonance.

(Reproduced with permission) [74]

7.4.2 Fourier Transform Infrared (FTIR) Spectroscopy

FTIR detects functional groups by observing their peak regions in the spectra. This analysis can provide insights for nanoparticle capping and stability. FTIR spectroscopy demonstrates the adsorption, reflection, and emission of IR light through a substance [76]. It is significantly more sensitive and provides faster analysis than the other spectroscopic methods. Figure 7.5 illustrates that the interferometer and optical system are the fundamental elements of FTIR.

Banerjee et al. reported employing the FTIR technique to describe silver nanoparticles generated from three Indian medicinal herbs: *O. tenuiflorum* (black tulsi), *A. indica* (neem), and *M. balbisiana* (banana) [78].

Various absorbance bands have been observed at around 1025, 1074, 1320, 1382, 1610, and 2262 cm^{-1} in all three cases. The study shows that these peaks are geminal methyl groups, -C=C- moieties from aromatic rings, -C-O-C- ether linkages, -C-O- bonds, and alkyne bonds, respectively. It was proposed that the efficient capping and stabilization of the synthesized silver nanoparticles depend heavily on these functional groups. Sadeghi and Gholamhoseinpoor observed significant variations in the FTIR spectra of *Ziziphora tenuior* (Zt) extract of leaves before and after the bioreduction process. Structural analysis shows that silver nanoparticles are reduced and stabilized by associating with nitrogen atoms.

7.4.3 X-Ray Diffraction (XRD)

It is a highly regarded non-destructive analysis technique for investigating crystal structure and atomic spacing of crystalline materials. Moreover, it gives significant

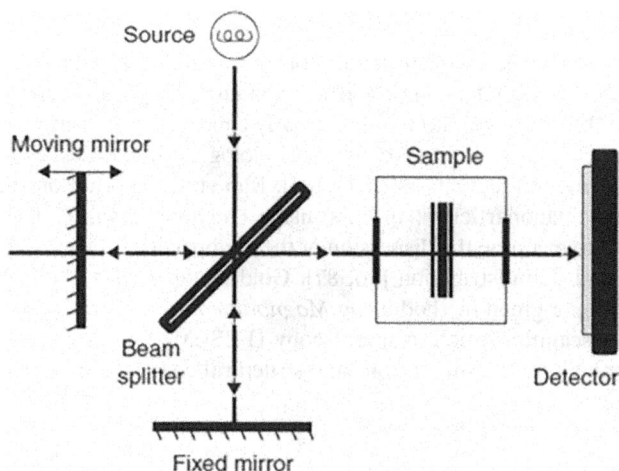

FIGURE 7.5 FTIR spectroscopy.

Source: [77].

TABLE 7.4

Applications of XRD

Applications of XRD	Peak position	Pattern indexing
		Crystal phase analysis
		Unit cell dimensions
	Intensity	Reaction kinetics
		Phase abundance
		Structural analysis
	Shape	Crystallite size
		Particle size distribution
		Crystalline growth kinetics

information regarding other structural properties, such as crystal phases, orientations, crystallinity, and crystal defects [79]. The various applications of XRD are shown in Table 7.4. The crystallinity of the silver nanoparticles (Ag NPs) is determined by XRD, and its peak analysis verified that the primary component of the nanoparticles was silver, and no contaminants were detectable [80]. Likewise, XRD diagrams for iron oxide powder have been investigated to evaluate its crystallinity and possible biomedical applications [81]. Research on the extracellular production of Ag NPs employing the fungus *Fusarium oxysporum* has been reported, and XRD was employed to characterize the produced nanoparticles. Table 7.4 outlines the key applications of X-ray diffraction (XRD), including phase identification, structural analysis, and crystallite size, strain, and purity assessment in nanomaterials [82].

7.4.4 Scanning Electron Microscopy

It is widely used for the characterization of green nanomaterials. It gives detailed insights into their morphological properties, including shape, size, and surface composition [83, 84]. A skinny coating of electrically conducting substances, such as gold, is essential for characterizing non-conductive biological molecules, as they generate insufficient signals for accurate imaging [85]. Rao et al. reported on the green production of silver nanoparticles (Ag NPs) using *Ocimum sanctum*. The SEM study was performed to examine the dispersion of the nanoparticles, providing information on their size and size distribution [86, 87]. Gold nanoparticles (Au NPs) were also synthesized using a green method using *Magnolia kobus* and *Diospyros kaki* plants. Field-emission scanning electron microscopy (FESEM) was applied to investigate synthesized nanoparticles' dimensions and structural characteristics at varying reaction temperatures.

7.4.5 Transmission Electron Microscopy

It is a widely employed electron microscopy technique for the detailed characterization of nanomaterials. TEM is a more advanced technique that produces a direct image with excellent spatial resolution at the atomic scale (<1 nm). Additionally, it provides valuable information about the chemical composition of the sample [88]. TEM is an essential analytical technique for characterizing nanomaterials' morphological and compositional properties [89]. When coupled with energy-dispersive X-ray spectroscopy (EDS), both SEM and TEM facilitate elemental analysis of these materials. Both techniques enable the identification of nanomaterial size, morphology, and aggregation. TEM offers a significant advantage over SEM. It provides superior spatial resolution and detailed insight into the internal structural features of the sample [90]. Sheny et al. investigated green synthesis of palladium nanoparticles using the dried leaf extract of *Anacardium occidentale* [91]. TEM was used to analyze the morphology of the produced nanoparticles and assess their degree of aggregation. Additionally, high-resolution TEM (HRTEM) was utilized to examine the crystalline structure of nickel and nickel oxide nanoparticles [92, 93].

7.4.6 Particle Size and Zeta Potential

A zeta potential instrument is employed to analyze the surface charge of nanoparticles. The zeta potential is crucial for evaluating the behavior of dispersed particles in liquid systems, particularly in determining significant colloidal stability and analyzing particle deposition in a water-cooling system. Colloidal stability is primarily influenced by the surface charge of nanoparticles and their distribution within the surrounding environment. The properties and behavior of a given pH in a solution and the presence of an electrolyte can be characterized. The electrical potential of the interfacial surface is measured to determine this quantitatively. A high zeta potential value shows that nanoparticles (NPs) are stable in dispersion due to electrostatic interactions. This parameter is crucial for calculating the size distributions of nanoparticles, with dynamic light scattering (DLS) being the widely employed method for the size distribution of nanoparticles [94].

7.5 Applications of Green Nanoparticles

Nanoparticles are currently in great demand on the commercial scale because of their diverse applications in electronics, industries, the environment, agriculture, and the biomedical industry. The most commonly used nanoparticles in this field, Ag and Au NPs, have been extensively studied and are highly beneficial for biological applications. Some of the important applications are discussed as follows:

7.5.1 Biomedical Applications

Nanoparticles synthesized using green synthesis have numerous medical applications. Research demonstrates that numerous synthesized green nanoparticles have been utilized based on their antibacterial, anticancer, and drug delivery systems in both therapeutic and diagnostic uses. Furthermore, the applications of green nanoparticles are covered under the following sections:

7.5.1.1 Drug Delivery Systems

An effective drug dose is the one that can be delivered to a specific target site. However, the development of advanced drug delivery systems can achieve the distribution of the therapeutic molecules with enhanced efficacy in a specified period. Drug delivery methods might be facilitated by nanoparticles, especially when paired with other biomolecules. This approach enhances the effectiveness of drug delivery, ensuring that the drug reaches the body and functions effectively [95]. Among various metals, gold (Au) nanoparticles are generally considered non-immunogenic and non-toxic, making them safe and suitable for biological applications. Additionally, they can be easily functionalized, allowing for further modifications to suit specific medical purposes. These unique properties make Au nanoparticles highly suitable for developing advanced sites and transport for medication delivery systems. Aubin-Tam and Hamad-Schifferli recognized this potential and created a medication delivery method that combines Au nanoparticles with infrared light. This method might properly transport and effectively release different drug molecule doses. The effectiveness of this approach was due to the ability of Au nanoparticles to have various shapes to respond to varying wavelengths of infrared radiation, enabling controlled and targeted drug release [96].

Various nanostructures have been synthesized for drug delivery applications. Nanotechnology has contributed to the medical industry by enabling the applications of nanoparticles in medication delivery [97]. When supplied at the appropriate place and dose, nanoparticles can deliver medicine to targeted cells. There is a significant decrease in adverse effects and overall medication use [98]. This approach is less costly and has fewer adverse effects.

7.5.1.2 Antimicrobial Activities

Antibiotic resistance has become a primary concern worldwide, and the problem is expected to worsen over time. This resistance has developed because bacteria can rapidly evolve and adapt their genetic material, making them less affected by antimicrobial agents. As a result, finding new and effective treatments has become a key focus of research. Green-synthesized nanoparticles have shown promising results in combining results in countering multidrug-resistant bacteria and could offer an alternative in the fight against these resistant infections. Their ability to target and eliminate harmful bacteria suggests they play a crucial part in future antimicrobial therapies [99]. Nanoparticles (NPs) and other combinations have been mixed with organic and inorganic compounds to enhance antimicrobial activity. In recent studies, metal ions and metal-based compounds, including gold nanoparticles (Au NPs), selenium nanoparticles (Se NPs), silver nanoparticles (Ag NPs), copper oxide nanoparticles (CuO NPs), and zinc oxide nanoparticles, are effective as antimicrobial coatings. The effective antimicrobial properties of silver (Ag) against numerous bacterial types are widely recognized. Green-manufactured silver nanoparticles (Ag NPs) from *Carrissa carandas* leaf extract have demonstrated notable antibacterial action against several human pathogenic microorganisms, including Shigella flexneri, which causes shigellosis and is more affected by inhibition [100].

7.5.2 Environmental Applications

Green-synthesized nanoparticles provide environmentally friendly solutions to environmental applications such as wastewater treatment, pollutant elimination, and soil cleansing. They are biocompatible and less toxic, making them suitable for sustainable environmental management.

7.5.2.1 Water Purification

Access to clean water is more important than ever due to rising pollution levels and a growing world population. Since 70% of the human body is made of water, it is necessary that we clean up the contaminated water sources. Water contamination is mostly caused by industrial effluents, domestic sewage, marine waste, and agricultural fertilizers and pesticides. Nanoparticles can act as water purifying agents, which can help to reduce or completely eradicate pollutants from the freshwater. Nanoparticles could be used for wastewater treatment by adsorption, nanofiltration, disinfection, and photocatalysis. Titanium oxide nanoparticles (TiO_2 NPs), iron oxide nanoparticles (Fe_2O_3 and Fe_3O_4 NPs), and CeO_2 nanostructures are studied to be able to remove heavy metal atoms like As and Cr [101]. Nanomaterials have a very tiny diameter and enormous dimension, as well as outstanding adsorption qualities and strong reduction capacity; these distinctive characteristics help to remove toxins from wastewater [102]. TiO_2 nanocomposites incorporating mesoporous silica have demonstrated the ability to effectively remove aromatic pollutants from wastewater [103]. A substantial method for eliminating contaminants from water bodies will be required in the future; nanotechnology is meeting this demand with an improved technique that guarantees the supply of potable water of superior quality.

7.5.2.2 Wastewater Treatment Process

The rapid growth of the global population, increasing industrial activities, and the usage of synthetic chemicals have significantly contaminated aquatic habitats through untreated wastewater discharge. Consequently, natural water sources may be unsafe for human consumption, because they contain many contaminants. These include organic contaminants such as pharmaceuticals, dyes, and pesticides; inorganic compounds like fluoride, arsenic, copper, and mercury; and biological contaminants such as algae and bacteria [104, 105]. Various physical, chemical, and biological methods can be employed for wastewater treatment. However, ongoing research is focused on developing more efficient and cost-effective purification technologies

Nanotechnology has also been recognized as a potential solution, providing high-efficiency pollutant removal from wastewater. Many strategies combining various nanoparticles have been investigated to increase contaminant removal, as shown in Figure 7.6.

Adsorption is an effective technique for treating wastewater because it is relatively inexpensive, simple to use, and produces only a few secondary contaminants [106]. As nanotechnology has developed, several nanostructured materials have become effective adsorbents for cleaning drinking water, surface water, and pollutants from industrial [107, 108]. The tiny particle size, high porosity, and huge surface area of nanoadsorbents allow them to have a higher adsorption capacity and quicker kinetics than traditional adsorbents [109]. Additionally, they have strong catalytic activity and chemical reactivity. Carbon nanotubes, graphene, and other metal oxides (e.g., Fe_3O_4, TiO_2, ZnO, MgO, MnO_2, and CeO_2) have shown promise in wastewater purification technologies based on their ability to effectively remove contaminants like dyes and

FIGURE 7.6 Nanoparticle-based methods applied for wastewater treatment.

heavy metals [110]. Das et al. successfully removed Cu (II) and Co (II) ions and methylene blue dye from water-based solutions using biosynthesized magnetite nanoparticles derived by the crude latex of *Jatropha curcas* and *Cinnamomum tamala* leaf extract [111].

7.5.3 Food Industry

When combined with the latest advances, nanoparticles significantly improve food production, storage, packaging, and transportation. Nanotechnology plays an important part in sustaining functional characteristics through the use of colloids, emulsions, and biopolymer solutions based on nanoparticles. It offers new opportunities for developing nanostructures that improve food ingredients and enable sensor-based monitoring in the food industry. Nanostructures are employed as antimicrobials, nanocarriers, nanoadditives, and nanocomposites in food production and as nanodetectors for quality control [112]. When compared to traditional materials, the high surface-to-volume ratio of nanomaterials allows them to be efficient enzyme supports. Nano-carriers allow food additives to be delivered precisely without changing their physicochemical characteristics. For bioactive delivery, particle size is essential as certain types of cells are not necessarily able to absorb microparticles. By increasing surface area, encapsulation with nanocomposites or nanoemulsions increases bioavailability. Challenges concerning the performance and toxicity of nanomaterials must be resolved to advance their development and applications. In the food sector, it is crucial to have effective regulations for their production, usage, and disposal. Furthermore, raising public knowledge and acceptance of nano-enabled food and agricultural goods is important.

7.5.4 Agriculture

Nanotechnology is bringing significant changes to agriculture and food production. It has the ability to greatly transform modern farming practices by improving efficiency and sustainability [113, 114].

In conventional farming, a large proportion of agrochemicals used on crops fail to reach their intended target. This loss occurs due to various factors, including leaching into soil, breakdown by sunlight (photolysis), drifting away from the target area, chemical decomposition in water (hydrolysis), and degradation by microorganisms. By incorporating nanotechnology, these challenges can be addressed, leading to more effective and precise agricultural practices [115, 116].

When agricultural pathogens are specially targeted, the antimicrobial properties discussed above could play an essential role in protecting crops. Zinc oxide nanoparticles (ZnO NPs) have attained significant popularity in agriculture, owing to their capability to counter plant diseases. Their antibacterial properties have been demonstrated by ZnO NPs produced from lemon fruit, which show effectiveness despite the soft rot bacterial pathogen *Dickeya dadantii*. Additionally, their antifungal activity has been observed in ZnO NPs synthesized using Eucalyptus globulus extract, which exhibited strong fungicidal effects against major fungal pathogens affecting apple or chards [117, 118].

7.6 Major Challenges and Future Directions

Current developments in nanoparticle investigation have significantly expanded their potential applications. Extensive research work has been conducted on the green synthesis of metallic nanoparticles using biological sources, including plants, bacteria, fungi, and yeast. However, large-scale production remains constrained by several challenges, which are outlined as follows:

i. Optimization of reactants (e.g., plant extracts, microbial inocula, fermentation media) and process conditions (e.g., temperature, pH, agitation) is essential for precise control over nanoparticle size and morphology.
ii. For specialized applications, research should concentrate on improving the physicochemical characteristics of nanoparticles.
iii. The function of metabolites produced from plants and microbial cellular components in the creation of nanoparticles has to be further investigated.
iv. Scaling up green synthesis methods for commercial nanoparticle production must be prioritized.
v. Optimizing process parameters is required to maximize nanoparticle production and stability while reducing reaction time.

To advance the future development of green-synthesized nanoparticles (NPs), it is essential to overcome current challenges to enable economic and efficient large-scale manufacturing, comparable to conventional techniques. The process of extraction and cleaning of nanoparticles from the reaction mixture is a crucial process, ensuring product quality, and is a significant field that needs more research. Furthermore, thorough toxicity evaluations of plants and animals are required to enable the safe use of these nanoparticles in a variety of industries. Although the majority of research so far has been conducted in laboratories on a small scale, growing endeavors are being made to explore the broader potential of green-synthesized nanoparticles. Researchers are focusing on increasing their application in agriculture, environmental management, and other fields to satisfy the growing requirements of the worldwide community and to enhance the well-being of humans [119].

7.7 Conclusion

Green synthesis offers an environmentally friendly, non-toxic substitute for traditional physical and chemical techniques for metal nanoparticle (NP) production. Various plant-derived materials, including leaf, fruit, seed, and bark extracts, along with microorganisms like actinomycetes, fungi, and bacteria, have demonstrated efficacy in producing metal and metal oxide nanoparticles (NPs) such as Au, Ag, Pt, Pd, Ni, Se, Cu, CuO, and TiO_2. Nanoparticle size, morphology, and reaction kinetics are influenced by factors such as temperature, aeration, pH, concentration, reaction time, and salt concentration. Advanced characterization methods, including UV–vis spectroscopy, FTIR, XRD, SEM, TEM, EDX, and AFM, are employed to analyze structural and morphological properties of biosynthesized NPs. Due to their cost efficiency, non-toxicity, accessibility, and environmental compatibility, plant extract-mediated nanoparticle synthesis has been extensively investigated for diverse industrial applications. Plant-based green nanoparticle synthesis is a promising advancement in nanotechnology, promoting environmental sustainability and long-term progress in nanoscience. These nanoparticles demonstrate diverse applications across various industries such as biomedical (medicine, cosmetics, biotechnology, etc.), electronics, optics, textiles, environmental applications (water treatment, dye degradation), and agriculture. Notably, they hold significant potential for revolutionizing drug delivery systems in biomedical sciences.

REFERENCES

[1] Yap, Y.H., Azmi, A.A., Mohd, N.K., Yong, F.S.J., Kan, S.Y., Thirmizir, M.Z.A., & Chia, P.W. (2020). Green synthesis of silver nanoparticle using water extract of onion peel and application in the acetylation reaction. *Arabian Journal for Science and Engineering*, 45, 4797–4807.

[2] National Nanotechnology Initiative. (2021). Benefits and applications of nanotechnology. Retrieved August 17, 2021, from www.nano.gov/you/nanotechnology-benefits

[3] Wegner, K., Schimmöller, B., Thiebaut, B., Fernandez, C., & Rao, T.N. (2011). Pilot plants for industrial nanoparticle production by flame spray pyrolysis. *KONA Powder and Particle Journal*, 29, 251–265.

[4] Ion, J.C. (2006). *Laser processing of engineering materials: Principal procedure and industrial application.* Elsevier.

[5] Zeng, H., Du, X.-W., Singh, S.C., Kulinich, S.A., Yang, S., He, J., & Cai, W. (2012). Nanomaterials via laser ablation/irradiation in liquid: A review. *Advanced Functional Materials*, 22, 1333–1353.

[6] Amendola, V., & Meneghetti, M. (2013). What controls the composition and the structure of nanomaterials generated by laser ablation in liquid solution? *Physical Chemistry Chemical Physics*, 15, 3027–3046.

[7] Kubota, K., Dahbi, M., Hosaka, T., Kumakura, S., & Komaba, S. (2018). Towards K-ion and Na-ion batteries as "beyond Li-ion." *The Chemical Record*, 18, 459–479.

[8] Su, D., Ahn, H.-J., & Wang, G. (2013). Hydrothermal synthesis of α-MnO2 and β-MnO2 nanorods as high-capacity cathode materials for sodium-ion batteries. *Journal of Materials Chemistry A*, 1, 4845–4850.

[9] Hosono, E., Saito, T., Hoshino, J., Okubo, M., Saito, Y., Nishio-Hamane, D., Kudo, T., & Zhou, H. (2012). High power Na-ion rechargeable battery with single-crystalline Na0.44MnO2 nanowire electrode. *Journal of Power Sources*, 217, 43–46.

[10] Song, H. (2018). One-step convenient hydrothermal synthesis of MoS2/RGO as a high-performance anode for sodium-ion batteries. *International Journal of Electrochemical Science*, 13, 4720–4730.

[11] Lin, B., Zhu, X., Fang, L., Liu, X., Li, S., Zhai, T., Xue, L., Guo, Q., Xu, J., & Xia, H. (2019). Birnessite nanosheet arrays with high K content as a high-capacity and ultrastable cathode for K-ion batteries. *Advanced Materials*, 31, 1900060.

[12] Walter, J.G., Petersen, S., Stahl, F., Scheper, T., & Barcikowski, S. (2010). Laser ablation-based one-step generation and bio-functionalization of gold nanoparticles conjugated with aptamers. *Journal of Nanobiotechnology*, 8, 21.

[13] Salmaso, S., Caliceti, P., Amendola, V., Meneghetti, M., Magnusson, J.P., Pasparakis, G., & Alexander, C. (2009). Cell uptake control of gold nanoparticles functionalized with a thermoresponsive polymer. *Journal of Materials Chemistry*, 19, 1608–1615.

[14] Leng, J., Wang, Z., Wang, J., Wu, H.-H., Yan, G., Li, X., Guo, H., Liu, Y., Zhang, Q., & Guo, Z. (2019). Advances in nanostructures fabricated via spray pyrolysis and their applications in energy storage and conversion. *Chemical Society Reviews*, 48, 3015–3072.

[15] Aboulouard, A., Gultekin, B., Can, M., Erol, M., Jouaiti, A., Elhadadi, B., Zafer, C., Demic, S. (2020). Dye sensitized solar cells based on titanium dioxide nanoparticles synthesized by flame spray pyrolysis and hydrothermal sol-gel methods: A comparative study on photovoltaic performances. *Journal of Materials Research and Technology*, 9, 1569–1577.

[16] Joshi, D., & Adhikari, N. (2019). An overview on common organic solvents and their toxicity. *Journal of Pharmaceutical Research International*, 28, 1–18.

[17] Tobiszewski, M., Namieśnik, J., & Pena-Pereira, F. (2017). Environmental risk-based ranking of solvents using the combination of a multimedia model and multi-criteria decision analysis. *Green Chemistry*, 19, 1034–1042.

[18] Akinyemi, P.A., Adegbenro, C.A., Ojo, T.O., & Elugbaju, O. (2019). Neurobehavioral effects of organic solvents exposure among wood furniture makers in Ile-Ife, Osun State, Southwestern Nigeria. *Journal of Health Pollution*, 9, 190604.

[19] Mueller, R., Jossen, R., Pratsinis, S.E., Watson, M., & Akhtar, M.K. (2004). Zirconia nanoparticles made in spray flames at high production rates. *Journal of the American Ceramic Society*, 87, 197–202.

[20] Strobel, R., Baiker, A., & Pratsinis, S.E. (2006). Aerosol flame synthesis of catalysts. *Advances in Powder Technology*, 17, 457–480.

[21] Teoh, W.Y., Amal, R., & Mädler, L. (2010). Flame spray pyrolysis: An enabling technology for nanoparticles design and fabrication. *Nanoscale*, 2, 1324–1347.

[22] Devatha, C.P., & Thalla, A.K. (2018). Green synthesis of nanomaterials. In *Synthesis of inorganic nanomaterials* (pp. 169–184). Woodhead Publishing.

[23] Jegadeeswaran, P., Shivaraj, R., & Venckatesh, R. (2012). Green synthesis of silver nanoparticles from extract of *Padina tetrastromatica* leaf. *Digest Journal of Nanomaterials and Biostructures*, 7(3), 991–998.

[24] Huston, M., DeBella, M., DiBella, M., & Gupta, A. (2021). Green synthesis of nanomaterials. *Nanomaterials*, 11(8), 2130.

[25] Gurunathan, S., Kalishwaralal, K., Vaidyanathan, R., Venkataraman, D., Pandian, S.R.K., Muniyandi, J., Hariharan, N., & Eom, S.H. (2009). Biosynthesis, purification and characterization of silver nanoparticles using *Escherichia coli*. *Colloids and Surfaces B: Biointerfaces*, 74, 328–335.

[26] Kalimuthu, K., Suresh Babu, R., Venkataraman, D., Bilal, M., & Gurunathan, S. (2008). Biosynthesis of silver nanocrystals by *Bacillus licheniformis*. *Colloids and Surfaces B: Biointerfaces*, 65, 150–153.

[27] Sweeney, R.Y., Mao, C., Gao, X., Burt, J.L., Belcher, A.M., Georgiou, G., Iverson, B.L. (2004). Bacterial biosynthesis of cadmium sulfide nanocrystals. *Chemistry & Biology*, 11, 1553–1559.

[28] Bai, H.J., Zhang, Z.M., Guo, Y., & Yang, G.E. (2009). Biosynthesis of cadmium sulfide nanoparticles by photosynthetic bacteria *Rhodopseudomonas palustris*. *Colloids and Surfaces B: Biointerfaces*, 70, 142–146.

[29] Labrenz, M., Druschel, G.K., Thomsen-Ebert, T., Gilbert, B., Welch, S.A., Kemner, K.M., Logan, G.A., Summons, R.E., Stasio, G.D., Bond, P.L., & Lai, B. (2000). Formation of sphalerite (ZnS) deposits in natural biofilms of sulfate-reducing bacteria. *Science*, 290, 1744–1747.

[30] Sinha, R., Karan, R., Sinha, A., & Khare, S.K. (2011). Interaction and nanotoxic effect of ZnO and Ag nanoparticles on mesophilic and halophilic bacterial cells. *Bioresource Technology*, 102, 1516–1520.

[31] Thakkar, K.N., Mhatre, S.S., & Parikh, R.Y. (2010). Biological synthesis of metallic nanoparticles. *Nanomedicine*, 6(2), 257–262.

[32] Boroumand Moghaddam, A., Namvar, F., Moniri, M., Md. Tahir, P., Azizi, S., & Mohamad, R. (2015). Nanoparticles biosynthesized by fungi and yeast: A review of their preparation, properties, and medical applications. *Molecules*, 20(9), 16540–16565.

[33] Siddiqi, K.S., & Husen, A. (2016). Fabrication of metal nanoparticles from fungi and metal salts: Scope and application. *Journal of Nanoscale Research Letters*, 11(1), 1311–1312.

[34] Mukherjee, P., Ahmad, A., Mandal, D., Senapati, S., Sainkar, S.R., Khan, M.I., Parishcha, R., Ajaykumar, P.V., Alam, M., Kumar, R., & Sastry, M. (2001). Fungus-mediated synthesis of silver nanoparticles and their immobilization in the mycelial matrix: A novel biological approach to nanoparticle synthesis. *Nano Letters*, 1, 515–519.

[35] Ahmad, A., Senapati, S., Khan, M., Kumar, R., & Sastry, M. (2005). Extra-/intracellular biosynthesis of gold nanoparticles by an alkalotolerant fungus, *Trichothecium* sp. *Journal of Biomedicine and Nanotechnology*, 1, 47–53.

[36] Gericke, M., & Pinches, A. (2006). Microbial production of gold nanoparticles. *Gold Bulletin*, 39, 22–28.

[37] Birla, S.S., Tiwari, V.V., Gade, A.K., Ingle, A.P., Yadav, A.P., & Rai, M.K. (2009). Fabrication of silver nanoparticles by *Phoma glomerata* and its combined effect against *Escherichia coli*, *Pseudomonas aeruginosa*, and *Staphylococcus aureus*. *Letters in Applied Microbiology*, 48, 173–179.

[38] Mukherjee, P., Senapati, S., Mandal, D., Ahmad, A., Khan, M.I., Kumar, R., Sastry, M. (2002). Extracellular synthesis of gold nanoparticles by the fungus *Fusarium oxysporum*. *ChemBioChem*, 3, 461–463.

[39] Bansal, V., Rautaray, D., Bharde, A., Ahire, K., Sanyal, A., Ahmad, A., & Sastry, M. (2005). Fungus-mediated biosynthesis of silica and titania particles. *Journal of Materials Chemistry*, 15, 2583–2589.

[40] Sanghi, R., & Verma, P. (2009). Biomimetic synthesis and characterization of protein-capped silver nanoparticles. *Bioresource Technology*, 100, 501–504.

[41] Bansal, V., Rautaray, D., Ahmad, A., & Sastry, M. (2004). Biosynthesis of zirconia nanoparticles using the fungus Fusarium oxysporum. *Journal of Materials Chemistry*, 14, 3303–3305.

[42] Kowshik, M., Ashtaputre, S., Kharrazi, S., Vogel, W., Urban, J., Kulkarni, S.K., & Paknikar, K.M. (2002). Extracellular synthesis of silver nanoparticles by a silver-tolerant yeast strain MKY3. *Nanotechnology*, 14, 95–100.

[43] Khan, Z., Bhadouria, P., & Bisen, P.S. (2005). Nutritional and therapeutic potential of Spirulina. *Current Pharmaceutical Biotechnology*, 6, 373–379.

[44] Carmichael, W.W., Mahmood, N.A., & Hyde, E.G. (1990). Natural toxins from cyanobacteria (blue-green algae). In *Marine toxins* (pp. 87–106). American Chemical Society.

[45] Singaravelu, G., Arockiamary, J.S., Kumar, V.G., & Govindaraju, K. (2007). A novel extracellular synthesis of monodisperse gold nanoparticles using marine alga, *Sargassum wightii* Greville. *Colloids and Surfaces B: Biointerfaces*, 57(1), 97–101.

[46] Bhagyaraj, S., Oluwafemi, O.S., Kalarikkal, N., & Thomas, S. (Eds.). (2018). *Synthesis of inorganic nanomaterials: Advances and key technologies*. Elsevier.

[47] Dikshit, P.K., Kumar, J., Das, A.K., Sadhu, S., Sharma, S., Singh, S., Gupta, P.K., & Kim, B.S. (2021). Green synthesis of metallic nanoparticles: Applications and limitations. *Catalysts*, 11(8), 902.

[48] Sharma, D., Kanchi, S., & Bisetty, K. (2019). Biogenic synthesis of nanoparticles: A review. *Arabian Journal of Chemistry*, 12(8), 3576–3600.

[49] Jayandran, M., Haneefa, M., & Balasubramanian, V. (2015). Green synthesis and characterization of manganese nanoparticles using natural plant extracts and its evaluation of antimicrobial activity. *Journal of Applied Pharmaceutical Science*, 5(1), 105–110.

[50] Thunugunta, T., Reddy, A.C., & Lakshman Reddy, D.C. (2015). Green synthesis of nanoparticles: Current prospectus. *Nanotechnology Reviews*, 4(4), 303–323.

[51] Herlekar, M., Barve, S., & Kumar, R. (2014). Plant-mediated green synthesis of iron nanoparticles. *Journal of Nanoparticles*, 2014(1), 140614.

[52] Amin, M., Anwar, F., Janjua, M.R.S.A., Iqbal, M.A., & Rashid, U. (2012). Green synthesis of silver nanoparticles through reduction with *Solanum xanthocarpum* L. berry extract: Characterization, antimicrobial, and urease inhibitory activities against *Helicobacter pylori*. *International Journal of Molecular Sciences*, 13(8), 9923–9941.

[53] Krithiga, N., Rajalakshmi, A., & Jayachitra, A. (2015). Green synthesis of silver nanoparticles using leaf extracts of *Clitoria ternatea* and *Solanum nigrum* and study of its antibacterial effect against common nosocomial pathogens. *Journal of Nanoscience*, 2015(1), 928204.

[54] Sharmila, G., Thirumarimurugan, M., & Sivakumar, V.M. (2016). Optical, catalytic and antibacterial properties of phytofabricated CuO nanoparticles using *Tecoma castanifolia* leaf extract. *Optik*, 127(19), 7822–7828.

[55] Mazumder, J.A., Khan, E., Perwez, M., Gupta, M., Kumar, S., Raza, K., & Sardar, M. (2020). Exposure of biosynthesized nanoscale ZnO to Brassica juncea crop plant: Morphological, biochemical and molecular aspects. *Scientific Reports*, 10, 8531.

[56] Sangeetha, G., Rajeshwari, S., & Venckatesh, R. (2011). Green synthesis of zinc oxide nanoparticles by *Aloe barbadensis* miller leaf extract: Structure and optical properties. *Materials Research Bulletin*, 46(12), 2560–2566.

[57] Kharissova, O.V., Dias, H.V.R., Kharisov, B.I., Perez, B.O., & Perez, V.M.J. (2013). The greener synthesis of nanoparticles. *Trends in Biotechnology*, 31(4), 240–248.

[58] Vadlapudi, V., & Kaladhar, D.S.V.G.K. (2014). Review: Green synthesis of silver and gold nanoparticles. *Middle-East Journal of Scientific Research*, 19(6), 834–842.

[59] Rai, A., Singh, A., Ahmad, A., & Sastry, M. (2006). Role of halide ions and temperature on the morphology of biologically synthesized gold nanotriangles. *Langmuir*, 22(2), 736–741.

[60] Gardea-Torresdey, J.L., Tiemann, K.J., Gamez, G., Dokken, K., & Pingitore, N.E. (1999). Recovery of gold (III) by alfalfa biomass and binding characterization using X-ray microfluorescence. *Advances in Environmental Research*, 3(1), 83–93.

[61] Armendariz, V., Herrera, I., Peralta-Videa, J.R., Jose-Yacaman, M., Troiani, H., Santiago, P., & Gardea-Torresdey, J.L. (2004). Size controlled gold nanoparticle formation by *Avena sativa* biomass: Use of plants in nanobiotechnology. *Journal of Nanoparticle Research*, 6, 377–382.

[62] Soni, N., & Prakash, S. (2011). Factors affecting the geometry of silver nanoparticles synthesis in *Chrysosporium tropicum* and *Fusarium oxysporum*. *American Journal of Nanotechnology*, 2(1), 112–121.

[63] Darroudi, M., Ahmad, M.B., Zamiri, R., Zak, A.K., Abdullah, A.H., & Ibrahim, N.A. (2011). Time-dependent effect in green synthesis of silver nanoparticles. *International Journal of Nanomedicine*, 6(1), 677–681.

[64] Kuchibhatla, S.V., Karakoti, A.S., Baer, D.R., Samudrala, S., Engelhard, M.H., Amonette, J.E., Thevuthasan, S., & Seal, S., (2012). Influence of aging and environment on nanoparticle chemistry: Implication to confinement effects in nanoceria. *Journal of Physical Chemistry C*, 116(26), 14108–14114.

[65] Mudunkotuwa, I.A., Pettibone, J.M., & Grassian, V.H. (2012). Environmental implications of nanoparticle aging in the processing and fate of copper-based nanomaterials. *Environmental Science & Technology*, 46(13), 7001–7010.

[66] Dubey, S.P., Lahtinen, M., & Sillanpää, M. (2010). Tansy fruit mediated greener synthesis of silver and gold nanoparticles. *Process Biochemistry*, 45, 1065–1071.

[67] Dubey, R. (2024). A review of UV-visible spectroscopy: Techniques and applications. *IJNRD – International Journal of Novel Research and Development*, 9(10), b412–b423. (www.IJNRD.org), ISSN: 2456-4184. https://ijnrd.org/viewpaperforall?paper=IJNRD2410147

[68] Khalid, K., Ishak, R., & Chowdhury, Z.Z. (2024). UV – Vis spectroscopy in nondestructive testing. In *Non-destructive material characterization methods* (pp. 391–416). Elsevier.

[69] Ahmad, P., Singh, A.K., Chauhan, D.K., Dubey, N.K., Sharma, S., & Tripathi, D.K. (2018). *Nanomaterials in plants, algae, and micro-organisms: Concepts and controversies*. Elsevier Science & Technology.

[70] Rocha, F.S., Gomes, A.J., Lunardi, C.N., Kaliaguine, S., & Patience, G.S. (2018). Experimental methods in chemical engineering: Ultraviolet visible spectroscopy – UV-Vis. *The Canadian Journal of Chemical Engineering*, 96(12), 2512–2517.

[71] De Oliveira, O., Marystela, F., de Lima Leite, F., & Da Róz, A.L. (2017). Nanocharacterization techniques. In M. Deans (Ed.), *Nanocharacterization techniques* (pp. 1–10). Elsevier.

[72] Kouvaris, P., Delimitis, A., Zaspalis, V., Papadopoulos, D., Tsipas, S.A., & Michailidis, N. (2012). Green synthesis and characterization of silver nanoparticles produced using *Arbutus unedo* leaf extract. *Materials Letters*, 76, 18–20.

[73] Singhal, G., & Bhavesh, R. (2011). Biosynthesis of silver nanoparticles using *Ocimum sanctum* (Tulsi) leaf extract and screening its antimicrobial activity. *Journal of Nanoparticle Research*, 13(7), 2981–2988.

[74] Parida, U.K., Bindhani, B.K., & Nayak, P. (2011). Green synthesis and characterization of gold nanoparticles using onion (*Allium cepa*) extract. *World Journal of Nano Science and Engineering*, 1, 93–98.

[75] Silveira, A.P., Bonatto, C.C., Lopes, C.A.P., Rivera, L.M.R., & Silva, L.P. (2018). Physicochemical characteristics and antibacterial effects of silver nanoparticles produced using the aqueous extract of Ilex paraguariensis. *Materials Chemistry and Physics*, 216, 476–484.

[76] Lin, P.C., Lin, S., Wang, P.C., & Sridhar, R. (2014). Techniques for physicochemical characterization of nanomaterials. *Biotechnology Advances*, 32(4), 711–726.

[77] Subramanian, A., & Rodriguez-Saona, L. (2009). Fourier transform infrared (FTIR) spectroscopy. In D.W. Sun (Ed.), *Infrared spectroscopy for food quality analysis and control* (pp. 145–178). Academic Press.

[78] Banerjee, P., Satapathy, M., Mukhopadhaya, A., & Das, P. (2014). Leaf extract-mediated green synthesis of silver nanoparticles from widely available Indian plants: Synthesis, characterization, antimicrobial property and toxicity analysis. *Bioresources and Bioprocessing*, 1, 3.

[79] Bunaciu, A.A., Udriştioiu, E.G., & Aboul-Enein, H.Y. (2015). X-ray diffraction: Instrumentation and applications. *Critical Reviews in Analytical Chemistry*, 45(4), 289–299.

[80] Shameli, K., Ahmad, M.B., Jazayeri, S.D., Sedaghat, S., Shabanzadeh, P., Jahangirian, H., & Abdollahi, Y. (2012). Synthesis and characterization of polyethylene glycol mediated silver nanoparticles by the green method. *International Journal of Molecular Sciences*, 13(6), 6639–6650.

[81] Cheng, F., Su, C., Yang, Y., & Yeh, C. (2005). Characterization of aqueous dispersions of Fe_3O_4 nanoparticles and their biomedical applications. *Biomaterials*, 26, 729–738.

[82] Bishnoi, A., Kumar, S., & Joshi, N. (2017). Wide-angle X-ray diffraction (WXRD): Technique for characterization of nanomaterials and polymer nanocomposites. In *Microscopy methods in nanomaterials characterization* (pp. 313–337). Elsevier.

[83] Luo, F., Yang, D., Chen, Z., Megharaj, M., & Naidu, R. (2016). One-step green synthesis of bimetallic Fe/Pd nanoparticles used to degrade Orange II. *Journal of Hazardous Materials*, 303, 145–153.

[84] Henning, S., & Adhikari, R. (2017). Scanning electron microscopy, ESEM, and X-ray microanalysis. In *Microscopy methods in nanomaterials characterization* (pp. 1–30). Elsevier.

[85] Suzuki, E. (2002). High-resolution scanning electron microscopy of immunogold-labelled cells by the use of thin plasma coating of osmium. *Journal of Microscopy*, 208(3), 153–157.

[86] Rao, Y.S., Kotakadi, V.S., Prasad, T.N.V.K.V., Reddy, A.V., & Gopal, D.S. (2013). Green synthesis and spectral characterization of silver nanoparticles from Lakshmi tulasi (*Ocimum sanctum*) leaf extract. *Spectrochimica Acta Part A: Molecular and Biomolecular Spectroscopy*, 103, 156–159.

[87] Song, J.Y., Jang, H., & Kim, B.S. (2009). Biological synthesis of gold nanoparticles using *Magnolia kobus* and *Diospyros kaki* leaf extracts. *Process Biochemistry*, 44, 1133–1138.

[88] Wang, Z.L. (1999). Transmission electron microscopy and spectroscopy of nanoparticles. In *Characterization of nanophase materials* (pp. 37–80). Springer.

[89] Mayeen, A., Shaji, L.K., Nair, A.K., & Kalarikkal, N. (2018). Morphological characterization of nanomaterials. In *Characterization of nanomaterials* (pp. 335–364). Elsevier.

[90] Hall, J.B., Dobrovolskaia, M.A., Patri, A.K., & McNeil, S.E. (2007). Characterization of nanoparticles for therapeutics. *Nanomedicine*, 2(3), 333–349.

[91] Sheny, D.S., Philip, D., & Mathew, J. (2012). Rapid green synthesis of palladium nanoparticles using the dried leaf of *Anacardium occidentale*. *Spectrochimica Acta Part A: Molecular and Biomolecular Spectroscopy*, 91, 35–38.

[92] Medina-Ramirez, I., Bashir, S., Luo, Z., & Liu, J.L. (2009). Green synthesis and characterization of polymer-stabilized silver nanoparticles. *Colloids and Surfaces B: Biointerfaces*, 73(2), 185–191.

[93] Davar, F., Fereshteh, Z., & Salavati-Niasari, M. (2009). Nanoparticles Ni and NiO: Synthesis, characterization and magnetic properties. *Journal of Alloys and Compounds*, 476, 797–801.

[94] Shaheen, T.I., Salem, S.S., & Fouda, A. (2021). Current advances in fungal nanobiotechnology: Mycofabrication and applications. In A. Lateef, E.B. Gueguim-Kana, N. Dasgupta, & S. Ranjan (Eds.), *Microbial Nanobiotechnology: Principles and Applications* (pp. 113–143). Springer Singapore.

[95] Ghosh, P., Han, G., De, M., Kim, C.K., & Rotello, V.M. (2008). Gold nanoparticles in delivery applications. *Advanced Drug Delivery Reviews*, 60, 1307–1315.

[96] Aubin-Tam, M.-E., & Hamad-Schifferli, K. (2008). Structure and function of nanoparticle – protein conjugates. *Biomedical Materials*, 3, 034001.

[97] Lo, S., & Fauzi, M.B. (2021). Current update of collagen nanomaterials – fabrication, characterisation and its applications: A review. *Pharmaceutics*, 13, 316.

[98] Zeinali, M., Abbaspour-Ravasjani, S., Ghorbani, M., Babazadeh, A., Soltanfam, T., Santos, A.C., Hamishehkar, H., & Hamblin, M.R. (2020). Nanovehicles for co-delivery of anticancer agents. *Drug Discovery Today*, 25(8), 1416–1430.

[99] Nadeem, M., Tungmunnithum, D., Hano, C., Abbasi, B.H., Hashmi, S.S., Ahmad, W., & Zahir, A. (2018). The current trends in the green syntheses of titanium oxide nanoparticles and their applications. *Green Chemistry Letters and Reviews*, 11, 492–502.

[100] Singh, R., Hano, C., Nath, G., & Sharma, B. (2021). Green biosynthesis of silver nanoparticles using leaf extract of *Carissa carandas* L. and their antioxidant and antimicrobial activity against human pathogenic bacteria. *Biomolecules*, 11, 299.

[101] Orge, C.A., Orfao, J.M.M., Pereira, M.M.F.R., Duarte de Farias, A.M., Neto, R.C.R., & Fraga, M.A. (2011). Ozonation of model organic compounds catalysed by nanostructured cerium oxides. *Applied Catalysis B: Environmental*, 103(1–2), 190–199.

[102] Theron, J., Walker, J.A., & Cloete, T.E. (2008). Nanotechnology and water treatment: Applications and emerging opportunities. *Critical Reviews in Microbiology*, 34, 43–69.

[103] Zhong, L.S., Hu, J.S., Cao, A.M., Liu, Q., Song, W.G., & Wan, L.J. (2007). 3D flowerlike ceria micro/nanocomposite structure and its application for water treatment and CO removal. *Chemistry of Materials*, 19(7), 1648–1655.

[104] Anjum, M., Miandad, R., Waqas, M., Gehany, F., & Barakat, M.A. (2019). Remediation of wastewater using various nanomaterials. *Arabian Journal of Chemistry*, 12, 4897–4919.

[105] Sharma, S., & Bhattacharya, A. (2017). Drinking water contamination and treatment techniques. *Applied Water Science*, 7, 1043–1067.

[106] Punia, P., Bharti, M.K., Chalia, S., Dhar, R., Ravelo, B., Thakur, P., & Thakur, A. (2021). Recent advances in synthesis, characterization, and applications of nanoparticles for contaminated water treatment – a review. *Ceramics International*, 47, 1526–1550.

[107] Gautam, R.K., & Chattopadhyaya, M.C. (2016). *Nanomaterials for wastewater remediation*. Elsevier.

[108] Sadegh, H., Ali, G.A., Gupta, V.K., Makhlouf, A.S.H., Shahryari-ghoshekandi, R., Nadagouda, M.N., Sillanpää, M., & Megiel, E. (2017). The role of nanomaterials as effective adsorbents and their applications in wastewater treatment. *Journal of Nanostructure in Chemistry*, 7, 1–14.

[109] Dil, E.A., Ghaedi, M., & Asfaram, A. (2017). The performance of nanorods material as adsorbent for removal of azo dyes and heavy metal ions: Application of ultrasound wave, optimization and modeling. *Ultrasonics Sonochemistry*, 34, 792–802.

[110] Santhosh, C., Velmurugan, V., Jacob, G., Jeong, S.K., Grace, A.N., & Bhatnagar, A. (2016). Role of nanomaterials in water treatment applications: A review. *Chemical Engineering Journal*, 306, 1116–1137.

[111] Das, C., Sen, S., Singh, T., Ghosh, T., Paul, S.S., Kim, T.W., Jeon, S., Maiti, D.K., Im, J., & Biswas, G. (2020). Green synthesis, characterization and application of natural product coated magnetite nanoparticles for wastewater treatment. *Nanomaterials*, 10, 1615.

[112] Ezhilarasi, P.N., Karthik, P., Chhanwal, N., & Anandharamakrishnan, C. (2013). Nanoencapsulation techniques for food bioactive components: A review. *Food and Bioprocess Technology*, 6, 628–647.

[113] Eid, A.M., Fouda, A., Abdel-Rahman, M.A., Salem, S.S., Elsaied, A., Oelmüller, R., Hijri, M., Bhowmik, A., Elkelish, A., & El-Din Hassan, S. (2021). Harnessing bacterial endophytes for promotion of plant growth and biotechnological applications: An overview. *Plants*, 10.

[114] Hofmann, T., Lowry, G.V., Ghoshal, S., Tufenkji, N., Brambilla, D., Dutcher, J.R., Gilbertson, L.M., Giraldo, J.P., Kinsella, J.M., Landry, M.P., Wilkinson, K.J. (2020). Technology readiness and overcoming barriers to sustainably implement nanotechnology-enabled plant agriculture. *Nature Food*, 1, 416–425.

[115] Pramanik, P., Krishnan, P., Maity, A., Mridha, N., Mukherjee, A., & Rai, V. (2020). Application of nanotechnology in agriculture. *Environmental Nanotechnology*, 4, 317–348.

[116] Salama, D.M., Abd El-Aziz, M.E., Rizk, F.A., & Abd Elwahed, M.S.A. (2021). Applications of nanotechnology on vegetable crops. *Chemosphere*, 266, 129026.

[117] Hossain, A., Abdallah, Y., Ali, M.A., Masum, M.M.I., Li, B., Sun, G., Meng, Y., Wang, Y., & An, Q. (2019). Lemon-fruit-based green synthesis of zinc oxide nanoparticles and titanium dioxide nanoparticles against soft rot bacterial pathogen *Dickeya dadantii*. *Biomolecules*, 9, 863.

[118] Ahmad, H., Venugopal, K., Rajagopal, K., De Britto, S., Nandini, B., Pushpalatha, H.G., Konappa, N., Udayashankar, A.C., Geetha, N., & Jogaiah, S. (2020). Green synthesis and characterization of zinc oxide nanoparticles using Eucalyptus globules and their fungicidal ability against pathogenic fungi of apple orchards. *Biomolecules*, 10, 425.

[119] Kamran, U., Bhatti, H.N., Iqbal, M., & Nazir, A. (2019). Green synthesis of metal nanoparticles and their applications in different fields: A review. *Zeitschrift für Physikalische Chemie*, 233(9), 1325–1349.

8

Environmental Fate, Transport, and Health Hazards of Nanomaterials

Vijay Kumar Vishvakarma and Gyanendra Kumar

8.1 Introduction

Nanotechnology has revolutionized industries by leveraging the unique properties of materials at the nanoscale. Nanomaterials, defined as particles with at least one dimension between 1 and 100 nm, exhibit high surface area-to-volume ratios, enhanced reactivity, and tunable properties.[1] These characteristics enable applications in drug delivery, renewable energy, and environmental remediation. However, the same properties that make nanomaterials valuable can lead to unintended environmental and health consequences.[2] Released through manufacturing, consumer products, or disposal, nanomaterials interact with environmental media and biota in complex ways, necessitating a thorough understanding of their fate, transport, and toxicity.[3]

This chapter synthesizes current knowledge on the environmental fate, transport, and health hazards of nanomaterials. It addresses their transformations, mobility across air, water, and soil, and impacts on human and ecological health. Case studies provide practical examples, while visual aids, including figures and tables, illustrate key concepts. In-text citations and a reference list provide a robust evidence base. This chapter synthesizes knowledge on nanomaterial fate, transport, and hazards, incorporating case studies, emerging nanomaterials, and regulatory perspectives, supported by visual aids and citations

8.2 Environmental Fate of Nanomaterials

The environmental fate of nanomaterials encompasses their behavior, transformation, and persistence after release into the environment. Nanomaterials enter ecosystems via intentional applications (e.g., nanopesticides), accidental spills, or end-of-life disposal (e.g., electronics).[4] Their fate is governed by intrinsic properties (size, shape, surface chemistry) and environmental factors (pH, ionic strength, organic matter).[5]

8.2.1 Transformations in Environmental Media

Nanomaterials undergo physical, chemical, and biological transformations that alter their behavior and toxicity.

DOI: 10.1201/9781003632498-8

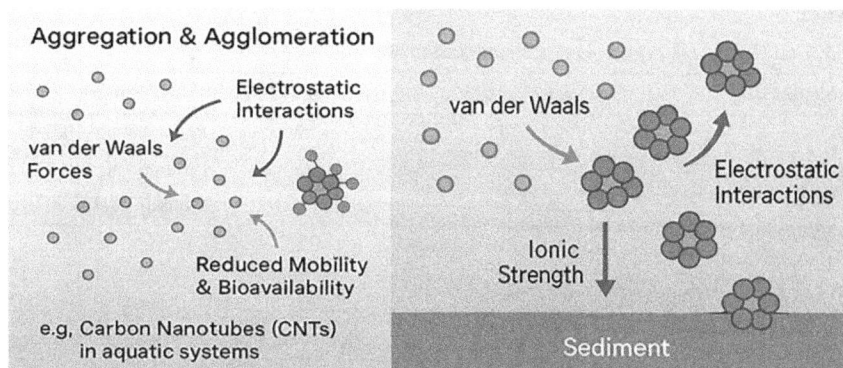

FIGURE 8.1 Diagram of nanomaterial aggregation and sedimentation.

8.2.1.1 Physical Transformations

The physical transformation of the nanomaterials is governed by aggregation and sedimentation. These two processes involve several key factors to transform them.[6] A schematic representation of these is given in Figure 8.1.

- **Aggregation and Agglomeration:** van der Waals forces and electrostatic interactions drive nanomaterial aggregation, reducing mobility and bioavailability. For instance, carbon nanotubes (CNTs) agglomerate in aquatic systems, limiting their transport.[7]
- **Sedimentation:** Aggregates settle in water bodies, concentrating nanomaterials in sediments. Titanium dioxide (TiO_2) nanoparticles settle rapidly in high-ionic-strength environments like seawater.[8]

8.2.1.2 Chemical Transformations

Chemical transformation requires several chemical reactions, surface modifications, and dissolution to transform the nanomaterials from one form to the other forms.[9] A few transformations of these nanomaterials are given in Table 8.1.

- **Oxidation and Reduction:** Zero-valent iron (nZVI) nanoparticles, used in groundwater remediation, oxidize to iron oxides, reducing toxicity but also efficacy.[10]
- **Surface Modifications:** Natural organic matter (NOM) coats nanomaterials, altering stability. Silver nanoparticles (AgNPs) coated with NOM exhibit reduced dissolution and toxicity.[11]
- **Particle Solubility:** Metal oxide nanoparticles, like ZnO, tend to aggregate in freshwater, forming larger flocs rather than remaining as individual nanoparticles. These particles rapidly dissolve, releasing zinc ions, with solubility similar to bulk ZnO. Toxicity observed in algae was primarily due to dissolved zinc, not the nanoparticle form itself, highlighting the need for careful interpretation in nanoparticle toxicity studies.[12,13]

TABLE 8.1

Common Chemical Transformations of Nanomaterials

Nanomaterial	Transformation	Environmental Impact	Reference
nZVI	Oxidation to Fe_2O_3	Reduced reactivity	Phenrat et al., 2007[14]
AgNPs	Sulfidation	Decreased toxicity	Levard et al., 2012[15]
ZnO NPs	Dissolution to Zn^{2+}	Increased toxicity	Franklin et al., 2007[16]

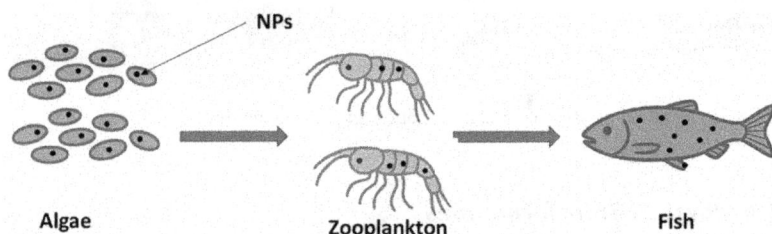

FIGURE 8.2 Bioaccumulation of NP concentrations in algae, zooplankton, and fish.

8.2.1.3 Biological Transformations

Microbial or enzymatic processes can degrade or modify nanomaterials. For example, bacteria reduce graphene oxide, altering its environmental behavior.[17]

8.2.2 Persistence and Bioaccumulation

Nanomaterial persistence varies by composition. Biodegradable nanomaterials, like polymeric nanoparticles, degrade over time, while persistent ones, like CNTs or quantum dots, may remain for decades.[18] Bioaccumulation occurs when nanomaterials accumulate in organisms, potentially transferring through food chains. AgNPs bioaccumulate in fish, leading to trophic transfer as in Figure 8.2.[19]

8.3 Transport of Nanomaterials

Nanomaterial transport determines their distribution across environmental compartments and the potential to reach sensitive receptors.[20]

8.3.1 Transport in Air

Airborne nanomaterials, released as aerosols during manufacturing, remain suspended due to their small size, enabling long-range transport. Deposition depends on atmospheric conditions and aggregation. TiO_2 nanoparticles may travel hundreds of kilometers before settling, posing inhalation risks as in Figure 8.3.[21,22]

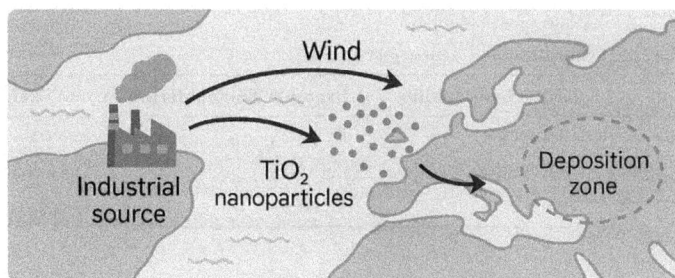

FIGURE 8.3 Atmospheric transport of TiO₂ nanomaterials.

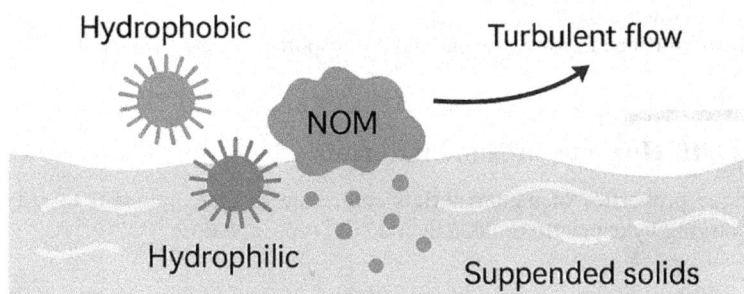

FIGURE 8.4 Influence of NOM on nanoparticle transport.

8.3.2 Transport in Water

In aquatic systems, transport depends on hydrodynamic conditions, surface chemistry, and environmental factors. Here, NOM plays a key role in the transport of these nanomaterials in the aquatic system.[23] Figure 8.4 summarizes these key effects.

- **Hydrodynamic Conditions:** Turbulent flow enhances dispersion, while stagnant conditions promote sedimentation.[24]
- **Surface Chemistry:** Hydrophobic nanomaterials (e.g., fullerenes) associate with suspended solids, reducing mobility, whereas hydrophilic nanoparticles remain dispersed.[25]
- **Environmental Factors:** NOM stabilizes nanoparticles, increasing transport distance. ZnO nanoparticles travel further in NOM-rich rivers.[26]

8.3.3 Transport in Soil

In terrestrial environments, nanomaterials interact with soil components. These components are porosity, permeability, pH, organic matter, and bioturbation.[27] Table 8.2 summarizes the key effects of these factors in the nanomaterials.

TABLE 8.2

Factors Affecting Nanomaterial Transport in Soil

Factor	Effect on Mobility	Example Nanomaterial	Reference
High porosity	Increases	TiO$_2$ NPs	Lv et al., 2016[31]
High organic matter	Decreases	CuO NPs	Peng et al., 2017[32]
Bioturbation	Increases	Ag NPs	Baccaro et al., 2019[33]

- **Porosity and Permeability:** Sandy soils allow greater mobility than clay-rich soils, where nanomaterials bind to fine particles.[28]
- **Organic Matter and pH:** High organic content or acidic pH immobilizes nanomaterials. CuO nanoparticles are retained in organic-rich soils, limiting groundwater contamination.[29]
- **Bioturbation:** Earthworms facilitate transport by mixing soil layers.[30]

8.4 Health Hazards of Nanomaterials

The unique properties of nanomaterials contribute to their potential toxicity, with effects varying by exposure route, nanomaterial type, and dose.

8.4.1 Human Health Impacts

Nanomaterials have numerous health impacts on human beings. Being small in size, they easily enter the human body and cause hazardous effects. There are five routes to enter the human body, i.e., oral, inhalation, subcutaneous, intravenous, and intramuscular. However, oral, inhalation, and subcutaneous are more prominent in the case of nanomaterials.[34-37]

8.4.1.1 Inhalation

Nanomaterials like CNTs or silica nanoparticles penetrate deep into the lungs, causing inflammation, oxidative stress, or fibrosis, as in Figure 8.5. Long, rigid CNTs are particularly concerning due to their asbestos-like behavior. Occupational exposure to TiO$_2$ nanoparticles is linked to respiratory irritation.[38-41]

8.4.1.2 Dermal Exposure

Skin is a robust barrier, but nanomaterials in sunscreens (e.g., ZnO) may penetrate damaged skin, causing localized toxicity. Systemic absorption is typically limited.[42]

8.4.1.3 Ingestion

Nanomaterials in food packaging or water may enter the gastrointestinal tract. AgNPs disrupt gut microbiota, leading to immune or metabolic effects.[43]

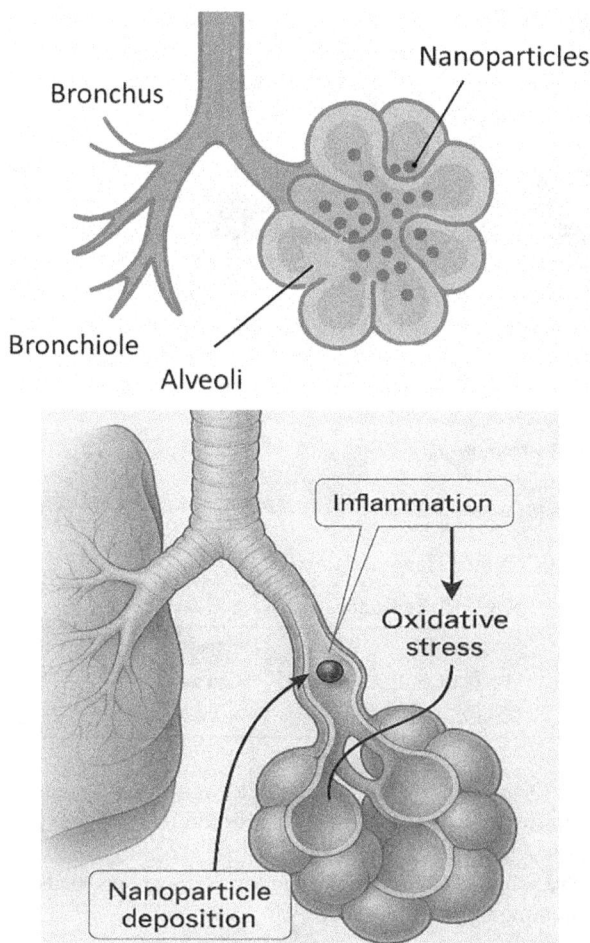

FIGURE 8.5 (a) Lung penetration by nanomaterials, (b) inflammation and oxidative stress.

8.4.1.4 Systemic Effects

Absorbed nanomaterials may translocate to organs like the liver, spleen, or brain. Quantum dots containing cadmium pose risks of neurotoxicity or genotoxicity.[44]

8.4.2 Environmental Health Impacts

Since they are small, their transportation is quite easy. So, they reach all parts of the ecosystem, and their accumulation results in biomagnification. So, they harm all parts and levels of the ecosystem.[45] Nanomaterials harm ecosystems by affecting non-target organisms, as in Figure 8.6:

- **Aquatic Organisms:** AgNPs are toxic to algae and fish, disrupting photosynthesis and reproduction. Concentrations as low as 1 µg/L impair *Daphnia magna*.[46]

Aquatic Organisms	Terrestrial Organisms	Trophic Transfer

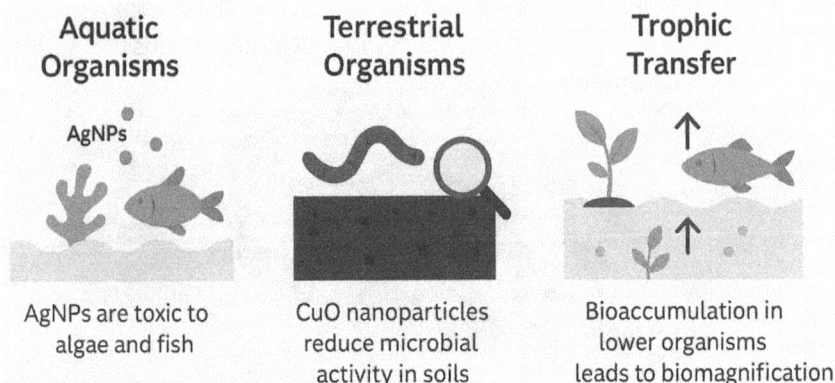

AgNPs are toxic to algae and fish	CuO nanoparticles reduce microbial activity in soils	Bioaccumulation in lower organisms leads to biomagnification

FIGURE 8.6 Toxicity of NPs to aquatic organisms.

TABLE 8.3

Toxicity Factors for Common Nanomaterials

Nanomaterial	Key Toxicity Factor	Effect	Reference
CNTs	Shape (rod-like)	Fibrosis	Poland et al., 2008[53]
Ag NPs	Size (<10 nm)	High toxicity	Navarro et al., 2008[54]
ZnO NPs	Surface charge	Cytotoxicity	Franklin et al., 2007[55]

- **Terrestrial Organisms:** CuO nanoparticles reduce microbial activity in soils, affecting nutrient cycling. Earthworms exposed to TiO_2 nanoparticles show reduced reproductive success.[47,48]
- **Trophic Transfer:** Bioaccumulation in lower organisms leads to biomagnification, threatening higher trophic levels.[49]

8.4.3 Factors Influencing Toxicity

Several factors influence the toxicity caused by nanomaterials. These are shape, size, surface, exposure, and concentration. So, the effect of these factors is summarized in Table 8.3.

- **Size and Shape:** Smaller nanoparticles have higher reactivity and toxicity. Rod-shaped nanomaterials are often more toxic than spherical ones.[50]
- **Surface Chemistry:** Functionalized surfaces mitigate or enhance toxicity. PEGylated nanoparticles are less toxic than uncoated ones.[51]
- **Dose and Exposure Duration:** Chronic low-dose exposure causes cumulative effects, complicating risk assessments.[52]

8.5 Case Studies

To illustrate the real-world implications of nanomaterials, this section presents three case studies focusing on their environmental fate, transport, and health hazards.[56]

8.5.1 Silver Nanoparticles in Wastewater Treatment Plants

Silver nanoparticles (AgNPs) are widely used in consumer products like textiles and coatings for their antimicrobial properties. Significant quantities are released into wastewater treatment plants (WWTPs), where their fate and ecological impacts are critical concerns. Studies show that 90% of AgNPs entering WWTPs partition to biosolids, with the remainder released into receiving waters. In biosolids, AgNPs undergo sulfidation, forming less toxic Ag_2S, but partial dissolution releases Ag^+ ions, which are highly toxic to aquatic organisms.[57-60]

Transport of AgNPs from WWTP effluents to rivers increases their environmental reach. Navarro et al. (2008) found that AgNPs at 1 µg/L impair *Daphnia magna* reproduction, disrupting aquatic food webs. Biosolids applied to agricultural fields introduce AgNPs into soils, where they reduce microbial diversity, affecting nutrient cycling.[29] This case highlights the need for advanced WWTP technologies to capture nanomaterials and regulations to limit their release.[61-63]

8.5.2 Titanium Dioxide Nanoparticles in Sunscreens

TiO_2 nanoparticles are common in sunscreens due to their UV-blocking properties. Their release into aquatic environments during recreational activities like swimming raises concerns about human and ecological health. TiO_2 NPs are relatively inert but can generate reactive oxygen species (ROS) under UV light, causing oxidative stress in aquatic organisms.[64-66]

Human exposure occurs via dermal contact, with limited penetration through intact skin but potential risks through damaged skin. In aquatic systems, TiO_2 NPs settle in sediments, reducing bioavailability but impacting benthic organisms. Concentrations in coastal waters can reach 10 µg/L during peak tourism, affecting coral reefs and fish. This case underscores the need for eco-friendly sunscreen formulations and monitoring of recreational waters.[67,68]

8.5.3 Carbon Nanotubes in Occupational Settings

Carbon nanotubes (CNTs) are valued for their strength and conductivity but pose significant inhalation risks in occupational settings like manufacturing facilities. Their fibrous, asbestos-like structure can cause lung inflammation and fibrosis, particularly for multi-walled CNTs. Workplace air concentrations of CNTs can exceed 10 µg/m³, surpassing safe exposure limits.[69,70]

CNTs released into the environment during production can become airborne, facilitating long-range transport. Their persistence in soils and water bodies raises concerns about long-term ecological impacts, though data are limited. This case emphasizes the importance of engineering controls, personal protective equipment, and life cycle assessments to minimize CNT exposures.[71,72]

8.5.4 Zero-Valent Iron Nanoparticles in Groundwater Remediation

Zero-valent iron nanoparticles (nZVI) are used to remediate contaminated groundwater by reducing pollutants like chlorinated solvents. Their environmental fate involves oxidation to iron oxides, reducing not only efficacy but also toxicity. Transport is

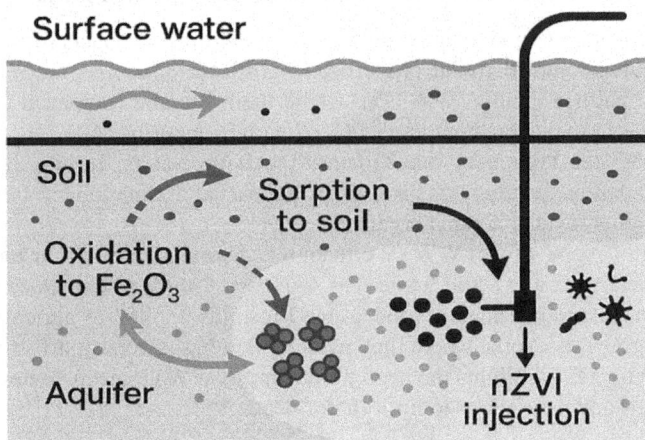

FIGURE 8.7 nZVI fate in groundwater remediation.

limited in soils due to aggregation and sorption to clay particles, but unintended migration to surface waters can occur in sandy aquifers. Ecologically, nZVI at high concentrations (>100 mg/L) can reduce microbial activity, disrupting soil ecosystems.[36] A case study in a California Superfund site showed that nZVI effectively degraded trichloroethylene but required careful monitoring to prevent nanoparticle leaching into nearby streams. A schematic representation of nZVI in groundwater remediation is depicted in Figure 8.7.

8.5.5 Quantum Dots in Biomedical Applications

Quantum dots (QDs), used in medical imaging, contain heavy metals like cadmium, posing toxicity risks if released during disposal. In landfills, QDs may leach into groundwater, with transport enhanced in high-porosity soils. Their small size (<10 nm) enables cellular uptake, causing oxidative stress and genotoxicity in aquatic organisms and humans. A hospital waste study showed QD accumulation in wastewater sludge, with potential trophic transfer in aquatic systems. Safe disposal and biodegradable QD designs are critical to mitigate risks.[73,74] The risk arises due to the quantum dots in the environment are covered in Table 8.4.

8.6 Emerging Nanomaterials

Emerging nanomaterials, such as MXenes and graphene-based materials, are gaining attention for energy storage, catalysis, and biomedical applications. Their environmental and health impacts are less studied, posing new challenges.[77]

8.6.1 MXenes

MXenes, two-dimensional transition metal carbides/nitrides, exhibit high conductivity and hydrophilicity. Their environmental fate involves oxidation in water,

TABLE 8.4

Environmental Risks of Quantum Dots

Pathway	Risk	Mitigation Strategy	Reference
Landfill leaching	Groundwater contamination	Secure containment	Hardman et al., 2006[75]
Wastewater release	Trophic transfer	Biodegradable QDs	Lovrić et al., 2005[76]

forming metal oxides, but their persistence is unknown. Transport in aquatic systems is enhanced by their colloidal stability, potentially reaching distant ecosystems. Toxicity studies show that MXenes (e.g., Ti_3C_2) cause oxidative stress in zebrafish at concentrations >50 μg/L. Human inhalation risks are a concern during manufacturing.[78,79]

8.6.2 Graphene-Based Materials

Graphene oxide (GO) and reduced graphene oxide (rGO) are used in composites and sensors. GO undergoes microbial reduction in water, altering its mobility. In soils, GO aggregates limit transport but impact microbial communities. Inhalation of GO nanosheets causes lung inflammation, similar to CNTs. Their large-scale production raises concerns about environmental releases.[80–82] The environmental fate of MXene and graphene-based material is depicted in Figure 8.8

8.7 Risk Assessment and Mitigation

8.7.1 Challenges in Risk Assessment

Assessing nanomaterial risks requires integrating data on fate, transport, and toxicity. Key challenges include:[83–87]

- **Standardization:** Lack of standardized testing protocols complicates comparisons.
- **Modeling:** Predictive transport models are limited by data gaps in transformation kinetics.
- **Regulation:** Regulatory frameworks lag behind nanomaterial development, with few jurisdictions mandating specific safety assessments.

8.7.2 Global Regulatory Frameworks

The regulation of nanomaterials varies significantly across regions, reflecting differences in legislative priorities, industrial practices, and risk assessment capabilities. Below is an overview of key global regulatory approaches, and it is also summarized in Table 8.5.

- **United States:** The Environmental Protection Agency (EPA) regulates nanomaterials primarily under the Toxic Substances Control Act (TSCA). However, the guidance for nanomaterials remains limited and largely general, lacking specific provisions tailored to the unique behaviors and risks associated with nanoscale substances.[88]

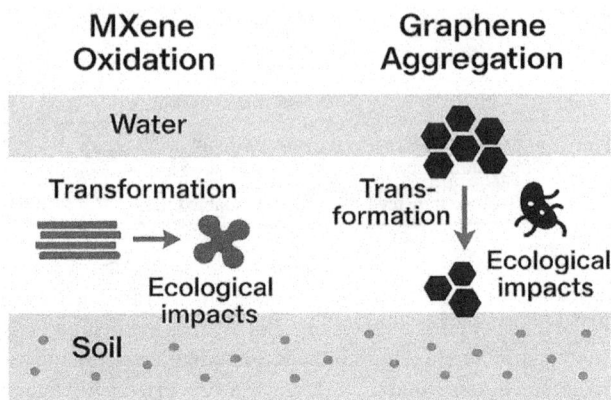

FIGURE 8.8 Environmental fate of MXenes and graphene.

TABLE 8.5

Global Nanomaterial Regulations

Region	Framework	Key Features	Reference
USA	TSCA	Limited nanomaterial specificity	EPA, 2017[88]
EU	REACH	Mandatory safety data	EC, 2020[89]
China	Occupational	Focus on worker safety	Wang et al., 2018[91]

- **European Union:** The Registration, Evaluation, Authorisation and Restriction of Chemicals (REACH) framework mandates the registration of nanomaterials. The EU has implemented stricter and more comprehensive safety assessments, requiring detailed physicochemical characterization and toxicity profiling of nanomaterials.[89]
- **China:** Emerging regulations focus on occupational safety but lack environmental standards. Harmonizing standards and mandating life cycle assessments are critical. Case studies highlight the need for tailored regulations, e.g., WWTP controls for AgNPs and disposal protocols for QDs[90].

8.7.3 Mitigation Strategies

The mitigation strategies are highly important in risk assessment. There are several branches of science that deals with mitigation strategies for nanomaterials. These are summarized in Table 8.6.

- **Green Nanotechnology:** Designing biodegradable nanomaterials, such as cellulose-based nanoparticles, reduces environmental impact.
- **Exposure Controls:** Engineering controls in manufacturing minimize releases.
- **Monitoring:** Sensitive detection methods, like single-particle ICP-MS, improve tracking in complex matrices.

TABLE 8.6

Mitigation Strategies for Nanomaterial Risks

Strategy	Description	Example Application	Reference
Green nanotechnology	Biodegradable designs	Cellulose NPs	Klemm et al., 2011[92]
Exposure controls	Ventilation, filtration	Manufacturing	NIOSH, 2009[93]
Monitoring	Advanced detection methods	SP-ICP-MS	Mitrano et al., 2014[94]

8.8 Conclusion

The environmental fate, transport, and health hazards of nanomaterials are driven by their unique properties and complex interactions with environmental and biological systems. Physical, chemical, and biological transformations govern their persistence and bioavailability, while transport mechanisms determine their distribution across air, water, and soil. Health hazards, ranging from respiratory toxicity to ecosystem disruption, highlight the need for careful risk management. Case studies on AgNPs, TiO_2 NPs, and CNTs illustrate real-world challenges and underscore the importance of life cycle assessments. Advances in standardization, modeling, and green nanotechnology offer pathways to mitigate risks, but regulatory gaps remain. Interdisciplinary research and robust policy frameworks are essential to balance the benefits of nanotechnology with environmental and public health safety.

REFERENCES

[1] Singh, N. B.; Kumar, B.; Usman, U. L.; Susan, Md. A. B. H. Nano Revolution: Exploring the Frontiers of Nanomaterials in Science, Technology, and Society. *Nano-Struct Nano-Objects* **2024**, *39*, 101299. https://doi.org/10.1016/j.nanoso.2024.101299.

[2] Kurul, F.; Turkmen, H.; Cetin, A. E.; Topkaya, S. N. Nanomedicine: How Nanomaterials Are Transforming Drug Delivery, Bio-Imaging, and Diagnosis. *Next Nanotechnology* **2025**, *7*, 100129. https://doi.org/10.1016/j.nxnano.2024.100129.

[3] Vineeth Kumar, C. M.; Karthick, V.; Kumar, V. G.; Inbakandan, D.; Rene, E. R.; Suganya, K. S. U.; Embrandiri, A.; Dhas, T. S.; Ravi, M.; Sowmiya, P. The Impact of Engineered Nanomaterials on the Environment: Release Mechanism, Toxicity, Transformation, and Remediation. *Environ Res* **2022**, *212*, 113202. https://doi.org/10.1016/j.envres.2022.113202.

[4] Lead, J. R.; Batley, G. E.; Alvarez, P. J. J.; Croteau, M.-N.; Handy, R. D.; McLaughlin, M. J.; Judy, J. D.; Schirmer, K. Nanomaterials in the Environment: Behavior, Fate, Bioavailability, and Effects—An Updated Review. *Environ Toxicol Chem* **2018**, *37* (8), 2029–2063. https://doi.org/10.1002/etc.4147.

[5] Abbas, Q.; Yousaf, B.; Amina; Ali, M. U.; Munir, M. A. M.; El-Naggar, A.; Rinklebe, J.; Naushad, M. Transformation Pathways and Fate of Engineered Nanoparticles (ENPs) in Distinct Interactive Environmental Compartments: A Review. *Environ Int* **2020**, *138*, 105646. https://doi.org/10.1016/j.envint.2020.105646.

[6] Shrestha, S.; Wang, B.; Dutta, P. Nanoparticle Processing: Understanding and Controlling Aggregation. *Adv Colloid Interface Sci* **2020**, *279*, 102162. https://doi.org/10.1016/j.cis.2020.102162.

[7] Xiong, Y.; Liu, X.; Xiong, H. Aggregation Modeling of the Influence of PH on the Aggregation of Variably Charged Nanoparticles. *Sci Rep* **2021**, *11* (1), 17386. https://doi.org/10.1038/s41598-021-96798-3.

[8] Markus, A. A.; Parsons, J. R.; Roex, E. W. M.; de Voogt, P.; Laane, R. W. P. M. Modeling Aggregation and Sedimentation of Nanoparticles in the Aquatic Environment. *Sci Total Environ* **2015**, *506–507*, 323–329. https://doi.org/10.1016/j.scitotenv.2014.11.056.

[9] Moon, G. D.; Ko, S.; Min, Y.; Zeng, J.; Xia, Y.; Jeong, U. Chemical Transformations of Nanostructured Materials. *Nano Today* **2011**, *6* (2), 186–203. https://doi.org/10.1016/j.nantod.2011.02.006.

[10] Zafar, A. M.; Javed, M. A.; Hassan, A. A.; Mohamed, M. M. Groundwater Remediation Using Zero-Valent Iron Nanoparticles (NZVI). *Groundw Sustain Dev* **2021**, *15*, 100694. https://doi.org/10.1016/j.gsd.2021.100694.

[11] Sharma, V. K.; Siskova, K. M.; Zboril, R.; Gardea-Torresdey, J. L. Organic-Coated Silver Nanoparticles in Biological and Environmental Conditions: Fate, Stability and Toxicity. *Adv Colloid Interface Sci* **2014**, *204*, 15–34. https://doi.org/10.1016/j.cis.2013.12.002.

[12] Escorihuela, L.; Fernández, A.; Rallo, R.; Martorell, B. Molecular Dynamics Simulations of Zinc Oxide Solubility: From Bulk down to Nanoparticles. *Food Chem Toxicol* **2018**, *112*, 518–525. https://doi.org/10.1016/j.fct.2017.07.038.

[13] Reed, R. B.; Ladner, D. A.; Higgins, C. P.; Westerhoff, P.; Ranville, J. F. Solubility of Nano-Zinc Oxide in Environmentally and Biologically Important Matrices. *Environ Toxicol Chem* **2012**, *31* (1), 93–99. https://doi.org/10.1002/etc.708.

[14] Phenrat, T.; Saleh, N.; Sirk, K.; Tilton, R. D.; Lowry, G. V. Aggregation and Sedimentation of Aqueous Nanoscale Zerovalent Iron Dispersions. *Environ Sci Technol* **2007**, *41* (1), 284–290. https://doi.org/10.1021/es061349a.

[15] Levard, C.; Hotze, E. M.; Lowry, G. V.; Brown, G. E. Environmental Transformations of Silver Nanoparticles: Impact on Stability and Toxicity. *Environ Sci Technol* **2012**, *46* (13), 6900–6914. https://doi.org/10.1021/es2037405.

[16] Franklin, N. M.; Rogers, N. J.; Apte, S. C.; Batley, G. E.; Gadd, G. E.; Casey, P. S. Comparative Toxicity of Nanoparticulate ZnO, Bulk ZnO, and ZnCl 2 to a Freshwater Microalga (Pseudokirchneriella Subcapitata): The Importance of Particle Solubility. *Environ Sci Technol* **2007**, *41* (24), 8484–8490. https://doi.org/10.1021/es071445r.

[17] Babaniyi, B. R.; Ogundele, O. D.; Thompson, S. O.; Aransiola, S. A. *Microbial Nanomaterial Synthesis: Types and Applications*; 2023; pp. 3–28. https://doi.org/10.1007/978-981-99-2808-8_1.

[18] Dilsad, A. M.; Ahuja, A.; Gupta, N.; Kumar Bachala, S.; Kumar Rastogi, V. Analysing the Impact of Nanomaterials on the Degradation Behaviour of Biopolymers: A Comprehensive Review. *Eur Polym J* **2024**, *214*, 113189. https://doi.org/10.1016/j.eurpolymj.2024.113189.

[19] Isibor, P. O.; Oyegbade, S. A.; Oni, J. G.; Ahmed, W. W.; Abiodun, E. O.; Oyewole, O. A. Nanoparticles in Food Chains: Bioaccumulation and Trophic Transfer. In *Environmental Nanotoxicology*; Springer Nature Switzerland: Cham, 2024; pp. 203–233. https://doi.org/10.1007/978-3-031-54154-4_11.

[20] Lowry, G. V.; Casman, E. A. Nanomaterial Transport, Transformation, and Fate in the Environment. In *Nanomaterials: Risks and Benefits*; Springer Netherlands: Dordrecht; pp. 125–137. https://doi.org/10.1007/978-1-4020-9491-0_9.

[21] Portugal, J.; Bedia, C.; Amato, F.; Juárez-Facio, A. T.; Stamatiou, R.; Lazou, A.; Campiglio, C. E.; Elihn, K.; Piña, B. Toxicity of Airborne Nanoparticles: Facts and Challenges. *Environ Int* **2024**, *190*, 108889. https://doi.org/10.1016/j.envint.2024.108889.

[22] Yin, X.-H.; Xu, Y.-M.; Lau, A. T. Y. Nanoparticles: Excellent Materials Yet Dangerous When They Become Airborne. *Toxics* **2022**, *10* (2), 50. https://doi.org/10.3390/toxics10020050.

[23] Peiffer, S.; Kappler, A.; Haderlein, S. B.; Schmidt, C.; Byrne, J. M.; Kleindienst, S.; Vogt, C.; Richnow, H. H.; Obst, M.; Angenent, L. T.; Bryce, C.; McCammon, C.; Planer-Friedrich, B. A Biogeochemical–Hydrological Framework for the Role of Redox-Active Compounds in Aquatic Systems. *Nat Geosci* **2021**, *14* (5), 264–272. https://doi.org/10.1038/s41561-021-00742-z.

[24] Mangal, D.; Conrad, J. C.; Palmer, J. C. Nanoparticle Dispersion in Porous Media: Effects of Hydrodynamic Interactions and Dimensionality. *AIChE Journal* **2021**, *67* (3). https://doi.org/10.1002/aic.17147.

[25] Xiao, Y.; Wiesner, M. R. Characterization of Surface Hydrophobicity of Engineered Nanoparticles. *J Hazard Mater* **2012**, *215–216*, 146–151. https://doi.org/10.1016/j.jhazmat.2012.02.043.

[26] Jones, E. H.; Su, C. Transport and Retention of Zinc Oxide Nanoparticles in Porous Media: Effects of Natural Organic Matter versus Natural Organic Ligands at Circumneutral PH. *J Hazard Mater* **2014**, *275*, 79–88. https://doi.org/10.1016/j.jhazmat.2014.04.058.

[27] Ogunkunle, C. O.; Oyedeji, S.; Okoro, H. K.; Adimula, V. Interaction of Nanoparticles with Soil. In *Nanomaterials for Soil Remediation*; Elsevier, 2021; pp. 101–132. https://doi.org/10.1016/B978-0-12-822891-3.00006-2.

[28] Tiede, K.; Boxall, A. B. A.; Tear, S. P.; Lewis, J.; David, H.; Hassellöv, M. Detection and Characterization of Engineered Nanoparticles in Food and the Environment. *Food Addit Contam A* **2008**, *25* (7), 795–821. https://doi.org/10.1080/02652030802007553.

[29] Cornelis, G.; Hund-Rinke, K.; Kuhlbusch, T.; van den Brink, N.; Nickel, C. Fate and Bioavailability of Engineered Nanoparticles in Soils: A Review. *Crit Rev Environ Sci Technol* **2014**, *44* (24), 2720–2764. https://doi.org/10.1080/10643389.2013.829767.

[30] Schlich, K.; Klawonn, T.; Terytze, K.; Hund-Rinke, K. Effects of Silver Nanoparticles and Silver Nitrate in the Earthworm Reproduction Test. *Environ Toxicol Chem* **2013**, *32* (1), 181–188. https://doi.org/10.1002/etc.2030.

[31] Lv, X.; Gao, B.; Sun, Y.; Dong, S.; Wu, J.; Jiang, B.; Shi, X. Effects of Grain Size and Structural Heterogeneity on the Transport and Retention of Nano-TiO2 in Saturated Porous Media. *Sci Total Environ* **2016**, *563–564*, 987–995. https://doi.org/10.1016/j.scitotenv.2015.12.128.

[32] Peng, C.; Shen, C.; Zheng, S.; Yang, W.; Hu, H.; Liu, J.; Shi, J. Transformation of CuO Nanoparticles in the Aquatic Environment: Influence of PH, Electrolytes and Natural Organic Matter. *Nanomaterials* **2017**, *7* (10), 326. https://doi.org/10.3390/nano7100326.

[33] Baccaro, M.; Harrison, S.; van den Berg, H.; Sloot, L.; Hermans, D.; Cornelis, G.; van Gestel, C. A. M.; van den Brink, N. W. Bioturbation of Ag2S-NPs in Soil Columns by Earthworms. *Environ Pollut* **2019**, *252*, 155–162. https://doi.org/10.1016/j.envpol.2019.05.106.

[34] Sharma, V. K.; Yngard, R. A.; Lin, Y. Silver Nanoparticles: Green Synthesis and Their Antimicrobial Activities. *Adv Colloid Interface Sci* **2009**, *145* (1–2), 83–96. https://doi.org/10.1016/j.cis.2008.09.002.

[35] Fadeel, B.; Garcia-Bennett, A. E. Better Safe than Sorry: Understanding the Toxicological Properties of Inorganic Nanoparticles Manufactured for Biomedical Applications. *Adv Drug Deliv Rev* **2010**, *62* (3), 362–374. https://doi.org/10.1016/j.addr.2009.11.008.

[36] Nel, A.; Xia, T.; Mädler, L.; Li, N. Toxic Potential of Materials at the Nanolevel. *Science (1979)* **2006**, *311* (5761), 622–627. https://doi.org/10.1126/science.1114397.

[37] Oberdörster, G.; Oberdörster, E.; Oberdörster, J. Nanotoxicology: An Emerging Discipline Evolving from Studies of Ultrafine Particles. *Environ Health Perspect* **2005**, *113* (7), 823–839. https://doi.org/10.1289/ehp.7339.

[38] Donaldson, K.; Murphy, F. A.; Duffin, R.; Poland, C. A. Asbestos, Carbon Nanotubes and the Pleural Mesothelium: A Review and the Hypothesis Regarding the Role of Long Fibre Retention in the Parietal Pleura, Inflammation and Mesothelioma. *Part Fibre Toxicol* **2010**, *7* (1), 5. https://doi.org/10.1186/1743-8977-7-5.

[39] Shi, H.; Magaye, R.; Castranova, V.; Zhao, J. Titanium Dioxide Nanoparticles: A Review of Current Toxicological Data. *Part Fibre Toxicol* **2013**, *10* (1), 15. https://doi.org/10.1186/1743-8977-10-15.

[40] Johnston, H. J.; Hutchison, G.; Christensen, F. M.; Peters, S.; Hankin, S.; Stone, V. A Review of the in Vivo and in Vitro Toxicity of Silver and Gold Particulates: Particle Attributes and Biological Mechanisms Responsible for the Observed Toxicity. *Crit Rev Toxicol* **2010**, *40* (4), 328–346. https://doi.org/10.3109/10408440903453074.

[41] Poland, C. A.; Duffin, R.; Kinloch, I.; Maynard, A.; Wallace, W. A. H.; Seaton, A.; Stone, V.; Brown, S.; MacNee, W.; Donaldson, K. Carbon Nanotubes Introduced into the Abdominal Cavity of Mice Show Asbestos-like Pathogenicity in a Pilot Study. *Nat Nanotechnol* **2008**, *3* (7), 423–428. https://doi.org/10.1038/nnano.2008.111.

[42] Gulson, B.; McCall, M.; Korsch, M.; Gomez, L.; Casey, P.; Oytam, Y.; Taylor, A.; McCulloch, M.; Trotter, J.; Kinsley, L.; Greenoak, G. Small Amounts of Zinc from Zinc Oxide Particles in Sunscreens Applied Outdoors Are Absorbed through Human Skin. *Toxicol Sci* **2010**, *118* (1), 140–149. https://doi.org/10.1093/toxsci/kfq243.

[43] Chen, Y.; Wang, R.; Xu, M. Metabolomics Analysis for Unveiling the Toxicological Mechanism of Silver Nanoparticles Using an *In Vitro* Gastrointestinal Digestion Model. *ACS Nanosci Au* **2024**, *4* (5), 327–337. https://doi.org/10.1021/acsnanoscienceau.4c00012.

[44] Bai, C.; Tang, M. Progress on the Toxicity of Quantum Dots to Model Organism-zebrafish. *J Appl Toxicol* **2023**, *43* (1), 89–106. https://doi.org/10.1002/jat.4333.

[45] Zhang, L.; Cui, Y.; Xu, J.; Qian, J.; Yang, X.; Chen, X.; Zhang, C.; Gao, P. Ecotoxicity and Trophic Transfer of Metallic Nanomaterials in Aquatic Ecosystems. *Sci Total Environ* **2024**, *924*, 171660. https://doi.org/10.1016/j.scitotenv.2024.171660.

[46] Xiang, Q.-Q.; Li, Q.-Q.; Wang, P.; Yang, H.-C.; Fu, Z.-H.; Liang, X.; Chen, L.-Q. Metabolomics Reveals the Mechanism of Persistent Toxicity of AgNPs at Environmentally Relevant Concentrations to *Daphnia Magna*. *Environ Sci Nano* **2025**, *12* (1), 563–575. https://doi.org/10.1039/D4EN00350K.

[47] Simonin, M.; Cantarel, A. A. M.; Crouzet, A.; Gervaix, J.; Martins, J. M. F.; Richaume, A. Negative Effects of Copper Oxide Nanoparticles on Carbon and Nitrogen Cycle Microbial Activities in Contrasting Agricultural Soils and in Presence of Plants. *Front Microbiol* **2018**, *9*. https://doi.org/10.3389/fmicb.2018.03102.

[48] Cañas, J. E.; Qi, B.; Li, S.; Maul, J. D.; Cox, S. B.; Das, S.; Green, M. J. Acute and Reproductive Toxicity of Nano-Sized Metal Oxides (ZnO and TiO2) to Earthworms (Eisenia Fetida). *J Environ Monit* **2011**, *13* (12), 3351. https://doi.org/10.1039/c1em10497g.

[49] Zheng, Y.; Nowack, B. Meta-Analysis of Bioaccumulation Data for Nondissolvable Engineered Nanomaterials in Freshwater Aquatic Organisms. *Environ Toxicol Chem* **2022**, *41* (5), 1202–1214. https://doi.org/10.1002/etc.5312.

[50] Nieves Lira, C.; Carpenter, A. P.; Baio, J. E.; Harper, B. J.; Harper, S. L.; Mackiewicz, M. R. Size- and Shape-Dependent Interactions of Lipid-Coated Silver Nanoparticles: An Improved Mechanistic Understanding through Model Cell Membranes and *In Vivo* Toxicity. *Chem Res Toxicol* **2024**, *37* (6), 968–980. https://doi.org/10.1021/acs.chemrestox.4c00053.

[51] Patlolla, A. K.; Shinde, A. K.; Tchounwou, P. B. A Comparison of Poly-Ethylene-Glycol-Coated and Uncoated Gold Nanoparticle-Mediated Hepatotoxicity and Oxidative Stress in Sprague Dawley Rats. *Int J Nanomedicine* **2019**, *Volume 14*, 639–647. https://doi.org/10.2147/IJN.S185574.

[52] Squillacioti, G.; Charreau, T.; Wild, P.; Bellisario, V.; Ghelli, F.; Bono, R.; Bergamaschi, E.; Garzaro, G.; Guseva Canu, I. Worse Pulmonary Function in Association with Cumulative Exposure to Nanomaterials. Hints of a Mediation Effect via Pulmonary Inflammation. *Part Fibre Toxicol* **2024**, *21* (1), 28. https://doi.org/10.1186/s12989-024-00589-3.

[53] Poland, C. A.; Duffin, R.; Kinloch, I.; Maynard, A.; Wallace, W. A. H.; Seaton, A.; Stone, V.; Brown, S.; MacNee, W.; Donaldson, K. Carbon Nanotubes Introduced into the Abdominal Cavity of Mice Show Asbestos-like Pathogenicity in a Pilot Study. *Nat Nanotechnol* **2008**, *3* (7), 423–428. https://doi.org/10.1038/nnano.2008.111.

[54] Navarro, E.; Piccapietra, F.; Wagner, B.; Marconi, F.; Kaegi, R.; Odzak, N.; Sigg, L.; Behra, R. Toxicity of Silver Nanoparticles to *Chlamydomonas Reinhardtii*. *Environ Sci Technol* **2008**, *42* (23), 8959–8964. https://doi.org/10.1021/es801785m.

[55] Franklin, N. M.; Rogers, N. J.; Apte, S. C.; Batley, G. E.; Gadd, G. E.; Casey, P. S. Comparative Toxicity of Nanoparticulate ZnO, Bulk ZnO, and ZnCl 2 to a Freshwater Microalga (Pseudokirchneriella Subcapitata): The Importance of Particle Solubility. *Environ Sci Technol* **2007**, *41* (24), 8484–8490. https://doi.org/10.1021/es071445r.

[56] Tran, T.-K.; Nguyen, M.-K.; Lin, C.; Hoang, T.-D.; Nguyen, T.-C.; Lone, A. M.; Khedulkar, A. P.; Gaballah, M. S.; Singh, J.; Chung, W. J.; Nguyen, D. D. Review on Fate, Transport, Toxicity and Health Risk of Nanoparticles in Natural Ecosystems: Emerging Challenges in the Modern Age and Solutions toward a Sustainable Environment. *Sci Total Environ* **2024**, *912*, 169331. https://doi.org/10.1016/j.scitotenv.2023.169331.

[57] Kent, R. D.; Oser, J. G.; Vikesland, P. J. Controlled Evaluation of Silver Nanoparticle Sulfidation in a Full-Scale Wastewater Treatment Plant. *Environ Sci Technol* **2014**, *48* (15), 8564–8572. https://doi.org/10.1021/es404989t.

[58] Kaegi, R.; Voegelin, A.; Ort, C.; Sinnet, B.; Thalmann, B.; Krismer, J.; Hagendorfer, H.; Elumelu, M.; Mueller, E. Fate and Transformation of Silver Nanoparticles in Urban Wastewater Systems. *Water Res* **2013**, *47* (12), 3866–3877. https://doi.org/10.1016/j.watres.2012.11.060.

[59] Kraas, M.; Schlich, K.; Knopf, B.; Wege, F.; Kägi, R.; Terytze, K.; Hund-Rinke, K. Long-Term Effects of Sulfidized Silver Nanoparticles in Sewage Sludge on Soil Microflora. *Environ Toxicol Chem* **2017**, *36* (12), 3305–3313. https://doi.org/10.1002/etc.3904.

[60] Jesmer, A. H.; Velicogna, J. R.; Schwertfeger, D. M.; Scroggins, R. P.; Princz, J. I. The Toxicity of Silver to Soil Organisms Exposed to Silver Nanoparticles and Silver Nitrate in Biosolids-Amended Field Soil. *Environ Toxicol Chem* **2017**, *36* (10), 2756–2765. https://doi.org/10.1002/etc.3834.

[61] Xiang, Q.-Q.; Li, Q.-Q.; Wang, P.; Yang, H.-C.; Fu, Z.-H.; Liang, X.; Chen, L.-Q. Metabolomics Reveals the Mechanism of Persistent Toxicity of AgNPs at Environmentally Relevant Concentrations to *Daphnia Magna*. *Environ Sci Nano* **2025**, *12* (1), 563–575. https://doi.org/10.1039/D4EN00350K.

[62] Forstner, C.; Orton, T. G.; Wang, P.; Kopittke, P. M.; Dennis, P. G. Wastewater Treatment Processing of Silver Nanoparticles Strongly Influences Their Effects on Soil Microbial Diversity. *Environ Sci Technol* **2020**, *54* (21), 13538–13547. https://doi.org/10.1021/acs.est.0c01312.

[63] Li, L.; Stoiber, M.; Wimmer, A.; Xu, Z.; Lindenblatt, C.; Helmreich, B.; Schuster, M. To What Extent Can Full-Scale Wastewater Treatment Plant Effluent Influence the Occurrence of Silver-Based Nanoparticles in Surface Waters? *Environ Sci Technol* **2016**, *50* (12), 6327–6333. https://doi.org/10.1021/acs.est.6b00694.

[64] Soler de la Vega, A. C.; Cruz-Alcalde, A.; Sans Mazón, C.; Barata Martí, C.; Diaz-Cruz, M. S. Nano-TiO2 Phototoxicity in Fresh and Seawater: Daphnia Magna and Artemia Sp. as Proxies. *Water (Basel)* **2020**, *13* (1), 55. https://doi.org/10.3390/w13010055.

[65] Botta, C.; Labille, J.; Auffan, M.; Borschneck, D.; Miche, H.; Cabié, M.; Masion, A.; Rose, J.; Bottero, J.-Y. TiO2-Based Nanoparticles Released in Water from Commercialized Sunscreens in a Life-Cycle Perspective: Structures and Quantities. *Environ Pollut* **2011**, *159* (6), 1543–1550. https://doi.org/10.1016/j.envpol.2011.03.003.

[66] Sendra, M.; Sánchez-Quiles, D.; Blasco, J.; Moreno-Garrido, I.; Lubián, L. M.; Pérez-García, S.; Tovar-Sánchez, A. Effects of TiO2 Nanoparticles and Sunscreens on Coastal Marine Microalgae: Ultraviolet Radiation Is Key Variable for Toxicity Assessment. *Environ Int* **2017**, *98*, 62–68. https://doi.org/10.1016/j.envint.2016.09.024.

[67] Ragnarsdóttir, O.; Abdallah, M. A.-E.; Harrad, S. Dermal Uptake: An Important Pathway of Human Exposure to Perfluoroalkyl Substances? *Environ Pollut* **2022**, *307*, 119478. https://doi.org/10.1016/j.envpol.2022.119478.

[68] Anderson, S. E.; Meade, B. J. Potential Health Effects Associated with Dermal Exposure to Occupational Chemicals. *Environ Health Insights* **2014**, *8s1*, EHI. S15258. https://doi.org/10.4137/EHI.S15258.

[69] Bergamaschi, E.; Garzaro, G.; Wilson Jones, G.; Buglisi, M.; Caniglia, M.; Godono, A.; Bosio, D.; Fenoglio, I.; Guseva Canu, I. Occupational Exposure to Carbon Nanotubes and Carbon Nanofibres: More Than a Cobweb. *Nanomaterials* **2021**, *11* (3), 745. https://doi.org/10.3390/nano11030745.

[70] Gupta, S. S.; Singh, K. P.; Gupta, S.; Dusinska, M.; Rahman, Q. Do Carbon Nanotubes and Asbestos Fibers Exhibit Common Toxicity Mechanisms? *Nanomaterials* **2022**, *12* (10), 1708. https://doi.org/10.3390/nano12101708.

[71] Nowack, B.; David, R. M.; Fissan, H.; Morris, H.; Shatkin, J. A.; Stintz, M.; Zepp, R.; Brouwer, D. Potential Release Scenarios for Carbon Nanotubes Used in Composites. *Environ Int* **2013**, *59*, 1–11. https://doi.org/10.1016/j.envint.2013.04.003.

[72] Petersen, E. J.; Zhang, L.; Mattison, N. T.; O'Carroll, D. M.; Whelton, A. J.; Uddin, N.; Nguyen, T.; Huang, Q.; Henry, T. B.; Holbrook, R. D.; Chen, K. L. Potential Release Pathways, Environmental Fate, And Ecological Risks of Carbon Nanotubes. *Environ Sci Technol* **2011**, *45* (23), 9837–9856. https://doi.org/10.1021/es201579y.

[73] Gómez-Sagasti, M. T.; Epelde, L.; Anza, M.; Urra, J.; Alkorta, I.; Garbisu, C. The Impact of Nanoscale Zero-Valent Iron Particles on Soil Microbial Communities Is Soil Dependent. *J Hazard Mater* **2019**, *364*, 591–599. https://doi.org/10.1016/j.jhazmat.2018.10.034.

[74] Fate, Transport, and Toxicity of Nanoscale Zero-Valent Iron (NZVI) Used During Superfund Remediation; Washington, 2009.

[75] Hardman, R. A. Toxicologic Review of Quantum Dots: Toxicity Depends on Physicochemical and Environmental Factors. *Environ Health Perspect* **2006**, *114* (2), 165–172. https://doi.org/10.1289/ehp.8284.

[76] Lovrić, J.; Bazzi, H. S.; Cuie, Y.; Fortin, G. R. A.; Winnik, F. M.; Maysinger, D. Differences in Subcellular Distribution and Toxicity of Green and Red Emitting CdTe Quantum Dots. *J Mol Med* **2005**, *83* (5), 377–385. https://doi.org/10.1007/s00109-004-0629-x.

[77] Lin, H.; Buerki-Thurnherr, T.; Kaur, J.; Wick, P.; Pelin, M.; Tubaro, A.; Carniel, F. C.; Tretiach, M.; Flahaut, E.; Iglesias, D.; Vázquez, E.; Cellot, G.; Ballerini, L.; Castagnola, V.; Benfenati, F.; Armirotti, A.; Sallustrau, A.; Taran, F.; Keck, M.; Bussy, C.; Vranic, S.; Kostarelos, K.; Connolly, M.; Navas, J. M.; Mouchet, F.; Gauthier, L.; Baker, J.; Suarez-Merino, B.; Kanerva, T.; Prato, M.; Fadeel, B.; Bianco, A. Environmental and Health Impacts of Graphene and Other Two-Dimensional Materials: A Graphene Flagship Perspective. *ACS Nano* **2024**, *18* (8), 6038–6094. https://doi.org/10.1021/acsnano.3c09699.

[78] Ye, W.; Lao, W.; Wu, L.; Guan, D.-X.; Zhang, C.; Feng, Y.; Mao, L. Enzyme-Driven Biodegradation of Ti 3 C 2 MXene: Unveiling Peroxidase-Mediated Pathways and Enhanced Bioaccumulation Risks in Aquatic Systems. *Environ Sci Nano* **2025**, *12* (6), 3357–3369. https://doi.org/10.1039/D5EN00124B.

[79] Nasrallah, G. K.; Al-Asmakh, M.; Rasool, K.; Mahmoud, K. A. Ecotoxicological Assessment of Ti 3 C 2 T $_x$ (MXene) Using a Zebrafish Embryo Model. *Environ Sci Nano* **2018**, *5* (4), 1002–1011. https://doi.org/10.1039/C7EN01239J.

[80] Peng, Z.; Liu, X.; Zhang, W.; Zeng, Z.; Liu, Z.; Zhang, C.; Liu, Y.; Shao, B.; Liang, Q.; Tang, W.; Yuan, X. Advances in the Application, Toxicity and Degradation of Carbon Nanomaterials in Environment: A Review. *Environ Int* **2020**, *134*, 105298. https://doi.org/10.1016/j.envint.2019.105298.

[81] Forstner, C.; Orton, T. G.; Skarshewski, A.; Wang, P.; Kopittke, P. M.; Dennis, P. G. Graphene Oxide Affects Soil Bacterial and Fungal Diversity Even at Parts-per-Trillion Concentrations. January 26, 2019. https://doi.org/10.1101/530485.

[82] Zhao, Y.; Liu, Y.; Zhang, X.; Liao, W. Environmental Transformation of Graphene Oxide in the Aquatic Environment. *Chemosphere* **2021**, *262*, 127885. https://doi.org/10.1016/j.chemosphere.2020.127885.

[83] Tang, W.; Zhang, X.; Hong, H.; Chen, J.; Zhao, Q.; Wu, F. Computational Nanotoxicology Models for Environmental Risk Assessment of Engineered Nanomaterials. *Nanomaterials* **2024**, *14* (2), 155. https://doi.org/10.3390/nano14020155.

[84] Longhin, E. M.; Rios-Mondragon, I.; Mariussen, E.; Zheng, C.; Busquets, M.; Gajewicz-Skretna, A.; Hofshagen, O.-B.; Bastus, N. G.; Puntes, V. F.; Cimpan, M. R.; Shaposhnikov, S.; Dusinska, M.; Rundén-Pran, E. Hazard Assessment of Nanomaterials: How to Meet the Requirements for (next Generation) Risk Assessment. *Part Fibre Toxicol* **2024**, *21* (1), 54. https://doi.org/10.1186/s12989-024-00615-4.

[85] Longhin, E. M.; Rios-Mondragon, I.; Mariussen, E.; Zheng, C.; Busquets, M.; Gajewicz-Skretna, A.; Hofshagen, O.-B.; Bastus, N. G.; Puntes, V. F.; Cimpan, M. R.; Shaposhnikov, S.; Dusinska, M.; Rundén-Pran, E. Hazard Assessment of Nanomaterials: How to Meet the Requirements for (next Generation) Risk Assessment. *Part Fibre Toxicol* **2024**, *21* (1), 54. https://doi.org/10.1186/s12989-024-00615-4.

[86] Zhang, R.; Zheng, X.; Fan, W.; Wang, X.; Zhao, T.; Zhao, X.; Peijnenburg, W. J. G. M.; Vijver, M. G.; Wang, Y. Fate Models of Nanoparticles in the Environment: A Critical Review and Prospects. *Environ Sci Nano* **2025**, *12* (7), 3394–3412. https://doi.org/10.1039/D5EN00342C.

[87] Schwirn, K.; Voelker, D.; Galert, W.; Quik, J.; Tietjen, L. Environmental Risk Assessment of Nanomaterials in the Light of New Obligations Under the REACH Regulation: Which Challenges Remain and How to Approach Them? *Integr Environ Assess Manag* **2020**, *16* (5), 706–717. https://doi.org/10.1002/ieam.4267.

[88] EPA 2017. https://www.legislation.vic.gov.au/in-force/acts/environment-protection-act-2017/004.

[89] EC. https://commission.europa.eu/law/law-topic/data-protection_en.

[90] Xu, X.; Ren, P.; Shi, W.; Deng, F.; Wang, Q.; Shi, S.; Shen, J.; Dong, G.; Han, J. Occupational Health and Safety Regulations in China: Development Process, Enforcement Challenges, and Solutions. *Front Public Health* **2025**, *13*. https://doi.org/10.3389/fpubh.2025.1522040.

[91] Wang, B.; Wu, C.; Kang, L.; Huang, L.; Pan, W. What Are the New Challenges, Goals, and Tasks of Occupational Health in China's Thirteenth Five-Year Plan (13th FYP) Period? *J Occup Health* **2018**, *60* (3), 208–228. https://doi.org/10.1539/joh.2017-0275-RA.

[92] Klemm, D.; Kramer, F.; Moritz, S.; Lindström, T.; Ankerfors, M.; Gray, D.; Dorris, A. Nanocelluloses: A New Family of Nature-Based Materials. *Angew Chem Int Ed* **2011**, *50* (24), 5438–5466. https://doi.org/10.1002/anie.201001273.

[93] NIOSH. https://www.cdc.gov/niosh/docs/2009-125/pdfs/2009-125.pdf.

[94] Mitrano, D. M.; Barber, A.; Bednar, A.; Westerhoff, P.; Higgins, C. P.; Ranville, J. F. Silver Nanoparticle Characterization Using Single Particle ICP-MS (SP-ICP-MS) and Asymmetrical Flow Field Flow Fractionation ICP-MS (AF4-ICP-MS). *J Anal At Spectrom* **2012**, *27* (7), 1131. https://doi.org/10.1039/c2ja30021d.

9

Nanomaterials for the Construction Industry

Vaibhav Sharma and Piyush Gupta

9.1 Introduction

Nanotechnology is a major area of research for civil engineering applications. The construction industry is a sector that might significantly benefit from nanomaterials (NMs) with novel functions and improved performance. The quality of construction materials is enhanced by using NMs, and the real estate sector takes advantage of the usage of NMs (Ahadi 2011). The incorporation of NMs in construction structures would not only extend their lifespan but also monitor energy consumption while assessing their responses to various agents such as fire, corrosion, water infiltration, fractures, and cracks. NMs accelerate the hydration rate during the first phase, enhance water absorption due to their surface characteristics, and diminish the availability of free water (Pedro Muñoz-Pérez et al. 2022). The integration of diverse NMs, including nano-silica (nano-SiO_2), nano-titanium dioxide (nano-TiO_2), carbon nanotubes (CNTs), nano-clay (NCly), and nano-alumina (nano-Al_2O_3), can substantially improve the mechanical properties of cementitious construction materials, especially in high-strength concrete applications (Sujitha, Ramesh, and Xavier 2024). Their incorporation into concrete can enhance compressive and tensile strength, as well as abrasion resistance (Jones et al. 2019). Among the NMs utilized in concrete, nano-silica has superior pozzolanic properties. It reacts with free lime generated during cement hydration to form additional C-S-H gel, which improves the concrete's strength, reduces its permeability, and increases its durability (Gopinath et al. 2012). NMs are utilized in the construction of bridges or platforms, hospitals, bricks, tiles, and ceramics (Mohtasham Moein et al. 2024). The incorporation of NMs significantly improves the qualities of asphalt, such as viscosity, rigidity, flexibility, and resistance to fatigue. It also increases the resistance to aging and UV radiation, thereby enhancing the quality of pavements and prolonging their lifespan (Yao et al. 2025). Nanoparticles (NPs), including silica fume (amorphous SiO_2), diminish concrete weight, increase strength and flexibility, and enhance insulation, hence reducing energy consumption in buildings. They enhance weather resistance, facilitate self-cleaning surfaces, purify vehicular emissions, augment crack resistance in polymers, provide biocidal surfaces for surgical environments, and improve fire resistance. Nano-TiO_2 improves concrete durability and preserves whiteness, utilizing its photocatalytic characteristics to decompose organic contaminants, bacteria, and nitrogen oxides (NO_x) (Broekhuizen et al. 2011). Nano-TiO_2 is employed in self-cleaning coatings for glass and surfaces,

DOI: 10.1201/9781003632498-9

which improves the performance of building materials. Nano-SiO$_2$ is frequently used in concrete to augment its compressive strength and diminish permeability, resulting in more durable constructions. CNTs have been added to enhance the mechanical strength and flexibility of composites. These NMs provide substantial benefits, such as better durability, lower maintenance expenses, improved energy efficiency, and superior environmental resistance, thereby advancing and sustaining construction practices. Thus, this chapter summarizes the applications of various NMs that can increase the overall performance of the construction sector.

9.2 NMs Used in Construction Industries

The integration of NMs in the building sector can enhance efficiency, sustainability, and resilience, facilitating novel solutions to modern difficulties. NMs, including nano-SiO$_2$, nano-TiO$_2$, CNT, nano-Fe$_2$O$_3$, and nano-cellulose (NCe), can enhance the endurance of structures by augmenting their mechanical and thermal properties. This may lead to a decrease in energy use and overall costs within the concrete sector (Saleem, Zaidi, and Alnuaimi 2021). The production of concrete has significantly improved by the incorporation of diverse supplemental cementitious materials and NMs. Due to their small size and enhanced surface area, NMs provide significant potential for augmenting the durability properties of concrete (Goel et al. 2022). NMs concrete is a next-generation concrete composed of materials with nanoscale grain size, which utilizes cement type I, nano-SiO$_2$ ranging from 10 to 150 nm, quartz powder sized between 0.3 and 25.0 μm, fine sand (quartz) measuring 50–650 μm, coarse aggregate sized 5–10 mm, and a super-plasticizer (Safiuddin et al. 2014). The various NMs are utilized in asphalt modification (Fang et al. 2013). The incorporation of NMs into asphalt binder enhances its ability to resist aging and moisture damage, along with its effectiveness under high and moderate temperatures (Ashish and Singh 2021). Diverse nanostructures of NMs demonstrated varying properties. NMs, owing to their reduced size and exceptionally large surface area, engage more vigorously with other particles within the soil matrix. The engineering properties of soil could be changed by using a small amount of NMs (Majeed and Taha 2013). The physical and mechanical qualities of soil were enhanced when mixed with NMs, which depended on the type, quality, and ratio of the NMs used, as well as the features of the native soil (Khudher et al. 2020).

9.2.1 Nano-SiO$_2$ and Nano-TiO$_2$

The incorporation of nano-SiO$_2$ in normal-strength concrete has been found to enhance compressive strength while reducing split tensile strength and flexural strength. Chloride permeability was reduced in nano-coated concrete (Gopinath et al. 2012). Lightweight coarse aggregate (LWA) was partially replaced with natural coarse aggregate at different proportions. The limitations of LWAs may be alleviated through the mixing of nano-SiO$_2$ particles. Incorporating 3% nano-SiO$_2$ into high-strength lightweight concrete (HSLWC) effectively reduced the drawbacks of LWA and led to a notable improvement in mechanical performance, while sorptivity and gas permeability decreased by as much as 25% and 40%, respectively, when compared to equivalent replacement levels of LWAs. Furthermore, nano-SiO$_2$ particles demonstrated superior performance in conventional concrete relative to LWCs

(Atmaca, Abbas, and Atmaca 2017). The interfacial transition zone became smaller and denser with the addition of nano-SiO_2. Nano-SiO_2 enhanced compressive strength and reduced porosity in lightweight concrete. Both water and chloride ion resistance were increased with 1% nano-SiO_2 (Du et al. 2015). Adding 2% nano-SiO_2 significantly reduced water sorptivity, permeable void content, chloride penetration, and porosity in high-volume fly ash (HVFA) concrete. The durability characteristics of concrete containing 38% Class F fly ash and 2% nano-SiO_2 as partial cement substitutes surpassed those of conventional cement concrete (Supit and Shaikh 2015). The addition of 1% nano-TiO_2 in concrete could achieve the 64.65 N/mm^2 (28 days) of compressive strength and 7.27 N/mm^2 (28days) of flexural strength (Patel and Mishra 2018). The ideal concentration of polyacrylonitrile was approximately 0.05 wt% in mortar for flexural and compressive strength, whereas the highest NO_x degradation achieved was approximately 77% in samples containing 1 wt% Nano-TiO_2, irrespective of the polyacrylonitrile concentration (Senff et al. 2015). Raw black rice husk ash (BRHA) was used as a partial replacement for cement, with nano-TiO₂ added at concentrations of 0.5%, 1.0%, and 1.5% in the blended cement. The inclusion of nano-TiO₂ improved both the mechanical properties and microstructural quality of the BRHA-based mortars (Noorvand et al. 2013). The use of 0.5% nano-Al_2O_3 combined with 1% nano-TiO₂ in cement led to increases of 42%, 34%, and 28% in compressive, splitting tensile, and flexural strengths, respectively. These NMs could improve the pore structure (Atiq Orakzai 2021). Nano-SiO_2 and nano-TiO_2 were responsible for achieving better compressive strength as compared to the control mortar. The compressive strength increased by 50% and 26% when 3% nano-SiO_2 and 1% nano-TiO_2 were added after 56 days. Moreover, the corrosion resistance of mortar was also enhanced by adding NMs (Vitharana et al. 2020). Reinforcing materials might be combined with MNs, which improve the performance and lifetime of concrete constructions. The mixture of nano-TiO_2 and CNTs in cement could endow structures with self-cleaning and self-sensing properties. These benefits may facilitate the photocatalytic degradation of contaminants and the structural health assessment of concrete constructions. NMs possess significant promise for applications in intelligent infrastructure utilizing high-strength concrete constructions (Bautista-Gutierrez et al. 2019). The shrinkage values of alkali-activated mortar diminished with the incorporation of polypropylene fibers (PPF) and NMs. When PPF were used in conjunction with NMs, there was a notable decrease in shrinkage and expansion, along with improved strength characteristics. This combination also achieved the lowest observed shrinkage and expansion levels, as well as the highest strength, which were observed in the alkali-activated mortar mix containing 2% nano-SiO_2, 1% nano-Al_2O_3, and 0.5% PPF (Dheyaaldin, Mosaberpanah, and Alzeebaree 2022). The addition of NMs notably enhanced the mechanical properties of concrete. The optimal dosages were found to be 3.54% for nano-SiO₂ and 4.24% for nano-Fe₂O₃. Concrete with these proportions achieved 67.27 MPa compressive strength, 8.95 MPa splitting tensile, and 9.56 MPa flexural strength. Nano-SiO_2 served as a supplementary cementing agent, while nano-Fe_2O_3 enhanced the hydration reaction rate of cementitious materials (Du et al. 2023). The addition of nano-SiO_2, nano-TiO_2, and nano-zinc oxide in cement was used to increase the anti-UV aging property. The composite NMs modifiers had a positive impact on the properties of asphalt (Zhang et al. 2021a). The base asphalt was modified with nano-zinc oxide, nano-SiO_2, and nano-TiO_2. NMs could reduce the length of linear chains of unaged materials but the phase angle aging index increased

(Zhang et al. 2021b). The pore size was increased when nano-SiO_2 and nano-Al_2O_3 were added to cement under low humidity and pressure conditions. The mass loss and porosity also improved (Zhang et al. 2020). NC and nano-SiO_2 were used to decrease the drying shrinkage up to 1% and increasing 28 days compressive strength and early strength in recycled mortar (Hou et al. 2020).

9.2.2 Nanoclays

Nanoclays (NCly) have emerged as potential additives to construction materials, serving as additional cementitious components. These NPs have novel characteristics that can increase the efficiency of concrete. The compressive strength, sustainability, and durability of concrete structures are improved by using NCly. Large surface area and high reactivity facilitate better chemical bonding between cement particles, resulting in improved mechanical characteristics (Bantie et al. 2024). The mixing of NCly in high-performance concrete could improve compressive, flexural strength, and enhance durability. Additionally, it helps reduce porosity, permeability, and water absorption (Mansi et al. 2022). At high temperature, the addition of NCly could improve the tensile modulus (Al-Safy et al. 2012). Two varieties of NCly were utilized: nanofil-15 and cloisite-15A. NCly modified the rheological characteristics of bitumen, enhancing stiffness, reducing phase angle, and improving resistance to aging (Jahromi and Khodaii 2009). Nano-Al_2O_3-modified cement pastes showed improvement in compressive strength at all ages due to a decrease in the number of macropores and the catalytic effect of nano-Al_2O_3 on cement hydration. However, the strength gain at 28 days was less significant (Zhan et al. 2019). The incorporation of 0.5% Al_2O_3 nano-powder enhanced the adherence of the cement mortar overlay to the concrete substrate. The cement mortar utilized for the overlay, incorporating 0.5% Al_2O_3 nano-powder, exhibited reduced porosity compared to the reference mortar at the interphase (Szymanowski and Sadowski 2019). Nano-metakaolin (NMK) is utilized in concrete applications. The incorporation of NMK enhanced some characteristics of concrete, particularly its mechanical qualities at early stages (Zhan et al. 2020). The use of 25% NMK as a substitute for cement enhanced compressive, flexural, split, and tensile strengths, as well as durability. The incorporation of NMK significantly enhanced water permeability and absorption, resulting in increased concrete density. The incorporation of NMK in the formulation of acid-resistant concrete demonstrated favorable outcomes in terms of chloride permeability and sulfate resistance. The use of NMK demonstrated enhanced flowability in concrete and cement mortar. It improved the workability of concrete. It mitigated shrinking resulting from particle compaction (Chandak and Pawade 2018). The coastal and marine concrete structures might be protected from chloride attack when NMK is used as a supplementary cementitious material. 10% NMK and 5% NMK helped to achieve a compressive strength of 49.8 MPa and 8.35 MPa, respectively (Pillay et al. 2022). The reference concrete M30 was formulated with 53-grade ordinary Portland cement, whereas the other mixtures had a partial substitution of ordinary Portland cement with metakaolin. The replacement levels for metakaolin were 5%, 10%, 15%, and up to 20% (by weight). The incorporation of metakaolin as a substitute for cement in concrete enhanced its performance by up to 15% (John 2013). Substituting up to 15% of metakaolin for ordinary Portland cement enhanced the mechanical characteristics of concrete. The addition of metakaolin up to 15% led to an increase in the corrosion of carbon steel (Parande et al. 2008). Nano-TiO$_2$ improves the compressive strength of cementitious materials and reduces moisture loss in cement-based materials during the drying process. It minimizes drying shrinkage in cement-based products (Zhang et al. 2015). Incorporating NMs such as nano-copper, NCly, and nano-magnesium at concentrations below 1% of the soil's dry weight could enhance the soil's properties. Geotechnical properties of soil could be improved by the addition of NMs (Majeed et al. 2014). Micro-sand, micro-cement, nano-SiO$_2$, NCly, and naphthalene sulphonate work as superplasticizers in nano-cement mortar. Compressive strength as well as tensile strength of the mortar were increased during the early stages of hardening, with a maximum increase of 22% in compressive strength, while a 3.7-fold increase in compressive strength was noted compared to traditional concrete strength levels (Al-Rifaie and Ahmed 2016).

9.2.3 Carbon Nanotubes

Carbon Nanotubes (CNTs) are highly promising NMs, offering exceptional chemical stability, mechanical, electrical, and thermal properties. Incorporating CNTs can substantially improve the mechanical, electrical, and thermal performance of composites (Kim et al. 2019). There are two distinct types of CNTs, i.e., single-walled carbon nanotubes and multi-walled carbon nanotubes (MWCNTs). CNTs are unique nanoscale materials that decrease the rate of cracking and permeability while enhancing resistance, strength, durability, and self-sensing properties. The incorporation of minimal amounts of CNTs enhanced fracture energy and toughness (Yesudhas Jayakumari, Swaminathan, and Partheeban 2023). In cases of concrete cracking, adding 0.05%–0.1% CNTs by weight of cement significantly increased mechanical strength up to 21% and durability up to 25%. The best performance was seen in concrete with higher CNT content and a lower aspect ratio (Carriço et al. 2018). The compressive and flexural properties of the cementitious binder are increased by the utilization of CNTs. The initial stiffness of textile-reinforced mortar (TRM) was improved when CNTs were mixed with the cementitious binder. The debonding of the TRM layers was also observed (Irshidat and Al-Shannaq 2018). CNTs below the critical incorporation concentration enhanced the mechanical qualities of ultra-high-performance concrete by facilitating pore filling, bridging, and creating a denser calcium–silicate–hydrate structure. CNTs significantly enhanced electrical conductivity and the resultant electromagnetic shielding efficacy up to the percolation threshold (Jung et al. 2020). Composite specimens with MWNTs showed a dramatic increase in electrical conductivity (over three orders of magnitude) and notable enhancements in compressive (36%) and flexural strength (18%) (Naeem et al. 2017). CNTs enhanced soil reinforcement by augmenting compressive strength. Nano-bentonite served as an additive in drilling fluids and composite materials to enhance soil mechanical properties. Colloidal silica and laponite were employed to mitigate or prevent soil liquefaction due to their unique rheological characteristics (Huang and Wang 2016). Nano-ZrO$_2$, nickel-coated MWCNTs, nano-boron nitride were the nanofillers used in cement mortar. The bond strength was increased up to 35.1%, 38.8%, and 42.8% due to the addition of 2 wt% nano-ZrO$_2$, 0.3 wt% Ni-coated MWCNTs, and 0.3 wt% nano-boron nitride in cement mortar (Wang et al. 2020). Four NMs are predominantly utilized to enhance concrete properties: nano-SiO$_2$, nano-TiO$_2$, CNTs, and carbon nanofibers. All four of these NMs have demonstrated enhancements in various concrete characteristics. Nano-TiO₂ and nano-SiO₂ both reduce bleeding and segregation, while also improving mechanical characteristics. Both carbon nanofibers and CNTs markedly enhance the mechanical characteristics of concrete (Safiuddin et al. 2014). Adding 1% oxidized multi-walled CNTs significantly boosts the compressive and flexural strength of Portland cement (Rattan, Sachdeva, and Chaudhary 2016). Nano-SiO_2, CNT, and nano-TiO_2 improved the 28-day strength of cement mortar when incorporated in optimum amounts, and the same optimization applied to concrete yielded favorable outcomes (Chakraborty, Ghosal, and Chakraborty 2019).

9.2.4 Cenosphere

Fly ash and Cenosphere (Cs) are mixed with concrete to enhance the strength of the concrete. Cs works as a pozzolanic material (Kavinkumar, Priya, and Praneeth 2023). Cs can be used as a partial replacement for fine aggregates, achieving compressive strengths of 21.27 N/mm² at 7 days and 33.80 N/mm² at 28 days with a 20% replacement. A 30% Cs replacement yielded splitting tensile strengths of 2.97 N/mm² and 3.52 N/mm² at 7 and 28 days, respectively (Zanjad, Pawar, and Nayak 2024). The addition of Cs in cement-based composites helps in decreasing the shrinkage. The value of embodied energy decreased from 10.12 to 7.20 GJ/m³ when 30% Cs was added to cement-based composites. There was an increase of about 63.94 to 70.69 MPa and 9.50 to 14.71 MPa in the compressive and flexural strength, respectively (Chen et al. 2020). The mixture of 80% fly ash Cs and superplasticizer was regarded as sustainable lightweight structural concrete of M25 grade (Patel et al. 2020). In Cs-based high-strength concrete, 70 MPa strength and 1500 kg/m³ density were obtained, and Cs acts as a fine aggregate, which increased the strength (Souza et al. 2019). The maximum compressive strength was achieved when samples were treated with 7% Cs. The concrete's strength, with the incorporation of Cs, was observed to more than double compared to a sample free from additives (Tanirbergenova et al. 2023). Cementitious composites incorporating fly ash Cs as a lightweight filler material showed an enhancement in mechanical performance, with specific strength values between 34.69 and 24.11 kPa/kg/m³. At 28 days, compressive strength 55.92–30.38 MPa, flexural strength 9.29–5.38 MPa, and tensile strength 3.51–1.66 MPa were observed. The elevated pozzolanic activity of fly ash Cs enhanced the mechanical characteristics, even at a lesser density (Hanif et al. 2017). Fly ash Cs were incorporated into manufactured sand. Strength decreased as the fly ash Cs content increased from 0% to 100%. Nevertheless, when the replacement level is kept below 35% by volume, the strength stays within the target limit of 31.2 N/mm² for M25 grade conventional concrete. The optimal recommendation was 35% fine aggregate as fly ash Cs (Kowsalya et al. 2022). Cement composites including aluminosilicate Cs exhibited insulating properties of 0.25 W/mK compared to standard cement-based mortars, making them a promising alternative to conventional materials. With compressive strengths ranging from 37.68 to 17.74 MPa, they could be utilized as structural components (Strzałkowski et al. 2023).

9.2.5 Nano-Fe₂O₃

This metal oxide nanoparticle is frequently utilized in construction owing to its magnetic characteristics, thermal endurance, and chemical resistance. The incorporation of nano-Fe_2O_3 powder as an additive in concrete offers numerous advantages, including a reduction in waste disposal and substantial cost savings compared to new materials (Edward et al. 2021). Density increased despite a reduction in workability. At 28 days, the compressive, tensile, and flexural strengths of the nano-Fe_2O_3-incorporated reinforced concrete enhanced by 24.82%, 6.58%, and 15.45%, respectively (Saini, Chaturvedy, and Pandey 2024). In construction, nano-Fe_2O_3 is commonly utilized as a pigment in paints and coatings, while providing superior UV and corrosion protection. It serves as a nano-additive in concrete, enhancing mechanical strength, reducing porosity, and improving durability. The synergistic impact of nano-Fe_2O_3 and sisal fiber reinforcement resulted in improved performance of electronic plastic waste concrete, with optimal performance noted in the electronic plastic waste concrete amended with 4% nano-iron oxide particles and 1% sisal fiber. Furthermore, its magnetic and dense properties render it appropriate for radiation shielding materials and protective coatings that inhibit corrosion in steel structures (Amjad, Ahmad, and Irshad Qureshi 2023).

9.2.6 Nano-Al_2O_3

Nano-Al_2O_3 has gained popularity in the construction sector for its remarkable hardness, thermal conductivity, and resistance to abrasion and corrosion. It is a material that is electrically insulating yet thermally conducting, rendering it appropriate for both structural and finishing applications. Improved mechanical properties were observed with the addition of 4% silica fume and 0.5% nano-Al_2O_3 compared to concrete without additives and concrete containing only 4% silica fume (Alazemi 2018). Mixing of Al_2O_3 nanofibers could increase the strength of cement-based materials up to 195 MPa. Nanofibers can offer calcium silicate hydrate nano-reinforcement and diminish shrinking. In ultra-high-performance concrete, the integration of Al_2O_3 nanofibers can substitute silica fume or metakaolin (Muzenski, Flores-Vivian, and Sobolev 2019). In concrete, nano-Al_2O_3 functions as a reinforcing ingredient that markedly enhances compressive and flexural strength. Maximum flexural and compressive strengths were achieved using 1% nano-Al_2O_3 with 100% fly ash, while the highest splitting tensile strength was obtained with 1.5% nano-Al_2O_3. Additionally, the inclusion of nano-Al_2O_3 significantly extends the setting time of alkali-activated self-compacting concrete, supporting its production under ambient conditions (Younus, Mosaberpanah, and Alzeebaree 2023).

9.2.7 Graphene and Graphene Oxide

They are a novel category of carbon-based NMs with exceptional mechanical and multifunctional characteristics. Graphene is a monolayer of carbon atoms organized in a 2D honeycomb lattice, known for its strength, approximately 200 times that of steel, and its superior electrical and thermal conductivity (Chuah et al. 2014). GO is an NM characterized by 1D at the nanoscale, while the other 2D are larger in scale. One advantage of GO over other NPs is its oxygen functions, which facilitate easy dispersion in an aqueous medium (Akarsh et al. 2022). The addition of 0.25 wt.% GO increased the modulus of elasticity by 30.5% and compressive strength of GBFS-FA concrete by 37.5%, while chloride ion permeability reduced by 35.3% compared to the GO-free counterparts. The highest reduction of 33% in thermal conductivity was observed in the FA65-G35 wt% samples. Furthermore, they work as smart materials with self-sensing properties and are utilized in coatings and barriers that offer enhanced protection against UV radiation, moisture, and abrasion. Their distinctive qualities render them viable candidates for application in energy-efficient structures and sophisticated insulation systems (Maglad et al. 2022).

9.2.8 Nano-Cellulose

Nano-Cellulose (NCe) is an NM that is derived from biomass resources characterized by its exceptional strength, elevated specific surface area, and high surface energy. Moreover, it is non-toxic. The properties of NCe could be altered by various techniques. Due to its elevated elastic modulus, stiffness, and minimal thermal expansion coefficient, NCe serves as an exceptional material for polymer reinforcement (Shen et al. 2020). The utilization of NCe in cementitious materials has garnered increasing attention in recent years due to its renewability, sustainability, and distinctive features (Guo et al. 2020). NCe enhances the mechanical properties of cement systems, with improvements driven by advancements at the nano/microscale. It appears to boost the hydration process and improve the micromechanical properties of the C-S-H gel matrix. The addition of cellulose filaments (CF) led to increases of 15%–25% in macro-mechanical parameters, with a maximum increase of 74% in flexural toughness, a 15% increase in the degree of hydration, and a 12%–25% enhancement in the micromechanical properties of the C-S-H matrix (Hisseine et al. 2019).

9.3 Application of NMs as Additives in Concrete Sector

The incorporation of nanotechnology, i.e., NMs, in the construction sector, particularly in the formulation of asphalt, road construction, and concrete pavement materials, presents significant potential for improving infrastructure performance, strength, durability, and longevity (Imoni et al. 2023; Utsev et al. 2022). Pervious concrete is regarded as a pavement material and causes less environmental impact; therefore, it has gained more attention (Dash and Gupta 2022; Gupta and Itishree 2022). There was a significant enhancement in compressive strength due to the incorporation of NPs at various ages. Furthermore, the incorporation of Nano-SiO_2 and NMK has demonstrated a significant decrease in water absorption and porosity of high-strength concrete and high-performance concrete. Moreover, the incorporation of Nano-SiO_2 and NMK improved resistance to magnesium sulfate, as well as to both nitric and sulfuric acids (at a pH value of 1.0), across various concrete grades (Diab et al. 2019). Nano-SiO_2 exhibited the most substantial enhancement in durability and strength among the admixtures; an average improvement of 2%–10% was observed compared to micro-silica. Metakaolin had a minimal effect on durability and strength. The highest compressive strength was achieved with 4% nano-SiO_2 substitution, while the peak flexural strength was noted with 13% micro-silica substitution (Arel and Thomas 2017). The mixing of nano-SiO_2 with concrete may help in improving water resistance, which was confirmed by the water permeability test (Ji 2005). Nano-SiO_2 increased the viscosity and hydration process of cement paste. The compressive strengths and the bond strengths of cement paste in which nano-SiO_2 was added were significantly higher than those with silica fume, particularly at early ages. As the nano-SiO_2 content increased, the rate of bond strength enhancement surpassed that of compressive strength augmentation (Qing et al. 2007). The addition of nano-SiO_2 and silica fume improved the mechanical properties and durability of concrete. Additionally, incorporating macro-polymeric fibers into the mixture enhanced the mechanical characteristics, depending on the fiber volume content (Fallah and Nematzadeh 2017). Compressive strength of 42 MPa was achieved in 28 days, which was equivalent to conventional geopolymer. Furthermore, 1wt % of nano-SiO_2 could result in a 35% decrease in setting time, while thermal conductivity decreased from 0.709 W/mK to 0.505 W/mK (Mortada et al. 2023). Regression equations had been formulated to enhance the estimation of the ideal proportions of superplasticizer dosages. Mathematical models linking slump and slump flow parameters to the dosage of a nano-modified additive help design concrete mixes with the desired flowability. The use of this additive reduces the risk of segregation in the mix and promotes cement hydration during colder months (Svintsov et al. 2020).

Table 9.1 tabulates the utilization of various types of NMs and their advantages in the construction sector.

The different types of NMs and their applications in the field of construction are illustrated in Figure 9.1.

9.4 Challenges and Recommendations

There is a need to develop effective methods for the dispersion of NMs in cement samples. The modified cement-based materials will be produced on a large scale when an effective method is established for the dispersion of NMs that ensures stability and good integration within the cement matrix. Furthermore, an additional hurdle is in determining the ideal dosage of the NMs incorporated into the cement paste to achieve enhanced mechanical strength and increased durability. The toxicity of NMs should be determined by using scenario analysis, which may help in finding the factors that are responsible for the toxic nature of NMs. Findings indicate that energy consumption in supply chains, primarily attributed to sophisticated NMs manufacturing processes, may lead to higher ecotoxicity compared to the direct release of NMs in numerous practical scenarios. Nano-Ag exhibited greater toxicity, mostly attributable to upstream operations resulting in metal releases (silver mining), as well as the possibility of downstream silver releases in worst-case scenarios (Garvey et al. 2019). The challenges of NMs-based manufacturing encompass cost-effectiveness, product quality, desired functionality, process repeatability, and production scalability (Singh et al. 2022). To attain the ideal mechanical properties of cement-based materials, it is crucial to incorporate an appropriate amount of NMs into the mixture. Excessive amounts of NMs incorporated into cement can lead to diminished compressive, tensile, and flexural strength of cement-based constructions. Nonetheless, the incorporation of NMs in asphalt mixtures presents challenges owing to insufficient practical applications and inadequate understanding. There is a lack of comprehensive information on life cycle cost analysis, standard guidelines, and mixing procedures. There is a need for additional investigation for the comprehensive utilization of various NMs in asphalt modifications. The utilization of NMs in the construction sector has

TABLE 9.1

Utilization of Different Types of NMs and Their Advantages

Types of Nanomaterials	Area	Main Applications	References
Nano-SiO$_2$	High-strength lightweight concrete	25% of sorptivity reduced by 25% and gas permeability by 40%	(Atmaca, Abbas, and Atmaca 2017)
	Lightweight concrete	Water and chloride ion resistance were increased by 1% nano-SiO$_2$	(Du et al. 2015)
	HVFA concrete	2% nano-SiO$_2$ reduced moisture sorptivity, permeable voids, chloride permeability, and porosity in HVFA concretes	(Supit and Shaikh 2015)
NCly	Concrete	Improvement in tensile modulus	(Al-Safy et al. 2012)
	Bitumen binder	Modified the rheological properties of bitumen	(Jahromi and Khodaii 2009)
	Concrete	25% NMK could enhance water permeability and absorption	(Chandak and Pawade 2018)
	Metakaolin-based concrete	5% NMK: compressive strength 49.8 MPa, flexural strength 8.35 MPa; 10% NMK: compressive strength 49.8 MPa, flexural strength 8.35 MPa	(Pillay et al. 2022)
	Metakaolin blended concrete/ mortar	The corrosion resistance of carbon steel was enhanced by adding up to 15% metakaolin	(Parande et al. 2008)
Nano-Al$_2$O$_3$	Cement mortar	0.5% Al$_2$O$_3$ nanopowder exhibited reduced porosity	(Szymanowski and Sadowski 2019)
Combination of nano-Al$_2$O$_3$ and nano-TiO$_2$	Concrete	Compressive strength increased by up to 42%, splitting tensile by 34%, and flexural strength by 28%	(Atiq Orakzai 2021)
NCe	–	Special material for polymer reinforcement	(Shen et al. 2020)
CNTs	Concrete	0.05%–0.1% CNTs increased mechanical strength and durability by up to 21% and 25%, respectively	(Carriço et al. 2018)
	Cement concrete	Fracture energy and toughness were increased	(Yesudhas Jayakumari, Swaminathan, and Partheeban 2023)
	Cement composite	The composite specimens exhibited improvement in electrical conductivity, compressive and flexural strength	(Naeem et al. 2017)
	RC beams	CNTs were used in multi-layer TRM, which resulted in the debonding of the TRM layers	(Irshidat and Shannaq 2018)

TABLE 9.1 *(Continued)*

Utilization of Different Types of NMs and Their Advantages

Types of Nanomaterials	Area	Main Applications	References
Cs	Cement	Shrinkage property was decreased. 63.94–70.69 MPa and 9.50–14.71 MPa were the compressive and flexural strengths, respectively	(Chen et al. 2020)
	High-strength concrete	70 MPa strength and 1500 kg/m^3 density were achieved	(Souza et al. 2019)
	Lightweight structural concrete	Achieved M25-grade concrete when 80% fly ash Cs and superplasticizer were used in concrete	(Patel et al. 2020)

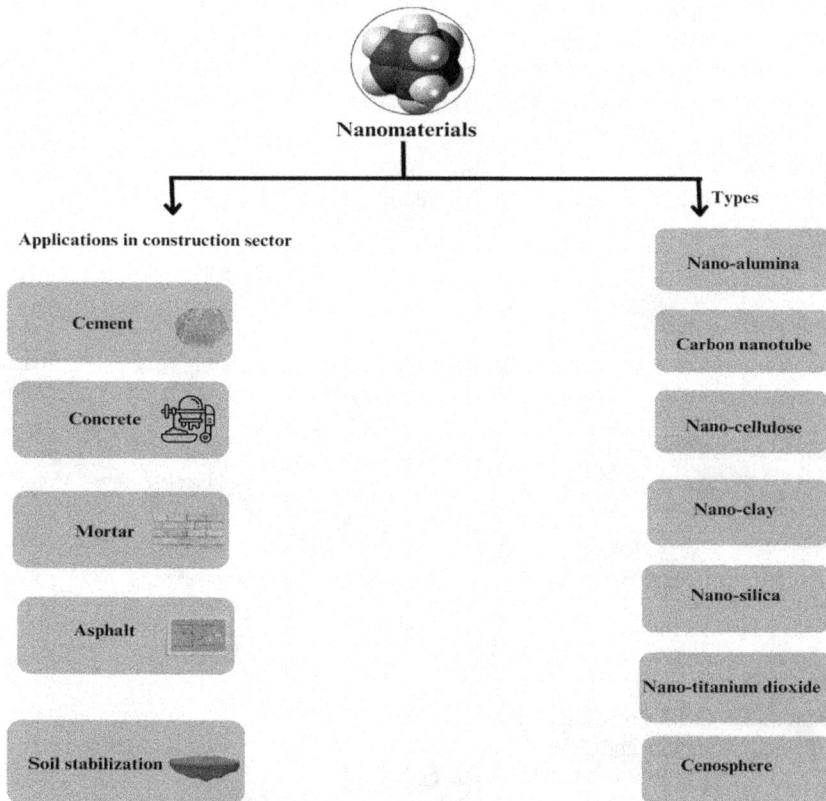

FIGURE 9.1 The different applications of NMs in the construction industry.

progressed from theoretical investigation to actual implementation. In this context, future research on NM-based construction will speed up and broaden the application of NMs in building materials. Developed nations significantly contribute to the research framework of NMs-based construction.

9.5 Conclusion

NMs in the construction sector present a sustainable alternative to mitigate the excessive consumption of non-renewable resources utilized in material production. Research indicates that the incorporation of NMs yields favorable outcomes in enhancing the mechanical properties of concrete, including augmented compressive and tensile strength. Additionally, in soil stabilization, improvements are observed in unconfined compressive strength and an increase in the plasticity limit. Furthermore, their application addresses asphalt issues such as grooving, cracking, and warping. The incorporation of diverse NMs into construction and composite materials has demonstrated considerable improvements in mechanical strength, durability, and functional efficacy. Nano-SiO_2 significantly decreases permeability and enhances durability at minimal doses, whereas NCly, such as NMK, augment strength, alter rheological characteristics, and bolster corrosion resistance. Nano-Al_2O_3, whether utilized independently or in conjunction with nano-TiO_2, diminishes porosity and enhances structural integrity. NCe serves as an eco-friendly reinforcement for polymers, whereas CNTs, even in low concentrations, markedly enhance strength, toughness, durability, and electrical conductivity. These findings underscore the efficacy and adaptability of NMs in enhancing conventional materials, providing sustainable and outstanding performance that will be used in engineering.

REFERENCES

Ahadi, Parisa. 2011. "Applications of nanomaterials in construction with an approach to energy issue." *Advanced Materials Research* 261–263: 509–14. https://doi.org/10.4028/WWW.SCIENTIFIC.NET/AMR.261-263.509.

Akarsh, P. K., D. Shrinidhi, Shriram Marathe, and Arun Kumar Bhat. 2022. "Graphene oxide as nano-material in developing sustainable concrete – a brief review." *Materials Today: Proceedings* 60: 234–46. https://doi.org/10.1016/J.MATPR.2021.12.510.

Al-Rifaie, Wail N., and Waleed K. Ahmed. 2016. "Effect of nanomaterials in cement mortar characteristics." *Journal of Engineering Science and Technology* 11 (9): 1321–32.

Al-Safy, Rawaa, Riadh Al-Mahaidi, George P. Simon, and Jana Habsuda. 2012. "Experimental investigation on the thermal and mechanical properties of nanoclay-modified adhesives used for bonding CFRP to concrete substrates." *Construction and Building Materials* 28 (1): 769–78. https://doi.org/10.1016/J.CONBUILDMAT.2011.09.009.

Alazemi, Athbi. 2018. "Investigate the effects of nano aluminum oxide on compressive, flexural strength, and porosity of concrete." Graduate Theses and Dissertations. https://ecommons.udayton.edu/graduate_theses/6801

Amjad, Hassan, Farhan Ahmad, and Muhammad Irshad Qureshi. 2023. "Enhanced mechanical and durability resilience of plastic aggregate concrete modified with nano-iron oxide and sisal fiber reinforcement." *Construction and Building Materials* 401: 132911. https://doi.org/10.1016/J.CONBUILDMAT.2023.132911.

Arel, Hasan Şahan, and Blessen Skariah Thomas. 2017. "The effects of nano- and micro-particle additives on the durability and mechanical properties of mortars exposed to internal and external sulfate attacks." *Results in Physics* 7: 843–51. https://doi.org/10.1016/J.RINP.2017.02.009.

Ashish, Prabin Kumar, and Dharamveer Singh. 2021. "Use of nanomaterial for asphalt binder and mixtures: A comprehensive review on development, prospect, and challenges." *Road Materials and Pavement Design* 22 (3): 492–538. https://doi.org/10.1080/14680629.2019.1634634.

Atiq Orakzai, Muhammad. 2021. "Hybrid effect of nano-alumina and nano-titanium dioxide on mechanical properties of concrete." *Case Studies in Construction Materials* 14 (June): e00483. https://doi.org/10.1016/J.CSCM.2020.E00483.

Atmaca, Nihat, Mohammed Layth Abbas, and Adem Atmaca. 2017. "Effects of nano-silica on the gas permeability, durability and mechanical properties of high-strength lightweight concrete." *Construction and Building Materials* 147: 17–26. https://doi.org/10.1016/J.CONBUILDMAT.2017.04.156.

Bantie, Zenamarkos, Asmare Tezera, Desalegn Abera, and Tesfa Nega. 2024. "Nanoclays as fillers for performance enhancement in building and construction industries: State of the art and future trends." *Developments in Clay Science and Construction Techniques*. https://doi.org/10.5772/INTECHOPEN.1005147.

Bautista-Gutierrez, Karla P., Agustín L. Herrera-May, Jesús M. Santamaría-López, Antonio Honorato-Moreno, and Sergio A. Zamora-Castro. 2019. "Recent progress in nanomaterials for modern concrete infrastructure: Advantages and challenges." *Materials* 12 (21): 3548. https://doi.org/10.3390/MA12213548.

Broekhuizen, Pieter Van, Fleur Van Broekhuizen, Ralf Cornelissen, and Lucas Reijnders. 2011. "Use of nanomaterials in the European construction industry and some occupational health aspects thereof." *Journal of Nanoparticle Research* 13 (2): 447–62. https://doi.org/10.1007/S11051-010-0195-9/FIGURES/6.

Carriço, A., J. A. Bogas, A. Hawreen, and M. Guedes. 2018. "Durability of multi-walled carbon nanotube reinforced concrete." *Construction and Building Materials* 164: 121–33. https://doi.org/10.1016/J.CONBUILDMAT.2017.12.221.

Chakraborty, Arun Kumar, Mainak Ghosal, and Arun Kr Chakraborty. 2019. "Application of nanomaterials on cement mortar and concrete: A study application of nanomaterials on cement mortar and concrete: A Study." Accessed March 22, 2025. www.researchgate.net/publication/334231601.

Chandak, Mayuri A., and P. Y. Pawade. 2018. "Influence of metakaolin in concrete mixture: A Review." *The International Journal of Engineering and Science* 1 (1): 37–41.

Chen, Wenhua, Zhanfeng Qi, Lei Zhang, and Zhiyi Huang. 2020. "Effects of cenosphere on the mechanical properties of cement-based composites." *Construction and Building Materials* 261: 120527. https://doi.org/10.1016/J.CONBUILDMAT.2020.120527.

Chuah, Samuel, Zhu Pan, Jay G. Sanjayan, Chien Ming Wang, and Wen Hui Duan. 2014. "Nano reinforced cement and concrete composites and new perspective from graphene oxide." *Construction and Building Materials* 73: 113–24. https://doi.org/10.1016/J.CONBUILDMAT.2014.09.040.

Dash, Subhakanta, Itishree Mohanty, and Piyush Gupta. 2022. Durability and water absorption properties of fly ash based pervious concrete: A case study. 5 (4). https://doi.org/10.31031/SBB.2022.05.000617

Dheyaaldin, Mahmood Hunar, Mohammad Ali Mosaberpanah, and Radhwan Alzeebaree. 2022. "Shrinkage behavior and mechanical properties of alkali activated mortar incorporating nanomaterials and polypropylene fiber." *Ceramics International* 48 (16): 23159–71. https://doi.org/10.1016/J.CERAMINT.2022.04.297.

Diab, Ahmed M., Hafez E. Elyamany, Abd Elmoaty M. Abd Elmoaty, and Muftah M. Sreh. 2019. "Effect of nanomaterials additives on performance of concrete resistance against magnesium sulfate and acids." *Construction and Building Materials* 210: 210–31. https://doi.org/10.1016/J.CONBUILDMAT.2019.03.099.

Du, Hongjian, Suhuan Du, and Xuemei Liu. 2015. "Effect of nano-silica on the mechanical and transport properties of lightweight concrete." *Construction and Building Materials* 82: 114–22. https://doi.org/10.1016/J.CONBUILDMAT.2015.02.026.

Du, Xiaoqi, Yanlong Li, Binghui Huangfu, Zheng Si, Lingzhi Huang, Lifeng Wen, and Meiwei Ke. 2023. "Modification mechanism of combined nanomaterials on high performance concrete and optimization of nanomaterial content." *Journal of Building Engineering* 64: 105648. https://doi.org/10.1016/J.JOBE.2022.105648.

Edward, Edeh O., Job O. Fredrick, Shittu N. Olawale, and Anowai I. Solomon. 2021. "Effect of nano iron oxide (Fe2O3) on concrete subjected to physical sulphate attack (PSA)." Accessed April 9, 2025. www.ijert.org.

Fallah, Saber, and Mahdi Nematzadeh. 2017. "Mechanical properties and durability of high-strength concrete containing macro-polymeric and polypropylene fibers with nano-silica and silica fume." *Construction and Building Materials* 132: 170–87. https://doi.org/10.1016/J.CONBUILDMAT.2016.11.100.

Fang, Changqing, Ruien Yu, Shaolong Liu, and Yan Li. 2013. "Nanomaterials applied in asphalt modification: A review." *Journal of Materials Science & Technology* 29 (7): 589–94. https://doi.org/10.1016/J.JMST.2013.04.008.

Garvey, Therese, Elizabeth A. Moore, Callie W. Babbitt, and Gabrielle Gaustad. 2019. "Comparing ecotoxicity risks for nanomaterial production and release under uncertainty." *Clean Technologies and Environmental Policy* 21 (2): 229–42. https://doi.org/10.1007/S10098-018-1648-6/FIGURES/8.

Goel, Gaurav, Payal Sachdeva, Akshay Kumar Chaudhary, and Yadvendra Singh. 2022. "The use of nanomaterials in concrete: A review." *Materials Today: Proceedings* 69: 365–71. https://doi.org/10.1016/J.MATPR.2022.09.051.

Gopinath, S., P. Ch Mouli, A. R. Murthy, N. R. Iyer, and S. Maheswaran. 2012. "Effect of nano silica on mechanical properties and durability of normal strength concrete." Accessed March 19, 2025. https://doi.org/10.2478/v.10169-012-0023-y.

Guo, Aofei, Zhihui Sun, Noppadon Sathitsuksanoh, and Hu Feng. 2020. "A review on the application of nanocellulose in cementitious materials." *Nanomaterials* 10 (12): 2476. https://doi.org/10.3390/NANO10122476.

Gupta, Piyush, and Itishree Mohanty. 2022 "A multifunctional and sustainable pavement material: Pervious concrete." *The Journal of Solid Waste Technology and Management*: 576–587.

Hanif, Asad, Zeyu Lu, Su Diao, Xiaohui Zeng, and Zongjin Li. 2017. "Properties investigation of fiber reinforced cement-based composites incorporating cenosphere fillers." *Construction and Building Materials* 140: 139–49. https://doi.org/10.1016/J.CONBUILDMAT.2017.02.093.

Hisseine, Ousmane A., William Wilson, Luca Sorelli, Balázs Tolnai, and Arezki Tagnit-Hamou. 2019. "Nanocellulose for improved concrete performance: A macro-to-micro investigation for disclosing the effects of cellulose filaments on strength of cement systems." *Construction and Building Materials* 206: 84–96. https://doi.org/10.1016/J.CONBUILDMAT.2019.02.042.

Hou, Shaodan, Zhenhua Duan, Zhiming Ma, and Amardeep Singh. 2020. "Improvement on the properties of waste glass mortar with nanomaterials." *Construction and Building Materials* 254: 118973. https://doi.org/10.1016/J.CONBUILDMAT.2020.118973.

Huang, Yu, and Lin Wang. 2016. "Experimental studies on nanomaterials for soil improvement: A review." *Environmental Earth Sciences* 75 (6): 1–10. https://doi.org/10.1007/S12665-015-5118-8/TABLES/1.

Imoni, Samson, Michael Toryila Tiza, Mogbo Onyebuchi, Ebenezer Ogirima, and Akande Corresponding. 2023. "The use of nanomaterials for road construction." *Bincang Sains Dan Teknologi* 2 (03): 108–17. https://doi.org/10.56741/bst.v2i03.435.

Irshidat, Mohammad R., and Ammar Al-Shannaq. 2018. "Using textile reinforced mortar modified with carbon nano tubes to improve flexural performance of RC beams." *Composite Structures* 200: 127–34. https://doi.org/10.1016/J.COMPSTRUCT.2018.05.088.

Jahromi, Saeed Ghaffarpour, and Ali Khodaii. 2009. "Effects of nanoclay on rheological properties of bitumen binder." *Construction and Building Materials* 23 (8): 2894–2904. https://doi.org/10.1016/J.CONBUILDMAT.2009.02.027.

Ji, Tao. 2005. "Preliminary study on the water permeability and microstructure of concrete incorporating nano-SiO2." *Cement and Concrete Research* 35 (10): 1943–47. https://doi.org/10.1016/J.CEMCONRES.2005.07.004.

John, Nova. 2013. "Strength properties of metakaolin admixed concrete." *International Journal of Scientific and Research Publications* 3 (6): 1–7. www.ijsrp.org.

Jones, Wendy, Alistair Gibb, Chris Goodier, Phil Bust, Mo Song, and Jie Jin. 2019. "Nanomaterials in construction – what is being used, and where?" 172 (2): 49–62. https://doi.org/10.1680/JCOMA.16.00011.

Jung, Myungjun, Young soon Lee, Sung Gul Hong, and Juhyuk Moon. 2020. "Carbon nanotubes (CNTs) in ultra-high performance concrete (UHPC): Dispersion, mechanical properties, and electromagnetic interference (EMI) shielding effectiveness (SE)." *Cement and Concrete Research* 131: 106017. https://doi.org/10.1016/J.CEMCONRES.2020.106017.

Kavinkumar, V., A. K. Priya, and R. Praneeth. 2023. "Strength of light weight concrete containing fly ash cenosphere." *Materials Today: Proceedings.* https://doi.org/10.1016/J.MATPR.2023.04.094.

Khudher, Fawzi, Mohd Hafez, Mohammed Yousif Fattah, Marwan Samir Al-Shaikhli, Fawzi Kh Khalaf, Mohammed Y. Fattah, Marwan S. Al, and Marwan S. Al-Shaikli. 2020. "A review study on optimizing the performance of soil using nanomaterials a review study on the optimizing the performance of soil using nanomaterials." *Advances in Industrial Engineering and Management* 9 (2): 1–10. https://doi.org/10.7508/aiem.02.2020.01.10.

Kim, G. M., I. W. Nam, Beomjoo Yang, H. N. Yoon, H. K. Lee, and Solmoi Park. 2019. "Carbon nanotube (CNT) incorporated cementitious composites for functional construction materials: The state of the art." *Composite Structures* 227: 111244. https://doi.org/10.1016/J.COMPSTRUCT.2019.111244.

Kowsalya, M., S. Sindhu Nachiar, Anandh Sekar, and P. T. Ravichandran. 2022. "Study on mechanical and microstructural properties of concrete with fly ash cenosphere as fine aggregate – a sustainable approach." *Buildings* 12 (10): 1679. https://doi.org/10.3390/BUILDINGS12101679.

Maglad, Ahmed M., Osama Zaid, Mohamed M. Arbili, Guilherme Ascensão, Adrian A. Șerbănoiu, Cătălina M. Grădinaru, Rebeca M. García, Shaker M. A. Qaidi, Fadi Althoey, and Jesús de Prado-Gil. 2022. "A study on the properties of geopolymer concrete modified with nano graphene oxide." *Buildings* 12 (8): 1066. https://doi.org/10.3390/BUILDINGS12081066.

Majeed, Zaid Hameed, and Mohd Raihan Taha. 2013. "A review of stabilization of soils by using nanomaterials." *Australian Journal of Basic and Applied Sciences* 7 (2): 576–81.

Majeed, Zaid Hameed, Raihan Taha, and Ibtehaj Taha Jawad. 2014. "Stabilization of soft soil using nanomaterials." *Research Journal of Applied Sciences, Engineering and Technology* 8 (4): 503–9.

Mansi, Aseel, Nadhim Hamah Sor, Nahla Hilal, and Shaker M. A. Qaidi. 2022. "The impact of nano clay on normal and high-performance concrete characteristics: A review." *IOP Conference Series: Earth and Environmental Science* 961 (1): 012085. https://doi.org/10.1088/1755-1315/961/1/012085.

Mohtasham Moein, Mohammad, Komeil Rahmati, Ashkan Saradar, Jaeyun Moon, and Moses Karakouzian. 2024. "A critical review examining the characteristics of modified concretes with different nanomaterials." *Materials* 17 (2): 409. https://doi.org/10.3390/MA17020409.

Mortada, Youssef, Eyad Masad, Reginald B. Kogbara, Bilal Mansoor, Thomas Seers, Ahmad Hammoud, and Ayman Karaki. 2023. "Development of Ca(OH)2-based geopolymer for additive manufacturing using construction wastes and nanomaterials." *Case Studies in Construction Materials* 19: e02258. https://doi.org/10.1016/J.CSCM.2023.E02258.

Muzenski, Scott, Ismael Flores-Vivian, and Konstantin Sobolev. 2019. "Ultra-high strength cement-based composites designed with aluminum oxide nano-fibers." *Construction and Building Materials* 220: 177–86. https://doi.org/10.1016/J.CONBUILDMAT.2019.05.175.

Naeem, Faizan, H. K. Lee, H. K. Kim, and I. W. Nam. 2017. "Flexural stress and crack sensing capabilities of MWNT/cement composites." *Composite Structures* 175: 86–100. https://doi.org/10.1016/J.COMPSTRUCT.2017.04.078.

Noorvand, Hassan, Abang Abdullah Abang Ali, Ramazan Demirboga, Nima Farzadnia, and Hossein Noorvand. 2013. "Incorporation of nano tio2 in black rice husk ash mortars." *Construction and Building Materials* 47: 1350–61. https://doi.org/10.1016/J.CONBUILDMAT.2013.06.066.

Parande, Anand Kuber, B. Ramesh Babu, M. Aswin Karthik, K. K. Deepak Kumaar, and N. Palaniswamy. 2008. "Study on strength and corrosion performance for steel embedded in metakaolin blended concrete/mortar." *Construction and Building Materials* 22 (3): 127–34. https://doi.org/10.1016/J.CONBUILDMAT.2006.10.003.

Patel, Nikunj, and C. B. Mishra. 2018. "Laboratory investigation of nano titanium dioxide (TiO2) in concrete for pavement." *International Research Journal of Engineering and Technology* 5 (5). www.irjet.net.

Patel, Sudeep K., Hara P. Satpathy, Amar N. Nayak, and Chitta R. Mohanty. 2020. "Utilization of fly ash cenosphere for production of sustainable lightweight concrete." *Journal of the Institution of Engineers (India): Series A* 101 (1): 179–94. https://doi.org/10.1007/S40030-019-00415-6/FIGURES/25.

Pedro Muñoz-Pérez, Sócrates, Yoshida Mirella Gonzales-Pérez, and Tabita Elizabeth Pardo-Muñoz. 2022. "The use of nanomaterials in the construction sector: A literary review." *Revista DYNA* 89 (221): 101–9. https://doi.org/10.15446/dyna.v89n221.100210.

Pillay, Deveshan L., Oladimeji B. Olalusi, Moses W. Kiliswa, Paul O. Awoyera, John Temitope Kolawole, and Adewumi John Babafemi. 2022. "Engineering performance of metakaolin based concrete." *Cleaner Engineering and Technology* 6: 100383. https://doi.org/10.1016/J.CLET.2021.100383.

Qing, Ye, Zhang Zenan, Kong Deyu, and Chen Rongshen. 2007. "Influence of nano-SiO2 addition on properties of hardened cement paste as compared with silica fume." *Construction and Building Materials* 21 (3): 539–45. https://doi.org/10.1016/J.CONBUILDMAT.2005.09.001.

Rattan, A., P. Sachdeva, and A. Chaudhary. 2016. "Use of nanomaterials in concrete." *International Journal of Latest Research in Engineering and Technology (IJLRET)* 02: 81–84. www.ijlret.com.

Safiuddin, Md, Marcelo Gonzalez, Jingwen Cao, and Susan L. Tighe. 2014. "State-of-the-art report on use of nano-materials in concrete." *International Journal of Pavement Engineering* 15 (10): 940–49. https://doi.org/10.1080/10298436.2014.893327.

Saini, Raman, Gyanendra Kumar Chaturvedy, and Umesh Kumar Pandey. 2024. "Examining the effects of nano iron oxide on physical and mechanical characteristics of rubberized concrete." *Innovative Infrastructure Solutions* 9 (6): 1–18. https://doi.org/10.1007/S41062-024-01494-6/FIGURES/13.

Saleem, Haleema, Syed Javaid Zaidi, and Nasser Abdullah Alnuaimi. 2021. "Recent advancements in the nanomaterial application in concrete and its ecological impact." *Materials* 14 (21): 6387. https://doi.org/10.3390/MA14216387.

Senff, L., R. C. E. Modolo, D. M. Tobaldi, G. Ascenção, D. Hotza, V. M. Ferreira, and J. A. Labrincha. 2015. "The influence of TiO2 nanoparticles and poliacrilonitrile fibers on the rheological behavior and hardened properties of mortars." *Construction and Building Materials* 75: 315–30. https://doi.org/10.1016/J.CONBUILDMAT.2014.11.002.

Shen, Renjie, Shiwen Xue, Yanru Xu, Qi Liu, Zhang Feng, Hao Ren, Huamin Zhai, and Fangong Kong. 2020. "Research progress and development demand of nanocellulose reinforced polymer composites." *Polymers* 12 (9): 2113. https://doi.org/10.3390/POLYM12092113.

Singh, Subhash, Ashwani Kumar, Sanjay K. Behura, and Kartikey Verma. 2022. "Challenges and opportunities in nanomanufacturing." *Nanomanufacturing and Nanomaterials Design* 17–30. https://doi.org/10.1201/9781003220602-2.

Souza, Felipe Basquiroto De, Oscar Rubem Klegues Montedo, Rosielen Leopoldo Grassi, and Elaine Gugliemi Pavei Antunes. 2019. "Lightweight high-strength concrete with the use of waste cenosphere as fine aggregate." *Matéria (Rio de Janeiro)* 24 (4): e12509. https://doi.org/10.1590/S1517-707620190004.0834.

Strzałkowski, Jarosław, Agata Stolarska, Dominik Kożuch, and Joanna Dmitruk. 2023. "Hygrothermal and strength properties of cement mortars containing cenospheres." *Cement and Concrete Research* 174: 107325. https://doi.org/10.1016/J.CEMCONRES.2023.107325.

Sujitha, V. S., B. Ramesh, and Joseph Raj Xavier. 2024. "Effects of nanomaterials on mechanical properties in cementitious construction materials for high-strength concrete applications: A review." *Journal of Adhesion Science and Technology* 38 (20): 3737–3768. https://doi.org/10.1080/01694243.2024.2354093.

Supit, Steve Wilben Macquarie, and Faiz Uddin Ahmed Shaikh. 2015. "Durability properties of high volume fly ash concrete containing nano-silica." *Materials and Structures/Materiaux et Constructions* 48 (8): 2431–45. https://doi.org/10.1617/S11527-014-0329-0/FIGURES/12.

Svintsov, Alexander P., Evgeny L. Shchesnyak, Vera V. Galishnikova, Roman S. Fediuk, and Nadezhda A. Stashevskaya. 2020. "Effect of nano-modified additives on properties of concrete mixtures during winter season." *Construction and Building Materials* 237: 117527. https://doi.org/10.1016/J.CONBUILDMAT.2019.117527.

Szymanowski, Jacek, and Lukasz Sadowski. 2019. "The development of nanoalumina-based cement mortars for overlay applications in concrete floors." *Materials* 2 (21): 3465. https://doi.org/10.3390/MA12213465.

Tanirbergenova, Sandugash K., Balaussa K. Dinistanova, Nurzhamal K. Zhylybayeva, Dildara A. Tugelbayeva, Gulya M. Moldazhanova, Aizat Aitugan, Kairat Taju, and Meruyert Nazhipkyzy. 2023. "Synthesis of cenospheres from ash and their application." *Journal of Composites Science* 7 (7): 276. https://doi.org/10.3390/JCS7070276.

Utsev, Terlumun, Toryila Michael Tiza, Onyebuchi Mogbo, Sitesh Kumar Singh, Ankit Chakravarti, Nagaraju Shaik, and Surendra Pal Singh. 2022. "Application of nano-materials in civil engineering." *Materials Today: Proceedings* 62: 5140–46. https://doi.org/10.1016/J.MATPR.2022.02.480.

Vitharana, Menaka Gayan, Suvash Chandra Paul, Sih Ying Kong, Adewumi John Baba-femi, Md Jihad Miah, and Biranchi Panda. 2020. "A study on strength and corro-sion protection of cement mortar with the inclusion of nanomaterials." *Sustainable Materials and Technologies* 25: e00192. https://doi.org/10.1016/J.SUSMAT.2020. E00192.

Wang, Xinyue, Sufen Dong, Ashraf Ashour, Wei Zhang, and Baoguo Han. 2020. "Effect and mechanisms of nanomaterials on interface between aggregates and cement mor-tars." *Construction and Building Materials* 240: 117942. https://doi.org/10.1016/J. CONBUILDMAT.2019.117942.

Yao, Hui, Jiani Zeng, Yiran Wang, and Zhanping You. 2025. "Review of nanomaterials in pavement engineering." *ACS Applied Nano Materials* 8 (11). https://doi.org/10.1021/ ACSANM.4C06153/ASSET/IMAGES/MEDIUM/AN4C06153_0020.GIF.

Yesudhas Jayakumari, Breetha, Elangovan Nattanmai Swaminathan, and Pachaivannan Partheeban. 2023. "A review on characteristics studies on carbon nanotubes-based cement concrete." *Construction and Building Materials* 367: 130344. https://doi. org/10.1016/J.CONBUILDMAT.2023.130344.

Younus, Shimal Jameel, Mohammad Ali Mosaberpanah, and Radhwan Alzeebaree. 2023. "The performance of alkali-activated self-compacting concrete with and without nano-alumina." *Sustainability*15 (3): 2811. https://doi.org/10.3390/SU15032811.

Zanjad, Nilesh, Shantanu Pawar, and Chittaranjan Nayak. 2024. "Review on application of lightweight cenosphere in construction material." *Macromolecular Symposia* 413 (1): 2300006. https://doi.org/10.1002/MASY.202300006.

Zhan, Bao Jian, Dong Xing Xuan, and Chi Sun Poon. 2019. "The effect of nanoalumina on early hydration and mechanical properties of cement pastes." *Construction and Build-ing Materials* 202: 169–76. https://doi.org/10.1016/J.CONBUILDMAT.2019.01.022.

Zhan, Pei min, Zhi hai He, Zhi ming Ma, Chao feng Liang, Xiao xiang Zhang, Annulo Addi-sayehu Abreham, and Jin yan Shi. 2020. "Utilization of nano-metakaolin in concrete: A review." *Journal of Building Engineering* 30: 101259. https://doi.org/10.1016/J. JOBE.2020.101259.

Zhang, Ai, Wencui Yang, Yong Ge, and Penghuan Liu. 2020. "Effect of nanomateri-als on the mechanical properties and microstructure of cement mortar under low air pressure curing." *Construction and Building Materials* 249: 118787. https://doi. org/10.1016/J.CONBUILDMAT.2020.118787.

Zhang, Henglong, Haihui Duan, Chongzheng Zhu, Zihao Chen, and Huan Luo. 2021a. "Mini-review on the application of nanomaterials in improving anti-aging proper-ties of asphalt." *Energy & Fuels* 35 (14): 11017–11036. https://doi.org/10.1021/ACS. ENERGYFUELS.1C01035.

Zhang, Rui, Xin Cheng, Pengkun Hou, and Zhengmao Ye. 2015. "Influences of nano-TiO2 on the properties of cement-based materials: Hydration and drying shrink-age." *Construction and Building Materials* 81: 35–41. https://doi.org/10.1016/J. CONBUILDMAT.2015.02.003.

Zhang, Shuai, Haokai Hong, Henglong Zhang, and Zihao Chen. 2021b. "Investigation of anti-aging mechanism of multi-dimensional nanomaterials modified asphalt by FTIR, NMR and GPC." *Construction and Building Materials* 305: 124809. https:// doi.org/10.1016/J.CONBUILDMAT.2021.124809.

10

Nanomaterials as Fire Retardants

Seema

10.1 Introduction

Fire hazards pose significant risks to human life, infrastructure, and the environment. In 2023, 1,389,000 fires resulted in 3,670 civilian deaths and 13,350 injuries, and severe property loss.[1] These numbers are still increasing and are totally unacceptable. In recent years, the significant increase in fire cases around the globe, forced researchers to investigate and develop more efficient and effective fire-retardant materials. To overcome these concerns, several researchers have used nanotechnology to create advanced and innovative fire-retardant solutions. The development of effective fire-retardant materials is crucial to mitigating these risks. Traditional fire retardants often rely on halogenated compounds, which, despite their effectiveness, pose environmental and health concerns due to toxic byproducts. The emergence of nanomaterials has revolutionized fire retardancy by offering high efficiency at lower loading levels, improved thermal stability, and reduced environmental impact. This chapter discusses the role of nanomaterials in enhancing fire retardancy across various applications, focusing on their synergistic interactions with traditional flame retardants and their use in textiles, paints, and coatings.

Polymer nanocomposites represent one of the most extensively studied applications of nanomaterials for fire retardancy. Polymers are inherently flammable due to their hydrocarbon-rich composition, but incorporating nanomaterials into their matrices significantly improves fire resistance, thermal stability, and smoke suppression. Nanomaterials not only reduce the peak heat release rate (PHRR) and total heat release (THR) but also maintain or enhance the mechanical, optical, and thermal properties of the host polymer.

Nanomaterials contribute to flame retardancy in polymer nanocomposites through several mechanisms, including barrier effect (formation of a protective char or ceramic-like layer on the polymer surface during combustion, hindering heat and oxygen transfer), catalytic effect (promotion of carbonaceous char formation via catalysis, which suppresses the emission of flammable volatiles), radical trapping (scavenging of reactive free radicals during combustion, thus inhibiting flame propagation), and heat dissipation (distribution of thermal energy away from ignition sources, preventing localized heating and degradation).

Each polymer requires tailored nanomaterial incorporation strategies depending on polarity, processing methods, and end-use requirements (see Figure 10.1).

DOI: 10.1201/9781003632498-10

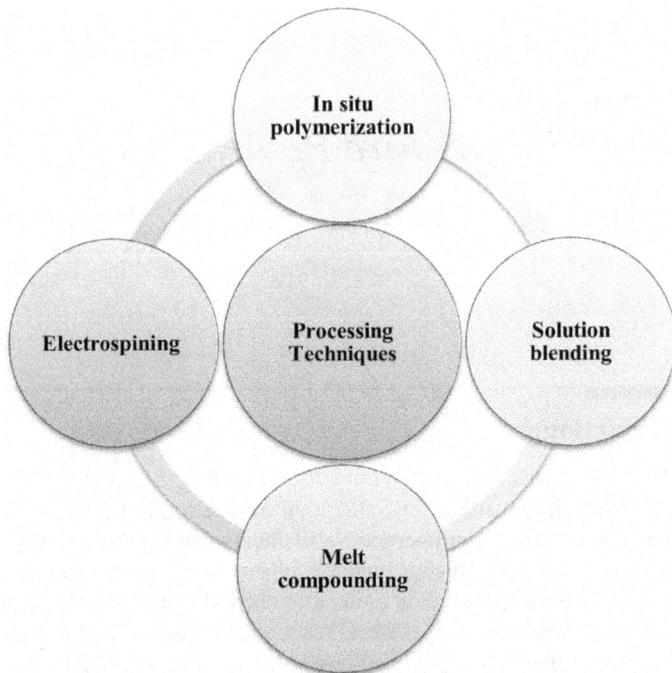

FIGURE 10.1   Processing techniques of polymer nanocomposites.

The dispersion and compatibility of nanofillers within the polymer matrix (see Figure 10.2) are critical for achieving uniform flame-retardant performance. Surface modification (e.g., silanization, functionalization with organic groups) of nanoparticles often enhances compatibility and interfacial bonding.

10.2 Types of Nanomaterials as Fire Retardant

Nanomaterials have revolutionized fire retardancy by providing highly efficient flame resistance at lower loading levels while maintaining or even enhancing the mechanical, thermal, and electrical properties of materials. The effectiveness of nanomaterials in fire retardancy depends on their composition, structure, and mode of action. Various nanomaterials (see Figure 10.3 and Table 10.1) have been explored for fire-retardant applications, including carbon-based nanomaterials, metal oxide nanoparticles, layered double hydroxides (LDHs), clay-based nanomaterials, and boron-based nanomaterials.[2]

10.2.1 Carbon-Based Nanomaterials

Carbon-based nanomaterials, including graphene, graphene oxide (GO), carbon nanotubes (CNTs), carbon nanofibers (CNFs), and fullerene-based structures, are highly

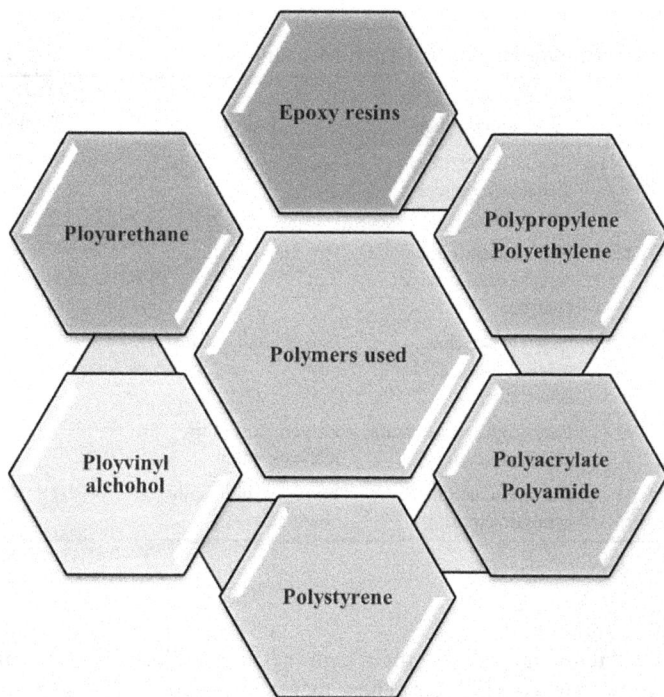

FIGURE 10.2 Common polymers studied for fire-retardant nanocomposites.

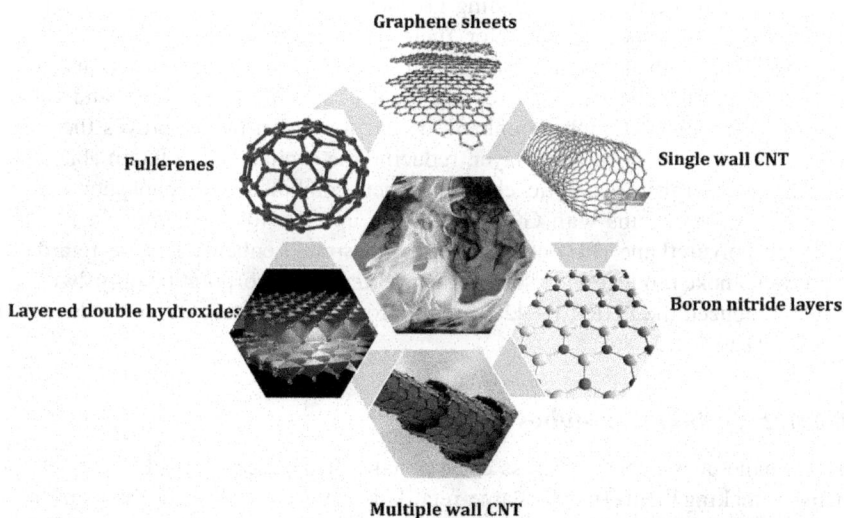

FIGURE 10.3 Types of fire-retardant nanomaterials.

TABLE 10.1

Nanomaterials as Fire Retardants and Their Mechanisms

Nanomaterial type	Key fire retardant mechanism	Example	Reference
Graphene, GO, CNTs	Barrier effect, char formation, radical scavenging	Epoxy, polyurethane	3,4,5,6,7,8,9
Metal oxide (TiO2, ZnO, Fe2O3)	Heat absorption, smoke suppression, catalytic charring	Polymers, coatings, foams	10,11,12,13
LDHs	Gas release, char enhancement, acid scavenging	Epoxy coatings, textiles	14,15,16,17,18,19
Clay based nanomaterials	Barrier effect, mechanical reinforcement	Polyamides, fabrics, fire resistant paints	20,21,22,23,24,25,26,27, 28,29
Boron based nanomaterials	Gas phase action, catalytic charring	Epoxy resins, foams, coatings	30,31,32,33,34

effective in fire retardancy due to their high thermal stability, excellent mechanical properties, and strong barrier effect. These materials improve char formation and prevent oxygen diffusion, thereby slowing down combustion.[35]

10.2.1.1 Graphene and Graphene Oxide

Graphene and its derivatives, including graphene oxide (GO) and reduced graphene oxide (rGO), are widely studied for their flame-retardant properties.[36] The key advantages of graphene-based materials include high aspect ratio, responsible for an effective physical barrier that prevents the diffusion of heat, oxygen, and volatile degradation products. Graphene-enhanced char formation that improves the stability and compactness of the char layer, reducing the availability of flammable gases, and shows synergistic catalytic charring when combined with phosphorus flame retardants. Epoxy resins with GO have shown up to a 40% reduction in PHRR.[37] Graphene-polyurethane (PU) composites demonstrated enhanced flame retardancy with lower smoke production.[38] Song et al. reported that incorporating only 2wt% GO in epoxy reduced the PHRR by 42% and improved the limiting oxygen index (LOI) from 20.4% to 27.2%.[39]

10.2.1.2 Carbon Nanotubes

Carbon nanotubes (CNTs) possess exceptional thermal stability and electrical conductivity, making them effective flame-retardant additives.[40] Their mechanisms of fire retardancy include heat dissipation, where CNTs act as a heat sink, reducing localized temperature increases. Char reinforcement by CNTs improves char integrity, making it more resistant to thermal decomposition. In another way, free radical scavenging by CNTs reduces the formation of combustion-promoting free radicals in the gas phase. CNT-enhanced polypropylene (PP) composites exhibited reduced flammability and improved thermal stability.[41] Multi-walled CNTs (MWCNTs) in epoxy resins improved char formation and reduced total smoke release by 50%.[42] CNTs enhance thermal conductivity and structural integrity of the char, acting as a scaffold during combustion. They also show synergistic effects with traditional flame retardants. Zhang et al. demonstrated that adding MWCNTs to polystyrene reduced PHRR and increased the time to ignition.[43]

10.2.1.3 Carbon Nanofibers and Fullerenes

Carbon nanofibers (CNFs) reinforce the char structure and improve thermal conductivity.[44]

Fullerene is a third allotropic form of carbon after diamond and graphite. Since the discovery of fullerenes, C_{60} has been widely explored by researchers due to its stability. C_{60} fullerene spherical structure comprises a sphere with 12 pentagons and 20 hexagons, where each carbon is attached to other atoms by sp^2 hybridization. C_{60} is named buckminsterfullerene due to its soccer-ball-like structure, in tribute to the geodesic dome architect, Buckminster Fuller. The impacts of fullerene (C_{60}) on the thermal stability of polymer–fullerene composites have been widely explored and well documented. Fullerenes (C_{60}) function as radical scavengers, suppressing flame propagation in polymeric materials.[45]

10.2.2 Metal Oxide Nanoparticles

Metal oxide nanoparticles, such as titanium dioxide (TiO_2), zinc oxide (ZnO), iron oxide (Fe_2O_3), aluminum oxide (Al_2O_3), and cerium oxide (CeO_2), contribute to fire retardancy through multiple mechanisms, including heat absorption, smoke suppression, and catalytic charring.[46] Metal oxide nanoparticles are commonly used in small loadings (1–5wt%) to catalyze char formation and suppress smoke. Their synergistic effect is often more pronounced when combined with organic flame retardants. Pan et al. (2015) demonstrated that the inclusion of 2wt% TiO_2 nanoparticles in PU foam resulted in a 28% reduction in HRR and significant suppression of smoke and toxic gas emissions.[47]

TiO_2 nanoparticles act as thermal insulators and smoke suppressants due to their high thermal stability and ability to reflect heat. They promote char formation, reduce the release of combustible gases, and work synergistically with phosphorus and nitrogen flame retardants. Polyester–TiO_2 nanocomposites exhibited improved fire resistance and reduced heat release rate (HRR).[48] TiO_2-enhanced polyurethane foams showed better thermal stability and self-extinguishing properties.[49]

ZnO nanoparticles function as smoke suppressants and fire retardants by acting as catalysts to promote charring, neutralizing free radicals that drive combustion, and reducing the emission of toxic smoke and carbon monoxide (CO).[50] ZnO-PP composites exhibited a 30% reduction in smoke density.[51] ZnO-coated textiles demonstrated improved flame resistance in fire-protective clothing.[52,53] Iron oxide (Fe_2O_3) enhances char formation and promotes the formation of a stable carbonaceous layer.[54] CeO_2 acts as a radical scavenger, inhibiting fire spread.[55]

10.2.3 Layered Double Hydroxides

An anionic clay with planar structure, layered double hydroxide (LDH), also known as a hydrotalcite-type material, is composed of a double layer of a cation layer and an anion capable of ion exchange.[56] They provide flame retardancy by releasing water and CO_2 upon decomposition, which cools the burning surface.[57] They enhance the char strength, prevent material degradation, and act as acid scavengers, reducing the formation of toxic gases. LDH-enhanced epoxy coatings showed up to 50% improvement in fire resistance. LDHs combined with phosphorus retardants exhibited enhanced synergistic effects.[58] They are particularly effective in combination with phosphorus- or nitrogen-based retardants. Wang et al. (2011) observed that adding LDH to EVA (ethylene vinyl acetate) enhanced LOI and reduced both THR and PHRR by over 50%.[59]

10.2.4 Clay-Based Nanomaterials

Phyllosilicate minerals, clay nanoplatelets, have been widely explored for many applications due to their unique structure, thermal resistance, barrier properties, mechanical strength, and natural abundance.[60] Nanoclays, such as montmorillonite (MMT), halloysite nanotubes (HNTs), and vermiculite, improve fire resistance by creating a tortuous path for flammable gases, reducing flame propagation by enhancing char stability and mechanical strength.[61] Nanoclays such as MMT are widely used due to their natural abundance, cost-effectiveness, and ease of modification. When exfoliated in polymer matrices, they increase the viscosity of the melt, delay volatile gas release, and act as a physical barrier against heat and oxygen. Gilman et al. showed that PP/MMT nanocomposites exhibited a 63% reduction in HRR and significant smoke suppression.[62] MMT-nanoclay in polyamide composites reduced the THR by 45%.[63] HNT-coated fabrics exhibited improved flame retardancy with minimal environmental impact.[64]

10.2.5 Boron-Based Nanomaterials

Boron compounds bear low toxicity, molecular diversity, and various mechanisms of fire-retardant action, and show high-potential flame-retardant properties.[65] Boron-containing nanomaterials, including boron nitride (BN) nanosheets, boric acid nanoparticles, and borate compounds, function as effective flame retardants through gas phase action by releasing boric oxide vapors that dilute flammable gases.[66]

They also promote catalytic charring with enhanced formation of a protective carbonaceous layer. BN-polyurethane foams showed improved fire resistance and heat insulation properties.[67] Boron-modified epoxy resins exhibited significant reductions in heat release rates.[68]

10.3 Synergistic Effects of Nanomaterials with Traditional Flame Retardants

The integration of nanomaterials with traditional flame retardants has emerged as a highly effective strategy to enhance flame retardancy while minimizing the total additive content required.[69] This synergistic approach exploits the individual advantages of both systems. Traditional flame retardants offer gas-phase flame inhibition and condensed-phase char formation, whereas nanomaterials provide barrier effects, catalytic activity, and thermal stability. Together, these components result in enhanced fire performance, mechanical integrity, and environmental compatibility. While nanomaterials alone can enhance fire resistance, their combination with traditional flame retardants such as phosphorus-based, nitrogen-based, and halogen-free intumescent systems can result in synergistic flame-retardant effects. These synergistic systems not only improve flame retardancy but also allow for lower overall additive loading, minimizing negative impacts on mechanical properties and processability.[70]

The synergistic behavior between nanomaterials and conventional flame retardants arises due to their complementary functions during combustion. Some common synergistic mechanisms include enhanced char formation, where nanomaterials catalyze the formation of stable, thermally resistant char, while traditional flame retardants promote intumescence or release fire-suppressing gases.[71] Nanomaterials reinforce the protective barrier formed by intumescent flame retardants, enhancing their strength, thermal insulation, and oxygen impermeability. By gas-phase radical trapping, nanomaterials can interact with flame-inhibiting species (e.g., PO•, HPO•) to stabilize them or prolong their activity in quenching combustion radicals. Metal oxide or layered nanomaterials catalyze carbonization reactions, increasing the yield and density of char.[72]

Synergistic systems achieve high flame retardancy with reduced material addition (often <10 wt%). Preserved mechanical properties with reduced filler content mean better retention of strength, ductility, and toughness. Multifunctionality-enabled coatings or composites also exhibit UV resistance, antimicrobial properties, or conductivity. An improved environmental profile is shown by many synergistic systems as they are halogen-free, minimizing toxicity and smoke.

10.3.1 Notable Nanomaterial–Traditional Retardant Combinations

Graphene oxide (GO) enhances the performance of IFR systems (typically containing ammonium polyphosphate (APP), pentaerythritol, and melamine) by improving char structure and strength. Zhou et al. showed that a GO/APP system in PP significantly improved the limiting oxygen index (LOI) and UL-94 ratings while reducing the PHRR by 65%.[73]

MMT is a type of layered silicate that enhances barrier properties and promotes char formation. When incorporated with phosphorus-based flame retardants, MMT forms a cohesive, reinforced char structure. Gilman et al. demonstrated that MMT, when used with triphenyl phosphate in a polystyrene matrix, led to more effective flame retardancy at reduced phosphorus content, highlighting a clear synergistic effect.[74]

CNTs provide electrical conductivity, thermal stability, and mechanical reinforcement. When combined with nitrogen-rich MPP, they synergistically improve the flame resistance of polymer matrices like polyamide 6 (PA6). Zhang et al. found that CNT/MPP hybrid systems in PA6 significantly improved LOI, char residue, and passed UL-94 V-0 rating with minimal dripping.[75]

LDHs decompose endothermically and release water, diluting flammable gases and supporting char formation. Their combination with APP results in higher LOI values and improved smoke suppression. Wang et al. reported that APP-LDH hybrid flame retardants in EVA copolymers enhanced LOI values, reduced smoke release, and increased char yield compared to either additive alone[76]

Nanoparticles like TiO_2, ZnO, and Fe_2O_3 act as char catalysts and reinforce the intumescent char structure. They can also improve UV resistance and other physical properties. Sun et al. (2023) incorporated Zn ferrite nanoparticles into a phosphorus-based system for epoxy resin, resulting in a 75% reduction in total smoke production and improved thermal stability.[77]

The synergy between nanomaterials and conventional flame retardants is based on several interactive mechanisms, including improved char formation, where nanoparticles (e.g., metal oxides, graphene) catalyze carbonization, enhancing the quality and quantity of char formed by intumescent systems.[78] Nanofillers such as layered silicates or graphene sheets create a barrier effect and act as a physical shield that slows down heat and mass transfer. Gas-phase radical trapping is another mechanism where some nanoparticles enhance the flame-inhibiting effect of phosphorus- or nitrogen-based compounds by stabilizing or promoting flame-quenching radicals. The high thermal stability of nanoparticles contributes to reduced heat penetration into the bulk polymer.[79,80]

A few challenges and considerations in obtaining synergism are the dispersion control, a highly required parameter because uniform distribution of nanomaterials is crucial to realize synergy; agglomeration can reduce performance. Processing compatibility is the other considerable factor, as some traditional flame retardants may chemically interact with or degrade nanomaterials during processing. Cost and scaling of hybrid systems can be more expensive due to the need for high-purity nanomaterials and complex formulations

10.4 Mechanisms of Fire Retardancy Using Nanomaterials

Nanomaterials enhance fire retardancy through a combination of physical and chemical mechanisms that slow down combustion, reduce heat release, suppress smoke production, and prevent material degradation. Unlike traditional fire retardants, nanomaterials function efficiently at lower loading levels while improving the mechanical and thermal properties of materials. The primary fire-retardant mechanisms of nanomaterials include the barrier effect, catalytic charring, heat absorption and energy dissipation, and gas phase action (Figure 10.4 and Table 10.2).[81,82]

10.4.1 Barrier Effect

The barrier effect is one of the most important mechanisms by which nanomaterials improve fire resistance. During combustion, nanomaterials form a protective layer on the material surface that acts as a physical barrier against heat transfer, oxygen diffusion, and volatile combustible gases. This effect significantly slows down the combustion process and enhances the material's thermal stability.

10.4.1.1 Formation of a Compact Char Layer

Certain nanomaterials, such as graphene, CNTs, nanoclays, and LDHs, enhance the formation of a stable char layer on the material surface during pyrolysis. The char

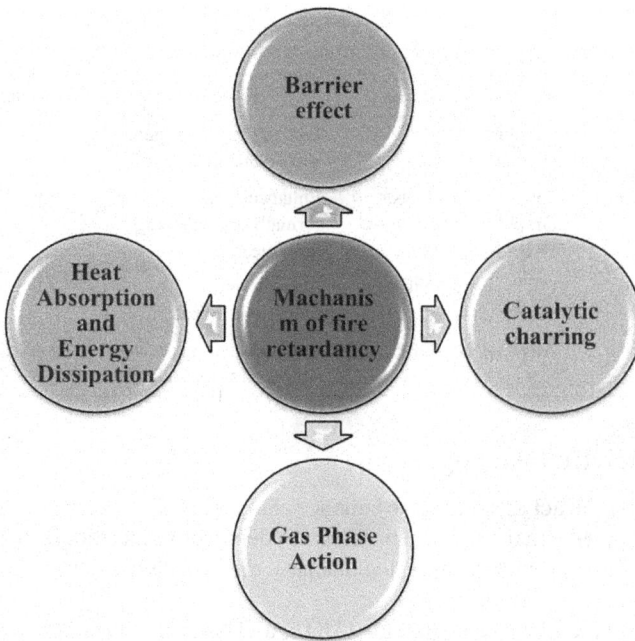

FIGURE 10.4 Mechanism of fire retardancy using nanomaterials.

layer protects the underlying material by reducing the release of flammable gases and preventing heat penetration, thereby delaying ignition and combustion.[100,101]

10.4.1.2 Physical Shielding of Heat and Oxygen

MMT clay, HNTs, and vermiculite form an intumescent barrier upon heating, which expands and shields the material from further heat exposure. Graphene- and carbon-based nanocomposites enhance the formation of dense, thermally stable coatings that prevent oxygen from reaching the burning surface.

10.4.1.3 Reduced Escape of Volatile Combustible Gases

In polymeric materials, thermal degradation produces volatile organic compounds (VOCs) that contribute to fire propagation. Nanoclays, layered silicates, and metal oxides (such as ZnO and TiO_2) create tortuous pathways that trap these volatile gases, delaying their release and reducing fuel availability for combustion.

A study on epoxy composites containing graphene oxide (GO) demonstrated a 50% reduction in heat release rate (HRR) due to the improved barrier effect.[102] Nanoclay-based coatings in polyurethane foams significantly increased time to ignition and reduced PHRRs.[103]

TABLE 10.2

Summary of the Mechanism of Fire-Retardant Nanomaterials

Mechanism	Key Nanomaterials	Function	Reference
Barrier Effect	Graphene, CNTs, Nanoclays, LDHs	Prevents heat, oxygen, and volatile gas transfer	83,84,85,86,87,88,89
Catalytic Charring	Fe_2O_3, TiO_2, LDHs, Graphene-Phosphorus Hybrids	Enhances char formation, preventing fuel release	90,91,92
Heat Absorption	CNTs, Boron Nitride, Silica Aerogels	Absorbs and redistributes heat	93,94,95,96,97
Gas Phase Action	$Al(OH)_3$, $Mg(OH)_2$,	Releases inert gases and suppresses free radicals	98,99

10.4.2 Catalytic Charring

Nanomaterials can act as catalysts to enhance char formation, which is crucial for preventing fire spread. Char is a carbonaceous residue that remains after thermal degradation, acting as a protective layer that insulates the underlying material from further heat exposure.

Iron oxide (Fe_2O_3), titanium dioxide (TiO_2), and cerium oxide (CeO_2) promote the catalytic decomposition of polymer chains into high-carbon-content residues rather than volatile gases. These nanoparticles lower the degradation temperature of polymers, ensuring that more carbon remains in the solid phase rather than being released as flammable gases.[104] LDHs release metal oxides upon heating, which catalyze char formation and promote cross-linking of polymeric structures, improving fire resistance. LDHs also act as acid scavengers, reducing the formation of combustible volatile compounds.[105]

Nanomaterials are often combined with phosphorus-based retardants (e.g., APP) to enhance charring effects. Studies have shown that graphene–phosphorus hybrid systems can double the char yield compared to conventional phosphorus flame retardants alone. The incorporation of TiO_2 nanoparticles in epoxy resins resulted in a 50% reduction in total smoke production, demonstrating enhanced char formation.[106]

10.4.3 Heat Absorption and Energy Dissipation

Some nanomaterials improve fire resistance by absorbing thermal energy and dissipating heat, reducing the overall temperature and slowing down combustion. Graphene, CNTs, boron nitride (BN), and metal oxides possess high thermal conductivity and stability, enabling them to absorb and redistribute heat away from the burning surface. This prevents localized overheating and delays the decomposition of materials. Silica aerogels and clay-based nanocomposites act as excellent thermal insulators by trapping air pockets within their porous structure, reducing heat transfer. Silicate nanofibers form high-temperature-resistant networks that slow down the penetration of heat into polymeric materials. CNTs incorporated into polyvinyl alcohol (PVA) composites significantly reduced the PHRR due to their ability to absorb and dissipate heat.[107] Boron nitride (BN) nanosheets in polyurethane foams improved fire resistance by reducing thermal conductivity and enhancing heat dissipation.[108]

10.4.4 Gas Phase Action

Certain nanomaterials function in the gas phase by releasing non-combustible gases that dilute flammable vapors, suppressing fire propagation.

Hydroxide-based nanoparticles ($Al(OH)_3$, $Mg(OH)_2$) decompose upon heating to release water vapor, which cools the surrounding area and dilutes oxygen concentration. Boron-containing nanomaterials release boric acid vapors, which further inhibit combustion reactions.[109]

10.4.4.1 Radical Scavenging Effect

Fire propagation occurs due to the formation of highly reactive free radicals (•OH, •H, and •O) during combustion.[110] Transition metal oxides (CeO_2, Fe_2O_3, MnO_2) function as radical scavengers, neutralizing these reactive species and slowing down fire spread.[111]

10.4.4.2 Smoke Suppression Mechanism

Nanomaterials such as clays,[112] zinc oxide (ZnO), LDHs, and silica nanoparticles trap soot and toxic gases, significantly reducing smoke release. Carbon-based materials (graphene, CNTs) promote complete combustion, reducing the production of toxic carbon monoxide (CO).[113] $Mg(OH)_2$ nanoparticles in polyethylene composites led to a 40% reduction in flammable gas release, improving fire safety. CeO_2 nanoparticles in polyurethane coatings acted as radical scavengers, reducing smoke generation and suppressing flame spread.

10.5 Applications of Fire-Retardant Nanomaterials

10.5.1 Coatings and Surface Treatments Using Nanomaterials

In addition to bulk incorporation into polymers, nanomaterials are increasingly used in fire-retardant coatings and surface treatments, especially for textiles, foams, wood, and plastic surfaces. These coatings, applied by various methods (see Figure 10.5), offer the advantage of post-processing fire resistance, enabling easy application on existing materials without the need for reformulation. Nanomaterial-based coatings can form intumescent char layers, promote heat shielding, and suppress smoke and toxic gas emissions during combustion.

Nanomaterial-based coatings typically function through one or more of the following mechanisms, including thermal insulation, where coatings form ceramic or carbonaceous layers that block heat transfer. The dense structure of nanomaterials impedes the ingress of oxygen and release of flammable volatiles. They catalyze the formation of a protective char that delays ignition and flame spread. Some nanomaterials release water or CO_2 upon decomposition, reducing surface temperature and diluting combustible gases.

MMT and LDHs are popular in intumescent coatings due to their lamellar structures, which offer excellent gas barrier properties and thermal resistance. When thermally degraded, they release non-combustible gases (e.g., water vapor, CO_2) and form a ceramic-like char that insulates the underlying material. Qian et al. developed

Layer by layer assembly	Dip Coating	Spray Coating	Sol–gel Technique
Precise and customizable multilayer structures	Simple method	Allows uniform deposition on large surfaces	Creates hybrid coatings with inorganic–organic frameworks
Uses oppositely charged polyelectrolytes and nanomaterials.	Involve immersion of substrates in nanomaterial-containing dispersions	Suitable for architectural applications.	Operational sat room or moderate temperatures.

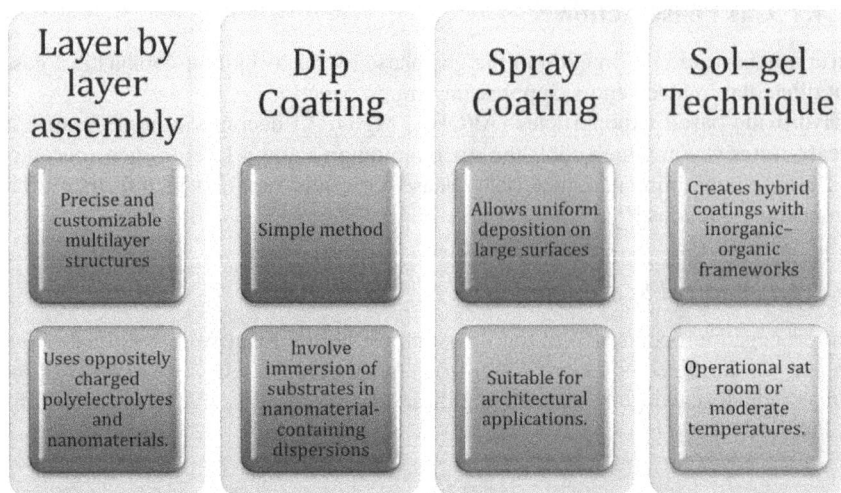

FIGURE 10.5 Methods of application of fire-retardant coatings.

flame-retardant coatings using MMT and LDH for cotton fabrics, achieving significant reductions in heat release and after-flame time.[114] GO-based coatings are increasingly favored due to their high thermal conductivity, gas barrier capability, and char-forming efficiency. They can be applied via layer-by-layer (LbL) assembly, spray coating, or dip-coating methods. Li et al. demonstrated that LbL GO-based coatings on cotton fabrics reduced the PHRR by over 60%, with improved self-extinguishing behavior.[115] CNTs can be incorporated into surface treatments to enhance mechanical strength, thermal conductivity, and char integrity. Their high aspect ratio contributes to barrier effects, reducing the rate of oxygen permeation and mass loss. Lee et al. applied CNT-based flame-retardant coatings on wood panels, resulting in higher LOI values and reduced smoke production.[116]

Metal oxide nanoparticles (TiO_2, ZnO, SiO_2) are used for their photocatalytic, UV-protective, and thermal barrier properties. TiO_2 and ZnO can enhance char formation and suppress toxic gas emissions when combined with intumescent agents. Chen et al. developed a flame-retardant coating of TiO_2 nanoparticles embedded in a polymer matrix, which improved thermal stability and fire performance of polyurethane foams.[117]

Hybrid nanocoatings (organic–inorganic synergistic coatings combining phosphorus-containing compounds, bio-based polymers, and nanomaterials have emerged as environmentally friendly solutions. These coatings offer multifunctional protection: fire resistance, antimicrobial activity, and weathering resistance.[118]

Nanomaterial-based coatings found wide applications in different areas, including textiles like flame-retardant uniforms, upholstery, and protective clothing, and offer numerous advantages (see Figure 10.6). Wood and paper industries with enhanced ignition resistance and charring for construction and packaging materials. Plastics and foams sector, where improved flame resistance in insulation and cushioning materials

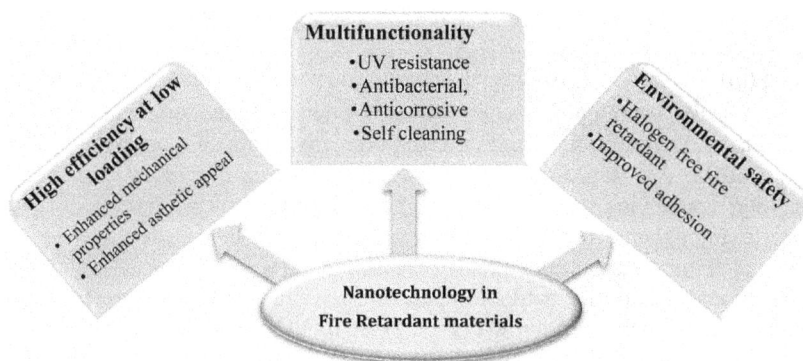

FIGURE 10.6 Benefits of nanotechnology in fire-retardant coatings.

is highly desirable, and metal surfaces with fire-resistive coatings for structural elements in aerospace and automotive sectors.

Nanomaterial-based coating materials offer various advantages, such as being environmentally friendly (often halogen-free), lightweight, and transparent options available, and they are customizable for multifunctionality (e.g., anti-static, UV resistant). However few challenges are associated are long-term durability and wash-fastness (especially in textiles), potential nanoparticle leaching issues, and majorly cost and scalability for large-scale production

10.5.2 Fire-Retardant Nanocoating for the Textile Industry

The textile industry faces growing demand for flame-retardant fabrics used in protective clothing, home furnishings, military gear, transportation, and public interiors. Traditional flame retardants used in textiles, like halogenated compounds, have raised health and environmental concerns. In response, nanotechnology-based coatings have emerged as a safer, more effective alternative, enabling durable, lightweight, and multifunctional flame-retardant finishes on textiles without significantly altering fabric properties.[119,120]

Nanocoatings enhance flame resistance through thermal shielding, where nanoparticles form a compact, heat-resistant layer on fabric surfaces, slowing thermal decomposition. Nanomaterials like graphene or nanoclays create a tortuous path and a barrier effect that limits oxygen ingress and flammable gas release. Some nanomaterials promote carbonaceous char formation that insulates and protects fibers from heat and flame. Certain nanoparticles absorb heat and release non-flammable gases (e.g., water vapor from LDHs), diluting fuel concentration via endothermic reactions.[121,122]

Graphene and graphene oxide (GO) are applied via LbL assembly, spray coating, or dip-coating. GO enhances thermal stability and facilitates char formation while maintaining fabric flexibility. Li et al. demonstrated that LbL GO coatings on cotton fabrics reduced PHRR by over 60%, achieving self-extinguishing behavior[123] Two-dimensional MMT and other nanoclays create gas barrier layers that delay heat and mass transfer. They are commonly used with intumescent agents in cotton and polyester textiles. Alongi et al. reported that cotton fabrics coated with MMT-based nanocoatings showed delayed ignition, improved char yield, and reduced flammability.[124]

LDHs release water upon decomposition, which cools the substrate and dilutes volatile gases. They are often used in waterborne coatings with biopolymers or phosphates. Carosio et al. applied an LbL coating containing LDH and phytic acid on cotton fabric, achieving self-extinguishing behavior after multiple flame exposures[125]

Metal oxide nanoparticles (TiO$_2$, ZnO) contribute to thermal resistance and char integrity of flame-retardant materials. TiO₂ also imparts UV protection, making it a dual-functional coating for outdoor applications. Wang et al. developed a TiO$_2$/APP hybrid coating for polyester fabric that increased its LOI from 18% to 28%.[126]

Carbon nanotubes (CNTs) reinforce the protective char and enhance the mechanical integrity of coated fabrics. They are usually dispersed in water-based binders or applied via sol–gel methods. Zou et al. applied CNT–silica hybrid coatings on polyester-cotton fabric, improving flame resistance while maintaining fabric flexibility.[127]

These coatings can be applied to the textiles via various methods, including layer-by-layer (LbL) assembly, where sequential adsorption of nanomaterials and polyelectrolytes is used to build uniform multilayered coatings. Pad–dry–cure is the conventional textile finishing process where fabrics are passed through a nanoparticle dispersion, squeezed, dried, and cured. The spray and dip coating method is used for flexible deposition of nanoparticles in solvent or water-based systems. Sol–gel coating, another important method, forms hybrid inorganic–organic coatings through controlled hydrolysis and condensation of metal alkoxides.

10.5.3 Nanotechnology in Paints and Coatings for Fire Retardants

Fire-retardant paints and coatings are critical components in the construction, transportation, and electronics industries, offering passive fire protection by slowing down ignition and reducing heat release.[128] The integration of nanotechnology into these systems significantly enhances their flame-retardant performance without compromising mechanical or aesthetic properties. Nanomaterials impart barrier properties, thermal stability, and char reinforcement, making them ideal additives in intumescent and non-intumescent fire-retardant coatings. In the construction industry, where steel structures, wood, and composite panels are used, fire-retardant paints with nanomaterials for passive fire protection have found wide applications. Transportation, including aircraft interiors, automotive parts, and train coaches, is coated with nanoparticle-enhanced intumescent layers. Electronics and Appliances bearing fire-retardant coatings for circuit boards and casings prevent fire spread due to overheating. Textile and furnishings utilizes fire-retardant coatings on curtains, carpets, and upholstery.

Nanomaterials in fire-retardant coatings function through various physical and chemical mechanisms, such as thermal barrier formation, where nanoparticles form a ceramic-like barrier layer upon heating, protecting the substrate; char reinforcement, where nanomaterials stabilize the carbonaceous char, reducing flammable volatile release; endothermic decomposition, some nanoparticles (e.g., LDHs) release water or inert gases, absorbing heat and diluting flammable vapors; gas barrier effect where plate-like nanostructures (e.g., nanoclays, graphene) restrict gas and heat diffusion through coatings; and catalytic char promotion, where metal oxides and carbon nano-structures catalyze crosslinking and aromatization of the polymer matrix.

Nanoclays (e.g., MMT) are well-dispersed in the polymer matrices, creating a tortu-ous path for heat and gas and improving the viscosity and mechanical strength of coat-ings. Giannelis et al. demonstrated that epoxy coatings with MMT nanoclay reduced HRR and smoke production.[129] Graphene and graphene oxide provide excellent bar-rier and thermal conductivity and promote char formation with reduce flammability at low loadings.[130]

LDHs release water upon heating, diluting combustibles, and cooling the surface. They are often combined with phosphorus-based systems for synergy. Wang et al. showed that acrylic coatings with Mg-Al LDHs improved the LOI and char residue, reducing the fire hazard significantly.[131] Metal oxide nanoparticles (e.g., TiO_2, ZnO, Fe_2O_3) act as catalysts for char formation and provide UV resistance, increasing coating durability and fire resistance. Chen et al. reported that TiO_2-based coatings enhanced fire retardancy while also providing photocatalytic self-cleaning properties.[132]

CNTs reinforce mechanical strength and electrical conductivity. They assist in forming a compact, stable char barrier. Zhao et al. developed a water-based fire-retardant paint with CNTs that reduced PHRR by over 50%.[133]

10.6 Future Prospects of Nanomaterials as Fire Retardants

While polymer nanocomposites with fire-retardant nanomaterials show significant promise, some challenges still remain, including achieving uniform nanoparticle dispersion at low loading levels, preventing nanoparticle agglomeration during pro-cessing, maintenance of transparency, mechanical strength, and flexibility, and man-agement of cost and scalability for industrial applications.

Future research directions include the use of bio-based or green nanomaterials (e.g., cellulose nanocrystals, lignin nanoparticles), hybrid nanofillers (e.g., graphene–metal oxide, clay–CNT composites), multifunctional composites combining fire retardancy with antimicrobial, electrical, or UV-shielding properties, and smart polymers that self-extinguish or indicate fire exposure through color change.

The future of fire retardancy lies in the fusion of advanced nanomaterials with emerging technologies such as AI, green chemistry, and additive manufacturing. As sustainability, performance, and multifunctionality become increasingly essential, nanotechnology offers a transformative approach to creating the next generation of intelligent and eco-friendly fire-retardant materials. The following trends and innova-tions are expected to shape the field over the coming decades. Next-generation fire-retardant materials aim to be intelligent and responsive, activating only under specific thermal or combustion conditions. These systems can expand, change color, or become insulating upon ignition, using thermo-responsive polymers or phase-change materials (PCMs) integrated with nanoparticles.[134] Bio-derived nanomaterials such as nano-cellulose, chitosan, and lignin-based nanoparticles are biodegradable and renewable, aimed at replacing toxic halogenated flame retardants.[135] Future materials will combine multiple nanomaterials (e.g., graphene + LDH + metal oxides) to offer synergistic fire protection, mechanical reinforcement, electrical/thermal conductiv-ity, and UV shielding.[136] Nano-encapsulation of flame-retardant additives ensures better dispersion, reduced reactivity under normal conditions, and targeted activation in fire scenarios.[137] Nanomaterials in 3D-printable fire-retardant inks and filaments allow rapid manufacturing of customized components with built-in fire resistance.[138] Machine learning helps predict fire behavior, optimize nanoparticle combinations, and simulate fire scenarios to accelerate flame-retardant material development.[139]

NOTES

[1] https://injuryfacts.nsc.org/home-and-community/safety-topics/fire-related-fatalities-and injuries/.

[2] Singh, M. K., Rangappa, S. M., Misra, M., Mohanty, A. K., & Siengchin, S. (2025). Recent advancements in nanostructured flame-retardants: Types, mechanisms, and applications in polymer composites. *Nano-Structures & Nano-Objects*, 42, 101468. https://doi.org/10.1016/j.nanoso.2025.101468.

[3] Liu, Q., Zhao, Y., Gao, S., Yang, X., Fan, R., Zhi, M., & Fu, M. (2021). Recent advances in the flame retardancy role of graphene and its derivatives in epoxy resin materials, *Composites Part A: Applied Science and Manufacturing*, 149, 106539, https://doi.org/10.1016/j.compositesa.2021.106539.

[4] Wang, X., Guo, W., Cai, W., & Hu, Y. (2022). Graphene-based polymer composites for flame-retardant application. In *Woodhead Publishing Series in Composites Science and Engineering, Innovations in Graphene-Based Polymer Composites*, pp. 61–89. Woodhead Publishing. ISBN 9780128237892, https://doi.org/10.1016/B978-0-12-823789-2.00004-2.

[5] Jamsaz, A., & Goharshadi, E. K. (2023). Graphene-based flame-retardant polyurethane: A critical review. *Polymer Bulletin*, 80, 11633–11669. https://doi.org/10.1007/s00289-022-04585-5.

[6] Yeoh, G. H., De Cachinho Cordeiro, I. M., Wang, W., Wang, C., Yuen, A. C. Y., Chen, T. B. Y., Vargas, J. B., Mao, G., Garbe, U., & Chua, H. T. (2024). Carbon-based Flame Retardants for Polymers: A bottom-up review. *Advanced Materials*, 36(42), e2403835.

[7] Kim, J., Jang, J., Yun, S., Kim, H. D., Byun, Y. Y., Park, Y. T., Song, J. I., & Cho, C. (2021). Synergistic flame retardant effects of carbon nanotube-based multilayer nanocoatings. *Macromolecular Materials and Engineering*, 306(9). https://doi.org/10.1002/mame.202100233.

[8] Sai,T., Ran, R.,Guo, Z., Song, P., & Fang, Z. (2022). Recent advances in fire-retardant carbon-based polymeric nanocomposites through fighting free radicals. *SusMat*, 2(4). https://doi.org/10.1002/sus2.73.

[9] Idumah, C. I. (2023). Recent advancements in fire retardant mechanisms of carbon nanotubes, graphene, and fullerene polymeric nanoarchitectures. *Journal of Analytical and Applied Pyrolysis*, 174, 106113, https://doi.org/10.1016/j.jaap.2023.106113.

[10] Vo, D. K., Do, T. D., Nguyen, B. T., Tran, C. K., Nguyen, T. A., Nguyen, D. M., Pham, L. H., Nguyen, T. D., Nguyen, T.-D., & Hoang, D. Q. (2022). Effect of metal oxide nanoparticles and aluminum hydroxide on the physicochemical properties and flame-retardant behavior of rigid polyurethane foam. *Construction and Building Materials*, 356, 129268. https://doi.org/10.1016/j.conbuildmat.2022.129268.

[11] Abdel-Salam, A. H., & Morsy, A. (2025). Novel high-efficiency nano metal oxide based on phosphorus as smart flame retardants with multiple reactive for sustainable cotton-polyester fabrics. *International Journal of Biological Macromolecules*, 294, 139502. https://doi.org/10.1016/j.ijbiomac.2025.139502.

[12] Li, J., Zhao, H., Liu, H., Sun, J., Wu, J., Liu, Q., Zheng, Y., & Zheng, P. (2023). Recent advances in metal-family flame retardants: A review. *RSC Advance*, 13, 22639–22662. https://doi.org/10.1039/D3RA03536K.

[13] Li, Q., & Pan, Y. T. (2024). Chapter 6 – Flame retardant properties of metal oxide/polymer nanocomposites. *Flame Retardant Nanocomposites. Woodhead Publishing Series in Composites Science and Engineering*, 2024, 201–224. https://doi.org/10.1016/B978-0-443-15421-8.00005-7.

[14] Wang, M., Xiao, G., Chen, C., Yang, Z., Zhong, F., Chen, C., Zou, R., Li, Y., & Li, R. (2022). Combining layered double hydroxides and carbon nanotubes to synergistically enhance the flame retardant properties of composite coatings. *Colloids and Surfaces A: Physicochemical and Engineering Aspects*, 638, 128315. https://doi.org/10.1016/j.colsurfa.2022.128315.

[15] Zhou, L.-L., Li, W.-X., Zhao, H.-B., & Zhao, B. (2022). Comparative Study of M(II) Al (M=Co, Ni) Layered Double Hydroxides for Silicone Foam: Characterization, Flame Retardancy, and Smoke Suppression. *International Journal of Molecular Sciences*, 23(19), 11049. https://doi.org/10.3390/ijms231911049.

[16] Xu, S., Liu, J., Liu, X., Li, H., Gu, X., Sun, J., & Zhang, S. (2022). Preparation of Ni-Fe layered double hydroxides and its application in thermoplastic polyurethane with flame retardancy and smoke suppression. *Polymer Degradation and Stability*, 202, 110043. https://doi.org/10.1016/j.polymdegradstab.2022.110043.

[17] Idumah, C. I., & Ogbu, J. E. (2024). Flame retardant mechanisms of Montmorillonites, layered double hydroxides and molybdenum disulfide polymeric nanoarchitectures for safety in extreme environments. *Polymer-Plastics Technology and Materials*, 63(6), 639–666. https://doi.org/10.1080/25740881.2023.2301294.

[18] Shi, X.-H., Li, X.-L., Shi, H., Liu, Q.-Y., Xie, W.-M., Wu, S.-J., Zhao, N., & Wang, D.-Y. (2024). Insight into the flame-retardant mechanism of different organic-modified layered double hydroxide for epoxy resin. *Applied Clay Science*, 248, 107233. https://doi.org/10.1016/j.clay.2023.107233.

[19] Wang, M., Xiao, G., Chen, C., Yang, Z., Zhong, F., Chen, C., Zou, R., Li, Y., & Li, R. (2022). Combining layered double hydroxides and carbon nanotubes to synergistically enhance the flame retardant properties of composite coatings. *Colloids and Surfaces A: Physicochemical and Engineering Aspects*, 638, 128315.

[20] Lee, I., Kim, S. J., Byun, Y. Y., Chang, I., Ju, Y.-W., Park, Y. T., Jang, J., & Cho, C. (2023). High flame retardancy enabled by dual clays-based multilayer nanocomposites. *Progress in Organic Coatings*, 183, 107784. https://doi.org/10.1016/j.porgcoat.2023.107784.

[21] Fu, Y., Wang, Z., Liu, H., Zhang, K., Zhang, Z., Song, Y., Wu, P., Li, Y., & Ling, Z. (2025). Flame-retardant effect of scalable all clay composite films. *Applied Clay Science*, 267, 107724. https://doi.org/10.1016/j.clay.2025.107724.

[22] Beryl, J. R., & Xavier, J. R. (2023). Halloysite for clay–polymer nanocomposites: Effects of nanofillers on the anti-corrosion, mechanical, microstructure, and flame-retardant properties—a review. *Journal of Materials Science*, 58, 10943–10974. https://doi.org/10.1007/s10853-023-08710-1.

[23] Rodrigues, J., & Shimpi, N. G. (2024). Nanostructured flame retardants: An overview. *Nano-Structures & Nano-Objects*, 39, 101253. https://doi.org/10.1016/j.nanoso.2024.101253.

[24] Temane, L. T., Orasugh, J. T., & Ray, S. S. (2023). Recent Advances and Outlook in 2D Nanomaterial-Based Flame-Retardant PLA Materials. *Materials*, 16(17), 6046. https://doi.org/10.3390/ma16176046.

[25] Attia, N. F., Elashery, S. E. A., El-Sayed, F., Mohamed, M., Osama, R., Elmahdy, E., Abd-Ellah, M., El-Seedi, H. R., Hawash, H. B., & Ameen, H. (2024). Recent advances in nanobased flame-retardant coatings for textile fabrics. *Nano-Structures & Nano-Objects*, 38, 101180. https://doi.org/10.1016/j.nanoso.2024.101180.

[26] Nguyen Thanh, T., Szolnoki, B., Vadas, D. Nacsa, M., Marosi G., & Bocz, K. (2023). Effect of clay minerals on the flame retardancy of polylactic acid/ammonium polyphosphate system. *Journal of Thermal Analysis and Calorimetry*, 148, 293–304. https://doi.org/10.1007/s10973-022-11712-x.

[27] Hu, Y., Liu, Y., Zheng, S., & Kang, W. (2024). Progress in application of silane coupling agent for clay modification to flame retardant polymer. *Molecule*, 29(17), 4143. https://doi.org/10.3390/molecules29174143.

[28] Sun, Y., Yang, P., & Sun, W. (2022). Effects of kaolinite on thermal, mechanical, fire behavior and their mechanisms of intumescent flame-retardant polyurea. *Polymer Degradation and Stability*, 197, 109842. https://doi.org/10.1016/j.polymdegradstab.2022.109842.

[29] Lama, N., Wilhite, J., Lvov, Y., Konnova, S., & Fakhrullin, R. (2023). Clay nanotube coating on cotton fibers for enhanced flame-retardancy and antibacterial properties. *Chemnanomat*, First published: 19 May 2023. https://doi.org/10.1002/cnma.202300106.

[30] Xia, L., Wang, X., Ren, T., Luo, L., Li, D., Dai, J., Xu, Y., Yuan, C., Zeng, B., & Dai, L. (2022). Green construction of multi-functional fire resistant epoxy resins based on boron nitride with core-shell structure. *Polymer Degradation and Stability*, 203, 110059.

[31] Tasi, T.-P., Hsieh, C.-T., Yang, H.-C., Liu, K.-C., Huang, Y.-R., Gandomi, Y. A., & Mallick, B. C. (2022). Enhanced fireproof performance of construction coatings by adding hexagonal boron nitride nanosheets. *Ceramics International*, 48(14), 20809–20816.

[32] Han, G., Zhao, X., Feng, Y., Ma, J., Zhou, K., Shi, Y., Liu, C., & Xie, X. (2021). Highly flame-retardant epoxy-based thermal conductive composites with functionalized boron nitride nanosheets exfoliated by one-step ball milling. *Chemical Engineering Journal*, 407, 127099.

[33] Tawiah, B., Yu, B., Yuen, R. K. K., Hu, Y., Wei, R., Xin, H. H., & Fei, B. (2019). Highly efficient flame retardant and smoke suppression mechanism of boron modified graphene Oxide/Poly(Lactic acid) nanocomposites. *Carbon*, 150, 8–20.

[34] Dogan, M., Dogan, S. D., Savas, L. A., Ozcelik, G., & Tayfun, U. (2021). Flame retardant effect of boron compounds in polymeric materials. *Composites Part B: Engineering*, 222, s109088.

[35] Araby, S., Philips, B., Meng, Q., Ma, J., Laoui, T., & Wang, C. H. (2021). Recent advances in carbon-based nanomaterials for flame retardant polymers and composites. *Composites Part B: Engineering*, 212, 108675.

[36] Liu, S., He, M., Qin, Q., Liu, W., Liao, L., & Qin, S. (2024). Expanded properties and applications of porous flame-retardant polymers containing graphene and its derivatives. *Polymers*, 16(14), 2053. https://doi.org/10.3390/polym16142053.

[37] Li, H., Liu, C., Zhu, J., Huan, X., Xu, K., Geng, H., Chen, X., Li, T., Deng, D., Ding, W., Zu, L., Ge, L., Jia, X., & Yang, X. (2024). Intrinsically reactive hyperbranched interface governs graphene oxide dispersion and crosslinking in epoxy for enhanced flame retardancy. *Journal of Colloid and Interface Science*, 672, 465–476.

[38] Yao, Y., Jin, S., Ma, X., Yu, R., Zou, H., Wang, H., Lv, X., & Shu, Q. (2020). Graphene-containing flexible polyurethane porous composites with improved electromagnetic shielding and flame retardancy. *Composites Science and Technology*, 200, 108457.

[39] Song, P., Liu, L., & Wu, Y. (2015). Flame retardant effect of graphene oxide and functionalized graphene oxide in epoxy resins. *RSC Advances*, 5(31), 24731–24739. https://doi.org/10.1039/C4RA16004A.

[40] Taib, M. N. A. M., Khairuddin, N. F. M., & Saleh, T. A. (2025). Functionalized carbon nanotubes: Synthesis, properties, and application in polymer for flame retardancy – A review. *Journal of Thermal Analysis and Calorimetry*. https://doi.org/10.1007/s10973-025-14016-y.

[41] Elbakyan, L., & Zaporotskova, I. (2025). Polypropylene modified with carbon nanomaterials: Structure, properties and application (a review). *Polymers*, 17(4), 517. https://doi.org/10.3390/polym17040517.

[42] Gu, L., Qiu, C., Qiu, J., Yao, Y., Sakai, E., & Yang, L. (2020). Preparation and characterization of DOPO-functionalized MWCNT and its high flame-retardant performance in epoxy nanocomposites. *Polymers*, 12(3), 613. https://doi.org/10.3390/polym12030613.

[43] Yan, L., Xu, Z., & Zhang, J. (2016). Flame retardant and smoke suppression mechanism of multi-walled carbon nanotubes on high-impact polystyrene nanocomposites. *Iranian Polymer Journal*, 25(7), 623–633.

[44] Baddam, Y., Ijaola, A. O., & Asmatulu, E. (2021). Fabrication of flame-retardant and superhydrophobic electrospun nanofibers. *Surfaces and Interfaces*, 23, 101017.

[45] Pan, Y., Guo, Z., Ran, S., & Fang, Z. (2020). Influence of fullerenes on the thermal and flame-retardant properties of polymeric materials. 137(1), s47538. https://doi.org/10.1002/app.47538.

[46] Harhoosh, A. A. (2023). Nontoxic flame retardants based on metal oxide nanoparticles. *Research in Medical & Engineering Sciences*, 10(4), RMES.000744. https://doi.org/10.31031/RMES.2023.10.000744.

[47] Pan, H., Wang, W., Pan, Y., Son, L., Hu, Y., & Liew, K. M. (2015). Formation of layer-by-layer assembled titanate nanotubes filled coating on flexible polyurethane foam with improved flame retardant and smoke suppression properties. *ACS Applied Materials & Interfaces*, 7(1), 101–111.

[48] Chen, X., Wan, M., Gao, M., Wang, Y., & Yi, D. (2020). Improved flame resistance properties of unsaturated polyester resin with TiO_2-MXOY solid superacid. *Chinese Journal of Chemical Engineering*, 28(9), 2474–2482. https://doi.org/10.1016/j.cjche.2020.06.018.

[49] Wang, L., Yan, W. J., Zhong, C.-Z., Chen, C.-R., Luo, Q., Pan, Y.-T., Tang, Z.-H., & Xu, S. (2024). Construction of TiO_2-based decorated with containing nitrogen-phosphorus bimetallic layered double hydroxides for simultaneously improved flame retardancy and smoke suppression properties of EVA. *Materials Today Chemistry*, 36, 101952. https://doi.org/10.1016/j.mtchem.2024.101952.

[50] Bartczak, P., Siwińska-Ciesielczyk, K., Haak, N., Parus, A., Piasecki, A., Jesionowski, T., & Borysiak, S. (2023). Closed-cell polyurethane spray foam obtained with novel TiO_2–ZnO hybrid fillers – mechanical, insulating properties and microbial purity. *Journal of Building Engineering*, 65, 105760. https://doi.org/10.1016/j.jobe.2022.105760.

[51] Cheng, C., Shuqian, S., Mingmei, S., Zhengwen, W., Xingrong, Z., & Linsheng, T. (2023). Synergistic flame retardancy of ZnO with piperazine pyrophosphate/melamine polyphosphate in PP. *Polymer Testing*, 117, 107878. https://doi.org/10.1016/j.polymertesting.2022.107878.

[52] Verbič, A., Gorjanc, M., & Simončič, B. (2019). Zinc oxide for functional textile coatings: Recent advances. *Coatings*, 9(9), 550. https://doi.org/10.3390/coatings9090550.

[53] Saleemi, S., Mannan, H. A., Riaz, T., Hai, A. M., Zeb, H., & Khan, A. K. (2024). Optimizing synergistic silica – zinc oxide coating for enhanced flammability resistance in cotton protective clothing. *Fibers*, 12(5), 44. https://doi.org/10.3390/fib12050044.

[54] Zhang, L., Yang, D., Li, Z., Zhai, Z., Li, X., de La Vega, J., & Wang, D.-Y. (2024). Ultrafine iron oxide decorated mesoporous carbon nanotubes as highly efficient flame retardant in epoxy nanocomposites via catalytic charring effect. *Sustainable Materials and Technologies*, 39, e00845. https://doi.org/10.1016/j.susmat.2024.e00845.

[55] Hobson, J., Yin, G.-Z., Yu, X., Zhou, X., Prolongo, S. G., Ao, X., & Wang, D.-Y. (2022). Synergistic effect of cerium oxide for improving the fire-retardant, mechanical and ultraviolet-blocking properties of EVA/magnesium hydroxide composites. *Materials*, 15(17), 5867. https://doi.org/10.3390/ma15175867.

[56] Altalhi, A. A., Mohamed, E. A., & Negm, N. A. (2024). Recent advances in layered double hydroxide (LDH)-based materials: Fabrication, modification strategies, characterization, promising environmental catalytic applications, and prospective aspects. *Energy Advances*, 3, 2136–2151.

[57] Liu, Y., Gao, Y., Wang, Q., & Lin, W. (2018). The synergistic effect of layered double hydroxides with other flame retardant additives for polymer nanocomposites: A critical review. *Dalton Transactions*, 47, 14827–14840.

[58] Li, P., Dang, L., Li, Y., Lan, S., & Zhu, D. (2022). Enhanced flame-retardant and mechanical properties of epoxy resin by combination with layered double hydroxide, $Mg_2B_2O_5$ whisker, and dodecyl dihydrogen phosphate. *Materials & Design*, 217, 110608. https://doi.org/10.1016/j.matdes.2022.110608.

[59] Wang, D. Y., Liu, Y., & Wang, Y. Z. (2011). Layered double hydroxides as flame retardant and smoke suppressant for polymers. *Journal of Materials Chemistry*, 21(26), 9456–9467. https://doi.org/10.1039/C1JM10577H.

[60] Lee, I., Kim, J., Yun, S., Jang, J., Cho, S. Y., Cho, J. S., Ryu, J. H., Choi, D., & Cho, C. (2024). Synergistic combination of dual clays in multilayered nanocomposites for enhanced flame retardant properties. *ACS Omega*, 9(6), 6606–6615. https://doi.org/10.1021/acsomega.3c07534.

[61] Kausar, A. (2020). Flame retardant potential of clay nanoparticles, in micro and nano technologies. *Clay Nanoparticles*, 2020, 169–184. ISBN 9780128167830, https://doi.org/10.1016/B978-0-12-816783-0.00007-4.

[62] Gilman, J. W., Jackson, C. L., & Morgan, A. B. (2000). Flammability properties of polymer-layered-silicate nanocomposites. polypropylene and polystyrene nanocomposites. *Chemistry of Materials*, 12(7), 1866–1873.

[63] Jang, B. N., & Wilkie, C. A. (2005, April). The effect of clay on the thermal degradation of polyamide 6 in polyamide 6/clay nanocomposites. *Polymer*, 46(10), 3264–3274.

[64] Jin, X., Xiang, E., Zhang, R., Qin, D., He, Y., Jiang, M., & Jiang, Z. (2022). An eco-friendly and effective approach based on bio-based substances and halloysite nanotubes for fire protection of bamboo fiber/polypropylene composites. *Journal of Materials Research and Technology*, 17, 3138–3149. https://doi.org/10.1016/j.jmrt.2022.02.051.

[65] Dogan, M., Dogan, S. D., Savas, L. A., Ozcelik, G., & Tayfun, U. (2021). Flame retardant effect of boron compounds in polymeric materials. *Composites Part B: Engineering*, 222, 109088. https://doi.org/10.1016/j.compositesb.2021.109088.

[66] Shen, K. K. Boron-based flame retardants in non-halogen based polymers. https://doi.org/10.1002/9781119752240.ch7.

[67] Owen Li, Sandeep Tamrakar, Zeynep Iyigundogdu, Debbie Mielewski, Kevin Wyss, James M. Tour, Alper Kiziltas. Flexible polyurethane foams reinforced with graphene and boron nitride nanofillers. (2022). *Polymer Composites*, 44(3). https://doi.org/10.1002/pc.27183.

[68] Ahmad, F., Zulkurnain, E. S. B., Ullah, S., Al-Sehemi, A. G., & Raza, M. R. (2020). Improved fire resistance of boron nitride/epoxy intumescent coating upon minor addition of nano-alumina. *Materials Chemistry and Physics*, 256, 123634. https://doi.org/10.1016/j.matchemphys.2020.123634.

[69] Xu, J., Niu, Y., Xie, Z., Liang, F., Guo, F., & Wu, J. (2023). Synergistic flame retardant effect of carbon nanohorns and ammonium polyphosphate as a novel flame retardant system for cotton fabrics. *Chemical Engineering Journal*, 451(Part 2), 138566. https://doi.org/10.1016/j.cej.2022.138566.

[70] Liu, S., Fang, Z., Yan, H., Chevali, V. S., & Wang, H. (2016). Synergistic flame retardancy effect of graphene nanosheets and traditional retardants on epoxy resin. *Composites Part A Applied Science and Manufacturing*, 89. https://doi.org/10.1016/j.compositesa.2016.03.012.

[71] Wang, Y., Liu, B., Chen, R., Wang, Y., Han, Z., Wang, C., & Weng, L. (2023, June 30). Synergistic effect of nano-silica and intumescent flame retardant on the fire reaction properties of polypropylene composites. *Materials (Basel)*, 16(13), 4759. https://doi.org/10.3390/ma16134759.

[72] Ai, L., Chen, S., Zeng, J., Yang, L., & Liu, P. (2019). Synergistic flame retardant effect of an intumescent flame retardant containing boron and magnesium hydroxide. *ACS Omega*, 4(2), 3314–3321. https://doi.org/10.1021/acsomega.8b03333.

[73] Zhou, K., Zhu, P., Zhang, J., & Tang, G. (2014). Enhanced flame retardancy and thermal stability of polypropylene with synergistic graphene oxide and intumescent flame retardant. *Journal of Hazardous Materials*, 264, 505–513. https://doi.org/10.1016/j.jhazmat.2013.11.028.

[74] Gilman, J. W., Jackson, C. L., Morgan, A. B., Harris, R., Manias, E., Giannelis, E. P., Wuthenow, M., Hilton, D., & Phillips, S. H. (2000). Flammability properties of polymer – Layered silicate nanocomposites. *Chemistry of Materials*, 12(7), 1866–1873. https://doi.org/10.1021/cm000176g.

[75] Zhang, S., Horrocks, A. R., & Price, D. (2016). Synergistic flame retardant systems based on carbon nanotubes and melamine polyphosphate for polyamide 6. *Polymer Degradation and Stability*, 134, 170–180. https://doi.org/10.1016/j.polymdegradstab.2016.09.014.

[76] Wang, D. Y., Liu, Y., & Wang, Y. Z. (2011). Layered double hydroxides as flame retardant and smoke suppressant for polymers. *Journal of Materials Chemistry*, 21(26), 9456–9467. https://doi.org/10.1039/C1JM10577H.

[77] Sun, X., Li, Z., Das, O., & Hedenqvist, M. S. (2023). Superior flame retardancy and smoke suppression of epoxy resins with zinc ferrite@polyphosphazene nanocomposites. *Composites Part A: Applied Science and Manufacturing*, 167, 7417. https://doi.org/10.1016/j.compositesa.2022.107417.

[78] Araby, S., Philips, B., Meng, Q., Ma, J., Laoui, T., & Wang, C. H. (2021). Recent advances in carbon-based nanomaterials for flame retardant polymers and composites. *Composites Part B: Engineering*, 212, 108675. https://doi.org/10.1016/j.compositesb.2021.108675.

[79] Kim, Y., Lee, S., & Yoon, H. (2021). Fire-safe polymer composites: Flame-retardant effect of nanofillers. *Polymers*, 13(4), 540. https://doi.org/10.3390/polym13040540.

[80] Kicko-Walczak, E., & Rymarz, G. (2015). Flame retardants nanocomposites-synergistic effect of combination conventional retardants with nanofillers of the flammability of thermoset resins. *Journal of Mechanics Engineering and Automation*, 5, 510–518. https://doi.org/10.17265/2159-5275/2015.09.005.

[81] Castrovinci, A., & Camino, G. (2007). Fire-retardant mechanisms in polymer nanocomposite materials. In Duquesne, S., Magniez, C., Camino, G. (eds) *Multifunctional Barriers for Flexible Structure. Materials Science*, vol. 97. Springer, Berlin, Heidelberg. https://doi.org/10.1007/978-3-540-71920-5_5.

[82] Shen, J., Liang, J., Lin, X., Lin, H., Yu, J., & Wang, S. (2021, December 27). The flame-retardant mechanisms and preparation of polymer composites and their potential application. *Construction Engineering. Polymers (Basel)*, 14(1), 82. https://doi.org/10.3390/polym14010082. PMID: 35012105; PMCID: PMC8747271.

[83] Yang Y, Díaz Palencia JL, Wang N, Jiang Y, & Wang DY. (2021). Nanocarbon-based flame retardant polymer nanocomposites. *Molecules*, 26(15), 4670. https://doi.org/10.3390/molecules26154670. PMID: 34361823; PMCID: PMC8348979.

[84] Liu, Q., Zhao, Y., Gao, S., Yang, X., Fan, R., Zhi, M., & Fu, M. (2021). Recent advances in the flame retardancy role of graphene and its derivatives in epoxy resin materials. *Composites Part A: Applied Science and Manufacturing*, 149, 06539.

[85] Dewaghe, C., Lew, C. Y., Claes, M., Belgium, S. A., & Dubois, P. (2011). Fire-retardant applications of polymer–carbon nanotubes composites: Improved barrier effect and synergism. In *Woodhead Publishing Series in Composites Science and Engineering, Polymer–Carbon Nanotube Composites*, pp. 718–745, Woodhead Publishing.

[86] Lee, I., Kim, S. J., Byun, Y. Y., Chang, I., Ju, Y.-W., Park, Y. T., Jang, J., & Cho, C. (2023). High flame retardancy enabled by dual clays-based multilayer nanocomposites. *Progress in Organic Coatings*, 183, 107784. https://doi.org/10.1016/j.porgcoat.2023.107784.

[87] Liu, Y., Hu, X., Han, S., Wang, Y., Wu, Z., & Qian, L. (2022). The fabrication of phytate and LDH-based hybrid nanosheet towards improved fire safety properties and superior smoke suppression of intumescent flame retardant LDPE. *Thermochimica Acta*, 714, 179271. https://doi.org/10.1016/j.tca.2022.179271.

[88] Dowbysz, A., Samsonowicz, M., Kukfisz, B., & Koperniak, P. (2024). Recent developments of nano flame retardants for unsaturated polyester resin. *Materials*, 17(4), 852. https://doi.org/10.3390/ma17040852.

[89] Huo, S., Guo, Y., Yang, Q., Wang, H., & Song, P. (2024). Two-dimensional nanomaterials for flame-retardant polymer composites: A mini review. *Advanced Nanocomposites*, 1(1), 240–247. ISSN 2949-9445, https://doi.org/10.1016/j.adna.2024.07.001.

[90] You, Z., Xiu, L., Fang, W., Jian-Wei, H., & Jian-Xin, D. (2014). Effect of metal oxides on fire resistance and char formation of intumescent flame retardant coating. *Journal of Inorganic Materials*, 29(9), 972–978. https://doi.org/10.15541/jim20130686.

[91] Shao, Z.-B., Cui, J., Lin, X.-B., Li, X.-L., Jian, R.-K., & Wang, D-Y. (2022). In-situ coprecipitation formed Fe/Zn-layered double hydroxide/ammonium polyphosphate hybrid material for flame retardant epoxy resin via synergistic catalytic charring. *Composites Part A: Applied Science and Manufacturing*, 155, 106841, https://doi.org/10.1016/j.compositesa.2022.106841.

[92] Chen, W., Liu, Y., Liu, P., Xu, C., Liu, Y., & Wang, Q. (2017). The preparation and application of a graphene-based hybrid flame retardant containing a long-chain phospha-phenanthrene. *Scientific Reports*, 7, 8759. https://doi.org/10.1038/s41598-017-09459-9.

[93] Zhang, Y., Ma, C., Yang, M., Yuan, D., Tang, K., Li, C., & Mao, B. (2023). Can additive carbon nanotubes reduce the PMMA fire risk? *Fire Safety Journal*, 136, 103757. https://doi.org/10.1016/j.firesaf.2023.103757.

[94] Han, G., Zhao, X., Feng, Y., Ma, J., Zhou, K., Shi, Y., Liu, C., & Xie, X. (2021). Highly flame-retardant epoxy-based thermal conductive composites with functionalized boron nitride nanosheets exfoliated by one-step ball milling. *Chemical Engineering Journal*, 407, 127099. https://doi.org/10.1016/j.cej.2020.127099.

[95] Guo, W., Yang, J., Li, L., Zhang, J., Wang, Z., Wang, J., & Zhang, J. (2024). Boron nitride based nanostructure inspired flame retarded thermoplastic polyurethane composites: Comprehensive evaluation and mechanism investigation. *Construction and Building Materials*, 457, 139478. https://doi.org/10.1016/j.conbuildmat.2024.139478.

[96] Sun, H., Pan, Y., He, S., Gong, L., Zhang, Z., Cheng, X., & Zhang, H. (2025). Hydrophobic silica aerogel with higher flame retardancy, thermal radiation shielding, and high-temperature insulation properties through introduction of TiO_2. *Gels*, 11(4), 249. https://doi.org/10.3390/gels11040249.

[97] Tao, J., Yang, F., Wu, T., Shi, J., Zhao, H.-B., & Rao, W. (2023). Thermal insulation, flame retardancy, smoke suppression, and reinforcement of rigid polyurethane foam enabled by incorporating a P/Cu-hybrid silica aerogel. *Chemical Engineering Journal*, 461, 142061,

[98] Zhu, K., Yang Y., Lin, C., Wang, Q., Ye, D., Jiang, H., & Wu, K. (2024). Effect of compounded aluminum hydroxide flame retardants on the flammability and smoke suppression performance of asphalt binders. *ACS Omega*, 9(2), 2803–2814, https://doi.org/10.1021/acsomega.3c08094.

[99] https://patents.google.com/patent/KR20080082135A/en

[100] Chen, J., & Han, J. (2019). Comparative performance of carbon nanotubes and nanoclays as flame retardants for epoxy composites. *Results in Physics*, 14, 102481. https://doi.org/10.1016/j.rinp.2019.102481.

[101] Guo, M. F., Li, J., & Wang, Y. W. (2021). Effects of carbon nanotubes on char structure and heat transfer in ethylene propylene diene monomer composites at high temperature. *Composites Science and Technology*, 211, 108852. https://doi.org/10.1016/j.compscitech.2021.108852.

[102] Wang, J. (2021). Functionalization of graphene oxide with polysilicone: Synthesis, characterization, and its flame retardancy in epoxy resin. *Polymers*, 13(21), 3857. https://doi.org/10.3390/polym13213857.

[103] Hejna, A. (2021). Clays as inhibitors of polyurethane foams' flammability. *Materials*, 14(17), 4826. https://doi.org/10.3390/ma14174826.

[104] Sharma, V., Agarwal, S., Mathur, A., Singhal, S., & Wadhwa, S. (2024). Advancements in nanomaterial based flame-retardants for polymers: A comprehensive overview. *Journal of Industrial and Engineering Chemistry*, 133, 38–52. https://doi.org/10.1016/j.jiec.2023.12.010.

[105] Sankeshi, S., Ganapathiraju, J., Bajaj, P., Mangali, M. K., Shaik, S. H., & Basak, P. (2024). 2D-nanostructures as flame retardant additives: Recent progress in hybrid polymeric coatings. *Nano-Structures & Nano-Objects*, 40, 101346. https://doi.org/10.1016/j.nanoso.2024.101346.

[106] Fan, H., Zhao, J., Zhang, J., Li, H., Zhang, S., Sun, J., Xin, F., Liu, F., Qin, Z., & Tang, W. (2022). TiO_2/SiO_2/kaolinite hybrid filler to improve the flame retardancy, smoke suppression and anti-aging characteristics of epoxy resin. *Materials Chemistry and Physics*, 277, 125576.

[107] Yang, Y., Díaz Palencia, J. L., Wang, N., Jiang, Y., & Wang, D.-Y. (2021). Nanocarbon-based flame retardant polymer nanocomposites. *Molecules*, 26(15), 4670. https://doi.org/10.3390/molecules26154670.

[108] Han, G., Zhao, X., Feng, G., Ma, J., Zhou, K., Shi, Y., Liu, C., & Xie, X. (2021). Highly flame-retardant epoxy-based thermal conductive composites with functionalized boron nitride nanosheets exfoliated by one-step ball milling. *Chemical Engineering Journal*, 407, 127099.

[109] Sharma, V., Agarwal, S., Mathur, A., Singhal, S., Wadhwa, S. (2024). Advancements in nanomaterial based flame-retardants for polymers: A comprehensive overview. *Journal of Industrial and Engineering Chemistry*, 133, 38–52. https://doi.org/10.1016/j.jiec.2023.12.010.

[110] Li, X., Kang, D., Zhang, L., Chen, J., Huang, S., Zou, Q., & He, Z. (2024). Effect of surface reaction on the distribution characteristics of temperature and OH radicals in microchannel combustion. *Fire*, 7(3), 71. https://doi.org/10.3390/fire7030071.

[111] Lin, M., Li, B., Li, Q., Li, S., Zhang, S. (2011). Synergistic effect of metal oxides on the flame retardancy and thermal degradation of novel intumescent flame-retardant thermoplastic polyurethanes. *Journal of Applied Polymer Science*, 121(4), 1951–1960. https://doi.org/10.1002/app.33759.

[112] Zhang, L., Niu, K., Wang, H., Wang, J., Liu, M., Lei, Y., & Yan, L. (2024). Char formation and smoke suppression mechanism of montmorillonite modified by ammonium polyphosphate/silane towards fire safety enhancement for wood composites. *Wood Science and Technology*, 58, 811–827. https://doi.org/10.1007/s00226-024-01546.

[113] Araby, S., Philips, B., Meng, Q., Ma, J., Laoui, T., & Wang, C. H. (2021). Recent advances in carbon-based nanomaterials for flame retardant polymers and composites. *Composites Part B: Engineering*, 212, 108675.

[114] Qian, L., & Sheng, K. (2011). Flame retardant and thermal properties of intumescent flame retardant coatings based on layered double hydroxide. *Applied Clay Science*, 53(4), 589–593. https://doi.org/10.1016/j.clay.2011.05.005.

[115] Li, Y., Wang, X., Song, L., & Hu, Y. (2015). Intumescent flame retardant cotton fabrics coated with graphene oxide and branched polyethylenimine via layer-by-layer assembly. *Carbohydrate Polymers*, 115, 229–235. https://doi.org/10.1016/j.carbpol.2014.08.062.

[116] Lee, Y. X., Wang, W., Lei, Y., Xu, L., Agarwal, V., Wang, C., & Yeoh, G. H. (2025, January). Flame-retardant coatings for wooden structures. *Progress in Organic Coatings*, 198, 108903. https://doi.org/10.1016/j.porgcoat.2024.108903.

[117] Chen, Y., Xu, W., & Wang, Y. (2016). Flame-retardant performance and thermal behavior of TiO_2–polyurethane nanocomposite coatings. *Progress in Organic Coatings*, 101, 150–156. https://doi.org/10.1016/j.porgcoat.2016.08.013.

[118] Patel, R., Chaudhary, M. L., Patel, Y. N., Chaudhari, K., & Gupta, R. K. (2025). Fire-Resistant Coatings: Advances in Flame-Retardant Technologies, Sustainable Approaches, and Industrial Implementation. *Polymers*, 17(13), 1814. https://doi.org/10.3390/polym17131814.

[119] Attia, N. F., Elashery, S. E. A., El-Sayed, F., Mohamed, M., Osama, R., Elmahdy, E., Abd-Ellah, M., El-Seedi, H. R., Hawash, H. B., & Ameen, H. (2024). Recent advances in nanobased flame-retardant coatings for textile fabrics. *Nano-Structures & Nano-Objects*, 38, 101180.

[120] Attia, N. F., Elashery, S. E. A., El-Sayed, F., Mohamed, M., Osama, R., Elmahdy, E., Abd-Ellah, M., El-Seedi, H. R., Hawash, H. B., & Ameen, H. (2024). Recent advances in nanobased flame-retardant coatings for textile fabrics. *Nano-Structures & Nano-Objects*, 38, 101180,

[121] Neha, & Madhu, M. (2024). Recent developments in the sustainable flame-retardants for textiles: A review. *JETIR*, 11(10). https://www.jetir.org/view?paper=JETIR2410350.

[122] Attia, N. F., Mohamed, A., Hussein, A., El-Demerdash, A.-G. M., & Kandil, S. H. (2023). Greener bio-based spherical nanoparticles for efficient multilayer textile fabrics nanocoating with outstanding fire retardancy, toxic gases suppression, reinforcement and antibacterial properties. *Surfaces and Interfaces*, 36, 102595.

[123] Li, Y., Wang, X., Song, L., & Hu, Y. (2015). Intumescent flame retardant cotton fabrics coated with graphene oxide and branched polyethylenimine via layer-by-layer assembly. *Carbohydrate Polymers*, 115, 229–235. https://doi.org/10.1016/j.carbpol.2014.08.062.

[124] Alongi, J., Ciobanu, M., & Malucelli, G. (2012). Thermal stability, flame retardancy and mechanical properties of cotton fabrics treated with inorganic coatings synthesized through sol–gel processes. *Carbohydrate Polymers*, 87(3), 2093–2099. https://doi.org/10.1016/j.carbpol.2011.10.032.

[125] Carosio, F., Alongi, J., & Malucelli, G. (2012). Layer-by-layer flame retardant coatings for cotton fabrics: Effect of nanoparticle composition. *RSC Advances*, 2, 1760–1767. https://doi.org/10.1039/C1RA00781D.

[126] Wang, Y., Lu, H., & Xu, J. (2017). TiO₂/ammonium polyphosphate coatings for enhancing flame retardancy of polyester textiles. *Surface & Coatings Technology*, 311, 278–284. https://doi.org/10.1016/j.surfcoat.2017.01.056.

[127] Zou, Y., Shi, Z., Sun, J., Luo, F., Dai, Y., Chen, D., & Li, H. (2024). Organic-inorganic hybrid-modified polyester/cotton fabric: Strategy for enhanced flame retardancy, chemical stability and amphiphobicity. *Industrial Crops and Products*, 222(1), 119441.

[128] Kolya, H., & Kang, C.-W. (2024). Eco-friendly polymer nanocomposite coatings for next-generation fire retardants for building materials. *Polymers*, 16(14), 2045. https://doi.org/10.3390/polym16142045.

[129] Giannelis, E. P. (2002). Polymer layered silicate nanocomposites. *Advanced Materials*, 8(1), 29–35. https://doi.org/10.1002/adma.200390081.

[130] El-Tantawy, A., Hassan, I. M., El-kady, O. A., Ali, A. I., Son, J. Y., & Ayoub, N. M. (2024). Low cost paints reinforced with an Al2O3/Y2O3/graphene nanocomposite for fire-resistant wood coating applications. *Materials Advances*, 5, 7377–7738. https://doi.org/10.1039/D4MA00552J.

[131] Wang, D. Y., Liu, Y., & Wang, Y. Z. (2011). Flame-retardant polymer nanocomposites with layered double hydroxides. *Journal of Materials Chemistry*, 21(26), 9456–9467. https://doi.org/10.1039/C1JM10577H.

[132] Jumiati, E., & Frida, E. (2025). Review of fire-retardant nanocomposite coating based on TiO2 (titanium dioxide)-MMT (montmorillonite)-CNF (cellulose nanofibers). *Polymers from Renewable Resources*, 16(2–3), 105–122. https://doi.org/10.1177/20412479251334021.

[133] Zhao, H., Wang, X., & Song, L. (2013). Carbon nanotube-reinforced flame-retardant coatings for flexible applications. *Composites Part B: Engineering*, 51, 36–42. https://doi.org/10.1016/j.compositesb.2013.02.037.

[134] Patel, R., Chaudhary, M. L., Patel, Y. N., Chaudhari, K., & Gupta, R. K. (2025). Fire-resistant coatings: Advances in flame-retardant technologies, sustainable approaches, and industrial implementation. *Polymers*, 17(13), 1814. https://doi.org/10.3390/polym17131814.

[135] Mukherjee, S., Uddin, K. M. A., Turku, I., Rohumaa, A., & Lipponen, J. (2025). Bio-based flame retardants derived from forest industry – An approach towards circular economy. *Resources, Environment and Sustainability*, 21, Article 100229. https://doi.org/10.1016/j.resenv.2025.100229.

[136] Murad, M. S., Hamzat, A. K., Asmatulu, E., & Smatulu, R. (2025). Flame-retardant fiber composites: Synergistic effects of additives on mechanical, thermal, chemical, and structural properties. *Advanced Composites and Hybrid Materials*, 8, 31. https://doi.org/10.1007/s42114-024-01111-1.

[137] Alongi, J., & Malucelli, G. (2015). State of the art and perspectives on the use of intumescent coatings for fire protection of materials. *Progress in Organic Coatings*, 76(4), 1297–1311.

[138] Babu, K., Das, O., Shanmugam, V., Mensah, R. A., Försth, M., Sas, G., Restás, A., & Berto, F. (2021). Fire behavior of 3D-printed polymeric composites. *Journal of Materials Engineering and Performance*, 30, 4745–4755. https://doi.org/10.1007/s11665-021-05627-1.

[139] Jafari, P., Zhang, R., Huo, S., Wang, Q., Yong, J., Hong, M., Deo, R., Wang, H., & Song, P. (2024). Machine learning for expediting next-generation of fire-retardant polymer composites. *Composites Communications*, 45, 101806. https://doi.org/10.1016/j.coco.2023.101806.

11

Nanotechnology: A Sustainable Solution for Climate Change Mitigation

Bhawani Shankar

11.1 Introduction

Climate change poses a serious hazard as seen by extreme weather, increasing sea levels, biodiversity loss, and catastrophic ecosystem effects. In the last 40–50 years, a rise in the surface temperature has been observed, which may further create harsh living conditions in the next 50–100 years, leading to catastrophic conditions [1]. Global warming and increased atmospheric CO_2 concentration are primary causes of climate change [2]. Greenhouse gases such as CO_2, nitrous oxide, methane, ozone, hydrofluorocarbons (HFCs), and volatile organic compounds contribute to atmospheric changes that drive global warming and other environmental concerns [3]. Since climate change is one of the most pressing issues facing humanity today, there is an urgent need for sustainable solutions that can minimize environmental deterioration, lower carbon emissions, and lessen adverse effects on ecological and human health.

Nanotechnology is one of the many suggested solutions that sticks out as a promising area with the potential to significantly impact the fight against climate change. Generally incorporating structures ranging from 1 to 100 nanometers (at least one dimension), nanotechnology is the manipulation of matter at the nanoscale [4]. It offers creative and sustainable solutions to climate problems, advances physical and chemical properties of the material, and biological effectiveness [5].

Nanotechnology may be essential in lowering our carbon footprint and reestablishing environmental equilibrium because of its capacity to improve material qualities, facilitate more effective energy usage, and offer innovative methods of pollution control [6]. Because of the special properties of nanoparticles (NPs) like increased surface area and reactivity, carbon capture, energy generation, waste management, and pollution remediation have advanced [7].

Nanostructured materials (e.g., nanocomposites, metal–organic frameworks, CNTs, nanocatalysts) offer sustainable solutions for greenhouse gas reduction, biofuel production, wastewater treatment, and environmental remediation. These materials enhance efficiency, reduce energy consumption, and minimize waste, providing innovative solutions for environmental challenges [8].

The use of nanotechnology as a sustainable means of reducing the negative effects of climate change on the environment is examined in this chapter. The basics of nanotechnology will be covered first, and then a thorough examination of the different

DOI: 10.1201/9781003632498-11

ways that nanotechnology might help address the environmental crisis around the world will follow. The risks and difficulties of implementing nanotechnologies will also be discussed, along with some successful case studies. In the last section, we will present our predictions on how nanotechnology will help fight climate change and promote sustainability in the future.

11.2 Fundamentals of Nanotechnology

Nanotechnology is an interdisciplinary field that brings together physics, chemistry, biology, and engineering. Fundamentally, it is the process of creating, synthesizing, and using materials whose structures and characteristics are revealed at the nanoscale [9]. Approximately 100,000 times smaller than the width of a human hair, a nanometer is one billionth of a meter. Materials display distinct behaviors at this scale that are not seen at bigger scales. Because of their superior mechanical strength, electrical conductivity, and chemical reactivity, NPs are perfect for a variety of uses, especially where conventional materials are inadequate [10]. The nanomaterials are classified as super nanomaterials (perfect geometry, very strong, and defect-free), smart nanomaterials (capable of changing size, shape, color, density, or any other physical property), and active nanomaterials (can change itself according to the surrounding environment) [6,11]. The characteristics of nanomaterials include:

11.2.1 Greater Surface Area

In comparison to bulk materials, nanomaterials frequently have a larger surface area in relation to their volume and have the extra advantage of better interaction with other materials [12]. They are perfect for catalysis, which allows for more active sites for chemical reactions, environmental cleanup by adsorbing greenhouse gases, energy storage, and transportation [13].

11.2.2 Quantum Effects

Materials with quantum mechanical characteristics at the nanoscale can behave differently. For instance, the electrical, optical, and magnetic characteristics of NPs may differ from those of their bulk counterparts. Some materials that are not magnetic in bulk, like palladium, platinum, and gold, become magnetic at the nanoscale when their size decreases because quantum effects become more apparent [14].

11.2.3 Strength and Durability

Because of the way atoms are structured at the nanoscale, nanomaterials are frequently stronger and more resilient than conventional materials. The reactivity, toughness, and other properties of nanomaterials are dependent on their unique size, shape, and structure, making them suitable for creating more durable materials for manufacturing and infrastructure [15].

11.3 The Effects of Climate Change on the Environment

A global environmental crisis is being fueled by climate change with far-reaching effects on infrastructure, human health, and ecosystems. Major changes are visible as increased temperature of land, air, and ocean, rising sea levels, changes in the hydrological cycle, melting of glaciers, changes in ocean currents, etc. [16]. Although the effects of climate change are intricate and varied, they can be broadly divided into the following categories:

11.3.1 The Effects of Climate Change on Ecosystems

The effects of climate change on ecosystems are among the most obvious and significant. Extreme weather events such as droughts, floods, and wildfires, along with rising temperatures and shifting precipitation patterns, damage to marine ecosystem, increase the number of glacial lakes, and warm rivers, lakes, oceans, etc., severely harming biodiversity and wildlife [17]. The delicate balance that supports natural systems is being upset more and more, and many species are in danger of going extinct as ecosystems become less stable.

11.3.2 Loss of Biodiversity

As temperatures rise, many species are finding it difficult to live in habitats that once supported a variety of ecosystems. Coral reefs that illustrate the close relationship between biodiversity and ecosystems are extremely sensitive to temperature fluctuations, ocean acidification, and warming waters [18]. Alterations in coral reef biodiversity have caused massive bleaching events and the devastation of important marine ecosystems, endangering the lives of hundreds of millions of people [19].

11.3.3 Habitat Destruction and Agricultural Impacts

Many animal and plant species must relocate or risk extinction because they cannot swiftly adjust to the changing climate. As a result, iconic species that depend on melting sea ice brought on by warming temperatures, like polar bears, have become less common. Extreme weather conditions have caused drought and increased salinity of soil and water, resulting in serious damage to the quality and quantity of crop yield [20-21]. Use of conventional agrochemicals is associated with adverse effects, including long-term stability issues, the development of resistance in pathogens from repeated overdosing, harm to beneficial microbial species, and exposure to humans and animals via the food chain [22]. Nanotechnology has provided various farming tools such as nanofertilizers, nanopesticides, precision farming tools, etc., for restoring ecosystem balance and better productivity [23].

11.3.4 Infrastructure and Human Health

The impact of climate change on human health is becoming more and more obvious. Particularly for susceptible groups like children and the elderly, higher temperatures worsen air quality problems and increase the prevalence of respiratory illnesses [24].

Since floods and storms have the potential to displace communities, spread diseases, and destroy infrastructure, extreme weather events also present serious risks to public health [25]. As mosquitoes prefer warmer climates, more frequent and severe heat-waves are aggravating the spread of vector-borne illnesses like dengue fever and malaria in addition to raising the risk of heat-related illnesses [26]. As sea levels rise, millions of people are being forced to migrate as coastal areas become inundated. Overcrowding, resource shortages, and social unrest result from the influx of displaced people, which strains the host nations and cities that have to house them.

11.3.5 Climate Change and Greenhouse Gases

The atmospheric release of greenhouse gases (GHGs) is the main cause of climate change. These gases, which trap solar heat and contribute to global warming, include carbon dioxide (CO_2), methane (CH_4), and nitrous oxide (N_2O). The concentration of these gases in the atmosphere has significantly increased over the past century due to human activities like burning fossil fuels, transportation, deforestation, and industrial agriculture, which have upset the natural carbon cycle [27]. Solar radiations in the form of visible light are absorbed by the land, water, and vegetation and re-emitted back to the atmosphere in the form of infrared radiation (longer wavelength), but GHGs trap a portion of emitted radiation, leading to global warming [27]. The overwhelming evidence of climate change's negative effects emphasizes the need for creative, scalable solutions to cut emissions, rebuild ecosystems, and increase resilience to inevitable effects. Nanotechnology can make promising contributions in this area.

11.4 The Use of Nanotechnology to Combat Climate Change

Many of the environmental issues brought on by climate change could be resolved with the help of nanotechnology. We may create solutions that increase energy efficiency, lower carbon emissions, clean up pollution, and strengthen our capacity to track and control environmental effects by utilizing the special qualities of nanomaterials.

11.4.1 Sequestration of Greenhouse Gases

GHGs contribute to global warming, causing irreversible damage to the ozone layer, health, and environment. Strategies to mitigate this include using non-conventional energy resources, carbon management, and increasing technology efficiency. Nano-technology offers sustainable solutions, leveraging nanomaterials' unique properties to capture GHGs. Functionalized nanomaterials like nanofilms, nanocomposites, carbon nanotubes (CNTs), graphene oxide, ceramics, zeolites, and metal–organic frameworks (MOFs) show great potential in GHG sequestration (GHGS) [28]. These materials have increased surface area, surface functional groups, and excellent properties, enabling better performance in environmental remediation. Various nanomaterials have been explored for GHGS, listed in Table 11.1. By harnessing nano-technology, we can develop clean and green alternatives to address global warming and climate change. Nanotechnology can transform technology to mitigate climate change consequences.

TABLE 11.1

Various Nanomaterials for Greenhouse Gas Sequestration Application (GHGS)

S. No.	Nanomaterial	Captured Greenhouse Gases (Adsorption Efficiency)	Examples	Reference
1.	Carbon nanotubes	CO_2 (70%) and CH_4 (90%)	MWCNTs	[29]
		CO_2 (93 mg/g)	Amino-substituted MWCNTs	[30]
		CO_2 (2.54 mmol/g), H_2S (0.22 mmol/g), SO_2 (5.5 mmol/g)	Double-walled CNTs	[31]
2.	Graphene	CO_2, CH_4, and H_2S	Nanoporous graphene	[32]
		CO_2, CH_4	NPs (Au, Pt, Ag, Ni, and Ru) @ graphene nanomaterial	[33]
		CO_2	MgO NPs @ graphene oxide	[34]
		CO_2	Reduced graphene oxide–MnO_2 composite	[35]
3.	Nano-zeolites	CO_2 (6.5 mmol/g)	Nano-zeolites	[36]
		CO_2 (7.5 mmol/g)	Ethylenediamine-modified nano-zeolite	[37]
		CH_4 (2.1 mmol/g)	Zeolite modified with Al	[38]
4.	Metal organic frameworks (MOFs)	CO_2, CH_4 (220 mg/g)	MOF-177	[39]
		CO_2	N-rich MOF	[40]
		CH_4	MOF-205	[41]
		CO_2, CH_4	Bimetallic In(III)/Pd(II)-based MOF	[42]
5.	Mesoporous silica NPs	CH_4 (412 mg/g)	Mesoporous carbon	[43]
		CO_2 (1.6 mmol/g)	Amine-grafted silica NPs	[44]
		CO_2 (202 mg/g)	Polyethyleneimine @ silica NPs	[45]

11.4.1.1 Carbon Capture and Storage

Nanotechnology can directly address climate change by developing sophisticated carbon capture and storage (CCS) technologies. The goal of these technologies is to absorb CO_2 emissions from power plants and other industrial sources and store them underground to keep them out of the atmosphere. Emissions from energy extraction are reduced and prevented from entering the atmosphere with the help of CCS technology. The CCS is divided into pre-combustion (capturing CO_2 prior to combustion), oxy-fuel combustion (capturing CO_2 within the combustion chamber), and post-combustion (capturing CO_2 as soon as the reactants are processed to products) captures [46]. Because of their high surface area and reactivity, nanomaterials can be used to create carbon capture systems that are more effective and economical. For example, MOFs and other nanoporous materials are very good at adsorbing CO_2, and they may be able to capture carbon at a far lower cost than is currently possible [46]. SiO_2 and TiO_2 NPs have been shown to be effective in changing wettability and lowering interfacial tension, while CNTs have been used to increase the adsorption capacity of CO_2 [47].

11.4.1.2 Nanocatalysts for Conversion

Using nanocatalysts to transform CO_2 into beneficial byproducts like chemicals or fuel is an additional strategy. This process, called carbon capture and utilization (CCU), is economically feasible because it mitigates the emission of CO_2 while converting it into fuels, chemicals, polymers, and other valuable products and lowers atmospheric CO_2 levels. This process, known as CCU, decreases atmospheric CO_2 levels by mitigating its emissions while transforming it into fuels, chemicals, polymers, and other valuable products [48]. The reduction in CO_2 to fuels and other useful chemicals has been documented using nanostructured metal oxides, metal sulfides, graphitic carbon nitride, graphene oxide, carbon nanofiber, carbon quantum dots, and MOFs [49].

11.4.2 Non-Conventional Energy Resources (Renewable Energy)

The development and optimization of renewable energy sources, such as solar, wind, and bioenergy, can also be greatly aided by nanotechnology. The shift away from fossil fuels (gasoline, petrol, and diesel) can be accelerated by nanotechnology by enhancing the performance and efficiency of renewable energy systems. Photovoltaic devices have utilized a variety of photovoltaic nanomaterials and nanocomposites to boost solar energy efficiency by directly converting photon energy into electrical energy [50].

11.4.2.1 Biofuels

Biofuels such as bioethanol, biogas, and biodiesel are green energy alternatives to combat the energy crisis and climate change [51]. Nanomaterials are generally used to immobilize enzymes onto metal oxide or magnetic NP-based matrices for higher efficiency during biofuel production [7].

11.4.2.2 Solar Cells

It has been demonstrated that nanostructured materials increase light absorption and electron transport, thereby increasing solar cell efficiency. Nanotechnology is also helping next-generation solar technologies like perovskite solar cells, polymer solar cells, quantum dots, etc., which have higher efficiency rates and cheaper production costs than conventional silicon-based solar panels [52].

11.4.2.3 Energy Storage

Systems for storing energy are essential to the dependability of renewable energy sources, and nanotechnology can improve them. The performance of energy storage devices, including capacitors and batteries (lithium-ion, sodium-sulfur, and redox flow), can be significantly enhanced by manipulating materials at the nanoscale, as nanomaterials are more efficient than traditional materials because of their special electrical, mechanical, and interfacial properties [53].

11.4.3 Water Purification and Desalination

Climate change is making it more difficult to find clean water in many parts of the world. Nanotechnology can offer sustainable solutions in the crucial fields of water purification by degrading dyes and other colorful pollutants and desalination, which are affected by the depletion or contamination of fresh water sources [54]. It is quite challenging to meet the need for clean and portable drinking water since treatment of water and wastewater by conventional methods has its limits. Nanotechnology offers a promising solution, enabling efficient water treatment through advanced processes like adsorption, membrane separation, disinfection, photocatalysis, and sensing [55]. These nanotechnology-based treatment methods can effectively remove pollutants by treating water with various nanomaterials, making water safe for use. By leveraging nanomaterials and nanotechnology, scientists aim to improve water treatment performance [56]. This approach can help address the growing demand for clean water and has the potential to revolutionize water treatment.

11.4.3.1 Nanofiltration

Water pollutants, salts, and other impurities can be eliminated by the extremely effective filtration capabilities of nanomaterials like graphene oxide (GO) membranes and CNTs [57-58]. Water treatment procedures using these technologies should be more economical and energy-efficient than those using conventional techniques.

11.4.3.2 Desalination

Freshwater content of the hydrosphere is only 30% fit for human consumption, and the remaining 70% of the planet is composed of ice caps, glaciers, and deep subterranean water. Since saline water makes up a significant amount, desalinating it to get rid of its salts and minerals presents a viable way to get fresh water out of it [59]. The process of desalinating seawater is a practical choice in areas with limited freshwater resources. Long-term sustainability of this solution can be achieved by using nanotechnology to increase the durability of desalination membranes and decrease the energy needed for desalination processes. Organic (CNT, graphene, etc.), inorganic (Ag NPs, Cu NPs, metal oxides, viz., ZnO, TiO, etc.), or hybrid (CNTs combined with metals or other support to improve adsorption) nanomaterials can be explored for a specific desalination process, although inorganic nanomaterials require a higher cost of production and have potential toxicities [60].

11.4.4 Monitoring of the Environment

To evaluate the effects of climate change and make sure mitigation measures are working, environmental conditions must be regularly monitored. By creating extremely sensitive sensors that can identify air-borne pollutants, heavy metals in water, pesticides in soil and foods, and measure climate variables more precisely, nanotechnology can enhance environmental monitoring systems [61]. Sensors detect and monitor physical parameters like concentration, temperature, and pressure. Nanosensors, made from nanomaterials, offer better sensitivity and performance due to improved optical properties [62]. Biosensors use biological elements to detect specific analytes or molecules. Both types of sensors enable precise detection and monitoring.

11.4.4.1 Monitoring of Air Quality

Real-time air quality monitoring is made possible by the use of nanomaterial-based sensors, particularly nanostructured metal-oxide gas sensors, to identify trace gases in the atmosphere, including CO_2, methane, and volatile organic compounds (VOCs) [63]. To better understand and control pollution levels, these sensors can be placed in industrial sites and urban areas.

11.4.4.2 Water Quality Sensors

Nanotechnology can also be used to create sensors that identify contaminants (pesticides, surfactants, dyes, phenolic compounds, etc.), heavy metals, and other pollutants in water [64]. Both developed and developing nations can use these sensors to monitor pollution levels and guarantee safe drinking water.

11.4.5 The Use of Nanotechnology to Reduce Pollution

Nanotechnology offers powerful tools for preventing, treating, and cleaning up soil, water, and air pollution [65]. Using new materials and methods, nanotechnology can help clean up contaminated areas and prevent further pollution.

11.4.5.1 Soil and Water Nanoremediation

Utilization of nanomaterials to purify contaminated soil and water is known as nanoremediation. Nanoremediation is more efficient and less expensive in comparison to conventional methods, and it uses engineered NPs to interact with pollutants to either break them down or capture them for removal [66].

11.4.5.2 Heavy Metal Removal

Communities are extremely concerned about the harmful effects of heavy metal pollution on human health and the long-term well-being of the Earth. Heavy metals like cadmium, chromium, arsenic, lead, and mercury do not degrade and can get into the food chain, leading to serious health issues [67]. Many nanomaterials are currently being used to remove pollutants from wastewater, but among conventional adsorbents, NPs, in particular, iron-based NPs (Fe NPs) are preferred due to their effective and efficient performance in wastewater remediation [68]. Carbonaceous nanomaterials, including graphene, CNTs, and chemically modified forms, have recently been developed to demonstrate exceptional adsorption efficacy for heavy metals, including arsenic and chromium [69].

11.4.5.3 Cleaning Up Oil Spills

Petroleum products and oil spills, which are considered a major threat to aquatic and terrestrial ecosystems, can be removed from water using nanomaterials. MWCNTs are widely used as adsorbent materials that have a high affinity for removing hydrocarbons from water. The functionalization of MWCNTs further increases the sorption capacity of adsorbents manifold to remove organic materials [70].

11.4.5.4 Control of Air Pollution

Nanomaterials are being developed to capture air pollutants that contribute to smog and respiratory diseases, such as sulfur dioxide (SO_2), nitrogen oxides (NO_x), VOCs, GHGs, and particulate matter (PM) [71].

11.4.5.5 Scrubbers with Nanocatalysts

To convert hazardous gases into less toxic forms, air pollution control systems can incorporate nanocatalysts [72]. Toxic gas molecules can be absorbed and detected by CNT sensors, while SO_2 and H_2S can be accurately detected by ultra-thin SnO_2 films in 3D structure nanosensors [73-74]. Ag NPs and Cu NPs detect the presence of NH_3 and H_2S gases, respectively, while Pt NPs in the CNT structure increase the sensitivity of the sensor to NO_2 [75].

11.4.5.6 Filter Media

Nanomaterials can be used to construct air filters that are extremely effective at trapping fine particles and other pollutants, which is vital for controlling indoor air pollution. Urban air quality could be improved by using these filters in personal protective equipment or industrial exhaust systems [76].

11.4.5.7 Sustainable Architecture

Sustainable architecture aims to reduce global warming and energy crises. Nanotechnology enables energy-efficient buildings through eco-friendly materials like coatings, composites, and solar cells [77]. Nano smart homes, buildings, and industries can achieve green energy goals. Shifting to clean energy sources, like solar energy, is crucial for industries and homes [78]. Nanomaterials can modernize conventional architecture, promoting sustainability. Sustainable architecture can mitigate climate change impacts [79]. Nanotechnology holds promise for a greener future.

11.4.6 Nanomaterials for Sustainable Agriculture

It is quite difficult to mitigate climate change associated with the agriculture sector because conventional methods of controlling weeds and pests, such as the excessive use of pesticides and herbicides, as well as improper use of mineral fertilizer, are the main cause of groundwater pollution and increased GHG emissions [80]. The application of nanotechnology has been extensively documented as an alternative solution provider, offering modern nano-agrochemicals such as nanofertilizers, nano-pesticides, nano-herbicides, and nano-emulsions, which enhance productivity and restore ecological balance [81]. There are two types of nanopesticides classified as type I, consisting of metal-based nanopesticides, namely, Ag, Ti, Cu, Fe, Al, and Zn; among these most used are Ag-, Cu-, and Ti-based NPs. Type II nano-pesticides made of polymers, clays, and zein particles, which act as nanocarriers for the active ingredients and include nanocapsule, nanosphere, nanogels, nanomicelle, nanoliposome, dendrimer, CNT, graphene, nanocomposite, mesoporous silica, etc. Type II nanopesticides are cost-effective, stimuli-responsive, and generally biocompatible [82]. These nanopesticides can control plant pathogens, including fungi and bacteria, which pose a threat to various crop plants. Nanofertilizers are designed to achieve controlled release of macro- and micro-nutrients to plants and are freely available currently in the market, containing nano zinc, silica, iron, titanium dioxide, and nano-urea [83]. In comparison to conventional NPK fertilizers, the use of nanofertilizers has shown a significant increase in crop yields with improved nutrient use efficiency [84].

11.5 Toxicological Aspects and Environmental Concerns of Nanomaterials

Although the physicochemical properties (electrical, optical, and magnetic) of nano-structured materials are unique, there are serious concerns regarding their toxicity [85]. Nanomaterials possess a large surface-to-volume ratio and can interact with a large contact area of the human body, making them super reactive to various biological tissues, organs, cells, etc., leading to severe nanotoxicity [86]. Because NPs are so small, they can pass through biological barriers and cause toxicity through a variety of routes, such as ingestion, inhalation, and dermal absorption, causing neurological toxicity and posing genetic risks [87]. Inflammation, fibrotic reactions, and oxidative stress can result from respiratory exposure to NPs [88]. NPs of heavy metals such as lead, mercury, and tin are stable and rigid, and these can bioaccumulate in the environment and potentially disturb ecological balance [89]. Aquatic ecosystems are negatively impacted by silver NPs, and in contrast to bulk gold, nano-gold can exhibit harmful in vivo long-term toxicity [90-91]. Graphene and fullerenes are examples of carbon-based NPs that might provide less hazardous substitutes [92]. Surface functionalization of nanomaterials with various moieties, like sodium citrate, polyvinylpyrrolidone, and/or surfactants, has been shown to considerably reduce their potential for nanotoxicity and enhance their safety profile [86]. Because the long-term health effects of NP exposure are still not well understood, risk analyses and standardized toxicological evaluations are required.

Although there are still methodological ambiguities and data gaps, life cycle assessment (LCA) can assess environmental impacts. Developing LCA techniques for nano-materials is crucial to guarantee safe and sustainable applications. A comprehensive understanding of NP interactions with biological systems and the environment is crucial for mitigating potential risks.

11.6 Conclusion

Nanotechnology presents a promising approach to mitigating climate change through various applications, including GHGS, CCS, renewable energy development, water purification, and pollution control. The unique physicochemical properties of nanomaterials enable enhanced performance in these applications. However, potential nanotoxicity and environmental impacts necessitate thorough risk assessments and standardized toxicological evaluations. Further research is required to develop safer nanomaterials and refine LCA methodologies. By advancing nanotechnology, we can harness its potential to create sustainable solutions for an environmentally conscious future.

REFERENCES

[1] N. Chausali; J. Saxena; R. Prasad; *J. Agric. Food Res.* **2023**, *12*, 100541.

[2] M. S. Apoorva; S. Kumar; A. Bhatia; *J. Agric. Phys.* **2023**, *23*, 1.

[3] G. V. Lowry; A. Avellan; L. M. Gilbertson; *Nat. Nanotechnol.* **2019**, *14*, 517.

[4] N. Chausali; J. Saxena; R. Prasad; *J. Agric. Food Res.* **2022**, *7*, 100257.

[5] T. Singh; S. Shukla; P. Kumar; V. Wahla; V. K. Bajpai; I. A. Rather; *Front. Microbiol.* **2017**, *8*, 1501.

[6] N. Chausali; J. Saxena; R. Prasad; *J. Agric. Food Res.* **2021**, *5*, 100191.

[7] M. Rai; J. C. dos Santos; M. F. Soler; P. R. F. Marcelino; L. P. Brumano; A. P. Ingle; S. Gaikwad; A. Gade; S. S. da Silva; *Nanotechnol. Rev.* **2016**, *5*, 231.

[8] J. L. Jones; D. M. Berube; M. Cuchiara; K. Grieger; E. A. Cohen Hubal; S. J. Kariko; P. Strader; Y. Theriault; *Environ. Systems Decisions* **2024**, *44*, 1039.

[9] N. Phogat; M. Kohl; I. Uddin; A. Jahan; *Precision Medicine*, Academic Press **2018**, p. 253–276.

[10] H. Saleem; S. J. Zaidi; N. A. Alnuaimi; *Materials* **2021**, *14*, 6387.

[11] A. D. Sontakke; Deepti; N. S. Samanta; M K. Purkait; *Advances in Smart Nanomaterials and Their Applications*, Elsevier **2023**, p. 23–50.

[12] N. Baig; I. Kammakakam; W. Falath; *Mater. Adv.* **2021**, *2*, 1821.

[13] U. Chadha; S. K. Selvaraj; H. Ashokan; S. P. Hariharan; V. M. Paul; V. Venkatarangan; V. Paramasivam; *Adv. Mater. Sci. Eng.* **2022**, *2022*, 1552334.

[14] N. Joudeh; D. Linke; *J. Nanobiotechnol.* **2022**, *20*, 262.

[15] I. Khan; K. Saeed; I. Khan; *Arabian J. Chem.* **2019**, *12*, 908.

[16] S. Bolan; L. P. Padhye; T. Jasemized; M. Govarthanan; K. Karmegam; H. Wijesekara; D. Amarasiri; D. Hou; P. Zhou; B. K. Biswal; R. Balasubramanian; H. Wang; K. H. M. Siddique; J. Rinklebe; M. B. Kirkham; N. Bolan; *Sci. Total Envir.* **2024**, *909*, 168388.

[17] J. E. Walsh; T. J. Ballinger; E. S. Euskirchen; E. Hanna; J. Mard; J. E. Overland: H. Tangen; T. Vihma; *Earth-Science Rev.* **2020**, *209*, 103324.

[18] O. Hoegh-Guldberg; E. S. Poloczanska; W. Skirving; S. Dove; *Front. Mar. Sci.* **2017**, *4*, 158.

[19] D. Allemand; D. Osborn; *Regional Studies Marine Sci.* **2019**, *28*, 100558.

[20] J. Schewe; S. N. Gosling; C. Reyer; F. Zhao; P. Ciais; J. Elliott; L. Francois; V. Huber; K. K. Lotze; S. I. Seneviratne; M. T. H. van Vliet; R. Vautard; Y. Wada; L. Breuer; M. Buchner; D. A. Carozza; J. Chang; M. Coll; D. Deryng; A. de Wit; T. D. Eddy; C. Folberth; K. Frieler; A. D. Friend; D. Gerten; L. Gudmundsson; N. Hanasaki; A. Ito; N. Khabarov; H. Kim; P. Lawrence; C. Morfopoulos; C. Muller; H. M. Schmied; R. Orth; S. Ostberg; Y. Pokhrel; T. A. M. Pugh; G. Sakurai; Y. Satoh; E. Schmid; T. Stacke; J. Steenbeek; J. Steinkamp; Q. Tang; H. Tian; D. P. Tittensor; J. Volkhoz; X. Wang; L. Warszawski; *Nat. Commun.* **2019**, 10, 1005.

[21] K. K. Nutan; R. S. Rathore; A. K. Tripathi; M. Mishra; A. Pareek; S. L. Singla-Pareek; *J. Exp. Bot.* **2020**, *71*, 490.

[22] A. Ahmed; P. He; P. He; Y. Wu; Y. He; S. Munir; *Envir. Int.* **2023**, *173*, 107819.

[23] M. Tripathi; S. Kumar; A. Kumar; P. Tripathi; S. Kumar; *Int. J. Curr. Microbiol. App. Sci.* **2018**, *7*, 196.

[24] K. N. Domingo; K. L. Gabaldon; M. N. Hussari; J. M. Yap; L. C. Valmadrid; K. Robinson; S. Leibel; *Eur. Respir. Rev.* **2024**, *33*, 230249.

[25] E. R. Parker; J. Mo; R. S. Goodman; *J. Climate Change and Health* **2022**, *8*, 100162.

[26] N. Mojahed; M. A. Mohammadkhani; A. Mohamadkhani; *Iran. J. Public Health* **2022**, *51*, 2664.

[27] M. Filonchyk; M. P. Peterson; L. Zhang; V. Hurynovich; Y. He; *Sci. Total Envir.* **2024**, 935, 173359.

[28] I. C. Nwuzor; *Advances in Nanofiber Research*, Intech Open, **2024** ch. 7, doi: 10.5772/intechopen.1007132.

[29] Z. Kang; M. Xue; D. Zhang; L. Fan; Y. Pan; S. Qiu; *Inorg. Chem. Commun.* **2015**, *58*, 79.

[30] M. Wang; Z. Wang; N. Li; J. Liao; S. Zhao; J. Wang; S. Wang; *J. Membr. Sci.* **2015**, *495*, 252.

[31] S. Majumdar; M. Maurya; J. K. Singh; *Energy Fuel.* **2018**, *32*, 6090.

[32] S. Gadipelli; Z. X. Guo; *Prog. Mater. Sci.* **2015**, *69*, 1.

[33] X. Wang; G. Ou; N. Wang; H. Wu; *ACS Appl. Mater. Interfaces* **2016**, *8*, 9194.

[34] C. A. Gunathilake; G. G. T. A. Ranathunge; R. S. Dassanayake; S. D. Illesinghe; A. S. Manchanda; C. S. Kalpage; R. M. G. Rajapakse; D. G. G. P. Karunaratne; *Envir. Sci. Nano* **2020**, 7, 1225.

[35] M. O. Aquatar; U. Bhatia; S. S. Rayalu; R. J. Krupadam; *Sci. Total Environ.* **2022**, *816*, 51522.

[36] Q. Jiang; J. Rentschler; G. Sethia; S. Weinman; R. Perrone; K. Liu; *Chem. Eng. J.* **2013**, *230*, 380.

[37] T. H. Pham; B. K. Lee; J. Kim; *J. Taiwan Inst. Chem. Eng.* **2016**, *66*, 239.

[38] M. Anbia; A. Eskandari; *Int. J. Eng.* **2016**, *29*, 1.

[39] D. Saha; Z. Bao; F. Jia; S. Deng; *Environ. Sci. Technol.* **2010**, *44*, 1820.

[40] J. S. Qin; D. Y. Du; W. L. Li; J. P. Zhang; S. L. Li; Z. M. Su; X. L. Wang; Q. Xu; K. Z. Shao; Y. Q. Lan; *Chem. Sci.* **2012**, 3, 2114.

[41] G. Xu; Z. Meng; Y. Liu; X. Guo; K. Deng; L. Ding; R. Lu; *Int. J. Energy Res.* **2019**, *43*, 7517.

[42] I. Bratsos; C. Tampaxis; I. Spanopoulos; N. Demitri; G. Charalambopoulou; D. Vourloumis; T. A. Steriotis; P. N. Trikalitis; *Inorg. Chem.* **2018**, 57, 7244.

[43] X. Liu; L. Zhou; J. Li; Y. Sun; W. Su; Y. Zhou; *Carbon* **2006**, *44*, 1386.

[44] L. Wang; R. T. Yang; *J. Phys. Chem.* **2011**, *115*, 21264.

[45] K. Li; J. Jiang; F. Yan; S. Tian; X. Chen; *Appl. Energy* **2014**, *136*, 750.

[46] O. O. Ayeleru; H. U. Modekwe; O. R. Onisuru; C. R. Ohoro; C. A. Akinnawo; P. A. Olubambi; *Sustainable Chem. Climate Action* **2023**, *3*, 100029.

[47] Y. T. Youns; A. K. Manshad; J. A. Ali; *Fuel* **2023**, *346*, 128680.

[48] O. Emmanuel; Rozina; T. C. Ezeji; *Next Sustainability* **2025**, *6*, 100108.

[49] D. Sharma; R. Sharma; D. Chand; A. Chaudhary; *Envir. Nanotech. Monitor. Manag.* **2022**, *18*, 100671.

[50] K. K. Jaiswal; C. R. Chowdhury; D. Yadav; R. Verma; S. Dutta; K. S. Jaiswal; Sangmesh; K. S. K. Karuppasamy; *Energy Nexus* **2022**, *7*, 100118.

[51] R. El-Araby; *Biotech. Biofuels Bioproducts* **2024**, *17*, 1.

[52] P. V. Tumram; R. Nafdey; P. R. Kautkar; S. V. Agnihotri; R. A. Khaparde; S. P. Wankhede; S. V. Moharil; *Mater. Sci. Eng.: B* **2024**, *307*, 117504.

[53] H. Mohammed; M. F. Mia; J. Wiggins; S. Desai; *Molecules* **2025**, *30*, 883.

[54] R. Yousefi; *Micro Nanosyst.* **2022**, *14*, 188.

[55] M. P. Ajith; M. Aswathi; E. Priyadarshini; P. Rajamani; *Bioresour. Tech.* **2021**, *342*, 126000.

[56] Nishu; S. Kumar; *Hybr. Adv.* **2023**, *3*, 100044.

[57] R. Yasmeen; F. S. Khan; W. U. Nisa; A. R. Saleem; M. Awais; M. Jameel; R. N. Dara; M. I. Khan; *Carbon Trends* **2025**, *19*, 100486.

[58] S. Elhenawy; M. Khraisheh; F. Al-Momani; M. Al-Ghouti; R. Selvaraj; A. Al-Muhtaseb; *Namomaterials* **2024**, *14*, 1707.

[59] M. B. Abid; R. A. Wahab; M. A. Salam; I. A. Moujdin; L. Gzara; *Heliyon* **2023**, *9*, e12810.

[60] R. Sirohi; Y. Kumar; A. Madhavan; N. A. Sagar; R. Sindhu; B. Bharathiraja; H. O. Pandey; A. Tarafdar *Envir. Tech. Innov.* **2023**, *30*, 103108.

[61] M. Sharma; P. Mahajan; A. S. Alsubaie; V. Khanna; S. Chahal; A. Thakur; A. Yadav: A. Arya; A. Singh; G. Singh; *Mater. Today Sust.* **2025**, *29*, 101068.

[62] M. Primozic; Z. Knez; M. Leitgeb; *Nanomaterials* **2021**, *11*, 292.

[63] X. Chen; M. Leishman; D. Bagnall; N. Nasiri; *Nanomaterials* **2021**, *11*, 1927.

[64] R. M. S. R. Mohamed; M. Morsin; N. Zainal; N. Nayan; C. Z. Zulkifli; N. H. Harun; *Envir. Tech. Innov.* **2021**, *24*, 102032.

[65] B. Karn; T. Kuiken; M. Otto; *Envir. Healt Perspectives* **2009**, *117*, 1823.

[66] M. L. D. Prado-Audelo; I. G. Kerdan; L. Escutia-Guadarrama; J. M. Reyna-Gonzalez; J. J. Magana; G. Leyva-Gomez; *Frontiers Envir. Sci.* **2021**, *9*, 793765.

[67] Q. Ali; M. A. Zia; M. Kamran; M. Shabaan; U. Zulfiqar; M. Ahmad; R. Iqbal; M. F. Maqsood; *Hybrid Adv.* **2023**, *4*, 100091.

[68] T. A. Aragaw; F. M. Bogale; B. A. Aragaw; *J. Saudi Chem. Soc.* **2021**, *25*, 101280.

[69] S. Lal; A. Singhal; P. Kumari; *J. Water Proc. Eng.* **2020**, *36*, 101276.

[70] T. A. Abdullah; T. Juzsakova; S. A. Hafad; R. T. Rasheed; N. Al-Jammal; M. A. Mallah; A. D. Salman; P. C. Le; E. Domokos; M. Aldulaimi; *Clean Tech. Envir. Policy* **2022**, *24*, 519.

[71] A. Soni; *Nanomed. Nanotech.* **2024**, *9*, 335.

[72] A. A. Beni; H. Jabbari; *Results Eng.* **2022**, *15*, 100467.

[73] Q. Meng; *Envir. Pollut.* **2018**, *240*, 848.

[74] C. Griessler; E. Brunet; T. Maier; S. Steinhauer; A. Kock; T. Jordi; F. Schrank; M. Schrems; *Microelect. Eng.* **2011**, *88*, 1779.

[75] I. Sharafeldin; S. Garcia-Rios; N. Ahmed; M. Alvarado; X. Vilanova; N. K. Allam; *J. Envir. Chem. Eng.* **2021**, *9*, 104534.

[76] D. S. de Almeida; L. D. Martins; M. L. Aguiar; *Chem. Eng. J. Adv.* **2022**, *11*, 100330.

[77] L. Pokrajac; A. Abbas; W. Chrzanowski; G. M. Dias; B. J. Eggleton; S. Maguire; E. Maine; T. Malloy; J. Nathwani; L. Nazar; A. Sips; J. Sone; A. van den Berg; P. S. Weiss; S. Mitra; *ACS Nano* **2021**, *15*, 18608.

[78] X. Huang; G. Xing; Y. Li; E. Nannen; *J. Nanomater.* **2015**, 524095.

[79] M. A. Macias-Silva; J. S. Cedeno-Munoz; C. A. Morales-Paredes; R. Tinizaray-Castillo; G. A. Perero-Espinoza; J. M. Rodriguez-Diaz; C. M. Jarre-Castro; *Case Stud. Chem. Envir. Eng.* **2024**, *10*, 100863.

[80] V. Quintarelli; M. B. Hassine; E. Radicetti; S. R. Stazi; A. Bratti; E. Allevato; R. Mancinelli; A. Jamal; M. Ahsan; M. Mirzaei; D. Borgatti; *Sustainability* **2024**, *16*, 9280.

[81] R. Prasad; A. Bhattacharyya; Q. D. Nguyen; *Front. Microbiol.* **2017**, *8*, 1014.

[82] D. Wang; N. B. Saleh; A. Byro; R. Zepp; E. Sahle-Demessie; T. P. Luxton; K. T. Ho; R. M. Burgess; M. Flury; J. C. White; C. Su; *Nat. Nanotechnol.* **2022**, *17*, 347.

[83] T. Adhikary; P. Basak; *Hybr. Adv.* **2024**, *6*, 100255.

[84] S. T. S. Valojai; Y. Niknejad; H. F. Amoli; D. B. Tari; *J. Plant Nutr.* **2021**, *44*, 1971.

[85] L. Xuan; Z. Ju; M. Skonieczna; P.-K. Zhou; R. Huang; *Med. Comm.* **2020**; *4*, e327.

[86] H. E. Thu; M. Haider; S. Khan; M. Sohail; Z. Hussain; *Open Nano* **2023**, *14*, 100190.

[87] M. Gao; Z. Yang; Z. Zhang; L. Chen; B. Xu; *Envir. Res.* **2024**, *259*, 119473.

[88] A. J. Ferreira; J. Cemlyn-Jones; C. R. Cordeiro; *Rev. Port. Pneumol.* **2013**, *19*, 28.

[89] R. K. Gupta; P. Guha; P. P. Srivastav; *J. Hazar. Mater. Lett.* **2024**, *5*, 100125.

[90] E. Navarro; F. Piccapietra; B. Wagner; F. Marconi; R. Kaegi; N. Odzak; L. Sigg; R. Behra; *Envir. Sci. Technol.* **2008**, *42*, 8959.

[91] M. Pozzi; S. J. Dutta; M. Kuntze; J. Bading; J. S. Rubbult; C. F. Abig; M. Langfeldt; F. Schulz; P. Horcajada; W. J. Parak; *J. Chem. Educ.* **2024**, *101*, 3146.

[92] A. Mahor; P. P. Singh; P. Bharadwaj; N. Sharma; S. Yadav; J. M. Rosenholm; K. K. Bansal; *C* **2021**, *7*, 19.

12

Applications of Nanotechnology for Eco-Friendly and Sustainable Economic Development

Surbhi Dhadda, Nikita Choudhary, and Sanjay K. Meena

12.1 Introduction

Nanotechnology, the forefront of modern scientific advancement, focuses on the study and manipulation of particles at the nanometer scale (sub-micron level). It plays a pivotal role in promoting sustainable development and addressing climate change. Among its diverse applications, nanotechnology contributes significantly to waste recycling and the decontamination of polluted soil and water. In various industrial sectors, environmentally benign bio-based nanoparticles are increasingly replacing hazardous metals and synthetic chemicals. Due to their exceptionally high surface-area-to-volume ratio, nanoparticles exhibit a range of unique and tunable properties. Their versatility enables transformative contributions toward achieving several global goals, such as eradicating hunger, ensuring access to clean water and sanitation, promoting good health and well-being, enabling affordable and clean energy, and supporting responsible production, consumption, and climate action. As global population growth intensifies pressure on natural resources, the development of pollution-free technologies becomes imperative. Nanotechnology offers promising avenues for fostering long-term, sustainable human development by facilitating clean energy solutions and environmental remediation. In particular, "green nanotechnology" – a subfield dedicated to eco-friendly and sustainable innovations – has emerged as a key player in minimizing environmental harm while maximizing benefits to human health. Researchers are actively exploring the potential of nanomaterials to manage, mitigate, and remediate pollution in air, water, and soil, while also enhancing the performance of conventional environmental technologies.

This chapter focuses on green nanotechnology as a critical tool for advancing sustainable development. The content is systematically organized around the diverse applications of green nanotechnology, with comprehensive discussions on nano-manufacturing techniques, the green synthesis of nanomaterials, and wastewater treatment, all examined through the lens of green chemistry principles. Key environmental applications of nanotechnology – such as reducing overall energy consumption during synthesis and production, facilitating the recyclability of products, and promoting the development and utilization of environmentally benign materials – are also explored in depth. While nanotechnology holds considerable potential to address contemporary

DOI: 10.1201/9781003632498-12

sustainability challenges, its adverse implications for environmental and human health cannot be overlooked. Despite the high efficiency and cost-effectiveness of nano-remediation technologies, there remains a pressing need for advanced research to fully understand and mitigate the potential long-term ecological risks they may pose. The chapter emphasizes the relevance of green chemistry concepts throughout the entire life cycle of nanoproducts, from design and development to eventual disposal. It also critically evaluates the applications and limitations of green nanotechnology in achieving genuine sustainability.

Although nanotechnology is widely regarded as a tool for enhancing human wellbeing and addressing global challenges, it also poses certain risks if not managed responsibly. When used appropriately, nanotechnology holds the potential to significantly contribute to the achievement of the 17 Sustainable Development Goals (SDGs) set for 2030. However, concerns over potential nanotoxicity have limited its advancement, despite its relevance as a transdisciplinary frontier technology capable of offering innovative solutions across all industrial sectors – primary, secondary, tertiary, and quaternary. To mitigate these risks and ensure that technological progress aligns with sustainable development, the adoption of green and environmentally responsible nanotechnological approaches is crucial. This chapter explores the concept, current research developments, and industrial applications of eco-friendly nanotechnology in the 21st century, highlighting its role in achieving the global SDGs and beyond.

One of the most pressing challenges of our time is to ensure sustainable development for future generations by integrating the principles of green chemistry and green engineering in the creation of materials and nanoproducts. This involves designing processes that avoid the generation of harmful byproducts and incorporating life cycle thinking at every stage of design and engineering.[1,2] Nanotechnology and advanced nanomaterials offer promising solutions for the remediation of groundwater, surface water, and wastewater contaminated with heavy metal ions, organic and inorganic pollutants, and various microorganisms.[3,4]

A number of nanomaterials are currently undergoing considerable research and development (R&D) for application in site remediation and environmental cleanup because of their special activity toward nondegradable pollutants. According to Allen and coworkers,[5] green manufacturing is the most effective strategy for minimizing and completely removing the discharge of harmful pollutants into the air, water, and land. Nanoadsorbents, including clays, zeolites, metals, metal oxides, polymeric membranes, porous nanofibers, and zero-valent iron, can be used to efficiently remove organic contaminants, dyes, and dye effluents from contaminated water.[6,7] According to Tratnyek and researchers,[8] semiconductor-based photocatalysts are employed in advanced oxidation processes (AOP) to break down and mineralize hazardous organic pollutants into ecologically benign chemicals under natural and artificial radiation.

Considered an emerging field, nanoscience offers a useful framework for investigating the implications and uses of green chemistry in a more comprehensive and secure manner. Overall, pollution reduction is a feature of nano-remediation in contrast to earlier remediation technologies.[9] Technologies based on nanotechnology can be utilized to minimize, avoid, and mitigate harm to the environment and human health.[10] The creation of inexpensive, nontoxic, and multipurpose, effective nanoproducts without the production of harmful byproducts is the goal of the green nano-manufacturing method.[11] The potential of nanotechnology to offer fresh approaches to controlling and eradicating pollution of the air, water, and land is now being investigated, and it is making significant strides in environmental protection technologies. According to Shapira and coauthors,[12] it also seeks to enhance the functionality of traditional technologies employed in cleanup operations.

Nanotechnology offers a fast, cost-effective solution for in situ environmental remediation, significantly reducing clean-up time, costs, and contamination risks.[13] Due to their effectiveness against stubborn pollutants, various nanomaterials – such as metal oxides, carbon nanotubes, and zeolites – are being researched for use in soil and water clean-up.[9,14] Nano-based methods can also address nonaqueous phase liquids (NAPLs), like heating oil spills, enhancing the efficiency and sustainability of environmental restoration efforts.[15–17] Nano-based remediation methods are more effective at reducing contaminants than traditional approaches.[18] Green nanotechnology focuses on creating eco-friendly products and processes, aiming to minimize environmental and health impacts by replacing harmful existing technologies.[19] While industrialization has advanced technology, it has not solved key social issues and has led to environmental harm. To address these challenges, the adoption of eco-friendly technologies is essential. Nanotechnology, with its unique properties at the nanoscale, offers solutions across various sectors such as healthcare, agriculture, and environmental remediation. It plays a vital role in promoting sustainability and achieving the United Nations' SDGs set for 2030.

Nanotechnology is expected to significantly transform society and improve the quality of life. However, its progress has been slower than initially predicted due to cautious attitudes from scientists, organizations, and governments toward commercialization. A revised timeline for nanotechnology development, based on trends from the past 15 years, was proposed by Aithal and team.[20] Encouragingly, the rise of green and eco-friendly nanotechnology processes is helping to accelerate growth and align progress with the updated projections.

12.1.1 Nanotechnology as Green Alternative

Nanotechnology, which operates at the nanometer (nm) scale, is a key research field driving innovation. It contributes to the development of advanced products, improved materials, and the replacement of traditional production methods with more efficient alternatives.[21] This chapter focuses on green nanotechnology in the context of green chemistry and sustainability. Green nanotechnology follows the principles of green chemistry, aiming to make chemical processes safer and more efficient.[22] The chapter explores the intersection of sustainability and nanotechnology, where terms like "sustainable" and "green chemistry" are commonly used.[23]

Green nanotechnology focuses on two main goals: process and product. It involves using multifunctional nanomaterials for renewable energy, pollution monitoring, green packaging, environmental remediation, and efficient wastewater treatment.[24] Green chemistry and sustainability principles help identify new research opportunities in this field.[25] Green chemistry ensures that nanotechnologies minimize unintended hazards by considering their broader implications, distinguishing itself from other technological trends that focus solely on products and processes.[14]

Green nanotechnology, guided by the 12 principles of green chemistry, is a vital and emerging branch of nanoscience focused on sustainability and safety in product manufacturing and process application. It offers significant potential for promoting occupational health and sustainable development. With a growing understanding of green nanotechnology, attention has shifted to its key application areas[26] that support greener growth and sustainability, as illustrated in Figure 12.1. Nano-based applications offer benefits like lower toxicity, cost-effectiveness, higher efficiency, and environmental reliability. Green nanotechnology enhances material and product design by minimizing pollution from the start, assessing life cycles, and promoting non-toxic, eco-friendly solutions. Its applications include photocatalysis, solar and fuel cells, cleaner production methods, and green synthesis of nanoparticles.[27] Additionally, it supports recycling industrial waste into useful nanomaterials, such as converting diesel soot or fly ash into valuable nano-products.

Green chemistry is a scientific approach aimed at designing chemical products and processes that reduce or eliminate the use and generation of hazardous substances. Introduced by Paul Anastas and John Warner in the late 1990s, the 12 Principles of Green Chemistry[25] serve as a foundational framework to guide chemists and engineers toward more sustainable practices in research, development, and manufacturing (Figure 12.2).

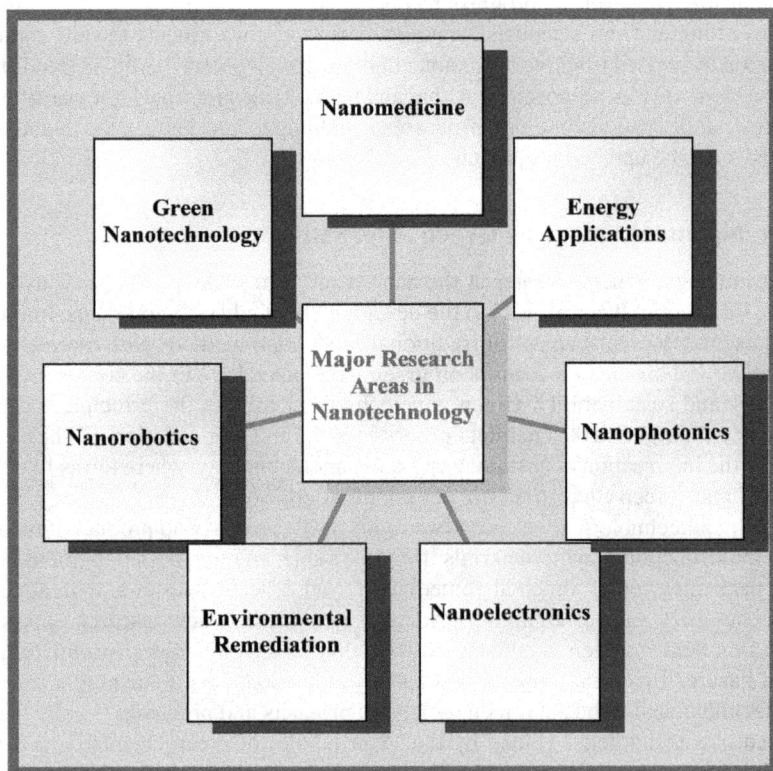

FIGURE 12.1   Major research areas in nanotechnology.

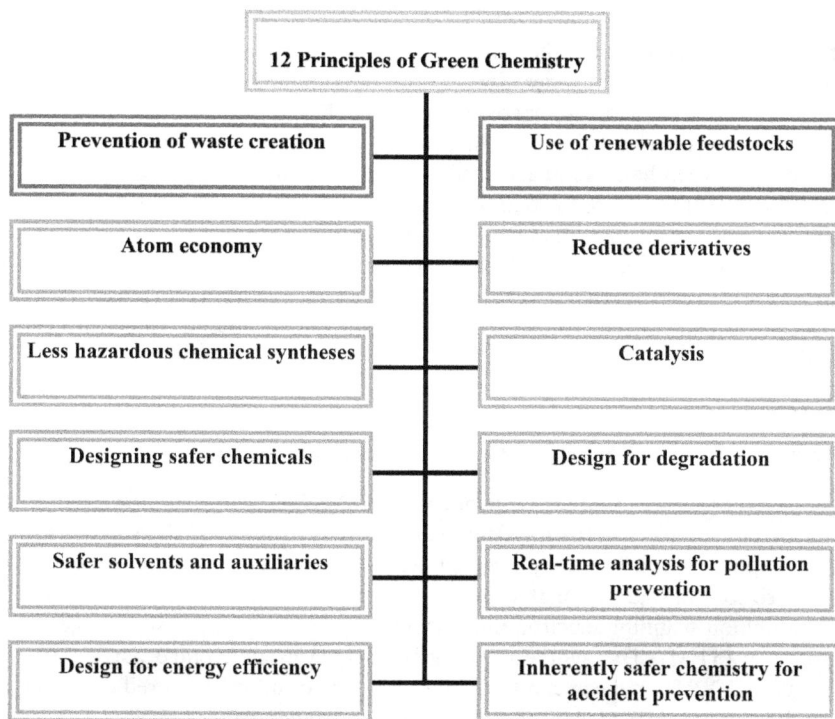

FIGURE 12.2 Twelve principles of green chemistry[25].

12.1.2 Green Nanotechnology: Eco-Friendly Synthesis and Applications

Green nanotechnology encompasses both the environmentally conscious synthesis of nanomaterials and their application in environmental remediation efforts. This dual focus not only enhances the feasibility of green nanotechnology but also contributes to its broader acceptance as a sustainable scientific approach. A prime example of green synthesis is the biological production of nanomaterials using plants, microbes, and other natural agents, which eliminates the need for toxic reagents and harsh processing conditions.[28] In the realm of chemical synthesis, the substitution of traditional organic solvents with less toxic, more environmentally benign alternatives significantly reduces the overall toxicity associated with nanomaterial production. Moreover, to qualify as "green," both the processes and resulting products must be evaluated through a life cycle assessment framework, considering the environmental impact from raw material sourcing to end-of-life disposal or recycling.

12.2 Eco-Friendly Synthesis Using Green Nanotechnology

A wide range of chemical and physical methods for nanomaterial synthesis are currently documented in the literature, offering high production rates and precise control over particle size and morphology. However, these conventional approaches are often deemed unsustainable due to their high energy and capital demands, reliance on hazardous chemicals, and the generation of substantial biological waste. Consequently, there is an increasing demand for cost-effective, eco-friendly, and biocompatible alternatives for nanomaterial synthesis. In contrast to traditional physical and chemical methods, green synthesis approaches offer safer, more sustainable, and economically viable pathways for nanoparticle production.[29] The core objective of nanotechnology is to establish reliable and efficient production techniques that allow precise control over chemical composition, morphology, and the development of well-dispersed systems on a large scale. A wide range of eco-friendly strategies utilizing plants, bacteria, and fungi for nanoparticle synthesis has been widely reported in the literature, owing to their economic viability, low toxicity, and inherent biocompatibility.

12.2.1 Synthesis of Nanoparticles from Plant Extracts

Jayprakash et al. synthesized silver nanoparticles (AgNPs) using a microwave-assisted green method, employing natural fruit extract of *Tamarindus indica* as both a reducing and capping agent. The morphological features of the resulting nanoparticles were characterized using various analytical techniques, including X-ray diffraction (XRD), high-resolution scanning electron microscopy (HR-SEM), and transmission electron microscopy (TEM). The synthesized AgNPs displayed an average particle size of 6–8 nm, and XRD analysis confirmed the presence of a face-centered cubic (FCC) crystalline structure of silver. The AgNPs demonstrated notable antibacterial activity, highlighting the potential of this simple, cost-effective, and environmentally benign synthesis route. Several studies have documented similar plant-based approaches for AgNPs synthesis, utilizing extracts from *Mangifera indica* leaves, *Garcinia mangostana* (mangosteen) leaves, *Jatropha curcas*, *Murraya koenigii* leaves, *Cinnamomum zeylanicum* leaves, *Aloe vera*, *Camellia sinensis*, honey, and mushrooms. Additionally, fruit extracts such as lemon, papaya, pear, tansy, and gooseberry have been effectively employed. Nanoparticles synthesized using plant or fruit extracts exhibit the added advantage of long-term stability by preventing agglomeration during storage.[30]

Aloe vera plant extract was employed in the synthesis of spinel-shaped polycrystalline nanopowders of $Ni_xCu_{0.25}Zn_{0.75-x}Fe_2O_4$ (where $x = 0.25$, 0.35, and 0.5), with an average particle size ranging from 15 to 40 nm. The synthesis was conducted via a straightforward solution-based method involving a mixture of metal nitrates and *Aloe vera* extract. The resulting nanomaterials exhibited clear ferromagnetic properties.[31] Similarly, coffee and tea extracts have been utilized for the green synthesis of stable noble metal nanoparticles, specifically palladium (Pd) and silver (Ag), with particle sizes ranging between 20 and 60 nm. These methods hold promise for the synthesis of other noble metals such as platinum (Pt) and gold (Au).[32] Metal oxides, including titanium dioxide, have also been produced via green routes. For instance, *Nyctanthes* leaf extract combined with titanium isopropoxide solution yielded titanium (IV) oxide nanoparticles with average sizes of 100–150 nm.[33] Furthermore, aqueous extracts from the manna of the *Hedysarum* plant and the soap-root (*Acanthophyllum bracteatum*) plant have been employed for nanoparticle synthesis, resulting in particles with diameters ranging from 29 to 68 nm.[34]

12.2.2 Synthesis of Nanoparticles from Natural Polysaccharides

Sulfated polysaccharides extracted from the marine red algae *Porphyra vietnamensis* were employed in the green synthesis of silver nanoparticles (Ag NPs). The resulting nanoparticles had an average size of approximately 13 ± 3 nm and exhibited a surface plasmon resonance peak at 404 nm. Spectroscopic analysis indicated that the reduction of silver nitrate was facilitated by the sulfate groups present in the polysaccharide structure.[35] In another eco-friendly approach, silver nanoparticles were synthesized using a silver (III)-containing rice wine and soda mixture, heated at temperatures ranging from 25 to 55°C and maintained at pH 6.5, without the addition of any external stabilizing agents. In this method, rice wine functioned both as a solvent and a reducing agent, while soda acted as a base catalyst and stabilizer. The resulting colloidal solution demonstrated excellent stability with minimal precipitation, even after extended storage over several months.[36]

In a separate study, Chen et al. developed chitosan-coated deformable liposomes containing flurbiprofen for ocular drug delivery, aiming to enhance transcorneal absorption and prolong pre-corneal drug residence time. These liposomes were prepared using a modified ethanol injection method, followed by chitosan coating. Gamma scintigraphy was employed *in vivo* to evaluate pre-corneal retention and drug drainage dynamics. The chitosan-coated deformable liposomes significantly extended the area under the remaining activity-time curve by 2.84-fold and 1.53-fold compared to flurbiprofen solution and uncoated deformable liposomes, respectively. No signs of ocular irritation or injury were observed following in vivo administration.[37]

Additionally, a novel in situ injectable nanocomposite hydrogel composed of curcumin, N,O-carboxymethyl chitosan, and oxidized alginate was formulated for dermal wound bandage applications. The nanocomposite was synthesized by incorporating methoxy poly(ethylene glycol)-β-poly(-caprolactone) copolymer into the hydrogel matrix. When applied to dorsal wounds in rats, the hydrogel demonstrated substantial enhancement in epidermal re-epithelialization and collagen deposition at the wound site.[38] Tian et al. (2012) formulated a drug delivery system based on glycyrrhetinic acid and modified sulfated chitosan for anticancer therapy. The resulting nanosystem exhibited a spherical morphology with an approximate size of 200 nm and showed marked anticancer activity.[39]

12.2.3 Synthesis of Nanoparticles from Microbial Source

Plant-derived extracts and microbial cultures have been extensively utilized across the globe for the eco-friendly or green synthesis of nanoparticles (NPs). Microorganisms are particularly well-suited for this purpose due to their rapid growth, cost-effective cultivation, and resilience under a wide range of environmental conditions, including variations in temperature, pressure, and pH. Their unique metabolic capabilities enable the synthesis of inorganic nanoparticles through both intracellular and extracellular reduction mechanisms. Microbes can thrive even in metal-contaminated environments, where they sequester metal ions and, through enzymatic activity, reduce them into their elemental nanoparticulate forms.[29]

Fungi-based green synthesis of nanomaterials is gaining significant global attention.[40] Compared to bacteria, fungal strains typically yield a higher quantity of nanoparticles due to their greater biomass. A wide variety of fungal species, including *Verticillium luteoalbum, Fusarium oxysporum, Colletotrichum* sp., *Trichothecium* sp., *Alternaria alternata, Trichoderma viride,* and *Aspergillus oryzae,* have been employed to synthesize nanoparticles of diverse shapes and sizes.[41] The use of such eco-friendly biological agents offers a sustainable alternative to conventional methods, allowing for the synthesis of biologically and pharmaceutically relevant nanomaterials without relying on toxic or hazardous chemicals.

12.2.4 Synthesis of Nanoparticles by Yeast

Kowshik and researchers[42] reported the extracellular synthesis of nanoparticles in large quantities, facilitated by simple downstream processing techniques. Their research focused on isolating a silver-tolerant yeast strain, MKY3, through inoculation with aqueous silver nitrate. Under environmentally stressed conditions, this strain facilitated the formation of silver nanoparticles ranging from 2 to 5 nm in size. Additionally, Dameron and coauthors[43] documented the biosynthesis of cadmium nanoparticles using *Candida glabrata* and *Schizosaccharomyces pombe.* Mourato and team[44] further investigated the biosynthesis of silver and gold nanoparticles utilizing an extremophilic yeast strain isolated from acid mine drainage. Moreover, the marine yeast *Rhodosporidium diobovatum* has been employed for the intracellular synthesis of stable lead sulfide nanoparticles.[45]

12.2.5 Applications of Nanotechnology for Sustainable Economic Development

Nanotechnology has emerged as a transformative force with the potential to address critical global challenges related to energy, healthcare, agriculture, water purification, and environmental sustainability. By enabling the design and manipulation of materials at the nanoscale, nanotechnology offers innovative solutions that are not only efficient and cost-effective but also environmentally benign. Its integration into various sectors contributes significantly to sustainable economic development by promoting resource conservation, reducing industrial waste, enhancing energy efficiency, and enabling the production of high-performance materials with minimal environmental impact. Furthermore, nanotechnology-driven advancements in green manufacturing, smart agriculture, and medical diagnostics are fostering inclusive growth and improving the quality of life, particularly in developing regions.

12.3 Conclusion

Nanoparticles offer a highly promising platform for a wide array of biological applications. Their biosynthesis through single-step processes has garnered increasing attention, encouraging further R&D in areas such as electrochemical sensors, biosensors, medicine, healthcare, and agriculture. This review highlights the synthesis of nanoparticles via biological methods, which are both environmentally sustainable and economically viable. It provides a comparative overview of the major synthesis techniques, namely, eco-friendly approaches with particular emphasis on biogenic methods. Continued advancements in this field are essential to translate the conceptual potential of nanotechnology into practical and scalable applications.

REFERENCES

[1] Galeazzo A, Furlan A, Vinelli A (2014) Lean and green in action: Interdependencies and performance of pollution prevention projects. *J Clean Prod* 85:191–200. https://doi.org/10.1016/j.jclepro.2013.10.015

[2] Qu X, Alvarez PJ, Li Q (2013) Applications of nanotechnology in water and wastewater treatment. *Water Res* 47(12):3931–3946. https://doi.org/10.1016/j.watres.2012.09.058

[3] Kumar A, Kumar A, Sharma G, Naushad M, Stadler FJ, Ghfar AA, Dhiman P, Saini RV (2017) Sustainable nano-hybrids of magnetic biochar supported g-C$_3$N$_4$/FeVO$_4$ for solar powered degradation of noxious pollutants – synergism of adsorption, photocatalysis & photo-ozonation. *J Clean Prod* 165:431–451. https://doi.org/10.1016/j.jclepro.2017.07.117

[4] Awual MR, Eldesoky GE, Yaita T, Naushad M, Shiwaku H, AlOthman ZA, Suzuki S (2015) Schiff based ligand containing nano-composite adsorbent for optical copper(II) ions removal from aqueous solutions. *Chem Eng J* 279:639–647. https://doi.org/10.1016/j.cej.2015.05.049

[5] Allen DT, Shonnard DR (2001) *Green engineering: Environmentally conscious design of chemical processes*. Pearson Education, London

[6] West JL, Halas NJ (2000) Applications of nanotechnology to biotechnology: Commentary. *Curr Opin Biotechnol* 11(2):215–217

[7] Seil JT, Webster TJ (2012) Antimicrobial applications of nanotechnology: Methods and literature. *Int J Nanomedicine* 7:2767. https://doi.org/10.2147/IJN.S24805

[8] Tratnyek PG, Johnson RL (2006) Nanotechnologies for environmental cleanup. *Nano Today* 1(2):44–48. https://doi.org/10.1016/S1748-0132(06)70048-2

[9] Bardos P, Bone B, Černík M, Elliott DW, Jones S, Merly C (2015) Nanoremediation and international environmental restoration markets. *Remediat J* 25(2):83–94. https://doi.org/10.1002/rem.21426

[10] Hood E (2004) Nanotechnology: Looking as we leap. *Environ Health Perspect* 112(13):A740. https://doi.org/10.1289/ehp.112-a740

[11] Dornfeld D, Yuan C, Diaz N, Zhang T, Vijayaraghavan A (2013) Introduction to green manufacturing. In: *Green manufacturing*. Springer, Boston, pp 1–23

[12] Shapira P, Youtie J (2015) The economic contributions of nanotechnology to green and sustainable growth. In: *Green processes for nanotechnology*. Springer, Cham, pp 409–434. https://doi.org/10.1007/978-3-319-15461-9_15

[13] Diallo MS, Fromer NA, Jhon MS (2013) Nanotechnology for sustainable development: Retrospective and outlook. In: *Nanotechnology for sustainable development*. Springer, Cham, pp 1–16. https://doi.org/10.1007/978-3-319-05041-6

[14] OECD (2011) *Fostering nanotechnology to address global challenges: Water*. Organisation for Economic Cooperation and Development, Paris. www.oecd.org/dataoecd/22/58/47601818.pdf

[15] Han C, Andersen J, Pillai SC, Fagan R, Falaras P, Byrne JA, Dunlop PSM, Choi H, Jiang W, O'Shea K, Dionysiou DD (2013) Chapter green nanotechnology: Development of nanomaterials for environmental and energy applications. *ACS Symp Ser* 1124:201–230. https://doi.org/10.1021/bk-2013-1124.ch012

[16] Deif AM (2011) A system model for green manufacturing. *J Clean Prod* 19(14):1553–1559. https://doi.org/10.1016/j.jclepro.2011.05.022

[17] Rusinko C (2007) Green manufacturing: An evaluation of environmentally sustainable manufacturing practices and their impact on competitive outcomes. *IEEE Trans Eng Manag* 54(3):445–454. https://doi.org/10.1109/TEM.2007.900806

[18] Sharma SK, Sanghi R (eds) (2012) *Advances in water treatment and pollution prevention.* Springer Science & Business Media, Dordrecht, The Netherlands

[19] Maksimović M, Omanović-Mikličanin E (2017) Towards green nanotechnology: Maximizing benefits and minimizing harm. In: *CMBEBIH 2017.* Springer, Singapore, pp 164–170. https://doi.org/10.1007/978-981-10-4166-2_26

[20] Aithal PS, Aithal S (2015) Ideal technology concept & its realization opportunity using nanotechnology. *IJAIEM* 4(2):153–164

[21] Dhingra R, Naidu S, Upreti G, Sawhney R (2010) Sustainable nanotechnology: Through green methods and life-cycle thinking. *Sustainability* 2(10):3323–3338. https://doi.org/10.3390/su2103323

[22] Andraos J (2005) Unification of reaction metrics for green chemistry: Applications to reaction analysis. *Org Process Res Dev* 9(2):149–163. https://doi.org/10.1021/op049803n

[23] Schulte PA, McKernan LT, Heidel DS, Okun AH, Dotson GS, Lentz TJ, Geraci CL, Heckel PE, Branche CM (2013) Occupational safety and health, green chemistry, and sustainability: A review of areas of convergence. *Environ Health* 12(1):31. https://doi.org/10.1186/1476-069X-12-31

[24] Wei Guo K (2011) Green nanotechnology of trends in future energy. *Recent Pat Nanotechnol* 5(2):76–88. https://doi.org/10.2174/187221011795909198

[25] Anastas PT, Warner JC (2000) *Green chemistry: Theory and practice*, vol. 30. Oxford University Press, Oxford

[26] Lazaro A, Quercia G, Brouwers HJH, Geus JW (2013) Synthesis of a green nano-silica material using beneficiated waste dunites and its application in concrete. *World J Nano Sci Eng* 3(3):41–51. https://doi.org/10.4236/wjnse.2013.33006

[27] Iavicoli I, Leso V, Ricciardi W, Hodson LL, Hoover MD (2014) Opportunities and challenges of nanotechnology in the green economy. *Environ Health* 13(1):78. https://doi.org/10.1186/1476-069X-13-78

[28] Ramya M, Subapriya MS (2012) Green synthesis of silver nanoparticles. *Int J Pharm Med Biol Sci* 1(1):54–61

[29] Fariq A, Khan T, Yasmin A (2017) Microbial synthesis of nanoparticles and their potential applications in biomedicine. *J Appl Biomed* 15(4):241–248. https://doi.org/10.1016/j.jab.2017.03.004

[30] Jayaprakash N, Vijaya JJ, Kaviyarasu K, Kombaiah K, Kennedy LJ, Ramalingam RJ, Munusamy MA, Al-Lohedan HA (2017) Green synthesis of Ag nanoparticles using Tamarind fruit extract for the antibacterial studies. *J Photochem Photobiol B* 169:178–185. https://doi.org/10.1016/j.jphotobiol.2017.03.013

[31] Laokul P, Maensiri S (2009) Aloe vera solution synthesis and magnetic properties of Ni-Cu-Zn ferrite nanopowders. *J Optoelectron Adv M* 11(6):857–862

[32] Nadagouda MN, Varma RS (2008) Green synthesis of silver and palladium nanoparticles at room temperature using coffee and tea extract. *Green Chem* (8):859–862. https://doi.org/10.1039/B804703K

[33] Sundrarajan M, Gowri S (2011) Green synthesis of titanium dioxide nanoparticles by Nyctanthes arbor-tristis leaves extract. *Chalcogenide Lett* 8(8):447–451

[34] Forough M, Farhadi K (2010) Biological and green synthesis of silver nanoparticles. *Turkish J Eng Environ Sci* 34(4):281–287. https://doi.org/10.3906/muh-1005-30

[35] Venkatpurwar V, Pokharkar V (2011) Green synthesis of silver nanoparticles using marine polysaccharide: Study of in-vitro antibacterial activity. *Mater Lett* 65(6):999–1002. https://doi.org/10.1016/j.matlet.2010.12.057

[36] Wu CC, Chen DH (2007) A facile and completely green route for synthesizing gold nanoparticles by the use of drink additives. *Gold Bull* 40(3):206–212. https://doi.org/10.1007/BF03215582

[37] Chen H, Pan H, Li P, Wang H, Wang X, Pan W, Yuan Y (2016) The potential use of novel chitosan-coated deformable liposomes in an ocular drug delivery system. *Colloids Surf B Biointerfaces* 143:455–462. https://doi.org/10.1016/j.colsurfb.2016.03.061

[38] Li X, Chen S, Zhang B, Li M, Diao K, Zhang Z, Li J, Xu Y, Wang X, Chen H (2012) In situ injectable nano-composite hydrogel composed of curcumin, N, O-carboxymethyl chitosan and oxidized alginate for wound healing application. *Int J Pharm* 437(1–2):110–119. https://doi.org/10.1016/j.ijpharm.2012.08.001

[39] Tian Q, Wang XH, Wang W, Zhang CN, Wang P, Yuan Z (2012) Self-assembly and liver targeting of sulfated chitosan nanoparticles functionalized with glycyrrhetinic acid. *Nanomedicine* 8(6):870–879. https://doi.org/10.1016/j.nano.2011.11.002

[40] Shamaila S, Sajjad AKL, Ryma N, Farooqi SA, Jabeen N, Majeed S, Farooq I (2016) Advancements in nanoparticle fabrication by hazard free eco-friendly green routes. *Appl Mater Today* 5:150–199. https://doi.org/10.1016/j.apmt.2016.09.009

[41] Jeevanandam J, Chan YS, Danquah MK (2016) Biosynthesis of metal and metal oxide nanoparticles. *ChemBioEng Rev* 3(2):55–67. https://doi.org/10.1002/cben.201500018

[42] Abdeen S, Praseetha PK (2013) Diagnostics and treatment of metastatic cancers with magnetic nanoparticles. *J Nanomedine Biotherapeutic Discov* 3(2):115. https://doi.org/10.4172/2155-983x.1000115

[43] Barabadi H, Honary S, Ebrahimi P, Mohammadi MA, Alizadeh A, Naghibi F (2014) Microbial mediated preparation, characterization and optimization of gold nanoparticles. *Braz J Microbiol* 45(4):1493–1501. https://doi.org/10.1590/s1517838220140004000046

[44] Sharma N, Pinnaka AK, Raje M, Fnu A, Bhattacharyya MS, Choudhury AR (2012) Exploitation of marine bacteria for production of gold nanoparticles. *Microb Cell Fact* 11:86. https://doi.org/10.1186/1475-2859-11-86

[45] Banu A, Rathod V, Ranganath E (2011) Silver nanoparticle production by *Rhizopus stolonifer* and its antibacterial activity against extended spectrum β-lactamase producing (ESBL) strains of Enterobacteriaceae. *Mater Res Bull* 46(9):1417–1423. https://doi.org/10.1016/j.materresbull.2011.05.008

13

Advancing Sustainable Development Goals through Nanotechnology

Himakshi Adhikari, Ishika Gupta, Sudhir Upadhyaya, Himanshi Goel, and Kumar Rakesh Ranjan

13.1 Introduction

Sustainable development stands at the forefront of global priorities. The reason is that we need to ensure long-term environmental stability, economic growth, and social well-being without depleting natural resources for future generations. The United Nations is responsible for the establishment of the Sustainable Development Goals (SDGs) [1]. The SDGs provide a comprehensive and multidimensional framework for tackling pressing global issues such as food security, healthcare, clean water access, renewable energy, and climate change mitigation [2]. In order to achieve these goals, we require the integration of cutting-edge scientific advancements and innovative technologies that can enhance efficiency, reduce resource consumption, and facilitate sustainable solutions across various sectors. One of the most promising and transformative fields in this regard is nanotechnology [3].

Nanotechnology, often regarded as a key enabler of the Fourth Industrial Revolution, involves the manipulation, engineering, and application of materials at the nanometer scale (1–100 nm) [4]. At this scale, materials exhibit unique physicochemical, electrical, mechanical, and optical properties, which differ significantly from their bulk counterparts, leading to novel applications in medicine, energy, water purification, environmental remediation, and advanced manufacturing [5]. The conceptual foundations of nanotechnology were first articulated by Richard P. Feynman in his seminal 1959 lecture, *"There's Plenty of Room at the Bottom,"* where he envisioned the possibility of atomic and molecular-level manipulation to create new materials and devices with unprecedented functionalities. However, the term "Nanotechnology" was formally introduced by Norio Taniguchi in 1974, in the context of ultra-precision machining and nanoscale material fabrication [6]. The field gained significant momentum in the 1980s, following the groundbreaking invention of the scanning tunneling microscope (STM) by Gerd Binnig and Heinrich Rohrer, which enabled direct visualization and manipulation of individual atoms. This discovery laid the foundation for molecular nanotechnology and paved the way for the development of atomic force microscopy (AFM), nanomaterials synthesis techniques, and molecular self-assembly approaches [7].

DOI: 10.1201/9781003632498-13

Since then, nanotechnology has emerged as a multidisciplinary domain, integrating principles from physics, chemistry, biology, and engineering to revolutionize diverse industries. Its applications are now being explored to accelerate progress toward achieving multiple SDGs [8]. It is important to note that all SDGs are interconnected, as progress in one area often influences multiple others. In this chapter, we specifically focus on the SDGs that have a direct and measurable link to nanotechnology, where nanoscale innovations are actively contributing to sustainable advancements in healthcare, clean energy, water purification, industrial development, and environmental conservation. By leveraging the unique properties of nanomaterials, researchers and industries are developing advanced solutions for achieving sustainability [9].

13.1.1 The Role of Technology in Achieving Sustainability

Technology plays a critical role in advancing sustainability by enabling smarter resource management, reducing environmental impact, and supporting circular economies. Nanotechnology – a branch of technology dealing with materials at the nanoscale – has emerged as a powerful extension of modern technology, offering precise control over matter. It plays a transformative role in sustainability by enhancing efficiency, reducing waste, and enabling eco-friendly innovations across multiple domains. Table 13.1 highlights all the SDGs and the direct/indirect role of nanotechnology in sustainable development.

13.2 Fundamentals of Nanotechnology

Nanotechnology is focused on designing, synthesizing, and manipulating materials at the nanoscale. At this scale, materials exhibit unique physical, chemical, and biological properties that differ significantly from their bulk counterparts, making nanotechnology essential in fields such as medicine, electronics, energy, and environmental science. In each of these fields, various types of nanomaterials have been employed.

13.2.1 Types of Nanomaterials and Their Properties

13.2.1.1 Carbon-Based Nanomaterials

Nanomaterials can be classified into four major types as shown in Figure 13.1. Carbon-based nanomaterials (NMs) include materials like graphene, carbon nanotubes (CNTs), carbon nanofibers (CNFs), and porous carbon. These materials exhibit properties such as high electrical conductivity, large surface area, excellent mechanical strength, and good electrochemical stability [10].

13.2.1.2 Metal-Based Nanomaterials

Metal oxide-based NMs are transition metal oxides such as TiO_2, MnO_2, Fe_2O_3, Co_3O_4, and NiO. These materials exhibit key properties like high theoretical

TABLE 13.1
SDGs and Role of Nanotechnology in Fulfilling Them

S. No.	Sustainable Development Goals	Role of Nanotechnology
1.	No poverty	While nanotechnology is not directly associated with SDG 1 (No Poverty), it plays a crucial indirect role. In agriculture, nano-based fertilizers and pest control enhance crop productivity, fostering economic stability for farmers. In healthcare, nanomedicine and advanced diagnostics enable cost-effective treatments, improving health outcomes in low-income populations. Furthermore, innovations in water purification and sustainable energy contribute to enhanced access to essential resources, thereby supporting poverty alleviation efforts.
2.	Zero Hunger	Nanotechnology supports SDG 2 (Zero Hunger) by enhancing agricultural productivity through nano-fertilizers, precision pest control, and soil health improvement. It also aids in food preservation and safety, reducing post-harvest losses and ensuring better nutrition.
3.	Good Health and Well-Being	Nanotechnology has transformed healthcare by facilitating precise diagnostics, targeted drug delivery, and advanced therapeutic approaches. Its significant role in disease prevention, early detection, and personalized medicine strengthens global health systems, enhancing the accessibility and effectiveness of quality healthcare.
4.	Quality Education	Nanotechnology indirectly supports SDG 4 by enhancing education through advanced learning tools, interactive simulations, and improved research capabilities. It modernizes teaching methods, fosters innovation, and strengthens educational infrastructure, creating a more effective and sustainable learning environment
5.	Gender Equality	Nanotechnology indirectly supports SDG 5 (Gender Equality) by improving women's healthcare through advanced diagnostics and reproductive treatments. It also enhances access to clean water and sanitation, reduces burdens on women, and creates new economic opportunities in technology-driven sectors, promoting gender inclusion.
6.	Clean Water and Sanitation	Nanotechnology supports SDG 6 (Clean Water and Sanitation) by enabling advanced water purification, desalination, and wastewater treatment through nanofilters and antimicrobial materials. It helps remove contaminants, making clean and safe water more accessible while improving sanitation systems.
7.	Affordable and Clean Energy	Nanotechnology supports SDG 7 (Affordable and Clean Energy) by enhancing solar panels' efficiency, improving energy storage in batteries and supercapacitors, and developing lightweight, high-performance materials for renewable energy technologies. It also helps in hydrogen production and energy conservation.
8.	Decent Work and Economic Growth	Nanotechnology indirectly supports SDG 8 (Decent Work and Economic Growth) by fostering industrial innovation, improving workplace safety with advanced materials, and enabling sustainable manufacturing. It also contributes to job creation in high-tech sectors, promoting economic development and technological advancement.

9.	Industry, Innovation, and Infrastructure	Nanotechnology plays a crucial role in *SDG 9* (*Industry, Innovation, and Infrastructure*) by enhancing material strength, improving manufacturing efficiency, and enabling advanced technologies. It contributes to sustainable industrial growth, smart infrastructure, and technological advancements across various sectors.
10.	Reduced Inequality	Nanotechnology indirectly supports SDG 10 (Reduced Inequalities) by improving access to affordable healthcare, clean water, and sustainable energy, benefiting marginalized communities. It also fosters economic opportunities through technological advancements, promoting inclusive growth and reducing the digital divide.
11.	Sustainable Cities and Communities	Nanotechnology contributes to SDG 11 (Sustainable Cities and Communities) by enhancing building materials for stronger and more sustainable infrastructure. It also improves air and water purification, energy efficiency, and waste management, making urban environments cleaner, safer, and more resilient.
12.	Responsible Consumption and Production	Nanotechnology contributes to SDG 12 (Responsible Consumption and Production) by enhancing resource efficiency, reducing waste through sustainable manufacturing, and developing eco-friendly materials. It also enables advanced recycling technologies and pollution control, promoting sustainable production practices.
13.	Climate Action	Nanotechnology plays an *indirect role* in *SDG 13 (Climate Action)* by enabling cleaner energy solutions, enhancing carbon capture processes, and developing eco-friendly materials that lower environmental impact. It also improves air and water purification technologies, supporting efforts to combat climate change.
14.	Life Below Water	Nanotechnology indirectly supports SDG 14 (Life Below Water) by helping in water pollution control, marine ecosystem protection, and sustainable aquaculture. It enables nanofilters for removing microplastics and heavy metals, nanocoatings to prevent biofouling on ships, and nano-sensors for monitoring ocean health, contributing to marine conservation.
15.	Life on Land	Nanotechnology indirectly contributes to SDG 15 (Life on Land) by enabling soil and water decontamination, promoting sustainable agriculture through nano-fertilizers and pesticides, and assisting in biodiversity conservation. It also aids in reforestation efforts with nano-enhanced plant growth stimulators and improves environmental monitoring for ecosystem protection.
16.	Peace, Justice, and Strong Institutions	Nanotechnology indirectly supports SDG 16 (Peace, Justice, and Strong Institutions) by enhancing cybersecurity through nano-encryption technologies, improving forensic science with advanced nano-sensors, and strengthening public safety with nano-surveillance tools. It also aids in anti-counterfeiting measures, ensuring transparency and security in governance systems.
17.	Partnerships for the Goals	Nanotechnology indirectly supports SDG 17 (Partnerships for the Goals) by fostering global collaboration in scientific research, technology transfer, and sustainable innovation. It enables partnerships among industries, governments, and research institutions to develop advanced solutions for healthcare, clean energy, and environmental sustainability, accelerating progress toward the SDGs.

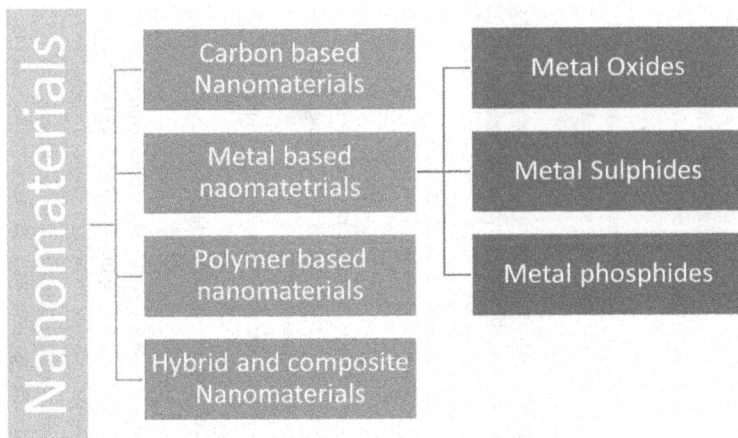

FIGURE 13.1 Classification of nanomaterials.

capacity, good structural stability, redox activity, and enhanced ion storage capabilities. They have wide applications in gas sensing, energy storage, and environmental decontamination. Metal sulfide-based NMs consist of materials such as MoS_2, WS_2, $CoSe_2$, and SnS_2, showcasing properties like high ion intercalation ability, good cycling stability, high energy density, and tunable electronic properties. They have shown applications in batteries, photocatalysis, and many others. Metal phosphides and nitrides comprise materials such as FeP, CoP, VN, and MoN. They exhibit properties like high electrical conductivity, good mechanical strength, and stable cycling performance, and they show good results in the field of batteries [11].

13.2.1.3 Polymer-Based Nanomaterials

Polymer-based NMs include polyaniline (PANI), polypyrrole (PPy), and poly(3,4-ethylenedioxythiophene) (PEDOT). They are widely used due to their properties of high flexibility, good conductivity, lightweight, and mechanical stability [12].

13.2.1.4 Hybrid and Composite Nanomaterials

Carbon-metal oxides, MXene-MOF hybrids, and metal sulfide–carbon composites show excellent synergy. Their combinations boost conductivity, stability, and electrochemical performance. These materials are promising for next-generation energy devices [13].

13.2.2 Synthesis and Fabrication Techniques

Several synthetic strategies and techniques are employed in nanotechnology. As synthetic strategies are concerned, there are, in general, three types of approaches for the production of NMs and fabrication of nanostructures [14].

13.2.3 A Bottom-Up Approach

Bottom-up approach builds NMs from atomic or molecular units, using self-assembly or chemical processes leading to the formation of nanostructures. This approach organizes atomic or molecular components in hierarchical nanocomplexes. A few of its examples include:

- **Sol–Gel Method:** Involves the transition of a system from a liquid "sol" into a solid "gel" phase to produce oxide NMs (e.g., silica, titania).
- **Chemical Vapor Deposition (CVD):** A gaseous precursor reacts or decomposes on a substrate to form a solid NM, widely used for thin films and CNTs.
- **Colloidal Synthesis:** Nanoparticles (e.g., gold, silver, QDs) are synthesized in solution using surfactants or stabilizers to control growth and prevent agglomeration.
- **Hydrothermal/Solvothermal Synthesis:** Reactions carried out in sealed vessels at high pressure and temperature to form nanocrystals with controlled morphology.
- **Atomic Layer Deposition (ALD):** Sequential, self-limiting chemical reactions allow for the layer-by-layer deposition of materials with atomic precision.
- **Microemulsion Method:** Utilizes water-in-oil or oil-in-water emulsions to confine the reaction environment and control nanoparticle size and shape.

13.2.4 Top-Down Approach

The top-down approach initiates with bulk material and breaks it down into nanoscale structures using mechanical or lithographic techniques. This approach adds or removes thin layers of bulk materials as do:

- **Ball Milling:** Mechanical grinding reduces bulk material into nanoscale particles through high-energy collisions.
- **Lithography (Photolithography and Electron Beam Lithography):** Used in nanofabrication to create precise patterns and structures on a substrate by etching away selective regions.
- **Etching (Wet and Dry Etching):** Chemical or plasma-based removal of material from a surface to form nanoscale features.
- **Laser Ablation:** A high-energy laser beam removes material from a solid target, creating nanoparticles in a liquid or gas medium.
- **Mechanical Milling and Attrition:** Involves the application of mechanical forces to break down particles into nanoscale dimensions.

13.2.4.1 Hybrid Techniques

A class of advanced fabrication methods now combines both approaches for improved control and scalability, such as:

- Nanoimprint lithography (NIL)
- Electrospinning (used to produce nanofibers)
- Template-assisted synthesis (e.g., using porous templates for nanowire or nanotube formation)

13.3 Nanotechnology and the SDGs

Addressing societal challenges through nanoscale chemistry is crucial for achieving the SDGs, paving the way for innovative and environmentally responsible solutions. By meticulously aligning nanoscale materials and processes with specific global needs, researchers are developing targeted interventions for areas such as clean energy, water purification, and sustainable agriculture. This approach not only leverages the unique properties of NMs to enhance performance and efficiency but also prioritizes minimizing environmental impact throughout the materials' life cycle. From designing biodegradable nanocarriers for targeted drug delivery to creating high-performance solar cells using earth-abundant NMs, the deliberate integration of sustainability principles into nanoscale chemistry fosters progress toward a healthier, more equitable, and environmentally sound future.

13.4 Application of Nanotechnology

Nanotechnology finds wide application in areas such as water purification, energy storage, medicine, and agriculture. Nano-based materials offer high surface area, enhanced reactivity, and selectivity, enabling efficient pollutant removal, improved battery and solar cell performance, targeted drug delivery, and controlled-release agrochemicals. Table 13.2 shows different nanotechnology approaches with various applications and advantages. These advancements support cleaner, more efficient systems aligned with SDGs.

TABLE 13.2

Nanotechnology in the Field of Medicine

Nano Strategy	Types	Application	Advantages
Nanocarriers	Liposomes, dendrimers, and polymeric nanoparticles, solid lipid nanoparticles (SLNs)	Targeted drug delivery, gene therapy, vaccine delivery.	Enhance drug delivery, improve solubility, bioavailability, and controlled release of drugs.
Nanoparticles	Cadmium-based QDs, carbon QDs, and perovskite QDs	Bioimaging, fluorescent labeling, photodynamic therapy	High photostability, tunable fluorescence, and better imaging contrast
Quantum dots	Nano-formulated drugs, theranostic nanoparticles, smart nanocarriers	Cancer therapy, neurological disorders, and regenerative medicine	Site-specific action, reduced side effects, and real-time monitoring possible
Nanomedicines	Gold nanoparticles, silver nanoparticles, iron oxide nanoparticles, silica nanoparticles	Drug delivery, biosensing, antimicrobial coatings, contrast agents	High surface-to-volume ratio, multifunctionality, and controlled interactions with biological systems

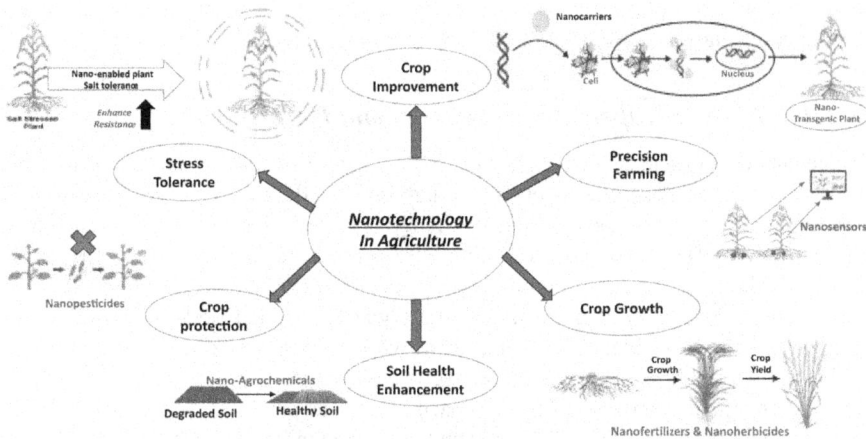

FIGURE 13.2 Nanotechnology in sustainable agriculture.

13.4.1 Nanotechnology in Sustainable Agriculture

In the field of sustainable agriculture, nanotechnology has been widely used. Figure 13.2 depicts some of the applications of nanotechnology in sustainable agriculture.

13.4.1.1 Nano-Sensors for Precision Agriculture

Nano-sensors play a vital role in detecting nutrient deficiencies in soil, monitoring environmental conditions, and alerting farmers to take corrective measures before damage occurs. These nano-biosensors collect chemical and biological signals from the soil and convert them into measurable electronic signals using transducers. Some nano-sensors are designed with specific nanoparticles (NPs) to detect toxic gases like nitrogen oxides (NO_x) and sulfur oxides (SO_x), which harm human health and contribute to environmental degradation [15].

The high sensitivity of nano-sensors is achieved through CNTs, which store energy, aid in filtration, and enhance soil quality. CNTs improve nutrient transfer to plants and accelerate growth rates. Additionally, metallic NPs, such as gold (Au), platinum (Pt), palladium (Pd), and earth metals, increase the potential and sensitivity of transducers, making them integral components of nano-sensors. These advanced sensing technologies enable real-time monitoring of soil fertility, reducing fertilizer waste and enhancing crop yields [16].

13.4.1.2 Nanoparticles for Plant Growth and Genetic Improvement

NPs are also applied directly to plants to modify genetic traits and enhance growth. Due to their physicochemical properties, NPs overcome several limitations of traditional agricultural biotechnology. For instance, silicon dioxide (SiO_2) NPs improve seedling development, root elongation, and nutrient uptake. Studies have shown that these NPs enhance photosynthesis rates, boost plant metabolism, and regulate gas exchange, ultimately leading to higher crop productivity [17].

13.4.1.3 Nano-Encapsulated Fertilizers and Herbicides

Nanotechnology provides innovative methods to improve fertilizer efficiency through nano-encapsulation. Nano-fertilizers, nano-herbicides, and nano-pesticides utilize nano-carriers that control the release of active ingredients, ensuring targeted and sustained nutrient delivery [18]. This precision minimizes overuse, reduces environmental pollution, and enhances plant absorption rates [19]. For example, poly(epsilon-caprolactone) nano-capsules are used for delivering atrazine, a widely used herbicide. These nano-carriers increase the herbicidal activity of atrazine while reducing its environmental impact. Similarly, NPs facilitate controlled pesticide release, improving pest resistance without excessive chemical application. This approach, known as precision farming, maximizes agricultural output while minimizing soil and water contamination [20].

13.4.1.4 Nano-Agrochemicals for Soil and Crop Improvement

Nanotechnology plays a crucial role in soil improvement and crop enhancement through nano-agrochemicals such as zeolites, which act as carriers for nutrients. These NMs ensure that essential plant nutrients are efficiently transported and absorbed, preventing nutrient loss and soil degradation [21]. Nano-herbicides offer a breakthrough in weed control by selectively targeting unwanted plants while preserving crops. Herbicides like triazines (e.g., chloro-s-triazines, atrazine) are encapsulated in nano-formulations, ensuring their efficient and controlled release into the soil. This enhances weed elimination without harming beneficial plants, improving overall farm productivity and sustainability [22].

13.4.2 Nanotechnology for Food Quality and Safety

Beyond agriculture, nanotechnology is transforming food safety and preservation. Nano-packaging materials enhance food shelf life by incorporating antimicrobial and oxygen-scavenging NPs, preventing spoilage and contamination. Nano-sensors in food packaging can detect bacterial contamination and chemical degradation, ensuring food quality from production to consumption [23]. Nanotechnology also improves food fortification and nutrient delivery. Nano-encapsulation techniques allow essential vitamins, minerals, and bioactive compounds to be incorporated into food products without altering their taste, texture, or stability. This technology is particularly beneficial in combating micronutrient deficiencies in malnourished populations [24].

13.4.3 Nanomaterials for Solar Cells

13.4.3.1 Metal NPs

Transition metal NPs have gained significant attention in solar cell applications due to their ability to enhance light absorption, improve charge transport, and reduce recombination losses. Among these, noble metal NPs such as gold (Au) and silver (Ag) are widely studied for their localized surface plasmon resonance (LSPR) effect, which enhances the optical absorption of solar cells by scattering and trapping incident light. This phenomenon is particularly beneficial in thin-film and perovskite solar cells, where optimizing light harvesting is critical for improving efficiency. Additionally, certain transition metal NPs, such as copper (Cu) and nickel (Ni), serve as effective charge transport materials, facilitating carrier mobility and reducing interfacial resistance in organic and dye-sensitized solar cells. Copper NPs, in particular, are considered a promising low-cost alternative to noble metals, offering comparable conductivity and stability in photovoltaic applications. Moreover, platinum (Pt) and palladium (Pd) NPs have been incorporated into perovskite solar cells to enhance interfacial charge transfer and catalytic activity, thereby improving overall device performance. The strategic incorporation of metal NPs in solar cells not only enhances optical and electronic properties but also contributes to the development of cost-effective and high-efficiency photovoltaic technologies [25].

13.4.3.2 Carbon Nanomaterials for Solar Cells

Carbon NMs are gaining significant traction as promising components in thin-film photovoltaic (PV) solar cells, largely due to their unique blend of properties, including tunable electronic structures, high electrical and thermal conductivity, robust chemical stability, and notable mechanical strength. Graphene, a two-dimensional sheet of sp^2-bonded carbon atoms, is explored for its potential as a transparent conducting electrode (TCE), offering an alternative to conventional materials like indium tin oxide (ITO) with the benefits of reduced material costs and enhanced device flexibility, although challenges related to achieving high conductivity and uniform large-area deposition remain. Strategies to overcome these challenges include doping with elements like nitrogen, stacking multiple graphene layers, and creating hybrid structures with other conductive materials. CNTs, available in single-walled (SWCNTs) and multi-walled (MWCNTs) forms, are also investigated for their high aspect ratios and excellent electrical conductivity, enabling their use as TCEs, charge transport layers, and active layer components in organic solar cells (OSCs), where they can facilitate efficient charge transport, reduce recombination losses, and enhance exciton dissociation. The performance of CNT-based solar cells is contingent on factors such as CNT chirality, network density, and alignment. The incorporation of these carbon NMs offers several key benefits, including enhanced light absorption, improved charge transport, and increased device stability. However, widespread adoption necessitates addressing challenges such as developing cost-effective synthesis methods, achieving precise doping control, ensuring long-term stability, and mitigating environmental impacts. Future research should focus on these areas and explore novel applications in advanced solar cell architectures like perovskite and tandem devices, ultimately leveraging the unique properties of carbon NMs to drive the next generation of solar energy technology [26].

13.4.3.3 2D Nanomaterials for Solar Cells

Two-dimensional (2D) NMs stand out as promising candidates to enhance solar cell performance due to their unique properties, including high surface area, quantum confinement effects, and tunable electronic structures. Several 2D materials are being actively explored, each offering specific advantages. Graphene's high electron mobility and conductivity make it suitable as a transparent electrode, although its zero bandgap necessitates modifications. Transition metal dichalcogenides (TMDs) like MoS_2 and WS_2 possess tunable band gaps, enabling efficient light absorption and strong light–matter interactions. MXenes, a class of transition metal carbides and nitrides, offer high electrical conductivity and mechanical strength, positioning them as excellent electrode materials. Black phosphorus (BP) presents a tunable direct bandgap and high carrier mobility, but its environmental instability requires protective strategies. Finally, quasi-2D perovskites, while not strictly 2D, offer improved stability and tunable optoelectronic properties. These materials are being integrated into diverse solar cell architectures, including silicon, thin-film, and perovskite-based devices. In silicon solar cells, graphene can replace traditional ITO electrodes, enhancing flexibility and reducing costs. In thin-film solar cells, TMDs and MXenes can serve as active layers or charge transport layers, boosting light absorption and charge collection. Quasi-2D perovskites are emerging as active absorbers in perovskite solar cells, whereas other 2D materials are used for interface engineering. Despite the potential, challenges remain in the large-scale production, stability, and integration of 2D materials into solar cells. Addressing these hurdles, particularly material degradation, and optimizing device architectures are crucial steps in realizing the full potential of 2D NMs for next-generation high-performance solar cells [27].

13.4.4 Energy Storage and Next-Generation Batteries

Rechargeable batteries, including lithium-ion (LIB), sodium-ion (SIB), and potassium-ion batteries (PIB), play a crucial role in energy storage but face challenges such as limited energy density, structural instability, and safety issues. Sodium- and potassium-based batteries offer cost-effective alternatives, but their larger ion sizes complicate electrode stability. The lithium–sulfur system provides high energy density but has limited energy density, shorter cycle life, structural degradation, safety concerns, and environmental impact [28, 29]. To overcome these challenges, researchers have developed NMs with enhanced conductivity, good ion transport, and improved electrode performance [30]. The nano-structuring of electrodes is being explored as a viable solution. Material engineering techniques are used to construct electrode architectures at the nanoscale, offering unique properties that enhance battery performance. The electrochemical performance of nanostructured electrodes depends not only on their nanoscale features but also on the design and synthesis methods, including phase fraction, phase purity, morphology, and particle/crystal size distribution. NMs are expected to provide innovative solutions for designing advanced electrode architectures. The use of NPs, in particular, helps to overcome issues related to ionic conductivity by reducing the length scale for ion transport. Additionally, NPs reduce excessive mechanical stress during charge and discharge cycles, which can cause repeated contraction and expansion. These mechanical changes can be better tolerated by NPs, preventing structural degradation and enhancing battery stability. Moreover, electrodes composed of NPs exhibit high porosity, providing ample internal space for expansion and contraction, leading to improved electrochemical performance. The increased surface area, numerous active sites, and short mass and charge diffusion distances contribute to superior energy storage capabilities.

Among various anode electrodes, nanostructured metal oxides have shown promise for use in LIBs, SIBs, and supercapacitors due to their high specific capacity/capacitance, typically two to three times higher than that of carbon/graphite-based materials. Metal oxides, particularly transition metal oxides (TMOs), have gained attention for their ability to undergo conversion reactions, allowing for greater ion storage capacity and improved battery performance. Combining metal oxide NPs like CeO_2, MnO_2, and ZrO_2 with carbon-based materials enhances reaction kinetics and electrode reversibility in batteries. Additionally, metal components such as TiO_2 nano-coating, Nb-doped WO_3, ZrO_2, and $NiCoO_2$ further improve battery performance. Decorating graphite nanoflakes with metal oxides (Li, Mn) significantly boosts anode stability and ensures long-term cycle life [31].

13.4.5 Nanotechnology in Medicine

The field of medicine is vast and well-developed, but faces many challenges. One of the major challenges in conventional medicine is the non-specific distribution of therapeutic agents, which often leads to systemic toxicity and reduced drug efficacy. Nanotechnology has revolutionized the healthcare field (as shown in Table 13.2) by enabling precise and effective drug delivery mechanisms. Developing nanocarriers, such as liposomes, dendrimers, and polymeric NPs, has facilitated enhanced drug delivery by improving solubility, bioavailability, and controlled release of drugs [32]. By engineering NPs with surface modifications, drugs can be directed specifically to diseased cells, thereby minimizing damage to healthy tissues. For instance, ligand-functionalized NPs can bind selectively to overexpressed receptors on cancer cells, allowing for precise tumor targeting [33]. This approach enhances therapeutic efficiency while reducing adverse effects, thus improving patient outcomes in oncological and chronic disease treatments. Furthermore, the impact of nanotechnology-driven drug delivery extends to treating infectious diseases, where nanoformulated antibiotics and antiviral drugs offer improved efficacy against resistant microbial strains [34].

Accurate and early diagnosis of diseases is crucial for effective treatment and disease management. Nanotechnology has significantly enhanced medical imaging and diagnostic techniques, contributing to the realization of SDG 3. NP-based contrast agents have improved imaging modalities such as magnetic resonance imaging (MRI), computed tomography (CT), and positron emission tomography (PET). Due to their high surface area and tunable properties, these NPs offer superior contrast enhancement, enabling the detection of pathological conditions at an early stage [35].

Quantum dot NPs represent a groundbreaking advancement in medical diagnostics. These semiconductor nanocrystals exhibit unique optical properties, including high fluorescence stability and tunable emission wavelengths, making them ideal for bioimaging applications. Quantum dots conjugated with biomolecules can specifically target diseased cells, allowing for real-time imaging and tracking of disease progression. Their application in detecting cancer biomarkers and neurodegenerative diseases holds immense potential in precision medicine, ultimately contributing to improved healthcare outcomes [36].

Nanomedicine integrates nanotechnology with medical science to develop novel therapeutic approaches for tissue engineering and regenerative medicine [37]. NMs, such as CNTs, graphene, and metallic NPs, play a crucial role in engineering biomimetic scaffolds that promote cell growth and tissue regeneration. These nanostructured biomaterials exhibit high mechanical strength, biocompatibility, and bioactivity, making them suitable for applications in bone, cartilage, and neural tissue repair. Additionally, nanosensors embedded in tissue scaffolds enable real-time monitoring of cellular responses, facilitating personalized treatment strategies [38].

The antimicrobial applications of NPs are particularly relevant in addressing healthcare-associated infections (HAIs), which pose significant challenges in clinical settings. Metallic NPs, such as silver, copper, and zinc oxide NPs, exhibit potent antimicrobial properties by disrupting microbial membranes and interfering with intracellular processes. Their incorporation into wound dressings, medical coatings, and implant materials provides effective protection against pathogenic bacteria and biofilm formation. The development of smart antimicrobial surfaces using nanotechnology is expected to reduce infection rates, improve patient safety, and contribute to global health advancements [39].

TABLE 13.3

Nanotechnology-Based Approach for water remediation

Aspects	Conventional Approach	Nanotechnology Approach
Filtration Method	Sand filtration, activated carbon, reverse osmosis	Nano-membranes, carbon nanotubes (CNTs), and graphene-based filters
Contamination Removal	Remove large particles and some microorganisms	Removes heavy metals, pathogens, and even nano-sized pollutants
Energy Efficiency	High energy consumption (especially in RO and distillation)	Lower energy consumption due to high surface area materials
Selectivity	Limited selectivity for specific pollutants	High selectivity using functionalized nanomaterials
Chemical Usage	Uses chlorine, alum, and other coagulants	Reduced chemical use with nanoadsorbents and photocatalysts
Cost and Maintenance	Frequent maintenance and filter replacement needed	Longer lifespan, self-cleaning membranes possible
Sustainability	Generates sludge, requires chemical disposal	Less sludge, more eco-friendly processes
Pathogen Removal	Kills bacteria with chlorine (leaves byproducts)	Nano-silver, TiO_2, and other nanomaterials offer antimicrobial effects

As we approach 2030, the integration of nanotechnology in healthcare is expected to drive significant advancements in disease prevention, diagnostics, and therapeutics. The emergence of personalized nanomedicine, where treatments are tailored based on an individual's genetic and molecular profile, will revolutionize patient care. Furthermore, the convergence of artificial intelligence with nanotechnology will enable the development of intelligent drug delivery systems, real-time health monitoring devices, and predictive analytics for disease management.

The widespread adoption of nanotechnology-driven healthcare solutions will contribute to achieving SDG by reducing disease burden, improving healthcare accessibility, and enhancing treatment outcomes. However, ethical considerations, regulatory frameworks, and safety assessments must be addressed to ensure the responsible application of nanotechnology in medicine.

13.4.6 Nanotechnology for Water Remediation

Nanotechnology offers solutions to a lot of problems, one of which is a sustainable alternative to conventional water treatment methods, such as the invention of nanomembranes, CNTs, and graphene-based filters [40–42]. As shown in Table 13.3, the conventional approach followed for this is sand filtration, activated carbon, and reverse osmosis, and it often involves high energy consumption, limited pollutant selectivity, and significant chemical usage. In contrast, nanotechnology-based approaches reduce chemical inputs, lower energy demands, and generate less sludge, contributing to more environmentally friendly and cost-effective solutions. The integration of functionalized NMs with antimicrobial properties further supports safe and clean water access,

---

13.5 Recovery of Nanomaterials

Although NMs are showing promising results in many fields, their recovery and recyclability remain crucial. Since NMs can accumulate in the environment and pose potential risks to human health and ecosystems, it is essential to develop efficient recovery, reuse, and disposal strategies. Ensuring proper recovery not only minimizes environmental contamination but also supports the principles of sustainability and circular economy by reducing material wastage and conserving resources. In order to recover these NMs various methodologies such as Hydrometallurgy, Pyrometallurgy, Nanohydrometallurgy, Thermal procedure (inert gas condensation), Vacuum separation, Oxidative polymerization, Precipitation method are being used. Out of which Hydrometallurgy aur Pyrometallurgy are the most effective processes that help in recovering NMs.

13.5.1 Nanohydrometallurgy

Nanohydrometallurgy is an innovative and eco-friendly process that uses nanotechnology to extract noble metals from batteries and electronic waste. This is an advanced version of hydrometallurgy, in which magnetic NPs are used for the separation of metals. A general outline of the process is depicted in Figure 13.3. The main advantages of this technique are as follows: i) simple operation – with the help of a magnet, metals can easily be separated; ii) highly efficient; iii) extracted NPs can be reused; and iv) eco-friendly.

Elements utilized in battery manufacturing pose significant concerns regarding toxicity and environmental contamination if improperly discarded. These risks underscore the critical need for efficient recycling and recovery strategies to mitigate potential environmental hazards.

Battery recycling has garnered substantial interest from the public, industry, and academia as it presents a viable solution for minimizing environmental risks associated with improper disposal while simultaneously facilitating the recovery of high-value metals for sustainable reuse. NMs used in nano-hydrometallurgy are mainly maghemite (y-Fe_2O_3), magnetite (Fe_3O_4) NPs, or nano zero-valent iron. They show superparamagnetic properties and are responsive to an external magnetic force induced by a magnet, helping with the fast and efficient separation of noble metal ions from the aqueous solution. The metals that are extracted can again be used in making batteries and other electronic devices [43].

13.5.2 Hydrometallurgy

Hydrometallurgy is an environmentally friendly and efficient technique widely used for the recovery of valuable metals and NMs from complex matrices, including industrial waste and spent NM-based products (see Figure 13.4). It involves the use of aqueous chemistry – typically leaching, solution concentration, and metal recovery through precipitation, solvent extraction, or electro-winning. This method is particularly advantageous for NMs due to its ability to operate at relatively low temperatures and to selectively separate target components. By integrating hydrometallurgical processes into the life cycle of NMs, industries can significantly reduce environmental pollution, promote material circularity, and enhance resource efficiency in line with green nanotechnology and SDGs [44].

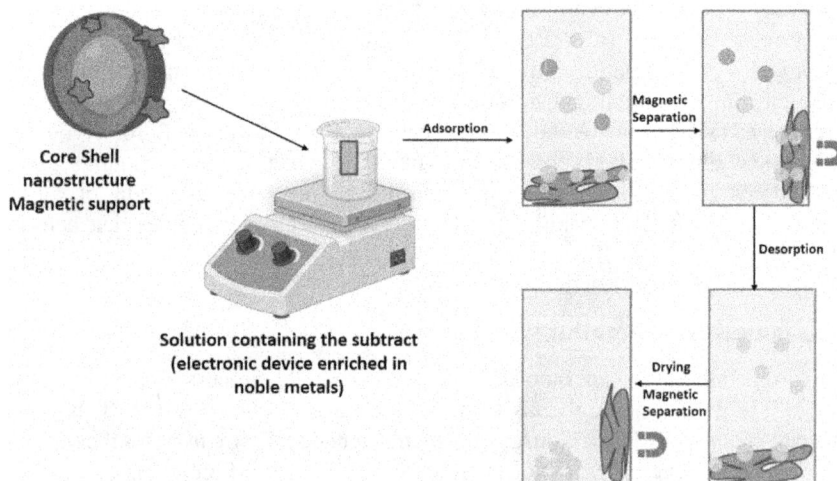

FIGURE 13.3 Schematic diagram of nano-hydrometallurgy.

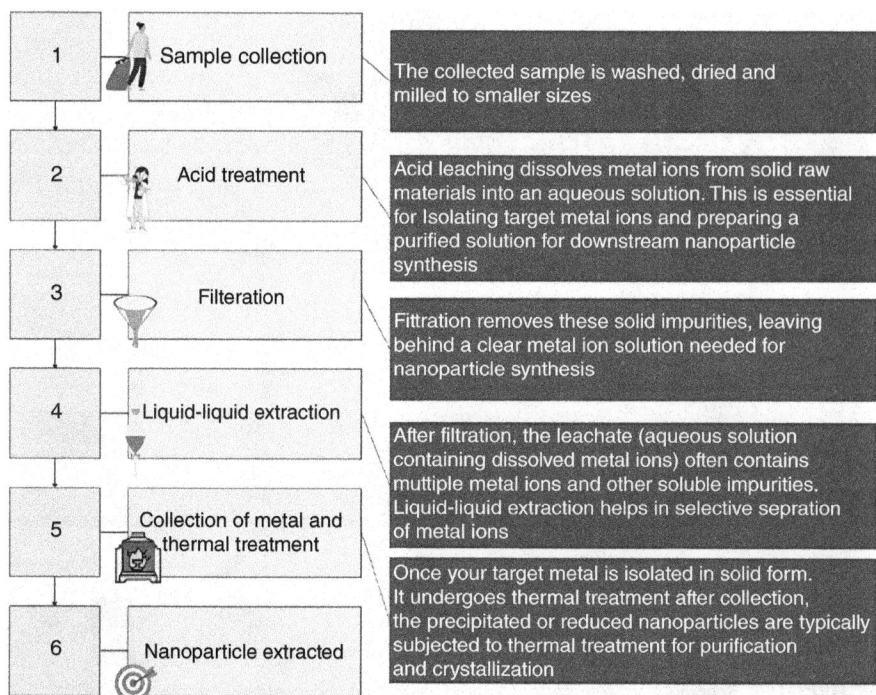

FIGURE 13.4 Schematic depiction of the hydrometallurgical pathway in nanotechnology.

13.5.3 Pyrometallurgical Pathway

Pyrometallurgy, a high-temperature extractive metallurgy branch, utilizes heat to drive chemical reactions and phase transformations, separating desired metals from their ores. This pathway typically begins with roasting (see Figure 13.5), where sulfide ores are heated in air to eliminate sulfur, forming metal oxides and sulfur dioxide. The resulting material undergoes smelting, involving melting with fluxing agents to create immiscible layers of molten slag and metal or matte; the slag dissolves impurities, while the denser metal phase settles. Subsequent converting oxidizes the matte, further purifying the target metal, such as producing blister copper from copper sulfide. Finally, refining techniques, like electrorefining, enhance metal purity to meet specific standards. Pyrometallurgy proves advantageous for high-grade ores, complex sulfide ores, and situations demanding rapid reaction kinetics, often with lower water consumption than hydrometallurgical alternatives. Furthermore, byproducts like sulfur dioxide can be harnessed for sulfuric acid production. However, the approach involves significant energy consumption, potential air pollution from emissions, and the generation of substantial slag. Consequently, while pyrometallurgy remains critical for certain applications, continuous efforts aim to mitigate its environmental impact through process optimization and sustainable technologies, ensuring its viability within a context-sensitive decision framework [44].

PYROMETALLURGICAL PATHWAY

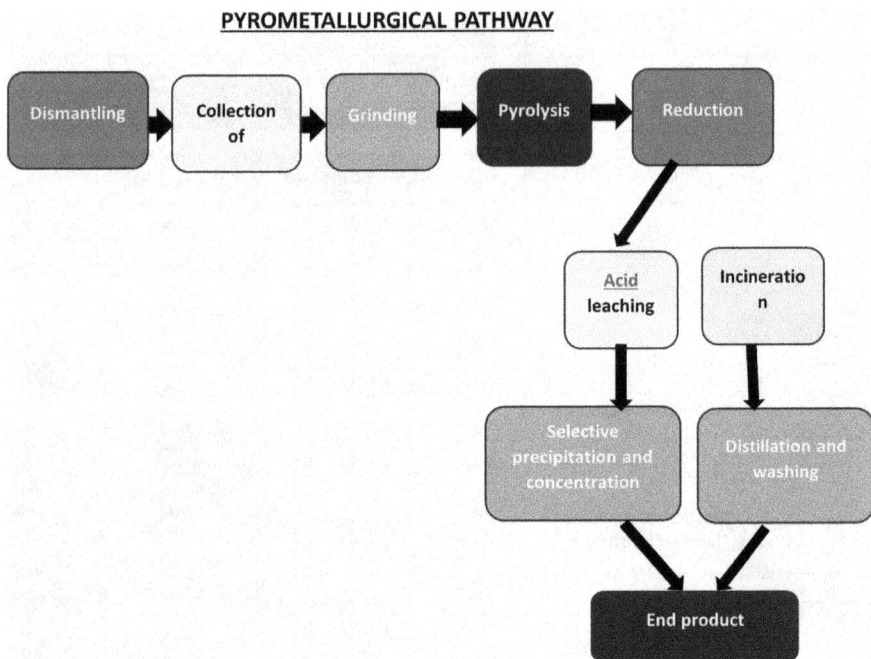

FIGURE 13.5 Schematic representation of the pyrometallurgical pathway of nanomaterials.

13.6 Challenges and Ethical Considerations

The rapid advancement of nanotechnology offers numerous benefits across various fields, yet it also raises significant challenges and ethical concerns. These include potential risks to human health and the environment, gaps in regulatory frameworks, and broader socio-economic implications. One of the primary concerns is the environmental and health risks associated with NMs. Due to their high surface area and unique physicochemical properties, NPs can interact with biological systems and ecosystems in unpredictable ways. Certain NMs, such as silver, titanium dioxide, and carbon-based NPs, have been shown to induce cytotoxicity, oxidative stress, and inflammatory responses in living organisms. Additionally, some NPs exhibit environmental persistence, accumulating in soil, water, and air, which may disrupt ecosystems. Exposure to engineered NMs through inhalation, ingestion, or dermal absorption is another concern, particularly for industrial workers and researchers handling these materials. Long-term effects remain uncertain, necessitating continuous study and comprehensive risk assessments.

To mitigate these risks, various regulatory frameworks and safety standards have been developed worldwide. The European Union's REACH (Registration, Evaluation, Authorisation, and Restriction of Chemicals) regulation requires manufacturers to assess and disclose potential hazards of NMs. In the US, the Environmental Protection Agency (EPA) has implemented guidelines for NM use in consumer products, while the International Organization for Standardization (ISO) and the Organisation for Economic Co-operation and Development (OECD) continue to develop technical standards for their safe assessment. However, despite these efforts, gaps persist in standardized toxicity testing, classification, and labeling of NMs, highlighting the need for further regulatory advancements.

Beyond regulatory concerns, nanotechnology introduces complex ethical, social, and economic challenges. One major issue is equitable access to nanotechnology, as high production costs may widen the technological divide between developed and developing nations. Additionally, the integration of nanotechnology in surveillance devices raises ethical concerns regarding personal privacy. From an economic perspective, while nanotechnology fosters innovation, it may lead to job displacement in traditional industries as automation and nano-enabled technologies replace conventional manufacturing methods. Public perception and acceptance of nanotechnology-based products also play a crucial role in determining their success. Misinformation and lack of awareness may lead to resistance, necessitating better science communication and public engagement. Furthermore, while nanotechnology holds promise for sustainable applications in areas such as energy storage and water purification, large-scale production could introduce unforeseen environmental consequences. Addressing these challenges requires a collaborative effort among scientists, policymakers, industry leaders, and the public to ensure that nanotechnology advances responsibly and sustainably.

13.7 Conclusion

The dominion of nanotechnology presents a compelling frontier in the pursuit of sustainable development. The diverse range of NMs, from carbon-based structures to metal oxides and innovative hybrid composites, offers a powerful toolkit for addressing critical challenges in energy, environmental protection, and resource management. These materials, synthesized through both bottom-up and top-down approaches, exhibit unique properties like enhanced conductivity, stability, and catalytic activity, enabling breakthroughs in energy storage, pollution remediation, and efficient resource utilization. However, realizing the full potential of nanotechnology for a sustainable future necessitates a holistic perspective. Rigorous LCAs are paramount to ensure that the benefits of NMs outweigh their potential environmental impacts. Furthermore, comprehensive research into their toxicity and environmental fate is essential to mitigate unforeseen risks. Finally, focusing on scalable, cost-effective, and ethically sound manufacturing practices is crucial for democratizing access to these technologies and ensuring their widespread adoption in the global quest for a more sustainable world. By carefully navigating these considerations, nanotechnology can emerge as a key enabler of a future where innovation and environmental stewardship go hand in hand.

REFERENCES

[1] L. Pokrajac *et al.*, "Nanotechnology for a Sustainable Future: Addressing Global Challenges with the International Network4Sustainable Nanotechnology," *ACS Nano*, vol. 15, no. 12, pp. 18608–18623, Dec. 2021, https://doi.org/10.1021/acsnano.1c10919.

[2] S. Morton, D. Pencheon, N. Squires, "Sustainable Development Goals (SDGs), and Their Implementation: A National Global Framework for Health, Development and Equity Needs a Systems Approach at Every Level," *British Medical Bulletin*, vol. 124, no. 1, pp. 81–90, Dec. 2017, https://doi.org/10.1093/bmb/ldx031.

[3] P. Berrone, H. E. Rousseau, J. E. Ricart, E. Brito, and A. Giuliodori, "How Can Research Contribute to the Implementation of Sustainable Development Goals? An Interpretive Review of SDG Literature in Management," *International Journal of Management Reviews*, John Wiley and Sons Inc, Apr. 2023, pp. 318–339. https://doi.org/10.1111/ijmr.12331.

[4] S. Malik, K. Muhammad, and Y. Waheed, "Nanotechnology: A Revolution in Modern Industry," *Molecules*, vol. 28, no. 2, p. 661, Jan. 2023, https://doi.org/10.3390/molecules28020661.

[5] J. O. Ogbuagu and P. C. Akubue, "Achieving Sustainable Development Goals Through Nanotechnology in Polymer and Textile," *European Journal of Engineering and Technology*, vol. 3, no. 7, 2015, Accessed: Mar. 04, 2025. [Online]. Available: www.idpublications.org

[6] W. S. Khan, E. Asmatulu, and R. Asmatulu, "Nanotechnology emerging trends, markets and concerns," *Nanotechnology Safety*, pp. 1–21, 1 Jan. 2025, https://doi.org/10.1016/B978-0-443-15904-6.00017-4.

[7] G. A. Silva, "Introduction to Nanotechnology and its Applications to Medicine," *Surgical Neurology*, vol. 61, no. 3, pp. 216–220, Mar. 2004, https://doi.org/10.1016/J.SURNEU.2003.09.036.

[8] S. Aithal and P. S. Aithal, "Green Nanotechnology Innovations to Realize UN Sustainable Development Goals 2030," Sep. 2021, Accessed: Apr. 01, 2025. [Online]. Available: https://papers.ssrn.com/abstract=3939567

[9] B. Singh, N. Chauhan, N. Agrawal, and A. Ayyagari, "Contribution of Microbial Nanotechnology to Sustainable Development Goals," *Nano-Microbiology for Sustainable Development*, pp. 1–23, 2025, https://doi.org/10.1007/978-3-031-78845-1_1.

[10] O. S. Ayanda *et al.*, "Recent Progress in Carbon-Based Nanomaterials: Critical Review," *Journal of Nanoparticle Research*, vol. 26, no. 5, p. 106, May 2024, https://doi.org/10.1007/s11051-024-06006-2.

[11] R. A. de Jesus *et al.*, "Metal Oxide Nanoparticles for Environmental Remediation," in *Biodegradation and Biodeterioration at the Nanoscale*, Elsevier, 2022, pp. 529–560. https://doi.org/10.1016/B978-0-12-823970-4.00025-7.

[12] G. M. Patel, V. Shah, J. Bhaliya, P. Pathan, and K. M. Nikita, "Polymer-Based Nanomaterials: An Introduction," in *Smart Polymer Nanocomposites*, Elsevier, 2023, pp. 27–59. https://doi.org/10.1016/B978-0-323-91611-0.00018-9.

[13] P. Palanisamy, M. Chavali, E. M. Kumar, and K. C. Etika, "Hybrid Nanocomposites and Their Potential Applications in the Field of Nanosensors/Gas and Biosensors," in *Nanofabrication for Smart Nanosensor Applications*, Elsevier, 2020, pp. 253–280. https://doi.org/10.1016/B978-0-12-820702-4.00011-8.

[14] M. S. Chavali and M. P. Nikolova, "Metal Oxide Nanoparticles and Their Applications in Nanotechnology," *SN Applied Sciences*, vol. 1, no. 6, p. 607, Jun. 2019, https://doi.org/10.1007/s42452-019-0592-3.

[15] M. Kaushal and S. P. Wani, "Nanosensors: Frontiers in Precision Agriculture," *Nanotechnology: An Agricultural Paradigm*, pp. 279–291, Jun. 2017, https://doi.org/10.1007/978-981-10-4573-8_13.

[16] S. A. Perdomo, J. M. Marmolejo-Tejada, and A. Jaramillo-Botero, "Review – Recent Advances in Nanosensors for Precision Agriculture," *Journal of Electrochemical Society*, vol. 170, no. 12, p. 127507, Dec. 2023, https://doi.org/10.1149/1945-7111/AD1306.

[17] A. Hosseinpour *et al.*, "Application of Zinc Oxide Nanoparticles and Plant Growth Promoting Bacteria Reduces Genetic Impairment Under Salt Stress in Tomato (*Solanum lycopersicum* L. 'Linda')," *Agriculture*, vol. 10, no. 11, p. 521, Nov. 2020, https://doi.org/10.3390/AGRICULTURE10110521.

[18] S. Singh *et al.*, "Nanoencapsulation of Fertilizers: Delivery and Applications," *Nanofertilizer Delivery, Effects and Application Methods*, pp. 97–110, Jan. 2024, https://doi.org/10.1016/B978-0-443-13332-9.00011-3.

[19] R. Sabarivasan and P. Murali Arthanari, "Application of Nanoencapsulation Technology in Agriculture for Effective and Sustainable Weed Management: A Critical Review," *Communications in Soil Science and Plant Analysis*, vol. 56, no. 2, pp. 277–291, 2024, https://doi.org/10.1080/00103624.2024.2413513.

[20] T. A. Wani, F. A. Masoodi, W. N. Baba, M. Ahmad, N. Rahmanian, and S. M. Jafari, "Nanoencapsulation of Agrochemicals, Fertilizers, and Pesticides for Improved Plant Production," *Advances in Phytonanotechnology: From Synthesis to Application*, pp. 279–298, Jan. 2019, https://doi.org/10.1016/B978-0-12-815322-2.00012-2.

[21] J. Pathak *et al.*, "Application of Nanoparticles in Agriculture: Nano-Based Fertilizers, Pesticides, Herbicides, and Nanobiosensors," *Molecular Impacts of Nanoparticles on Plants and Algae*, pp. 305–331, Jan. 2024, https://doi.org/10.1016/B978-0-323-95721-2.00012-9.

[22] M. Bhaskar, A. Kumar, and R. Rani, "Application of Nano Formulations in Agriculture," *Biocatalysis and Agricultural Biotechnology*, vol. 54, p. 102934, Nov. 2023, https://doi.org/10.1016/J.BCAB.2023.102934.

[23] M. Rai *et al.*, "Smart Nanopackaging for the Enhancement of Food Shelf Life," *Environmental Chemistry Letters*, vol. 17, no. 1, pp. 277–290, Mar. 2019, https://doi.org/10.1007/S10311-018-0794-8/METRICS.

[24] P. N. Ezhilarasi, P. Karthik, N. Chhanwal, and C. Anandharamakrishnan, "Nanoencapsulation Techniques for Food Bioactive Components: A Review," *Food and Bioprocess Technology*, no. 3, pp. 628–647, Aug. 2012, https://doi.org/10.1007/S11947-012-0944-0.

[25] L. V. Hublikar, S. V. Ganachari, F. A. Shilar, and N. Raghavendra, "Recent Advances in Transition Metal Oxide Nanomaterials for Solar Cell Applications: A Status Review and Technology Perspectives," *Materials Research Bulletin*, vol. 187, p. 113351, Jul. 2025, https://doi.org/10.1016/j.materresbull.2025.113351.

[26] W. A. Khan *et al.*, "Carbon Nanomaterials in Coatings: A Review Focusing Thin Film Photovoltaic Solar Cells," *Materials Science in Semiconductor Processing*, vol. 185, p. 108929, Jan. 2025, https://doi.org/10.1016/j.mssp.2024.108929.

[27] Shreya, P. Phogat, R. Jha, and S. Singh, "Emerging Advances and Future Prospects of Two Dimensional Nanomaterials Based Solar Cells," *Journal of Alloys and Compounds*, vol. 1001, p. 175063, Oct. 2024, https://doi.org/10.1016/j.jallcom.2024.175063.

[28] P. Phogat, S. Rawat, S. Dey, and M. Wan, "Advancements and Challenges in Sodium-Ion Batteries: A Comprehensive Review of Materials, Mechanisms, and Future Directions for Sustainable Energy Storage," *Journal of Alloys and Compounds*, vol. 1020, p. 179544, Mar. 2025, https://doi.org/10.1016/j.jallcom.2025.179544.

[29] P. Hong, C. Xu, C. Yan, Y. Dong, H. Zhao, and Y. Lei, "Prussian Blue and Its Analogues for Commercializing Fast-Charging Sodium/Potassium-Ion Batteries," *ACS Energy Letters*, vol. 10, no. 2, pp. 750–778, Feb. 2025, https://doi.org/10.1021/acsenergylett.4c02915.

[30] T. Emmerich *et al.*, "Enhanced Nanofluidic Transport in Activated Carbon Nanoconduits," *Nature Materials*, vol. 21, no. 6, pp. 696–702, Jun. 2022, https://doi.org/10.1038/s41563-022-01229-x.

[31] J. Zhu, Y. Ding, Z. Ma, W. Tang, X. Chen, and Y. Lu, "Recent Progress on Nano-structured Transition Metal Oxides as Anode Materials for Lithium-Ion Batteries," *Journal of Electronic Materials*, vol. 51, no. 7, pp. 3391–3417, Jul. 2022, https://doi.org/10.1007/s11664-022-09662-z.

[32] A. Kumar, P. K. Tyagi, S. Tyagi, S. *et al.* "Integrating green nanotechnology with sustainable development goals: A pathway to sustainable innovation," *Discover Sustainability*, vol. 5, no. 364, 2024. https://doi.org/10.1007/s43621-024-00610-x.

[33] R. Bajracharya *et al.*, "Functional Ligands for Improving Anticancer Drug Therapy: Current Status and Applications to Drug Delivery systems," *Drug Delivery*, vol. 29, no. 1, pp. 1959–1970, Dec. 2022, https://doi.org/10.1080/10717544.2022.2089296.

[34] O. M. Koo, I. Rubinstein, and H. Onyuksel, "Role of Nanotechnology in Targeted Drug Delivery and Imaging: A Concise Review," *Nanomedicine*, vol. 1, no. 3, pp. 193–212, Sep. 2005, https://doi.org/10.1016/J.NANO.2005.06.004.

[35] T. C. Jackson, B. O. Patani, D. E. Ekpa, T. C. Jackson, B. O. Patani, and D. E. Ekpa, "Nanotechnology in Diagnosis: A Review," *Advanced Nanoparticles*, vol. 6, no. 3, pp. 93–102, Aug. 2017, https://doi.org/10.4236/ANP.2017.63008.

[36] S. Y. Madani, F. Shabani, M. V. Dwek, and A. M. Seifalian, "Conjugation of Quantum Dots on Carbon Nanotubes for Medical Diagnosis and Treatment," *International Journal of Nanomedicine*, vol. 8, pp. 941–950, Mar. 2013, https://doi.org/10.2147/IJN.S36416.

[37] A. Zeeshan, M. Farhan, and A. Siddiqui, "Nanomedicine and Drug Delivery: A Mini Review," *International Nano Letters*, vol. 4, no. 1, pp. 1–7, Feb. 2014, https://doi.org/10.1007/S40089-014-0094-7.

[38] S. J. Kim, S. J. Choi, J. S. Jang, H. J. Cho, and I. D. Kim, "Innovative Nanosensor for Disease Diagnosis," *Accounts of Chemical Research*, vol. 50, no. 7, pp. 1587–1596, Jul. 2017, https://doi.org/10.1021/ACS.ACCOUNTS.7B00047/ASSET/IMAGES/LARGE/AR-2017-00047A_0006.JPEG.

[39] V. Raffa, O. Vittorio, C. Riggio, and A. Cuschieri, "Progress in Nanotechnology for Healthcare," *Minimally Invasive Therapy & Allied Technologies*, vol. 19, no. 3, pp. 127–135, Jun. 2010, https://doi.org/10.3109/13645706.2010.481095.

[40] N. Jain and N. Jee Kanu, "The Potential Application of Carbon Nanotubes in Water Treatment: A State-of-the-Art-Review," *Materials Today: Proceedings*, vol. 43, pp. 2998–3005, 2021, https://doi.org/10.1016/j.matpr.2021.01.331.

[41] S. J. Schmidt, W. Dou, and S. A. Sydlik, "Regeneratable Graphene-Based Water Filters for Heavy Metal Removal at Home," *ACS ES&T Water*, vol. 3, no. 8, pp. 2179–2185, Aug. 2023, https://doi.org/10.1021/acsestwater.3c00010.

[42] D. B. Tripathy and A. Gupta, "Nanomembranes-Affiliated Water Remediation: Chronology, Properties, Classification, Challenges and Future Prospects," *Membranes (Basel)*, vol. 13, no. 8, p. 713, Aug. 2023, https://doi.org/10.3390/membranes13080713.

[43] H. E. Toma, "Magnetic Nanohydrometallurgy: Principles and Concepts Applied to Metal Ion Separation and Recovery," *Chemical Engineering Research and Design*, vol. 216, pp. 251–269, Apr. 2025, https://doi.org/10.1016/j.cherd.2025.03.003.

[44] S. A. Mvokwe, O. O. Oyedeji, M. A. Agoro, E. L. Meyer, and N. Rono, "A Critical Review of the Hydrometallurgy and Pyrometallurgical Recovery Processes of Platinum Group Metals from End-of-Life Fuel Cells," *Membranes (Basel)*, vol. 15, no. 1, p. 13, Jan. 2025, https://doi.org/10.3390/membranes15010013.

14

Policy and Regulatory Issues in Achieving a Green Economy through Nanotechnology

Ikechukwu P. Ejidike, Opeyemi N. Avoseh, Adeniyi A. Adejare, Taiwo A. Akande, Toyib S. Oyewole, Folajimi T. Avoseh, Chinwendu B. Sunday, and Mercy O. Bamigboye

14.1 Introduction

The advancement of chemistry over the centuries has led to several divergent and convergent disciplines culminating in great discoveries and applications of its results in technology, medicine, agriculture, construction, food processing, cosmetics, lifestyles, and bioengineering (Anwar and Khitab, 2019; Ejidike et al., 2024a; Olaleru et al., 2025; Do et al., 2025). The advent of chemistry at the "nano" scale has opened rooms for the manipulation of materials at 10^{-9} m levels (Castillo, 2019; Ejidike et al., 2020; Gobinath et al., 2023; Keskin et al., 2023). In a famous 1959 lecture by Richard Feynman, the field of nanotechnology was foretold as "There's Plenty of Room at the Bottom" (Kolahalam et al., 2019), and the phrase was legally outlined in 1974 by Norio Taniguchi (Castillo, 2013; Anwar and Khitab, 2019). Thus, the study of materials or substances with at least one dimension at the nanometer (10^{-9} m).

Nanoscience gives the knowledge of the arrangement of atoms with their elementary properties at the nanoscale, whereas nanotechnology is the technology adopted in governing the material at the atomic stage for the amalgamation of novel nanomaterials with diverse characteristics (Korada and Hamid, 2017; Kolahalam et al., 2019; Ogunsile et al., 2024; Ejidike et al., 2024a, 2024b). The extensive dialogues about the technical definition of nanomaterials are that it cannot be applied to all regulations dealing with nanotechnology (Castillo, 2013; Korada and Hamid, 2017; Anwar and Khitab, 2019). However, nanomaterials are substances with sizes below 100 nm. Materials of the nanosized display characteristics with diverse physicochemical possessions than the bulk material, depending on the shape and size. For instance, gold, known as a yellow, shiny precious metal that does not corrode, possesses a face-centered cubic configuration. Its small size of about 10 nm absorbs green light and consequently appears red. These outstanding changes make manipulation and technological application possible (Roduner, 2006; Khitab et al., 2018; Shah et al., 2021).

Nanomaterials can also adopt different sizes and shapes, different from bulk materials like nanoparticles, nanorods, and nano-sheets, which can be categorized based on their dimensionality (Haleem et al., 2023; Gobinath et al., 2023; Ejidike et al., 2023; Do et al., 2025). Therefore, a clear understanding of nanomaterials is well

DOI: 10.1201/9781003632498-14

understood without any ambiguity. Further description of the term nanomaterials as described by the International Standardization Organisation (ISO) as a "material with any peripheral measurement in the nanoscale or possess a surface structure in the nano-range," having a "span range roughly 1 nm to 100 nm" (Ma et al., 2024; Haleem et al., 2023; Ejidike et al., 2023; Keskin et al., 2023). This definition has emerged as real and applicable. The ISO further classified nanomaterials as either nano-objects or nano-designed materials, with internal or surface structure in the nanoscale (Khitab et al., 2018; Ejidike and Clayton 2022; Keskin et al., 2023; Dimo et al. 2024).

Recently, there have been new concepts and concerns about advanced materials that respond to external stimuli or their surroundings to produce a dynamic and reversible change in specialized properties. They have been applied in drug delivery, cancer treatment, and self-healing materials for coating. These materials are called "smart materials" or "smart nanomaterials" (Yoshida and Lahann, 2008; Korada and Hamid, 2017; Haleem et al., 2023; Ejidike and Clayton, 2022; Onisuru and Onisuru, 2025).

14.2 Risk Assessment and Safety Testing

Recent scientific advancements in the use of nanomaterials have seen its viral applications in drug delivery, cosmetics, drug therapy, and other areas related to consumer products. Therefore, with the extensive exposure to these nanomaterials, there are extensive concerns about the toxicity and the risk assessment related to nanomaterials. The health risk assessment of nanomaterials has been evaluated in several reports (Warheit, 2018; Ghamarpoor et al., 2023; Stuetz et al., 2023; Ejidike et al., 2024a).

A gross amount of more than 300 million euros has been used by the European Commission to evaluate the health risks of nanomaterials. Since the Fifth Framework Programme (FP5), several white papers on issues addressing the health effects of nanomaterials have been produced in response to specified inducements. The International Risk Governance Council (IRGC) in 2007 raised concern about the dynamism of active nanostructures posing further or amplified risks and, at the same time, requiring a piece of advanced knowledge and competence to assess those risks (Cui et al., 2005; Ghamarpoor et al., 2023; Gottardo et al., 2021; Mech et al., 2022; Kostapanou et al., 2024). The European Union (EU) has been at the vanguard of formulating nanomaterials and their products on the safety and risk assessment regulations. The OECD's white paper identified nanomaterials with potential health risks, and these include nanomaterials like silicon dioxide, titanium dioxide, carbon nanotubes, and carbon black (Gottardo et al., 2021; Mech et al., 2022; Kostapanou et al., 2024). Consequently, a limited understanding of the potential toxicities and environmental impacts of various nanomaterials makes comprehensive risk assessment and appropriate safety testing challenging. This will be beneficial for policy and regulatory issues in achieving a green economy through nanotechnology (Danchin, 2022; Pandey et al., 2023; Ejidike et al., 2024a, 2024b).

14.3 International Harmonization

14.3.1 Lack of Global Standards and Regulations for Nanomaterials

The rapid advancement of nanotechnology presents significant opportunities for fostering a sustainable economy. However, the absence of globally harmonized standards and regulations for nanomaterials poses substantial challenges for companies operating across multiple jurisdictions (Iavicoli et al., 2014; Allan et al., 2021; Rodríguez-Gómez et al., 2023; Wasti et al., 2023; Elzein, 2024). This lack of uniformity can lead to regulatory uncertainties, increased compliance costs, and probable risks to public health and the environment (Allan et al., 2021; Elzein, 2024; Ejidike et al., 2024a).

One of the primary obstacles to international harmonization is the variability in regulatory definitions and testing protocols for nanomaterials. The complexity and diversity of nanomaterials make it difficult to establish a universal regulatory definition, complicating efforts to monitor and predict their environmental fate (Ray et al., 2009; Wahab et al., 2024; Baer et al., 2013). Additionally, there is a scarcity of reliable techniques for estimating exposures to nanomaterials, hindering accurate risk assessments. These challenges are compounded by knowledge gaps in toxicity and toxic mechanisms, making it difficult to predict the hazards associated with nanomaterials (Iavicoli et al., 2014; Allan et al., 2021). The lack of standardized nomenclature, test methods, and characterization protocols further exacerbates these challenges. This inconsistency hampers the development of shared standards, leading to fragmented regulatory approaches that can stifle innovation and compromise safety. Moreover, the absence of internationally agreed-upon definitions and terminology for nanotechnology, as well as standardized protocols for toxicity quantification and environmental impact evaluations, impedes effective regulation (Oualikene-Gonin et al., 2023).

The EPA regulates nanoscale materials under the Toxic Substances Control Act (TSCA), considering them as "chemical substances" with structures approximately 1–100 nanometers in size. Recognizing that these materials may exhibit unique properties compared to their larger-scale counterparts, the EPA has implemented a regulatory approach that includes information-gathering rules and premanufacture notifications. Manufacturers are required to report specific information, such as chemical identity, production volume, and health and safety data, to facilitate a comprehensive understanding of nanoscale materials in commerce (Chávez-Hernández et al., 2024; Petersen et al., 2022). This approach aims to guarantee that these materials are assembled and used in an approach that protects against arbitrary risks to human health and the environment.

The FDA's regulatory policy is product-focused and science-based, considering the unique properties of nanomaterials in the framework of each product's intended use. The agency emphasizes early consultation with manufacturers to address scientific and regulatory concerns related to nanotechnology products (Rodríguez-Gómez et al., 2025). The FDA's approach respects distinctions in legal standards for several product classes, leading to divergent regulatory outcomes based on the specific application of nanomaterials. For products not subject to premarket review, the FDA encourages manufacturers to consult with the agency before marketing their products to ensure compliance with safety standards (Allan et al., 2021; Rodríguez-Gómez et al., 2023).

14.3.2 Efforts Toward Harmonization

Despite these challenges, various international collaborations aim to harmonize nanotechnology regulations. Organizations such as the Organization for Economic Co-Operation and Development (OECD) have established working parties on manufactured nanomaterials to foster shared standards among member countries. These collaborations promote a harmonized approach to risk evaluation and management, enhancing public trust in nanoproducts across different markets (Rodríguez-Gómez et al., 2023).

Transnational research initiatives also play a crucial role in supporting these regulatory frameworks. Collaborative research between countries aids in developing best practices and regulatory tools while addressing key challenges posed by nanotechnology. Such partnerships facilitate knowledge sharing and technological advancements, ultimately shaping regulations that better protect public health and the environment.

14.3.3 Enhancing Harmonization

To improve international harmonization of nanotechnology regulations, several measures are recommended:

1. **Develop internationally agreed definitions and terminology:** Establishing clear and universally accepted definitions for nanomaterials is essential for consistent regulation across jurisdictions.
2. **Standardize testing protocols:** Creating standardized procedures for toxicity experimentation and environmental impact assessments will facilitate uniform risk evaluations and regulatory decisions.
3. **Enhance global collaboration:** Strengthening international collaborations through organizations like the OECD can promote the development of shared standards and best practices.
4. **Address knowledge gaps:** Investing in research to fill knowledge gaps in nanomaterial toxicity, exposure quantification, and environmental behavior is crucial for informed regulatory decisions.
5. **Implement reporting and labelling schemes:** Adopting compulsory reporting schemes for commercial products containing nanomaterials can enhance transparency and facilitate regulatory oversight (Isibor, 2024; Allan et al., 2021; Vega-Baudrit et al., 2023).

The rapid evolution and commercialization of nanotechnology have outstripped the establishment of comprehensive global regulatory frameworks, leading to a fragmented landscape that poses significant challenges for companies operating across multiple jurisdictions. The lack of harmonized international standards and regulations for nanomaterials not only hampers innovation but also raises concerns about environmental, health, and safety risks (Marchant et al., 2008).

Despite the global nature of nanotechnology applications, regulatory approaches to nanomaterials vary widely among countries. This divergence stems from differences in risk assessment methodologies, definitions of nanomaterials, and regulatory thresholds. For instance, the EU has implemented explicit requirements for nanomaterials under guidelines such as REACH (Registration, Evaluation, Authorization, and Restriction of Chemicals), while other regions may lack dedicated nanomaterial regulations, leading to inconsistencies in safety evaluations and market approvals (Rasmussen et al., 2025).

International organizations like the International Organization for Standardization (ISO) and the Organization for Economic Co-operation and Development (OECD) have initiated efforts to develop standardized testing methods and guidelines for nanomaterials (Rasmussen et al., 2019; Nielsen et al., 2023). The ISO Technical Committee 229 focuses on standardization in the field of nanotechnologies, addressing both scientific and regulatory needs (Sharifi et al., 2022). Similarly, Bleeker et al. (2023) reported that the OECD's operational delegation on assembled nanomaterials aims to promote international collaboration in human health and environmental protection aspects of nanomaterials. However, the adoption and implementation of these standards at the national level remain uneven, limiting their effectiveness in achieving true harmonization (Isibor, 2024).

14.3.4 Challenges in Achieving Harmonization

Several challenges impede the harmonization of nanotechnology regulations. Some of the identified challenges by Souto et al. (2024), Sarkar et al. (2023), and Doak et al. (2023) are highlighted below:

1. **Diverse definitions and classifications:** Countries often adopt different definitions for what constitutes a nanomaterial, leading to discrepancies in regulatory scope and applicability. For example, variations in size thresholds and consideration of specific properties can result in certain materials being regulated in one jurisdiction but not in another.
2. **Variability in risk assessment protocols:** The lack of standardized risk assessment methodologies leads to inconsistent evaluations of nanomaterial safety. Differences in testing protocols, exposure scenarios, and hazard identification can result in divergent regulatory decisions.
3. **Regulatory gaps and overlaps:** In some regions, existing chemical regulations are adapted to include nanomaterials, while others develop specific nanotechnology regulations. This patchwork approach can create regulatory gaps or overlaps, causing confusion for industry stakeholders and potentially compromising safety standards.
4. **Rapid technological advancements:** The swift pace of nanotechnology innovation challenges regulators' ability to keep up with emerging applications and associated risks. This lag can result in outdated regulations that do not adequately address current technological realities.

14.3.5 Initiatives Toward Harmonization

Several researchers have suggested ways to address the challenges of harmonization as stated above. Several initiatives have been proposed; below are some of the proposed or implemented initiatives (Halamoda-Kenzaoui et al., 2022; Bleeker et al., 2023; Oualikene-Gonin et al., 2023; Rodríguez-Gómez et al., 2023).

International regulatory cooperation: Bilateral and multilateral collaborations among regulatory agencies aim to share data, align safety assessments, and develop common regulatory approaches. For instance, the EU and the US have engaged in dialogues to harmonize nanomaterial regulations, focusing on mutual recognition of testing methods and standards.

Development of harmonized standards: Efforts to create universally accepted standards for nanomaterial characterization, testing, and risk assessment are underway. The ISO and OECD continue to play pivotal roles in this endeavor, working toward guidelines that can be adopted globally.

Capacity building and information sharing: Enhancing the capabilities of regulatory bodies, especially in developing countries, through training and resource sharing can promote more uniform regulatory practices. Platforms for exchanging information on best practices, regulatory updates, and scientific findings are essential for global harmonization.

Stakeholder engagement: Involving industry, academia, and civil society in the regulatory process ensures that diverse perspectives are considered, leading to more balanced and widely accepted regulations.

14.4 Stakeholder Engagement in Nanotechnology Governance

Achieving a sustainable economy through nanotechnology necessitates the energetic engagement of diverse stakeholders, including scientists, industry representatives, environmental groups, policymakers, and the public. Stakeholder engagement ensures that regulations are transparent, inclusive, and aligned with both scientific advancements and societal needs (Porcari et al., 2019).

A multi-stakeholder study by Porcari et al. (2019) examined perceptions of nanomaterials (NMs) and nano-related inventions among over 3,000 stakeholders across 15 countries. The study highlighted a growing harmony on the significance of scientific, unbiased, and trustworthy information concerning the potential environmental and health impacts of NMs. Stakeholders expressed a strong desire for internationally harmonized and robust regulations for nanomaterials, improved scientific evidence on nanomaterial hazards, and clear guidelines on their safe use. Engaging diverse stakeholders, including scientists, industry leaders, environmental organizations, and policymakers, is essential for the responsible development and regulation of nanotechnology. The inclusion of multiple perspectives helps to balance risks and benefits, ensuring that nanotechnology aligns with societal needs and sustainability goals (Amutha et al., 2024; Koltsov, 2019; Devasahayam, 2017). A risk governance methodology has been instrumental in Australia, where interdisciplinary collaborations among scientists, policymakers, and civil society have enhanced regulatory frameworks (Malakar et al., 2023). This model emphasizes the importance of transdisciplinary engagement in shaping policies that mitigate environmental and ethical concerns.

Additionally, stakeholder engagement fosters public trust and facilitates the integration of emerging technologies into regulatory frameworks (Schraven et al., 2021). The Responsible Innovation (RI) framework, encompassing anticipation, inclusion, reflexivity, and responsiveness, has been instrumental in addressing ethical, legal, and social implications (ELSI) of nanotechnology applications in diverse sectors, comprising food and agriculture (Grieger et al., 2022; Kokotovich et al., 2021; Malik et al., 2023). Regulatory policies must incorporate mechanisms for continuous stakeholder engagement to adapt to evolving technological advancements and societal expectations.

14.5 Risk Assessment and Management of Nanomaterials

14.5.1 Risk Governance and Regulatory Challenges

Risk assessment is a systematic process that combines data on hazards, exposures, and dose–response relationships to characterize potential risks within a population, such as workers handling nanomaterials. Despite advancements in occupational safety and health (OSH) research over the past decade, significant gaps persist in our understanding of nanomaterials' health effects and exposure levels. Risk governance is critical in nanotechnology regulation as it balances innovation with safety. Public and stakeholder perceptions of nanotechnology risks influence regulatory approaches, making it imperative to develop governance structures that are evidence-based and inclusive (Porcari et al., 2019). In Australia, regulatory bodies have adopted a dynamic risk governance approach, integrating interdisciplinary perspectives to assess and manage risks (Malakar et al., 2023). Unlike traditional risk governance models, which primarily focus on physical hazards, nanotechnology regulations must also address social, ethical, and environmental risks. The rapid pace of nanotechnology development often outpaces regulatory capabilities, creating gaps in risk assessment and policy implementation. Moreover, overlapping roles and mandates among regulatory agencies can lead to inefficiencies, necessitating a more coordinated and global regulatory framework (Amutha et al., 2024).

To effectively manage risks, nanotechnology industries must integrate safety-by-design approaches, comprehensive exposure monitoring, and regulatory compliance. Giubilato et al. (2020) propose a risk management framework (RMF) that integrates risk assessment with life cycle analysis to address hypothetical environmental and health risks linked with nano-biomaterials. This model supports decision-making for safer product development and regulatory alignment.

In workplace settings, engineering controls, personal protective equipment (PPE), and exposure reduction strategies are critical for protecting workers from nanoparticle exposure (Kuempel et al., 2012). Additionally, regulatory agencies must establish stricter pre-market testing guidelines to ensure the safe commercialization of nanomaterials.

14.5.2 The Role of Regulation in Sustainable Nanotechnology Innovation

Regulation portrays a dual role in nanotechnology invention. It can act as both a barricade and a propeller of responsible scientific advancements. In the food and agriculture sector, for example, stringent safety regulations have historically influenced the trajectory of innovation (Amenta et al., 2015). However, these regulations must evolve to encompass broader sustainability considerations, ensuring that nano-enabled solutions contribute to environmental and economic sustainability (Shukla et al., 2024).

Stakeholder studies in the US have revealed a spectrum of opinions vis-à-vis the role of guidelines in innovation, ranging from concerns about overregulation hindering technological advancements to perspectives that view guidelines as a necessary driver of responsible innovation (Grieger et al., 2022). A balanced regulatory approach that integrates RI principles can enhance the sustainability of nanotechnology applications, particularly in industries where environmental impact is a significant concern.

14.5.3 Ethical and Social Reflections in Nanotechnology Regulation

Ethical and social consequences of nanotechnology must be integrated into regulatory frameworks to promote responsible development. Governance approaches should prioritize transparency, public engagement, and ethical considerations to address societal concerns. In the context of nanotechnology risk governance, Australian research has demonstrated the importance of interdisciplinary approaches in bridging the gap among science, policy, and public perception (Malakar et al., 2023).

Moreover, international collaboration is essential in establishing standardized regulations that ensure the ethical use of nanotechnology. As governance mechanisms evolve, policymakers must consider public perception, predictive risk assessment tools, and global regulatory harmonization to facilitate responsible nanotechnology development (Wasti et al., 2023; Csóka et al., 2021; Amutha et al., 2024). Achieving a green economy through nanotechnology requires a proactive, adaptive, and globally coordinated regulatory framework. Stakeholder engagement, risk governance, and ethical considerations must be central to policy development to safeguard the safe and sustainable incorporation of nanotechnology. Regulation should not merely act as a compliance mechanism but should be leveraged as a tool to drive responsible innovation and societal benefits (Tripathy et al., 2024).

Future regulatory frameworks should prioritize interdisciplinary collaboration, transparency, and the use of advanced technologies for real-time risk assessment (Gottardo et al., 2021). Additionally, global regulatory harmonization can facilitate knowledge exchange and create standardized guidelines for the responsible governance of nanotechnology, and nanotechnology can contribute significantly to a sustainable economy by minimizing potential risks through the adoption of a comprehensive approach that integrates scientific, ethical, and policy perspectives (Kendal, 2022). Nanomaterials present unique challenges in risk assessment and administration due to their distinct physicochemical properties, like high reactivity, increased surface area, and potential for bioaccumulation. Traditional toxicological and risk assessment models might not be fully appropriate to nanomaterials, necessitating the development of specialized approaches.

In 2012, Kuempel et al. highlighted the importance of integrating hazard, exposure, and dose–response data in risk assessment for nanomaterials. The study emphasizes the need for workplace safety protocols, including engineering controls and PPE, to mitigate potential risks. However, challenges remain in developing standardized measures for workers' exposure and validating risk mitigation strategies. Research gaps also exist in chronic toxicity prediction and the long-term environmental impacts of nanomaterials (Kuempel et al., 2012).

Similarly, Schwirn et al. (2020) discuss the challenges associated with the EU's Registration, Evaluation, Authorization, and Restriction of Chemicals (REACH) regulation concerning nanomaterials. The study identifies key difficulties in classifying nanomaterials, assessing exposure concentrations in aquatic systems, and addressing the solubility of nanoforms. The authors propose pragmatic solutions, including improved testing methodologies and more refined exposure assessments, to enhance regulatory compliance (Schwirn et al., 2020).

14.6 Policy and Regulatory Approach in Achieving a Green Economy through Nanotechnology

Nanotechnology has appeared as a transformative field, suggesting innovative solutions for sustainable economic growth. However, achieving a green economy through nanotechnology requires robust policy and regulatory frameworks to ensure environmental safety, public health, and responsible innovation (Iavicoli et al., 2014; Kraegeloh et al., 2018). Governance mechanisms, including stakeholder engagement, risk governance, and ethical reflections, play a pivotal role in shaping the trajectory of nanotechnology (Amutha et al., 2024). Regulatory policies for nanomaterials vary globally, with frameworks such as the EU's REACH, the US Environmental Protection Agency's (EPA) Toxic Substances Control Act (TSCA), and guidelines from the World Health Organization (WHO) playing significant roles.

Under the REACH regulation, nanomaterials are classified based on their physicochemical properties, exposure potential, and toxicity profiles. The recent amendments, effective from January 2020, emphasize stricter risk assessment protocols, requiring industries to provide extensive safety data on nanomaterials (Schwirn et al., 2020). Despite these advancements, challenges persist in the harmonization of testing methods and regulatory classifications. In the medical sector, nano-biomaterials used in therapeutic devices and innovative therapy medicinal products (ITMP) are subject to regulatory scrutiny. The EU BIORIMA project developed an RMF to guide regulatory decisions, emphasizing integrated approaches to testing and assessment (Giubilato et al., 2020). This framework ensures that patient, occupational, and environmental risks are systematically evaluated throughout a product's life cycle.

14.6.1 Administering Frameworks for Nanomaterials

Several regulatory frameworks have been proposed to address the environmental and health risks of nanomaterials. A review by Grieger and co-workers evaluates different risk analysis frameworks and identifies a need for a multi-faceted approach. Most existing frameworks focus on occupational risk assessment while neglecting broader environmental implications. Additionally, these frameworks require further validation across various nanomaterials and application contexts (Grieger et al., 2022). The integration of nanotechnology into medical gadgets and pharmaceutical products also presents unique regulatory challenges. In 2020, Giubilato et al. proposed a risk management context for nanomaterials utilized in medical gadgets, accentuating the need for a widespread approach that reflects both human health and environmental safety. This framework is aligned with the EU's regulations and focuses on life cycle assessment, safe-by-design principles, and risk–benefit analysis (Giubilato et al., 2020).

Nanotechnology has the prospective to influence environmental sustainability via applications such as contamination remediation, water sanitization, and energy efficiency. However, the lack of standardized environmental risk assessments remains a challenge. Amorim et al. (2020) emphasize the need for updated environmental hazard testing methodologies under the Organization for Economic Co-operation and Development (OECD) guidelines. Their study outlines critical gaps in assessing the toxicity of nano-biomaterials and provides recommendations for refining testing protocols (Amorim et al., 2020).

Moreover, the European Medicines Agency (EMA) requires environmental risk assessments for pharmaceutical nanomaterials. This highlights the growing recognition of potential environmental hazards and the need for comprehensive management tactics to allay risks (Amorim et al., 2020). Achieving a green economy through nanotechnology requires international harmonization of regulations to safeguard consistency in risk assessment, safety standards, and ethical considerations. Regulatory groups such as the European Chemicals Agency (ECHA), the US Environmental Protection Agency (EPA), and the International Organization for Standardization (ISO) must collaborate to develop globally accepted guidelines.

Policy recommendations include:

1. **Standardized testing methods** – Developing harmonized testing methodologies for nanomaterials to facilitate risk assessment and regulatory compliance.
2. **Life cycle assessments (LCA)** – Encouraging industries to conduct LCAs to evaluate the environmental footprint of nanomaterials from production to disposal.
3. **Public transparency and stakeholder engagement** – Increasing transparency in nanotechnology research and involving stakeholders, including policymakers, industry leaders, and environmental advocates, in regulatory decision-making.
4. **Safe-by-design methods** – Promoting the advancement of nanomaterials with built-in safety features to minimize potential hazards.
5. **Enhanced monitoring and surveillance** – Establishing long-term monitoring programs to track the environmental and health impacts of nanomaterials

14.6.2 Stakeholder Engagement and Public Experience

Stakeholder engagement and public experience play a fundamental role in shaping nanotechnology policies and regulations in the world (Gaskell et al., 2005). A green economy helps with long-term growth, less damage to the environment, and sustainable development. Nanotechnology also offers new ways to solve problems such as cleaner energy technologies, advanced materials, and pollution remedies (Iavicoli et al., 2014; Kumar et al., 2024). Still, the application of nanotechnology raises questions regarding ethics, safety, and long-term environmental effects; therefore, public knowledge and stakeholder involvement become even more important (Babatunde et al., 2020; Verma et al., 2024). A study of stakeholder engagement and insights is vital to guaranteeing that adequate risk management systems are in place for nanotechnology and nano-related outputs or products. In reply, copious studies have been conducted to examine stakeholders' opinions of nanotechnology and nano-associated products across the past 15 years (Porcari et al., 2019). Excellent stakeholder involvement guarantees that the policy-making process considers the interests and worries of different stakeholders, including industry, government, civil society, and the public (Le Anh et al., 2019; Khatoon et al., 2025).

The way the public views nanotechnology is also effective since it can affect both the adoption of nanotechnology and policy decisions (Siegrist et al., 2007). A wide range of factors, including media coverage, cultural beliefs, and past technological experiences, affect how the public views nanotechnology (Rathore and Mahesh, 2021). According to several surveys, there is a great deal of concern regarding the unknown health and ecological risks associated with nanotechnology, even while many people are hopeful about its potential to address environmental challenges (Lee et al., 2005). Public skepticism may be increased in underdeveloped nations because of historical suspicion of industrial technologies and restricted access to scientific information (Rutjens et al., 2022). These perceptions are significantly shaped by media portrayals (Rathore and Mahesh, 2021).

Even when there is no proof of harm, sensational reports about nanoparticles in food or cosmetics might increase anxieties. For instance, public forums and educational outreach have been effectively employed by campaigns and surveys in the US and the UK to demystify nanotechnology and emphasize its role in pollution management and renewable energy (Pidgeon et al., 2009). Establishing trust with stakeholders and the public requires open and honest communication regarding the advantages and dangers of nanotechnology (Kearnes et al., 2004). This entails conducting inclusive and transparent decision-making procedures and disseminating factual and objective information regarding the possible drawbacks and advantages of nanotechnology (Renn, 2008).

For nanotechnology, several stakeholder involvement methods have been proposed. These models made it abundantly evident that early and ongoing stakeholder engagement is crucial, as is the requirement for transparency and participation in the decision-making processes (Jeffcoat et al., 2024; Roco et al., 1970). To increase public understanding of nanotechnology and its possible social effects, a few public engagement campaigns have also been started (Gaskell et al., 2005). Policies governing nanotechnology for a green economy are heavily influenced by public opinion and stakeholder participation (Iavicoli et al., 2014; Gaskell et al., 2005). Building confidence and trust among stakeholders and the public requires effective stakeholder engagement and open communication (Nygaard et al., 2021). Policymakers can pledge the responsible and sustainable development and handling of nanotechnology by interacting with stakeholders and the public (Nygaard et al., 2021; Jeffcoat et al., 2024).

14.7 Compliance and Enforcement

Responsible and sustainable development and application of nanotechnology depend on excellent compliance and enforcement measures (Gottardo et al., 2021). Compliance with relevant laws, rules, and industry codes is not merely a legal requirement but also essential in ensuring product safety and quality, environmental protection, and human health (Bolanle et al., 2024). The enforcement of public policies on compliance and implementation is critical in ensuring safe public health concerning the application of nanotechnology. Such compliance and enforcement involve a mix of setting up regulatory frameworks, together with monitoring compliance and implementing corrective actions in case there are breaches. Due to the nature of nanotechnology, which is well within the microscopic scale and spans from energy to waste management, compliance, and regulation, it offers both challenges and opportunities (Godwin et al., 1970; Gottardo et al., 2021).

The first step of compliance encompasses formulating comprehensive science-informed regulations. Most nations are at the outset of formulating protocols and policies for nanotechnology, and as such, the approaches differ from one country to another (Renn, 2008). Most, however, have a defined framework that emphasizes the precautionary principle concerning nanotechnology risk management. (O'Mathúna,

2011). For instance, the EU's REACH regulation is unique in the fact that it mandates manufacturers to register and evaluate for potential harm that nanomaterials pose to human health and the ecosystem (European Parliament, 2006). The EU has been proactive with its provision for REACH, as they have specific guidelines regarding the use of nanomaterials. REACH stipulates that producers and importers must obtain information regarding the properties and risks of substances to ensure that they are safe for use, including nanomaterials (European Parliament, 2006).

14.7.1 Compliance Mechanisms

Nanotechnology has compliance mechanisms, including self-regulation: business initiatives to create and apply standards and policies for nanotechnology use (National Nanotechnology Initiative, 2001); government regulation: laws and policies issued by state authorities and put into effect by state bodies (Reese, 2013; Renn, 2008; Khatoon et al., 2025); co-regulation: joint action of state and business in the formulation and enforcement of rules and policies (Kearnes et al., 2004; Reese, 2013). Enforcement in nanotechnology faces unique challenges due to the complexity of nanomaterials and the specialized methods of detecting and measuring those nanomaterials (Allan et al., 2021; Organization for Economic Co-operation and Development, 2020).

Enforcement mechanisms for nanotechnology are listed as follows:

> Inspections and audits: Consistent examination and accounting of compliance with policies and procedures (Sarahan, 2008). Penalties and fines: Financial charges for disobeying policies and standards (Vega-Baudrit et al., 2023). Public disclosure: Dissemination of information regarding the potential dangers of nanotechnology to the health of the people and the ecology of the environment (Vega-Baudrit et al., 2023). Compliance and enforcement mechanisms are fundamental in dealing with the expansion and application of nanotechnology (Gottardo et al., 2021). Constructive regulations, compliance mechanisms, and implementation mechanisms ought to be put in place to address the concerns posed by nanotechnology and put its innovation into use in achieving a green economy (Kumari et al., 2023).

14.7.2 International Cooperation

The development of nanotechnology has been much aided by international cooperation (Zheng et al., 2014). International cooperation is also vital for confronting the global challenges and prospects portrayed by nanotechnology (Cloete, 2024). The transition toward a green economy with the help of nanotechnology calls for cooperation on a global scale, as there are shared environmental issues, the impossibility of international borderless technology, and the need for an overarching law (Iavicoli et al., 2014; Babatunde et al., 2020; Devade et al., 2024).

Science and technology (S&T) developments from multiple nations are no longer limited to the S&T development of a single country. International cooperation has been actively encouraged by many nations to exchange ideas, create initiatives, and take part in research communities. According to recent research, both the quantity of intercontinental collaboration nanotechnology patents and their share of all nanotechnology patents have been gradually rising (Zheng et al., 2014). International collaboration is required as nanotechnology develops and spreads throughout the world to:

1. **Harmonize regulations:** Design and implement one regulation and standard in all countries, which may allow for greater investment and innovation trade (Marchant et al., 2010).
2. **Addressing global challenges:** Respond to climate change, lack of water, and food shortages by working together and encouraging diversity in research and development. (Anderson, 2024, Cloete, 2024; Pokrajac et al., 2021).
3. **Share best practices:** Engage governments in developing countries by sharing best practices, exposing them to training, and providing knowledge and skills on issues related to safe and sustainable expansion and the use of nanotechnology (Hankin and Read, 2016; Khatoon et al., 2025).

Several international initiatives have been launched to promote cooperation on nanotechnology:

1. **OECD Nanotechnology Inventiveness:** OECD has established a nanotechnology initiative to encourage global collaboration on nanotechnology safety, regulation, and innovation (Organisation for Economic Co-operation and Development (OECD, 2020).
2. **UNEP Nanotechnology Initiative:** To encourage the cautious and sustainable growth and application of nanotechnology, the United Nations Environment Programme (UNEP) has started a nanotechnology initiative (UNEP, 2025).
3. **ISO Nanotechnology Standards:** The International Organization for Standardization (ISO) has created standards for nanotechnology to promote global innovation and trade (ISO, 2005).

International cooperation is vital for promoting safe and sustainable growth and leveraging nanotechnology toward a green economy. By addressing the challenges and opportunities presented by international cooperation, we can promote innovation, trade, and investment in nanotechnology, and simultaneously support a green economy with a more sustainable future (Iavicoli et al., 2014; Pokrajac et al., 2021; Khatoon et al., 2025).

14.7.3 Public Engagement and Transparency

The expansion and application of nanotechnology in achieving a green economy require a comprehensive approach that involves public engagement and transparency. The subjects of public involvement, engagement, and debate have grown in importance in discussions about nanotechnology policy. Several social scientists in the field of Science and Technology Studies (STS) support upward public commitment, which has been incorporated into official discourse, particularly in the UK and throughout Europe (Laurent, 2008).

Over the past few decades, governmental and non-governmental organizations, as well as the local, scientific, and policy-making communities, have provided substantial support for public involvement and engagement initiatives (Khatoon et al., 2025).

Public involvement in policy-related issues promotes democratic participation and serves the public interest by balancing special interests in decision-making, particularly when all facets of society are represented (Suphattanakul, 2018; Bobbio, 2019; Scheinerman, 2023). Public engagement has provided a mechanism through which public confidence can be restored through good decision-making, fostering better transparency and legitimacy when both supportive and critical points are raised, and issues are deliberated broadly (Suphattanakul, 2018; Bobbio, 2019; Scheinerman, 2023; Davies et al., 2009).

For the public and stakeholders to make educated judgments regarding nanotechnology, transparency is essential to its progress (Morris et al., 2011; Liguori et al., 2016; Rasmussen et al., 2025). This can be accomplished by disclosing details on the study and advancement of nanotechnology, including any possible drawbacks or advantages. The emergence of interest in nanotechnology has sparked a heated discussion on responsibility and public involvement (Davies et al., 2009; Palencia et al., 2022). Numerous important studies have looked at various facets of how the populace views nanotechnology over the last six years. Crucially, and predictably, a survey revealed that many developed nations' populations have little knowledge about nanotechnology, with 80% of those surveyed saying they had little or no knowledge about nanotechnology (Davies et al., 2009; Suphattanakul, 2018; Bobbio, 2019).

Science policy has historically relied on key stakeholders, such as business interests, universities, or the healthcare industry. But now, the context is more multinational and multidisciplinary. A more integrated approach to research and society is emerging, with sustainability serving as a focal point for action (Morris et al., 2011; Ketprapakorn and Kantabutra, 2022; Palencia et al., 2022; Rasmussen et al., 2025).

Public engagement and transparency initiatives are as follows:

1. **Environmental Technologies Action Plan:** This plan aims to strengthen public recognition and understanding of nanotechnology and to involve citizens in the decision-making process (Bräutigam, 2007).
2. **National Nanotechnology Initiative (NNI):** The NNI is a US government initiative that aims to promote the development of nanotechnology and includes a public engagement component, which involves public consultations and outreach activities (Scheinerman, 2023; Bräutigam, 2007; Rodríguez-Gómez et al., 2023; Khatoon et al., 2025).

Public engagement and transparency are crucial components of a complete approach to building a green economy using nanotechnology. The NNI and the Environmental Technologies Action Plan are two examples of public involvement and transparency programs that highlight the significance of including citizens in the decision-making process (Bobbio, 2019; Scheinerman, 2023; Bräutigam, 2007; Laurent, 2008; Rodríguez-Gómez et al., 2023; Khatoon et al., 2025).

14.8 Conclusion

Achieving a green economy through nanotechnology faces policy and regulatory challenges, including ensuring the safety and sustainability of nanomaterials while fostering innovation, necessitating a robust regulatory framework that addresses potential risks and encourages responsible development. To enable society to have a green economy, the concept of "green nanotechnology" proposes exploiting nano-novelties in materials knowledge and engineering to create products and procedures that are energy resourceful as well as economically and ecologically sustainable. For the public and stakeholders to make educated judgments regarding nanotechnology, transparency is essential to its progress. This can be accomplished by disclosing details on the study and advancement of nanotechnology, including any possible drawbacks or advantages. However, the lack of uniformity has led to regulatory uncertainties, increased compliance costs, and hypothetical risks to public health and the environment. The complexity and diversity of nanomaterials have made it difficult to establish a universal regulatory definition, complicating efforts to monitor and predict their environmental fate. Therefore, international harmonization is mandatory for regulatory definitions and testing protocols for nanomaterials.

14.9 Authors' Contributions

All authors contributed to the development of the chapter.

14.10 Disclosure Statement

The authors of this chapter have no potential conflict of interest to report.

REFERENCES

Allan, J., Belz, S., Hoeveler, A., Hugas, M., Okuda, H., Patri, A., Rauscher, H., Silva, P., Slikker, W., Sokull-Kluettgen, B., Tong, W., Anklam, E. (2021). Regulatory landscape of nanotechnology and nanoplastics from a global perspective. *Regulatory Toxicology and Pharmacology*, 122, 104885. https://doi.org/10.1016/j.yrtph.2021.104885

Amenta, V., Aschberger, K., Arena, M., Bouwmeester, H., Botelho Moniz, F., Brandhoff, P., Gottardo, S., Marvin, H.J., Mech, A., Quiros Pesudo, L., Rauscher, H., Schoonjans, R., Vettori, M.V., Weigel, S., Peters, R.J. (2015). Regulatory aspects of nanotechnology in the agri/feed/food sector in EU and non-EU countries. *Regulatory Toxicology and Pharmacology*, 73(1), 463–476. https://doi.org/10.1016/j.yrtph.2015.06.016.

Amorim, M.J.B., Fernández-Cruz, M.L., Hund-Rinke, K., Scott-Fordsmand, J.J. (2020). Environmental hazard testing of nanobiomaterials. *Environmental Sciences Europe*, 32, 101. https://doi.org/10.1186/s12302-020-00369-8

Amutha, C., Gopan, A., Pushbalatatha, I., Ragavi, M., Reneese, J.A. (2024). Nano-technology and governance: Regulatory framework for responsible innovation. In: George, S.C., Tawiah, B. (eds) *Nanotechnology in societal development*. Advanced Technologies and Societal Change. Singapore: Springer. https://doi.org/10.1007/978-981-97-6184-5_14

Anderson, K. (2024). International cooperation key to combating the climate crisis. Leaf by Greenly. https://greenly.earth/en-us/blog/ecology-news/international-cooperation-key-to-combating-the-climate-crisis (Accessed on March 22, 2025).

Anwar, W., Khitab, A. (2019). Nanotechnology from engineers to toxicologists: Risks and remedial measures. *International Journal of Applied Nanotechnology Research*, 4(2), 1–25. https://doi.org/10.4018/IJANR.2019070101

Babatunde, D.E., Denwigwe, I.H., Babatunde, O.M., Gbadamosi, S.L., Babalola, I.P., Agboola, O. (2020). Environmental and societal impact of nanotechnology. *IEEE Access*, 8, 4640–4667. https://doi.org/10.1109/ACCESS.2019.2961513

Baer, D.R., Engelhard, M.H., Johnson, G.E., Laskin, J., Lai, J., Mueller, K., Munusamy, P., Thevuthasan, S., Wang, H., Washton, N., Elder, A., Baisch, B.L., Karakoti, A., Kuchibhatla, S.V., Moon, D. (2013, September). Surface characterization of nanomaterials and nanoparticles: Important needs and challenging opportunities. *Journal of Vaccum Science & Technology A*, 31(5), 50820. https://doi.org/10.1116/1.4818423.

Bleeker, E.A.J., Swart, E., Braakhuis, H., Fernández Cruz, M.L., Friedrichs, S., Gosens, I., Herzberg, F., Jensen, K.A., von der Kammer, F., Kettelarij, J.A.B., Navas, J.M, Rasmussen, K., Schwirn, K., Visser, M. (2023). Towards harmonisation of testing of nanomaterials for EU regulatory requirements on chemical safety – A proposal for further actions. *Regulatory Toxicology and Pharmacology*, 139, 105360. https://doi.org/10.1016/j.yrtph.2023.105360

Bobbio, L. (2019). Designing effective public participation. *Policy and Society*, 38(1), 41–57. https://doi.org/10.1080/14494035.2018.1511193

Bolanle, M., Ikre, W., Olabiyi, W. (2024). *Adhering to regulatory requirements and industry standards* (Vol. 1). Ladoke Akintola University of Technology.

Bräutigam, T. (2007). European Union Law. *European Journal of International Law*, 18(2), 377–378. https://doi.org/10.1093/ejil/chm018

Castillo, A.P.D. (2013). *Nanomaterials and workplace health & safety: What are the issues for workers?* Brussels: European Trade Union Institute, Pg. 01–44.

Castillo, A.P.D. (2019). Training for workers and safety representatives on manufactured nanomaterials. *New Solutions*, 29(1), 36–52. https://doi.org/10.1177/1048291119830085

Chávez-Hernández, J.A., Velarde-Salcedo, A.J., Navarro-Tovar, G., Gonzalez, C. (2024). Safe nanomaterials: From their use, application, and disposal to regulations. *Nanoscale Advances*, 6, 1583–1610. https://doi.org/10.1039/D3NA01097J

Cloete, K. (2024). *Nanotech diplomacy: Opportunities and challenges for international cooperation*. UNESCO-UNISA Africa Chair in Nanosciences and Nanotechnology Laboratories, University of South Africa (UNISA).

Csóka, I., Ismail, R., Jójárt-Laczkovich, O., Pallagi, E. (2021). Regulatory considerations, challenges and risk-based approach in nanomedicine development. *Current Medicinal Chemistry*, 28(36), 7461–7476. https://doi.org/10.2174/0929867328666210406115529

Cui, D., Tian, F., Ozkan, C.S., Wang, M., Gao, H. (2005). Effect of single wall carbon nanotubes on human HEK293 cells. *Toxicology Letters*, 155, 73–85. https://doi.org/10.1016/j.toxlet.2004.08.015.

Danchin, A. (2022). *In vivo, in vitro*, and *in silico*: An open space for the development of microbe-based applications of synthetic biology. *Microbial Biotechnology*, 15(1), 42–64. https://doi.org/10.1111/1751-7915.13937

Davies, S., Macnaghten, P., Kearnes, M. (eds). (2009). *Reconfiguring Responsibility: Deepening Debate on Nanotechnology*. Durham: DurhamUniversity.

Devade, K., Singh, P.K., Kumar, S., Kumar, H., Prasad, B., Rao, A.L.N., Sankhyan, A. (2024). Green nanotechnology based sustainable energy solutions and environmental impacts. *E3S Web of Conferences*, 511(1), 12. https://doi.org/10.1051/e3sconf/202451101031

Devasahayam, S. (2017). Overview of an internationally integrated nanotechnology governance. *International Journal of Metrology and Quality Engineering*, 8, 8. https://doi.org/10.1051/ijmqe/2017002

Dimo, S.N., Obidi, O.F., Nejo, A.O., Olaleru, S.A., Ejidike, I.P., Adetona, A.J. (2024). Biofabrication, spectroscopic, and photocatalytic studies of titania nanoparticles mediated by *Proteus mirabilis* strain NG-ABK-32 for smart applications. *Smart Science*, 12, 373–386. https://doi.org/10.1080/23080477.2024.2338651

Do, H.T.T., Nguyen, N.P.U., Saeed, S.I., Dang, N.T., Nguyen, T.T.H. (2025). Advances in silver nanoparticles: Unraveling biological activities, mechanisms of action, and toxicity. *Applied Nanoscience* 15, 1. https://doi.org/10.1007/s13204-024-03076-5

Doak, S.H., Andreoli, C., Burgum, M.J., Chaudhry, Q., Bleeker, E.A.J., Bossa, C., Domenech, J., Drobne, D., Fessard, V., Jeliazkova, N., Longhin, E., Rundén-Pran, E., Stępnik, M., El Yamani, N., Catalán, J., Dusinska, M. (2023). Current status and future challenges of genotoxicity OECD Test Guidelines for nanomaterials: A workshop report. *Mutagenesis*, 38(4), 183–191. https://doi.org/10.1093/mutage/gead017

Ejidike, I.P., Bamigboye, O.M., Ijimdiya, R.U., Seyinde, D.O., Ojo, O.O. (2020). Synthesis, characterization and biological evaluation of hexagonal wurtzite structured ZnO nanoparticle from Zn(II)-Schiff base complex. *Proceedings of the Nigerian Academy of Science*, 13, 136–147. https://doi.org/10.57046/YJHW2059

Ejidike, I.P., Clayton, H.S. (2022). Green synthesis of silver nanoparticles mediated by Daucus carota L.: Antiradical, antimicrobial potentials, *in vitro* cytotoxicity against brain glioblastoma cells. *Green Chemistry Letters and Reviews*, 15, 297–310. https://doi.org/10.1080/17518253.2022.2054290

Ejidike, I.P., Direm, A., Parlak, C., Bamigboye, M.O., Oluade, O., Adetunji, J.B., Ata, A., Eze, M.O., Hollett, J.W., Clayton, H.S. (2024b). Cadmium oxide nanoparticles from new organometallic Cd(II)-Schiff base complex and *in vitro* biological potentials: Dual *S. aureus* and *E. coli* DNA gyrase inhibition by the precursors via *in-silico* binding modes' study. *Research on Chemical Intermediates*, 50, 2763–2791. https://doi.org/10.1007/s11164-024-05291-9

Ejidike, I.P., Ijimdiya, R.U., Emmanuel-Akerele, H.A., Emmanuel, G.C., Ejidike, O.M., Bamigboye, M.O., Seyinde, D.O., Olaleru, A., Tanimowo, W.O., Awolope, R.O. (2023). Biosynthesis, characterization, and antimicrobial assessment of metal nanoparticles from *Dryopteris manniana* (HOOK.) C. Chr leaf extract. *Bulletin of Pharmaceutical Sciences Assiut University*, 46, 225–238. https://doi.org/10.21608/BFSA.2023.300937

Ejidike, I.P., Ogunleye, O., Bamigboye, M.O., Ejidike, O.M., Ata, A., Eze, M.O., Clayton, H.S., Nwankwo, V.U., Fatokun, J.O. (2024a). Role of nanotechnology in medicine: Opportunities and challenges. In: Shah, M.P., Bharadvaja, N., Kumar, L. (eds) *Biogenic nanomaterials for environmental sustainability: Principles, practices, and opportunities. Environmental science and engineering.* Cham: Springer, Pg. 353–375. https://doi.org/10.1007/978-3-031-45956-6_14

Elzein, B. (2024). Nano revolution: "Tiny tech, big impact: How nanotechnology is driving SDGs progress". *Heliyon*, 10(10), e31393. https://doi.org/10.1016/j.heliyon.2024.e31393.

European Parliament. (2006). REACH regulation – European Commission. https://environment.ec.europa.eu/topics/chemicals/reach-regulation_en

Gaskell, G., Eyck, T., Jackson, J., Veltri, G. (2005). Imagining nanotechnology: Cultural support for technological innovation in Europe and the United States. *Public Understanding of Science*, 14(1), 81–90. https://doi.org/10.1177/0963662505048949

Ghamarpoor, R., Fallah, A., Jamshidi, M (2023). Investigating the use of titanium dioxide (TiO2) nanoparticles on the amount of protection against UV irradiation. *Scientific Reports*, 13(1), 9793. https://doi.org/10.1038/s41598-023-37057-5.

Giubilato, E., Cazzagon, V., Amorim, M.J.B., Blosi, M., Bouillard, J., Bouwmeester, H., Costa, A.L., Fadeel, B., Fernandes, T.F., Fito, C., Hauser, M., Marcomini, A., Nowack, B., Pizzol, L., Powell, L., Prina-Mello, A., Sarimveis, H., Scott-Fordsmand, J.J., Semenzin, E., Stahlmecke, B., Stone, V., Vignes, A., Wilkins, T., Zabeo, A., Tran, L., Hristozov, D. (2020). Risk management framework for nano-biomaterials used in medical devices and advanced therapy medicinal products. *Materials (Basel)*, 13(20), 4532. https://doi.org/10.3390/ma13204532

Gobinath, E., Dhatchinamoorthy, M., Saran, P., Vishnu, D., Indumathy, R., Kalaiarasi, G. (2023). Synthesis and characterization of NiO nanoparticles using *Sesbania grandiflora* flower to evaluate cytotoxicity. *Results in Chemistry*, 6, 101043. https://doi.org/10.1016/j.rechem.2023.101043

Godwin, H.A., Angeles, L., Morris, J., States, U., Protection, E., Scott, N., Wiesner, M. (1970). Nanotechnology environmental health and safety issues. https://doi.org/10.1007/978-94-007-1168-6

Gottardo, S., Mech, A., Drbohlavová, J., Małyska, A., Bøwadt, S., Riego Sintes, J., Rauscher, H. (2021). Towards safe and sustainable innovation in nanotechnology: State-of-play for smart nanomaterials. *NanoImpact*, 21, 100297. https://doi.org/10.1016/j.impact.2021.100297.

Grieger, K., Merck, A., Kuzma, J. (2022). Formulating best practices for responsible innovation of nano-agrifoods through stakeholder insights and reflection. *Journal of Responsible Technology*, 10, 100030. https://doi.org/10.1016/j.jrt.2022.100030

Halamoda-Kenzaoui, B., Geertsma, R., Pouw, J., Prina-Mello, A., Carrer, M., Roesslein, M., Sips, A., Weltring, K.M., Spring, K., Bremer-Hoffmann, S. (2022). Future perspectives for advancing regulatory science of nanotechnology-enabled health products. *Drug Delivery and Translational Research*, 12(9), 2145–2156. https://doi.org/10.1007/s13346-022-01165-y

Haleem, A., Javaid, M., Singh, R.P., Rab, S., Suman, R. (2023). Applications of nanotechnology in medical field: A brief review. *Global Health Journal*, 7, 70–77. https://doi.org/10.1016/j.glohj.2023.02.008

Hankin, S.M., Read, S.A.K. (2016). Governance of nanotechnology: Context, principles and challenges. In: Murphy, F., McAlea, E., Mullins, M. (eds) *Managing risk in nanotechnology. Innovation, technology, and knowledge management*. Cham: Springer. https://doi.org/10.1007/978-3-319-32392-3_3

Iavicoli, I., Leso, V., Ricciardi, W., Hodson, L.L., Hoover, M.D. (2014). Opportunities and challenges of nanotechnology in the green economy. *Environmental Health*, 13, 78. https://doi.org/10.1186/1476-069X-13-78

International Organization for Standardization (ISO). (2005). ISO_TC 229 – nanotechnologies. www.iso.org/committee/381983.html (Accessed on March 10, 2025)

Isibor, P.O. (2024). Regulations and policy considerations for nanoparticle safety. In: Isibor, P.O., Devi, G., Enuneku, A.A. (eds). *Environmental nanotoxicology*. Cham: Springer, Pg 295–316. https://doi.org/10.1007/978-3-031-54154-4_14

Jeffcoat, P., Di Lernia, C., Hardy, C., New, E.J., Chrzanowski, W. (2024). (Re)imagining purpose: A framework for sustainable nanotechnology innovation. *NanoImpact*, 35, 100511. https://doi.org/10.1016/j.impact.2024.100511

Kearnes, M., Grove-white, R., Miller, P., Macnaghten, P., Wilsdon, J., Wynne, B. (2004). Bio-to-Nano? Learning the lessons, interrogating the comparison. *Nanotechnology, Risk and Sustainability*, 1, 1–27.

Kendal, E. (2022). Ethical, legal and social implications of emerging technology (ELSIET) symposium. *Bioethical Inquiry*, 19, 363–370. https://doi.org/10.1007/s11673-022-10197-5

Keskin, C., Ölçekçi, A., Baran, A., Baran, M.F., Eftekhari, A., Omarova, S., Khalilov, R., Aliyev, E., Sufianov, A., Beilerli, A., Gareev, I. (2023). Green synthesis of silver nanoparticles mediated *Diospyros kaki* L. (Persimmon): Determination of chemical composition and evaluation of their antimicrobials and anticancer activities. *Frontiers in Chemistry*, 11, 1187808. https://doi.org/10.3389/fchem.2023.1187808

Ketprapakorn, N., Kantabutra, S. (2022). Toward an organizational theory of sustainability culture. *Sustainable Production and Consumption*, 32, 638–654. https://doi.org/10.1016/j.spc.2022.05.020

Khatoon, U.T., Velidandi, A. (2025). An overview on the role of government initiatives in nanotechnology innovation for sustainable economic development and research progress. *Sustainability*, 17(3), 1250. https://doi.org/10.3390/su17031250

Khitab, A., Ahmad, S., Munir, M.J., Kazmi, S.M.S., Arshad, T., Khushnood, R.A. (2018). Synthesis and applications of nano titania particles: A review. *Reviews on Advanced Materials Science*, 53(1), 90–105. https://doi.org/10.1515/rams-2018-0007.

Kokotovich, A.E., Kuzma, J., Cummings, C.L. Grieger, K. (2021). Responsible innovation definitions, practices, and motivations from nanotechnology researchers in food and agriculture. *NanoEthics*, 15, 229–243. https://doi.org/10.1007/s11569-021-00404-9

Kolahalam, L.A., Kasi Viswanath, I.V., Diwakar, B.S., Govindh, B., Reddy, V., Murthy, Y.L.N. (2019). Review on nanomaterials: Synthesis and applications. *Materials Today: Proceedings*, 18, 2182–2190. https://doi.org/10.1016/j.matpr.2019.07.371.

Koltsov, D. (2019). Nanotechnology standards for industry, regulators and other stakeholders. Private communication at global Summit on regulatory science 2019 on nanotechnology and nanoplastics. EUR 30195 EN.

Korada, V.S., Hamid, N.H.B. (2017). *Engineering applications of nanotechnology from energy to drug delivery*. Springer International Publishing. https://doi.org/10.1007/978-3-319-29761-3

Kostapanou, A., Chatzipanagiotou, K.-R., Damilos, S., Petrakli, F., Koumoulos, E.P. (2024). Safe-and-sustainable-by-design framework: (Re-)designing the advanced materials lifecycle. *Sustainability*, 16(23), 10439. https://doi.org/10.3390/su162310439

Kraegeloh, A., Suarez-Merino, B., Sluijters, T., Micheletti, C. (2018). Implementation of safe-by-design for nanomaterial development and safe innovation: Why we need a comprehensive approach. *Nanomaterials (Basel)*, 8(4), 239. https://doi.org/10.3390/nano8040239

Kuempel, E.D., Geraci, C.L., Schulte, P.A. (2012). Risk assessment and risk management of nanomaterials in the workplace: Translating research to practice. *The Annals of Occupational Hygiene*, 56(5), 491–505. https://doi.org/10.1093/annhyg/mes040

Kumar, A., Tyagi, P.K., Tyagi, S., Ghorbanpour, M. (2024). Integrating green nanotechnology with sustainable development goals: A pathway to sustainable innovation. *Discover Sustainability*, 5(1). https://doi.org/10.1007/s43621-024-00610-x

Kumari, R., Suman, K., Karmakar, S., Mishra, V., Lakra, S.G., Saurav, G.K., Mahto, B.K. (2023). Regulation and safety measures for nanotechnology-based agri-products. *Frontiers in Genome Editing*, 5, 1–12. https://doi.org/10.3389/fgeed.2023.1200987

Laurent, B. (2008). Engaging the public in nanotechnology? Three visions of public engagement. CSI Working Papers Series 011, Centre de Sociologie de l'Innovation (CSI), Mines ParisTech.

Le Anh, N.L., Foster, M., Arnold, G. (2019). The impact of stakeholder engagement on local policy decision making. *Policy Sciences*, 52(4), 549–571. https://doi.org/10.1007/s11077-019-09357-z

Lee, C.-J., Scheufele, D.A., Lewenstein, B.V. (2005). Public attitudes toward emerging technologies. *Sage Journals*, 27(2). https://doi.org/10.1177/1075547005281474

Liguori, B., Hansen, S.F., Baun, A., Alstrup Jensen, K. (2016). Control banding tools for occupational exposure assessment of nanomaterials – ready for use in a regulatory context? *NanoImpact*, 2, 1–17. https://doi.org/10.1016/j.impact.2016.04.002

Ma, X., Tian, Y., Yang, R., Wang, H., Allahou, L.W., Chang, J., Williams, G., Knowles, J.C., Poma, A. (2024). Nanotechnology in healthcare, and its safety and environmental risks. *Journal of Nanobiotechnology*, 22(1), 715. https://doi.org/10.1186/s12951-024-02901-x

Malakar, Y., Lacey, J., Twine, N.A., Bauer, D.C. (2023). Applying a risk governance approach to examine how professionals perceive the benefits and risks of clinical genomics in Australian healthcare. *New Genetics and Society*, 42(1), e219247. https://doi.org/10.1080/14636778.2023.2192472

Malik, S., Muhammad, K., Waheed, Y. (2023). Nanotechnology: A revolution in modern industry. *Molecules*, 28(2), 661. https://doi.org/10.3390/molecules28020661

Marchant, G.E., Sylvester, D.J., Abbott, K.W. (2008). Risk management principles for nanotechnology. *Nanoethics*, 2(1), 43–60.10.1007/s11569-008-0028-9

Marchant, G.E., Sylvester, D.J., Abbott, K.W., Danforth, T.L. (2010). International harmonization of regulation of nanomedicine. *Studies in Ethics, Law, and Technology*, 3(3), 1–16. https://doi.org/10.2202/1941-6008.1120

Mech, A., Gottardo, S., Amenta, V., Amodio, A., Belz, S., Bøwadt, S., Drbohlavová, J., Farcal, L., Jantunen, P., Małyska, A., Rasmussen, K., Riego Sintes, J., Rauscher, H. (2022). Safe- and sustainable-by-design: The case of smart nanomaterials. A perspective based on a European workshop. *Regulatory Toxicology and Pharmacology*, 128, 105093. https://doi.org/10.1016/j.yrtph.2021.105093.

Morris, J., Willis, J., De Martinis, D., Hansen, B., Laursen, H., Sintes, J.R., Kearns, P., Gonzalez, M. (2011). Science policy considerations for responsible nanotechnology decisions. *Nature Nanotechnology*, 6(2), 73–77. https://doi.org/10.1038/nnano.2010.191

National Nanotechnology Initiative. (2001). Standards for nanotechnology. www.nano.gov/you/standards (Accessed on March 18, 2025)

Nielsen, M.B., Skjolding, L., Baun, A., Hansen, S.F. (2023). European nanomaterial legislation in the past 20 years – Closing the final gaps. *NanoImpact*, 32, 100487. https://doi.org/10.1016/j.impact.2023.100487

Nygaard, K., Graversgaard, M., Dalgaard, T., Jacobsen, B.H., Schaper, S. (2021). The role of stakeholder engagement in developing new technologies and innovation for nitrogen reduction in waters: A longitudinal study. *Water*, 13(22), 3313. https://doi.org/10.3390/w13223313

O'Mathúna, D.P. (2011). Taking a precautionary approach to nanotechnology. *Nanotechnology Development*, 1(1), 6. https://doi.org/10.4081/nd.2011.e6

Ogunsile, B.O., Okoh, O.S., Ejidike, I.P., Omolaja, O.R. (2024). Biosynthesis and optimization of AgNPs yield from *Chromolaena Odorata* leaf extract using response surface methodology (RSM). *Physical Chemistry Research*, 12, 21–31. https://doi.org/10.22036/pcr.2023.366212.2226

Olaleru, S.A., Molokwu, M.I., Mathew, S., Ejidike, I.P., Oyebamiji, O.O. (2025). Enhanced photocatalytic degradation of methylene blue dye using TiO2 nanoparticles obtained via chemical and green synthesis: A comparative analysis. *Pure and Applied Chemistry*. https://doi.org/10.1515/pac-2024-0326

Onisuru, O.R., Onisuru, O. (2025). Smart nanomaterials: Fundamentals, synthesis, and characterization. In: Ayeleru, O.O., Idris, A.O., Pandey, S., Olubambi, P.A. (eds) *Smart nanomaterials for environmental applications. Micro and nano technologies*. Elsevier, Pg. 117–140. ISBN 978-0-443-21794-4.

Organisation for Economic Co-operation and Development (OECD). (2020). Nanomaterials and advanced materials. www.oecd.org/en/topics/nanomaterials-and-advanced-materials.html (Accessed on March 16, 2025).

Oualikene-Gonin, W., Sautou, V., Ezan, E., Bastos, H., Bellissant, E., Belgodère, L., Maison, P., Ankri, J. (2023). Regulatory assessment of nano-enabled health products in public health interest. Position of the scientific advisory board of the French National Agency for the Safety of Medicines and Health Products. *Frontiers in Public Health*, 11, 1125577. https://doi.org/10.3389/fpubh.2023.1125577

Palencia, M., García-Quintero, A., Palencia Luna, V.J. (2022). Metrics for the sustainability analysis of nano-synthesis in the green chemistry approach. In: Shanker, U., Hussain, C.M., Rani, M. (eds) *Handbook of green and sustainable nanotechnology*. Cham: Springer. https://doi.org/10.1007/978-3-030-69023-6_85-1

Pandey, R.P., Vidic, J., Mukherjee, R., Chang, C.M. (2023). Experimental methods for the biological evaluation of nanoparticle-based drug delivery risks. *Pharmaceutics*, 15(2), 612. https://doi.org/10.3390/pharmaceutics15020612

Petersen, E.J., Ceger, P., Allen, D.G., Coyle, J., Derk, R., Garcia-Reyero, N., Gordon, J., Kleinstreuer, N.C., Matheson, J., McShan, D., Nelson, B.C., Patri, A.K., Rice, P., Rojanasakul, L., Sasidharan, A., Scarano, L., Chang, X. (2022). U.S. Federal Agency interests and key considerations for new approach methodologies for nanomaterials. *ALTEX*, 39(2), 183–206. https://doi.org/10.14573/altex.2105041

Pidgeon, N., Harthorn, B.H., Bryant, K., & Rogers-Hayden, T. (2009). Deliberating the risks of nanotechnologies for energy and health applications in the United States and United Kingdom. *Nature Nanotechnology*, 4(2), 95–98. https://doi.org/10.1038/nnano.2008.362

Pokrajac, L., Abbas, A., Chrzanowski, W., Dias, G.M., Eggleton, B.J., Maguire, S., Maine, E., Malloy, T., Nathwani, J., Nazar, L., Sips, A., Sone, J., Van Den Berg, A., Weiss, P.S., Mitra, S. (2021). Nanotechnology for a sustainable future: Addressing global challenges with the international network4Sustainable nanotechnology. *ACS Nano*, 15(12), 18608–18623. https://doi.org/10.1021/acsnano.1c10919

Porcari, A., Borsella, E., Benighaus, C., Grieger, K., Isigonis, P., Chakravarty, S., Kines, P., Jensen, K.A. (2019). From risk perception to risk governance in nanotechnology: A multi-stakeholder study. *Journal of Nanoparticle Research*, 21(11), 245. https://doi.org/10.1007/s11051-019-4689-9

Rasmussen, K., Rauscher, H., Kearns, P., González, M., Riego Sintes, J. (2019). Developing OECD test guidelines for regulatory testing of nanomaterials to ensure mutual acceptance of test data. *Regulatory Toxicology and Pharmacology*, 104, 74–83. https://doi.org/10.1016/j.yrtph.2019.02.008

Rasmussen, K., Sayre, P., Kobe, A., Gonzalez, M., Rauscher, H. (2025). 25 years of research and regulation: Is nanotechnology safe to commercialize? *Frontiers in Toxicology*, 7, 1629813. https://doi.org/10.3389/ftox.2025.1629813

Rathore, A., Mahesh, G. (2021). Public perception of nanotechnology: A contrast between developed and developing countries. *Technology in Society*, 67. https://doi.org/10.1016/j.techsoc.2021.101751

Ray, P.C., Yu, H., Fu, P.P. (2009, January). Toxicity and environmental risks of nanomaterials: Challenges and future needs. *Journal of Environmental Science Health. Part C, Environmental Carcinogenesis & Ecotoxicology Reviews*, 27(1), 1–35. https://doi.org/10.1080/10590500802708267.

Reese, M. (2013). Nanotechnology: Using co-regulation to bring regulation of modern technologies into the 21st century. *Health Matrix*, 23(2), 537–572. https://scholarlycommons.law.case.edu/healthmatrix/vol23/iss2/20

Renn, O. (2008). *Risk governance: Coping with uncertainty in a complex world* (1st ed., Issue 1). Earth Sciences, Law, Politics and International Relations. https://doi.org/10.4324/9781849772440

Roco, M.C., Hersam, M.C., Mirkin, C.A. (1970). Innovative and responsible governance of nanotechnology for societal development. In nanotechnology research directions for societal needs in 2020. 1(1), 561–617. Springer International Publishing. https://doi.org/10.1007/978-94-007-1168-6

Rodríguez-Gómez, F.D., Monferrer, D., Penon, O., Rivera-Gil, P. (2025). Regulatory pathways and guidelines for nanotechnology-enabled health products: A comparative review of EU and US frameworks. *Frontiers in Medicine (Lausanne)*, 12, 1544393. https://doi.org/10.3389/fmed.2025.1544393

Rodríguez-Gómez, F.D., Penon, O., Monferrer, D., Rivera-Gil, P. (2023). Classification system for nanotechnology-enabled health products with both scientific and regulatory application. *Frontiers in Medicine (Lausanne)*, 10, 1212949. https://doi.org/10.3389/fmed.2023.1212949

Roduner, E. (2006). Size matters: Why nanomaterials are different. *Chemical Society Reviews*, 35, 583–592. https://doi.org/10.1039/b502142c.

Rutjens, B.T., Sengupta, N., der Lee, R. van, van Koningsbruggen, G.M., Martens, J.P., Rabelo, A., Sutton, R.M. (2022). Science skepticism across 24 countries. *Social Psychological and Personality Science*, 13(1), 102–117. https://doi.org/10.1177/19485506211001329

Sarahan, P.C. (2008). Auditing as a tool for ensuring compliance with nanotechnology safety requirements. *Nanotechnology Law and Business*, 5(4), 417–428.

Sarkar, S., Pandey, A., Pant, A.B. (2023). Regulatory requirements for safety/toxicity assessment of cosmetics/nanocosmetic products: Challenges and opportunities. In: Pant, A.B., Dwivedi, A., Ray, R.S., Tripathi, A., Upadhyay, A.K., Poojan, S. (eds) *Skin 3-D models and cosmetics toxicity*. Singapore: Springer. https://doi.org/10.1007/978-981-99-2804-0_9

Scheinerman, N. (2023). Public engagement through inclusive deliberation: The Human Genome International Commission and Citizens' Juries. *The American Journal of Bioethics*, 23(12), 66–76. https://doi.org/10.1080/15265161.2022.2146786

Schraven, D., Joss, S., De Jong, M. (2021). Past, present, future: Engagement with sustainable urban development through 35 city labels in the scientific literature 1990–2019. *Journal of Cleaner Production*, 292, 125924. https://doi.org/10.1016/j.jclepro.2021.125924

Schwirn, K., Voelker, D., Galert, W., Quik, J., Tietjen, L. (2020). Environmental risk assessment of nanomaterials in the light of new obligations under the REACH regulation: Which challenges remain and how to approach them? *Integrated Environmental Assessment and Management*, 16(5), 706–717. https://doi.org/10.1002/ieam.4267

Shah, S., Shah, S.A., Faisal, S., Khan, A.A., Ullah, R., Ali, N., Bilal, M. (2021). Engineering novel gold nanoparticles using *Sageretia thea* leaf extract and evaluation of their biological activities. *Journal of Nanostructure in Chemistry*, 12, 129–140. https://doi.org/10.1007/s40097-021-00407-8

Sharifi, S, Mahmoud, N.N., Voke, E., Landry, M.P., Mahmoudi, M. (2022). Importance of standardizing analytical characterization methodology for improved reliability of the nanomedicine literature. *Nano-Micro Letters*, 14(1), 172. https://doi.org/10.1007/s40820-022-00922-5

Shukla, K., Mishra, V., Singh, J., Varshney, V., Verma, R., Srivastava, S. (2024). Nanotechnology in sustainable agriculture: A double-edged sword. *Journal of the Science of Food and Agriculture*, 104(10), 5675–5688. https://doi.org/10.1002/jsfa.13342

Siegrist, M., Keller, C., Kastenholz, H., Frey, S., Wiek, A. (2007). Laypeople's and experts' perception of nanotechnology hazards. *Risk Analysis*, 27(1), 59–69. https://doi. org/10.1111/j.1539-6924.2006.00859.x

Souto, E.B., Blanco-Llamero, C., Krambeck, K., Kiran, N.S., Yashaswini, C., Postwala, H., Severino, P., Priefer, R., Prajapati, B.G., Maheshwari, R. (2024). Regulatory insights into nanomedicine and gene vaccine innovation: Safety assessment, challenges, and regulatory perspectives. *Acta Biomaterialia*, 180, 1–17. https://doi.org/10.1016/j. actbio.2024.04.010

Stuetz, H., Reihs, E.I., Neuhaus, W., Pflüger, M., Hundsberger, H., Ertl, P., Resch, C., Bauer, G., Povoden, G., Rothbauer, M. (2023). The cultivation modality and barrier maturity modulate the toxicity of industrial zinc oxide and titanium dioxide nanoparticles on nasal, buccal, bronchial, and *alveolar mucosa* cell-derived barrier models. *International of Journal of Molecular Sciences*, 24(6), 5634. https://doi.org/10.3390/ ijms24065634.

Suphattanakul, O. (2018). Public Participation in decision-making processes: The concepts and tools. *Journal of Business and Social Review in Emerging Economies*, 4(2), 221–230. https://doi.org/10.26710/jbsee.v4i2.213

Tripathy, A., Patne, A.Y, Mohapatra, S., Mohapatra, S.S. (2024). Convergence of nanotechnology and machine learning: The state of the art, challenges, and perspectives. *International Journal of Molecular Sciences*, 25(22), 12368. https://doi.org/10.3390/ ijms252212368

UNEnvironment Programme (UNEP). (2025). UNEP – UN environment programme. www.unep.org/

Vega-Baudrit, J.-R., Camacho, M., Araya, A., Camacho, M. (2023). Regulating nanotechnology: Ensuring responsible and safe innovation in the advancement of science and technology. *Ciencia, Tecnologí-a y Salud*, 10(2), 177–191. https://doi. org/10.36829/63cts.v10i2.1581

Verma, V., Gupta, P., Singh, P., Pandey, N.K. (2024). Considerations in the development and deployment of nanotechnology. In: George, S.C., Tawiah, B. (eds) *Nanotechnology in societal development. Advanced technologies and societal change*. Singapore: Springer. https://doi.org/10.1007/978-981-97-6184-5_15

Wahab, Y.A., Al-Ani, L.A., Khalil, I., Schmidt, S., Tran, N.N., Escriba-Gelonch, M. Woo, M.W., Davey, K., Gras, S., Hessel, V., Julkapli, N.M. (2024). Nanomaterials: A critical review of impact on food quality control and packaging. *Food Control*, 163, 110466. https://doi.org/10.1016/j.foodcont.2024.110466.

Warheit, D.B. (2018). Hazard and risk assessment strategies for nanoparticle exposures: How far have we come in the past 10 years? *F1000Research*, 7, 376. https://doi. org/10.12688/f1000research.12691.1.

Wasti, S., Lee, I.H., Kim, S., Lee, J.H., Kim, H. (2023). Ethical and legal challenges in nanomedical innovations: A scoping review. *Frontiers in Genetics*, 14, 1163392. https://doi.org/10.3389/fgene.2023.1163392

Yoshida, M., Lahann, J. (2008). Smart nanomaterials. *ACS Nano*, 2, 1101–1107. https://doi. org/10.1021/nn800332g.

Zheng, J., Zhao, Z. yun, Zhang, X., Chen, D. zen, & Huang, M. hsuan. (2014). International collaboration development in nanotechnology: A perspective of patent network analysis. *Scientometrics*, 98(1), 683–702. https://doi.org/10.1007/s11192-013-1081-x

15

Nanotechnology and Its Role in Building a Sustainable Economy

Upendra Kumar Mishra, Satyendra Pratap Singh, S. Gaurav,
and Vishal Singh Chandel

15.1 Introduction

It is known to us that nanotechnology applications are in our daily lives and are easing our lives day by day. In recent conversations regarding the ways in which nanotechnology can address great societal issues, green nanotechnologies have garnered a lot of attention. In this study, a variety of specific applications of nanotechnology to green and sustainable growth are identified and discussed. Specific consideration is paid to the utilization of nanotechnology in the production and utilization of sustainable energy, the provision of water, and other environmental applications. It is anticipated that the world energy demand will shoot up to 30% from 2010 and 2035 [1], while at the same time, there are currently more than 800 million people around the world who do not have a clean drinking water facility [2]. It has never been more important to find ways to supply these needs that are both affordable and safe. This need is also connected with the fluctuations in markets, the stability of the supply chain, and the demands to minimize CO_2 emissions and other ecological and environmental concerns. Although the need for new-fangled energy sources and clean water is most acute in nations that are not members of the OECD, developed nations are nevertheless essential to find ways to provide technologies and logistics to further help and support their economies and populations. Nanotechnology applications are gaining interest because of these issues since they have the potential to both improve upon already existing technologies and provide new alternatives.

This study explores the potential scale and impact of nanotechnology in areas such as energy production and consumption, water supply, and various environmental and ecological applications. It also provides perspective on how nanotechnology's role in promoting green and sustainable development can be measured and economically assessed. Concerns are also voiced regarding the possible problems and dangers that may arise. The section begins by examining the range of definitions that have been suggested for both nanotechnology as a whole and specifically for green nanotechnology. It then moves on to explore sustainable uses of technology, examples of market projections, and signs of its economic influence. The conclusion emphasizes the importance of evaluating the economic implications of nanotechnology using a holistic life cycle approach – one that accounts for its full range of effects on the economy, environment, and society.

DOI: 10.1201/9781003632498-15

15.2 Nanotechnology: Path to Sustainable Environmental Performance and a Circular Economy

Altering the composition of matter on an atomic or molecular level, or nanotechnology, has transformed several fields, including resource management and environmental sustainability. Nanotechnology provides creative answers to crucial environmental problems by facilitating the creation of cutting-edge materials and procedures. Its uses are crucial for improving the quality of the environment, encouraging resource efficiency, and facilitating waste valorization, all of which lead to a more sustainable and circular economy that ultimately results in the development of numerous industries and jobs.

15.3 Environmental Quality through Nanotechnology

15.3.1 Water Purification

A basic human right and a fundamental component of sustainable development is having access to clean water. Because it makes it possible to create advanced filtration systems, nanotechnology is essential to enhancing the quality of water. For example, materials based on nano-cellulose have a large surface area and are chemically inert, which makes them useful for eradicating chemical and bacterial pollutants from contaminated water. Similarly, because of their special qualities, graphene-coated nanofilters have proven their potential in desalination and dye removal.

15.3.2 Air Pollution Control

There are substantial health and environmental hazards associated with air pollution. Air pollution monitoring and mitigation are made possible by nanotechnology. For better air quality, nanomaterials can be designed to absorb or neutralize hazardous chemicals, including nitrogen oxides and VOCs. Furthermore, nanotechnology makes it possible for scientists to develop sensors that can detect air pollutants in real time, allowing for immediate remedies.

15.3.3 Soil Remediation

Environmental health and agricultural productivity are at risk due to soil pollution brought on by manufacturing procedures and improper waste disposal. Nanoparticles are used in nanoremediation processes to break down or immobilize pollutants in the soil. For instance, heavy metals and organic contaminants in polluted soils have been treated using nanoscale zero-valent iron (nZVI). Such methods provide effective and affordable replacements for traditional soil remediation procedures.

15.4 Promoting Resource Efficiency and Circular Economy

15.4.1 Waste Valorization

Reducing waste and making optimal use of resources are essential to the shift from a linear to a circular economy. Waste valorization is made easier by nanotechnology, which makes it possible to turn waste materials into useful goods. For example, agro-industrial biowaste can be converted into green nanomaterials for wastewater treatment, following the circular economy and green chemistry concepts. This strategy promotes sustainable resource use in addition to waste reduction.

15.4.2 Recycling Rare Earth Metals

In many technological equipment and renewable energy systems, rare earth metals are a crucial part. For resource sustainability and to lessen the impact on the environment, these metals must be recycled effectively. Queen's University Belfast researchers have created techniques for recycling rare earth metals with ionic liquids, which provide useful components from industrial waste to dissolve selectively. By encouraging the reuse of essential resources, these developments aid in the development of a circular economy.

15.4.3 Sustainable Agriculture

Agriculture is both a contributor to and a victim of environmental degradation. Nanotechnology offers solutions for sustainable agricultural practices by enhancing resource efficiency and minimizing environmental impact. Nano-fertilizers and nano-pesticides enable targeted delivery of nutrients and protection, reducing excess use of chemicals. Additionally, nanosensors can monitor soil health and optimize irrigation, promoting sustainable farming practices.

15.5 Challenges in Adopting Nanotechnological Tools and Their Future Outlook

Even with its promising applications in environmental sustainability, nanotechnology continues to face several challenges. The potential toxicity of nanomaterials necessitates thorough risk assessments to ensure environmental and human safety. Moreover, the scalability of nanotechnology solutions requires further research and development to achieve cost-effectiveness and widespread adoption. Future efforts should focus on interdisciplinary collaboration, regulatory frameworks, and public awareness to harness the full potential of nanotechnology for a sustainable and circular economy.

Leading the way in innovations to achieve a circular economy and sustainable environmental performance is nanotechnology. Nanotechnology provides revolutionary answers to today's environmental problems through its wide range of uses in waste valorization, soil remediation, air pollution reduction, water purification, and sustainable agriculture. Nanotechnology can play a major role in creating a resilient and sustainable future by tackling current issues while encouraging teamwork.

Achieving sustainability in energy production and demand, sanitation, and the supply of potable or drinkable water, as well as mitigating the effects of changing environmental conditions, is one of the exciting areas that researchers and technologists are focusing on. Nanotechnology has the potential to fulfill these challenges.

Nanotechnology involves understanding and controlling materials at the nanoscale – typically between 1 and 100 nanometers – where unique properties emerge that enable novel applications. At this scale, substances often behave differently compared to their bulk counterparts or individual atoms and molecules, showing unusual physical, chemical, or biological traits. Some nanomaterials, for instance, are stronger or display unique magnetic behaviors not seen in other forms of the same material. Others offer improved thermal or electrical conductivity. They may exhibit increased chemical reactivity, enhanced light reflectivity, or color alteration as their size or structure is modified [3].

One nanometer is equal to one billionth of a meter since the word "nano" in the International System of Units implies one-billionth, or 10^{-9}. To help you visualize how little that is, here are a few examples:

- About 100,000 nanometers is the thickness of a standard sheet of paper.
- The diameter of a single human DNA strand is 2.5 nanometers.
- One inch is 25,400,000 nanometers.
- The width of a human hair is around 80,000–100,000 nanometers.
- The diameter of a single gold atom is approximately one-third of a nanometer.

One nanometer in diameter is the same as one softball in relation to the Earth when it comes to comparative scale. Things in the nanoscale are incredibly tiny, as shown by three visual representations of nanotechnology's size and scale in the figure on the right (Figure 15.1).

15.6 Superiority of Nano Over Bulk

By operating at the nanoscale, scientists can explore and harness the distinctive physical, chemical, mechanical, and optical properties that materials exhibit at this level. Nanotechnology is not solely about working with extremely small dimensions – it also involves leveraging the novel behaviors and interactions that arise at these scales.

Nanomaterials have many superior properties, such as chemical, physical, optical, and mechanical rheological properties, than their bulk counterparts, due to diverse

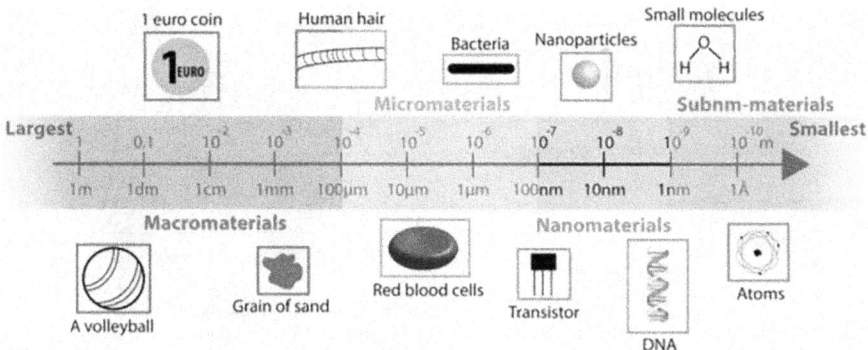

FIGURE 15.1 Scale line of nanomaterials.

Adopted from Brullot Ward [4]

reasons. These superior properties at the nanoscale draw the attention of scientists and technologists to explore them in various technological applications. Material scientists are working to synthesize low-cost, economical materials for different applications.

15.7 Quantum Confinement

The characteristics of materials can undergo substantial changes when particles with dimensions ranging from 1 to 100 nanometers are synthesized, in comparison to those at larger scales. At this scale, quantum processes govern the characteristics and behavior of particles (Figure 15.2). The notion of "tunability" of attributes is an intriguing and potent byproduct of nanoscale quantum effects. In other words, a scientist can essentially fine-tune a material quality by adjusting the particle's size. The thermal properties such as melting point, boiling point, freezing point, fluorescence, electrical conductivity and thermal conductivity, magnetic permeability and magnetic

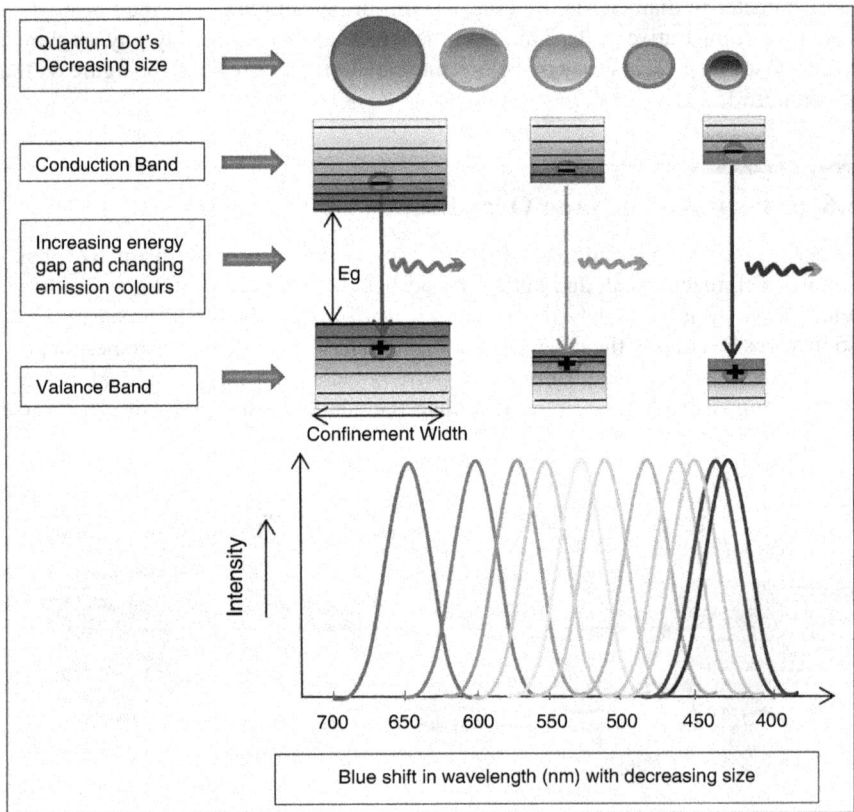

FIGURE 15.2   Representation of quantum confinement.

Adopted from S. Jagtap et al. [5]

susceptibility, chemical reactivity, and chemical stability of substances can vary at the nanoscale depending on their size.

The peculiarities of the nanoscale are exemplified by nanoscale gold. The size of the particle determines whether the nanoscale gold appears red or purple. Because of quantum phenomena, gold nanoparticles have a unique interaction with light compared to gold particles of a greater size.

15.8 Surface Properties of Nanomaterials

Nanoscale materials have a much higher surface area-to-volume ratio than their bulk counterparts. As the surface area-to-volume ratio increases, materials may exhibit intensified reactivity (Figure 15.3). A little thought experiment demonstrates the exceptionally high surface areas of nanoparticles. A solid cube with each side of 1 cm possesses a surface area of 6 cm^2, equivalent to one side of a half stick of gum. If a volume of 1 cm^3 were occupied by cubes measuring 1 millimeter per side, it would contain 1,000 such cubes (10 × 10 × 10), each possessing a surface area of 6 mm^2, resulting in a cumulative surface area of 60 cm^2 – marginally exceeding the dimensions of a simple credit card. When 1 cm^3 is occupied by micrometer-sized cubes – a

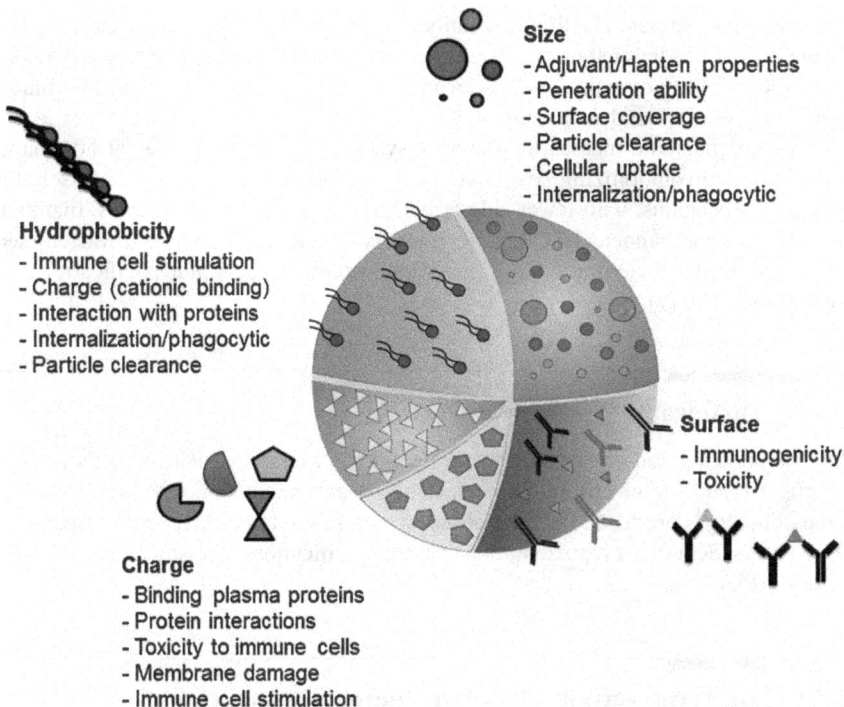

FIGURE 15.3 Representation of bacterial activity of nanoparticles due to large surface area.

Adopted from De Almeida, Letícia et al. [6]

trillion (10^{12}) of them, each with a surface area of 6 square micrometers – the cumulative surface area reaches 6 m^2. When a single cubic centimeter of volume is occupied by 1-nanometer-sized cubes – totaling 10^{21}, each with a surface area of 6 square nanometers – their cumulative surface area amounts to 6,000 square meters.

A 1 cm^3 of cubic nanoparticles possesses a surface area comparable to that of a football field. The increased surface area results in greater exposure of the substance to the surrounding environment, significantly accelerating the chemical reactions or reactivity of these materials. A notable advantage of increased surface area and enhanced reactivity in nanostructured materials is their contribution to the development of superior catalysts. Such a common illustration is the catalyst-based catalytic converter in an automobile, used for purifying or converting harmful exhaust into less harmful or non-harmful products to meet the standards fixed for automobile exhaust. The increased surface area of nanocatalysts has allowed contemporary catalytic converters to utilize significantly less precious metal than previously required to attain equivalent reductions in pollutant emissions. Engineers are improving the reactivity at the nano level to develop superior energy storage and converter devices (batteries, fuel cells), and catalysts of added efficiency.

15.9 Industrial Growth of Nanotechnology

The Indian government has financed numerous nanotechnology initiatives, start-ups, and enterprises. India's nanotechnology market is projected to grow at an annual compound rate of 19.1% over the coming decade, fueled by rising demand and continuous progress in research and development.

Nanomedicines and medication delivery systems are two examples of how nanotechnology is revolutionizing healthcare in India. Drugs are now more precisely targeted and efficacious, with fewer side effects and better patient outcomes, thanks to nanoparticles and nanocarriers. Nanotechnology is also being applied in various sectors across India, including energy generation and storage, water purification, and materials science [7].

15.10 Wastewater Treatment

Fluoride, arsenic, and nitrates are just a few of the toxins that Rite Water's filtering systems are capable of eliminating. Electrolysis and nanotechnology are combined in the company's products, like NanoBlue, which has a daily purification capacity of 50,000 liters. Schematic representation of water purification by photocatalyst is shown in Figure 15.4.

15.11 Catalytic Activity Toward Bacterial Detection

A strip that can detect bacterial contamination without the need for specialist equipment or skilled workers has been created by Module Innovation. The detecting capabilities of these strips are enhanced through the use of nanotechnology. When the strip

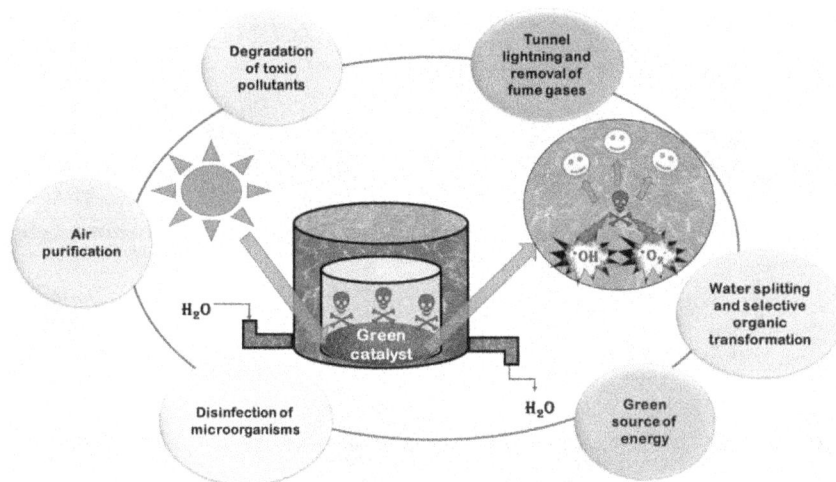

FIGURE 15.4 Representation of photocatalytic activity of nanomaterials.

Adopted from Szabolcs Bognar et al. [8]

comes into contact with germs, it changes color, making it easy to identify contamination. The business also sells nanofiber strips that are both reusable and designed to detect *Enterohaemorrhagic Escherichia coli*. A schematic representation for the action of silver nanoparticles towards bacterial activity is shown in Figure 15.5.

15.12 Li-Ion Batteries

A joint study conducted by the Indian Institute of Technology (IIT) and the Japan Advanced Institute of Science and Technology (JAIST) revealed that adding two-dimensional titanium diboride (TiB_2) nanosheets to lithium-ion batteries (Figure 15.6) can significantly boost their efficiency. This enhancement may lead to faster charging times and longer battery lifespan. Due to their nanoscale structure, TiB_2 nanosheets offer key benefits such as excellent high-rate performance, pseudocapacitive energy storage, and reliable functionality over repeated charge–discharge cycles. These attributes render it an excellent contender for application in lithium-ion batteries.

15.13 New Active Filaments Mimic Biology to Transport Nano-Cargo

Researchers at the Indian Institute of Technology Madras, in collaboration with the Institute of Mathematical Sciences, have developed an innovative method to rapidly transport nano-colloidal particles suspended in fluids or gels. Compared to traditional diffusion methods, this novel design can transport these nanoparticles more effectively and is inspired by micro-scale movements. By creating a workable active transport engine, the group made a substantial contribution to our knowledge of how

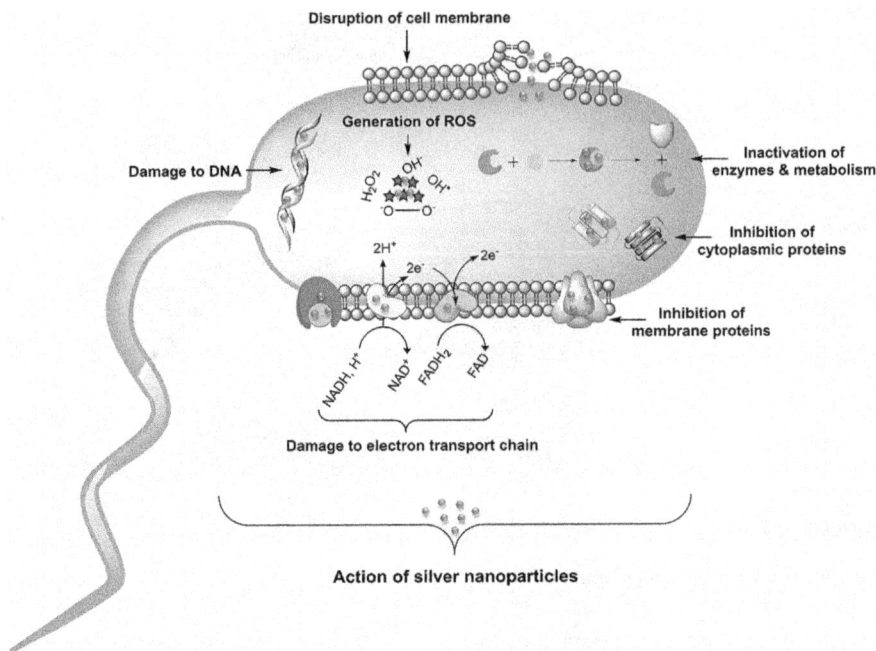

FIGURE 15.5   Antibacterial activity of silver nanoparticles.

Adopted from Priyanka Singh et al. [9]

momentum conservation functions in active systems. Potential uses for the novel nano-colloidal particle transport design include insemination, targeted drug delivery, and other fields where impaired motility is a problem. Additionally, the field of therapeutic therapies may be greatly impacted by this research.

15.14 India's Present Nanotechnology Market

With an initial budget of $12 million, the Nanoscience and Technology Mission (NSTM), under the direction of the Department of Science and Technology (DST), established the groundwork for nanotechnology in India during the 10th plan period (2002–2007). The Indian nanotechnology market is anticipated to increase significantly over the next ten years, with a predicted compound annual growth rate (CAGR) of 19.1%. The growing need for products based on nanotechnology in the energy and healthcare sectors is driving this expansion. Utilizing India's many natural resources, nanotechnology aims to satisfy the nation's expanding demands in a number of areas, such as electronics, healthcare, clean food and water, and premium textiles. The growth of nanotechnology market in Asia Pacific and global nanotechnology market is shown in Figures 15.7 and 15.8 respectively.

The following businesses and innovative start-ups are active in India's nanotechnology market and Table 15.1 represents the industrial growth of the India in the field of nanotechnology.

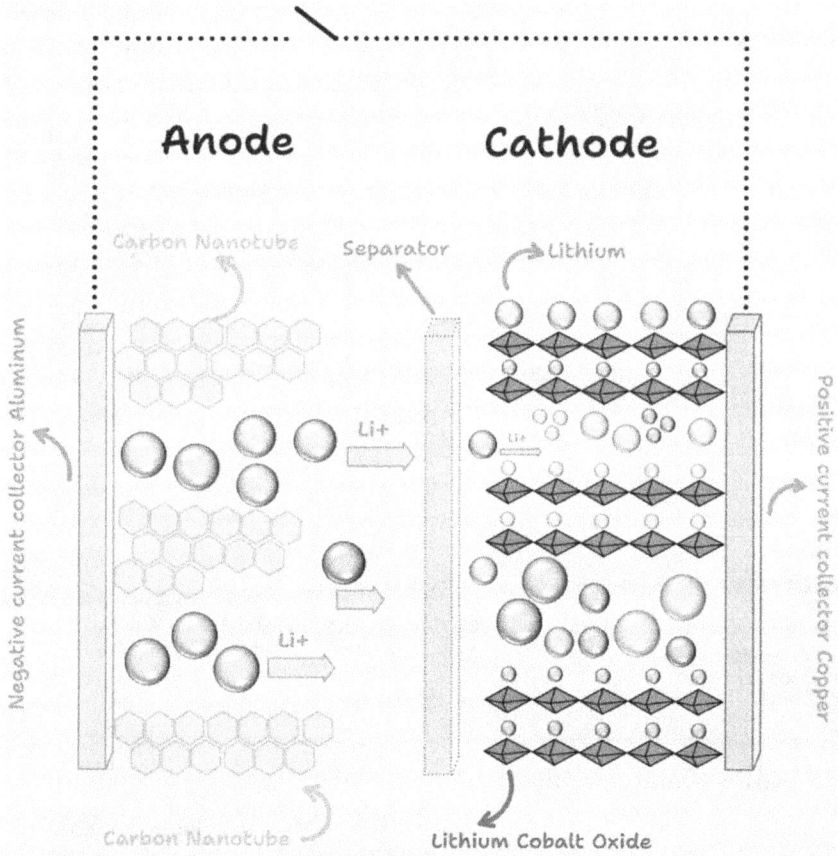

FIGURE 15.6   The representation of working of Li-ion battery.

Adopted from Ehsan Poorshakoor et al. [10]

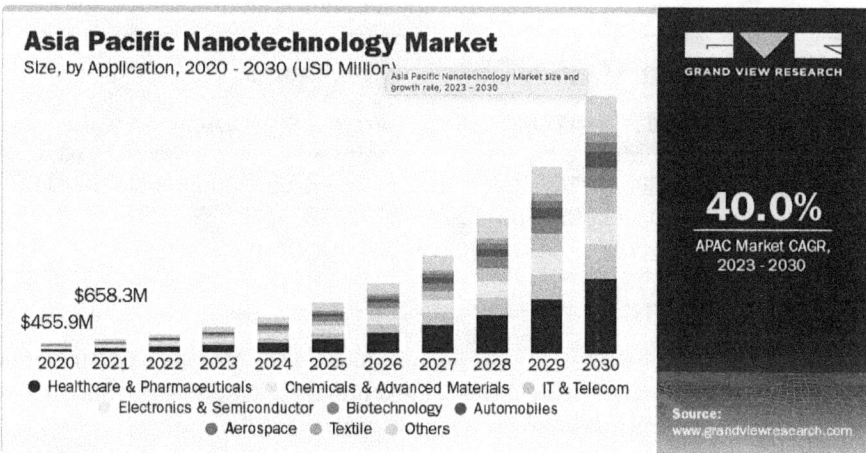

FIGURE 15.7   The representation of nanotechnology market in the Asia Pacific.

**Adopted from Nanotechnology Market Size, Share & Trends Analysis Report by Type
(Nanosensor, Nanodevice), by Application (Electronics & Semiconductor, Textile, Automobiles,
Aerospace), by Region, and Segment Forecasts, 2023–2030 [11]**

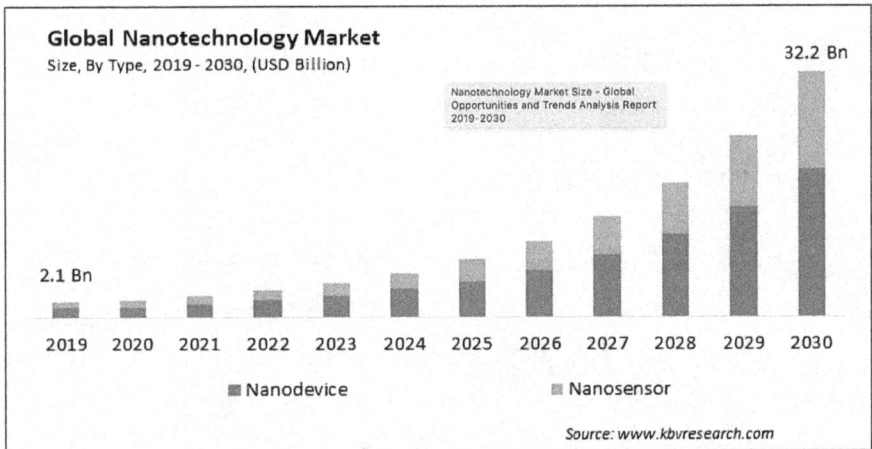

FIGURE 15.8 The status of global market of nanotechnology.

Adopted from Global Nanotechnology Market Size, Share & Industry Trends Analysis Report by Type (Nanodevice, and Nanosensor), by Application, by Regional Outlook and Forecast, 2023–2030 [12]

15.15 NoPo Nanotechnologies

Advanced nanomaterials, particularly single-walled carbon nanotubes, are produced by NoPo Nanotechnologies. The company makes its single-walled carbon nanotubes using a modified HiPCO technique. It states that these nanotubes can be utilized as heat conductors and in solar cells because they are 100 times stronger than steel and 1000 times more conductive than copper.

15.16 AMNIVOR Medicare

A medical firm called AMNIVOR Medicare extracts collagen from fish scales using nanotechnology. According to the corporation, this resource is inexpensive and plentiful. Numerous conditions, including burns, diabetes, and chronic wounds, can be treated using collagen and collagen-based products.

15.17 Nanosentrix

Conductive inks for printed electronics are manufactured by Nanosentrix. The ink is used in flexible electronics, RFID printing, and printable sensors. It is composed of conductive elements such as graphene, graphite, and carbon nanotubes.

15.18 Nanospan

Nanospan produces and distributes a range of graphene, carbon nanomaterials, and flexible electrodes and anodes based on nanosilicon for Li-ion batteries and

TABLE 15.1

Data Representation of Some Indian Companies Working in the Field of Nanotechnology [41]

Company Logo	Company Name	Working Field	Number of Employees
	Nanolife	Nanomedicine, oncology, ophthalmology, antibiotics, and cosmeceuticals	11–50
	Nanowatts Technologies	CMOS, RF MEMS, and wafer testing	1–10
	I Cube Nanotec India Pvt. Ltd.	Life sciences, specialty & fine chemicals, microelectronics, biofuels, petrochemicals, refining, and oil & gas	51–100
	NanoSniff Technologies	Microsensors and instrumentation with applications in sensing chemicals, particularly explosives, and bio-chemicals, focused on cardiac markers	11–50
	Nanofil Technologies Private Limited	Polypropylene compounds (including glass, talc, flame retardant), as well as ABS, SAN (glass fiber and flame retardant), and nylon compounds	251–500
	JK Nanosolutions	Innovative nanotech products for rapid and low-cost wastewater treatment	1–10

(Continued)

TABLE 15.1 (*Continued*)

Data Representation of Some Indian Companies Working in the Field of Nanotechnology

Company Logo	Company Name	Working Field	Number of Employees
	TechNanoIndia	Carbon nanotubes, graphene, quantum dots, metal nanoparticles, oxide nanoparticles, composites, alloys, and conductive inks	1–10
	Nanotech Soft-App IT Solution	FMCG, retail, automobile, e-commerce, manufacturing, pharmaceuticals, and IT	1–10
	Nanolitho Technologies Private Limited	Layout design services	11–50
	Navin Molecular	Complex small-molecule targets, offering high-quality starting materials and intermediates	501–1000

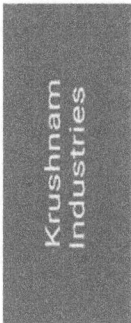	Krushnam Industries	Paint, ink, sealant, adhesive, polymer, paper, and rubber	51–100
	Ad-nanotech	Graphene (battery-grade, paint and coating-grade, composite-grade), carbon nanotubes (single-walled and multi-walled), and metal oxide nanopowders (such as zinc oxide, iron oxide, nano silica, TiO_2, Al_2O_3, MgO, CuO)	11–50
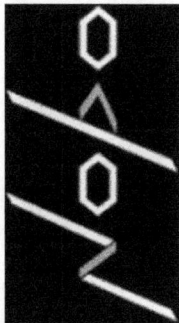	NoPo Nanotechnologies	Single-walled carbon nanotubes (SWCNT)	1–10
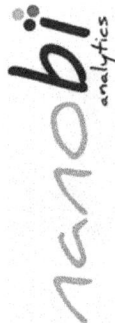	Nanobi Analytics	Microsoft SQL Server, PostgreSQL, and Oracle	51–100
	Nanotech Test & Measurement Solution	Automotive, electrical & electronics, aerospace, pharmaceutical, food & beverages, hospitals, and chemicals	11–50

(Continued)

TABLE 15.1 (*Continued*)

Data Representation of Some Indian Companies Working in the Field of Nanotechnology

Company Logo	Company Name	Working Field	Number of Employees
	EdgeTech Scientific Pvt. Ltd.	Service provider specializing in high technology equipment and materials related to various streams of research and industry, with a strong focus on nanotechnology	11–50
	Tata Nexarc	Service provider in the information technology	501–1000
	Nanomatrix Materials	Filtration, environmental clean-ups, and high-end applications	11–50
	NanOlife	Nanomedicine solutions targeted at areas such as oncology, ophthalmology, and antibiotic treatments, focusing on conditions like cancer and arthritis	51–100
	Nanoman	Multifunctional coatings that protect surfaces from liquids, environmental pollution, and weathering, suitable for both internal and external use	11–50
	New Nano Technologies	Health, energy efficiency, and environmental sustainability. Nano ceramic coatings for both hot and cold insulation, as well as innovative solutions for enhancing energy yield in solar power applications through self-cleaning and photocatalytic technologies	1–10

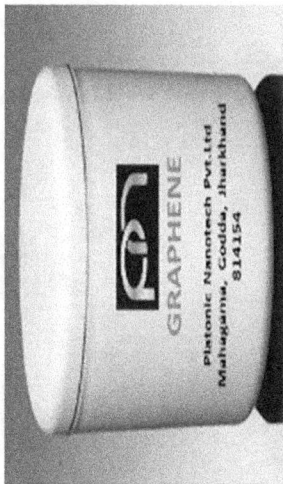

Company	Description	Size
Platonic Nanotech Pvt. Ltd.	High-quality graphene and graphene-based products	1–10
NanoLand Global	Clean-label technology, carbon reduction strategies	1–10
Nanosafe Solutions Private Limited	Antiviral high-filtration reusable masks (NSafe), antimicrobial water bottles (AqCure), and FDA-approved zero alcohol sanitizer (RubSafe)	11–50
RRCAT	Development and manufacturing of particle accelerators and laser technologies	1001–5000
NanoBios Lab	Healthcare research and development	11–50
20 Microns	Functional fillers, extenders, and tailored solutions	1001–5000

(Continued)

TABLE 15.1 (*Continued*)

Data Representation of Some Indian Companies Working in the Field of Nanotechnology

Company Logo	Company Name	Working Field	Number of Employees
	VVDN Technologies	Hardware development, software development, cloud applications, mechanical design, testing, automation, pre-compliance, and full-scale manufacturing	1001–5000
	Nanostuffs	Full stack (MERN) development solutions and end-to-end Vtex solutions,	251–500
	The Nano Labs	Utilizing Bluetooth, LoRaWAN, and hybrid technologies	11–50
	Clean Science and Technology	Chemicals, specializing in innovative, sustainable, and cost-effective catalytic manufacturing processes	501–1000

Company	Products/Services	Employees	Logo
Technhando Technologies LLP	Printed electrodes, microfluidic channels, ion-selective electrodes, miniaturized electrochemical readout electronics, and comprehensive data analysis solutions	1–10	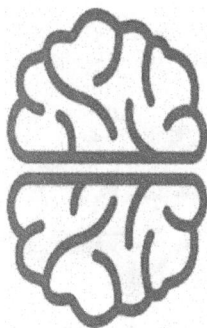
E-Spin Nanotech Pvt. Ltd.	High-quality, multifunctional nanofibers	11–50	
Sinsil International	Ellipsometers, potentiostats, galvanostats, solar simulators, and atomic force microscopy equipment	11–50	
Neon Laboratories Limited	Anesthesiology, critical care, gynecology, ophthalmology, orthopedics, oncology, and dermatology	1001–5000	NEON

(Continued)

TABLE 15.1 (*Continued*)

Data Representation of Some Indian Companies Working in the Field of Nanotechnology

Company Logo	Company Name	Working Field	Number of Employees
	NanoHealth	NanoHealth focuses on transforming the healthcare experience in India	11–50
	NexSemi Systems Pvt Ltd	Semiconductor technologies, ASIC design, and embedded software development	51–100
	Technanosoft Technologies	Salesforce implementation, IoT device integration, and reliable connectivity solutions	251–500

Naprod Life Sciences Pvt. Ltd.	Production of anti-cancer, general, and anesthetic formulations	251–500
Nature Technologies Pvt Ltd	Communication technologies	11–50
Molecular Connections	AI-powered data solutions in the informatics industry, specializing in Big Data, Data Science, and bioinformatics	1001–5000

(Continued)

TABLE 15.1 *(Continued)*

Data Representation of Some Indian Companies Working in the Field of Nanotechnology

Company Logo	Company Name	Working Field	Number of Employees
	Spinco Biotech	Chromatography, mass spectrometry, life science, Lyo & Lab Products, and chemistry and consumable products	501–1000
	Nan Technologies	IT software training, software development, and consulting services	11–50
	TriNANO Technologies	Nano coatings for solar panels	1–10
	Nano-Dye Technologies LLC	Cotton and cotton-blend textile dyeing industry	11–50

	Solar Industries India	Design, development, manufacture, and application of energy materials, specializing in industrial explosives and ammunition	1001–5000
	Detect Technologies	SaaS products that facilitate real-time intelligence and actionable insights	101–250
	Sun Pharma Advanced Research Company Ltd.	CMC development, preclinical translation, and clinical development and operations	251–500
	DRC Techno	Specializing in the gems & jewelry and healthcare sectors	51–100

(Continued)

TABLE 15.1 (*Continued*)

Data Representation of Some Indian Companies Working in the Field of Nanotechnology

Company Logo	Company Name	Working Field	Number of Employees
	Hyderabad Consulting Group	Aerospace and biomedical fields	11–50
	Gennova Biopharmaceuticals Limited	Healthcare sector, particularly in cardiovascular, neurology, nephrology, and oncology	501–1000
	Toshvin Analytical Pvt. Ltd.	High-technology analytical and laboratory instruments	251–500

Company	Description	Size
Enzene Biosciences	Biosimilars, phytopharmaceuticals, peptides, and regenerative products	101–250
NANOCOT	Nanotechnology-based coating systems designed for high-value assets in various sectors, including industrial, heavy machinery, marine, heavy automobiles, and architecture	1–10
MTAR Technologies Limited	Precision machining, assembly, testing, quality control, and specialized fabrication services, with a strong focus on high-stakes projects in the Indian Civilian Nuclear Power program, Indian Space program, Indian Defence, Global Defence, and Global Clean Energy sectors	1001–5000
Nilsan Nishotech Systems	Process vessels, piping systems, high-quality water systems, custom-built solutions, chromatography systems, and microfiltration membrane products	101–250
NexusNovus	Cleantech, agriculture, and food processing while leveraging a strong network, significant industry experience, and expertise in market entry, outsourcing, and management consulting	11–50

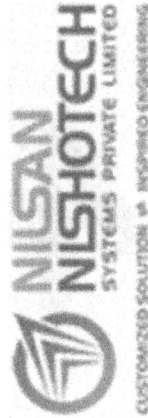

(Continued)

TABLE 15.1 (*Continued*)

Data Representation of Some Indian Companies Working in the Field of Nanotechnology

Company Logo	Company Name	Working Field	Number of Employees
	Industron Nanotechnology Pvt Ltd	Nanoindentation systems, complete mechanical testing solutions, and dynamic mechanical analysis tools	11–50
	Mirafra Technologies	ASIC design from specification to silicon, digital verification, physical design, design for test, post-silicon validation, and device driver development	501–1000

supercapacitors. Using cutting-edge tools like AFM, along with techniques like Fourier transform infrared spectroscopy and high-resolution transmission electron microscopy, it provides characterization and testing services for nanomaterials.

15.19 Kanopy Techno Solutions

Multiscale modeling, electrochemical instrumentation, nanofluidics, and nanofabrication technologies are among the electrochemistry and nanotechnology solutions provided by Kanopy Techno Solutions.

15.20 Saint-Gobain Glass India

SGG NANO, a different kind of glass made by Saint-Gobain Glass India, features sophisticated energy-efficient sun control and thermal insulation qualities. Using magnetically accelerated nanotechnology-based cathodic sputtering in a vacuum, several layers of specific nanometric metallic oxides and nitrides are applied to create the glass. This process improves the glass's insulation properties and overall energy efficiency.

Auto Fiber Craft (AFC) Powders manufactures specialized nanomaterials, such as nano-sized silver powder, which is utilized in RFID technology and electronic applications like conductive inks and pastes.

15.21 AVANSA Technology and Services

For businesses, organizations, and academic institutions that employ nanotechnology, AVANSA Technology & Services specializes in the synthesis and analysis of nanomaterials. Additionally, the company produces a variety of nanoparticles, carbon nanotubes, and graphene.

15.22 India's Nanotechnology Future

The growth of nanotechnology in India has been driven by its potential to address critical societal challenges, such as improving access to clean water and healthcare, while also contributing to economic development through the rise of nanotech-based industries. As public awareness continues to grow, India is in a strong position to harness nanotechnology for educational advancement and infrastructure development. This approach not only prepares the country to thrive in the global nanotech landscape but also helps build a skilled workforce essential for a knowledge-driven economy.

Nonetheless, more money is required, including long-term support for well-run research initiatives with noteworthy results. Furthermore, cooperation among research facilities nationwide may produce better outcomes.

15.23 Indicators of Economic Impact

The conceptualization of direct and indirect economic impacts. This study also highlights potential impact indicators and explores the related concerns and precautions associated with their use. The market analysis discussed in the previous section reveals substantial opportunities for green nanotechnology. Investment encompasses various domains such as research, technological development, prototype production, testing, and final product with marketing standards. Before these markets can emerge, it is essential to establish regulations, encourage user adoption, and implement monitoring systems.

It requires assessing the balance between the advantages and drawbacks of a new technology, which can vary based on the viewpoint – whether that of a manufacturer, competitor, consumer, employee, industry, region, or country. Policymakers frequently focus on the effects of economic development, new technologies, including green nanotechnologies, and their impact on employment wages. Green nanotechnology products and processes, along with their associated industries and services, will generate employment opportunities in research, manufacturing, implementation, and maintenance. However, predicting the exact number of new jobs created remains challenging. As traditional products are replaced, current employees may shift into roles within the green nanotechnology sector. Defining nanotechnology presents challenges, particularly employment or a sustainable job, as well as the definition of a sustainable nanotechnology position. This field encompasses various applications across multiple disciplines, including materials science, biology, and electronics, enabling advancements in areas such as drug delivery, energy storage, and the development of novel materials [13].

Nanotechnology that emphasizes environmentally sustainable practices is expected to account for a significant portion of future employment in the field. The extent to which these jobs are classified as "green" depends on the criteria used and may also include roles from other sectors considered environmentally beneficial, as evidenced by recent efforts to define and assess green employment [14–16].

As noted by Tassey [17], benefits can unfold across various time frames. In the near term, outcomes may include patent filings, research and development partnerships, and prototype creation, along with potential interest from venture capital investors. In the medium term, new products, processes, and business growth may become evident, while broader industry, economic, and societal advantages may materialize in the long term. Nanotechnology involves the control and alteration of matter at the atomic or molecular level, usually within the range of 1 to 100 nanometers. This field encompasses various applications across multiple disciplines, including materials science, biology, and engineering, facilitating advancements in areas such as drug delivery, electronics, and energy storage.

Research and development (R&D) experienced significant growth in the mid-1990s, driven by substantial increases in public investment. From the 2000s onward [18], after over a decade of substantial global public and private R&D and numerous scientific and technological achievements [19], nanotechnologies are primarily considered to be in the preliminary phases of development. Walsh et al. [16] examine the value of nanotechnologies in a study centered on the UK, commissioned by the Department for Environment, Food, and Rural Affairs (DEFRA). The research presents a methodology for estimating the net value added by nanotechnology to innovation.

Walsh and colleagues implement their methodology across multiple case studies on green nanotechnology, including applications such as nano-enabled food packaging, thin-film photovoltaics, fuel catalysts, and amperometric electrochemical gas sensors [20].

It is crucial to examine the potential and applications of green nanotechnologies. Life cycle assessment (LCA) analyzes a product's entire life cycle, covering resource extraction, production, use, recycling, and, eventually, the disposal of residual waste [21]. Within the framework of LCA, various approaches and tools are available, including methods that focus on economic inputs and outputs, their relationships, environmental impacts, and both direct and indirect energy demands throughout a product's life cycle [22, 23]. The use of nanotechnology in energy sector can reduce the energy cost and help to minimize carbon emissions in the environment [24]. Feasible methods for carbon nanotube production are energy-intensive, requiring high temperatures and pressures, which consequently lead to significant carbon dioxide emissions [25]. Recent estimates [22] reveal wide variations in the energy requirements for producing SWNTs, depending on the method used. Wender and Seager [26] present an argument utilizing a prospective LCA approach; the substantial energy demands for the large-scale production of SWCNT-enabled lithium-ion batteries are currently impractical.

Some applications of green nanotechnology have raised concerns about their environmental impact, as well as health and safety (EHS) risks to humans. Attention has been focused on the potential exposure risks to workers in laboratories and factories, particularly in the development and manufacturing of various nanostructure forms [27, 28]. The risks associated with nanostructures suggest a variety of ongoing actions and activities primarily in developed nations and by international organizations [29].

Groundwater contaminants have been utilized in various remediation [30, 31] projects in the US and in various European nations, including Germany and Italy, and the Czech Republic [32]. Quantum dots are tiny particles made from semiconductor materials, with customizable electrical and optical properties that offer promising applications in green technology, such as low-energy lighting and more efficient solar cells. Often made from cadmium and selenium, quantum dots may release toxic compounds under certain conditions during use or disposal [33, 34]. Numerous permutations and diverse applications of quantum dots, as well as their toxicology, are not yet addressed. Furthermore, at least one company, Nanoco, based in the UK, currently provides cadmium-free quantum dots that are being developed for LED lighting [35]. These examples highlight the uncertainties – especially over the long term – associated with the environmental, health, and safety (EHS) profiles of a broad and not yet clearly defined array of green nanotechnology applications. A number of recent and ongoing studies on nanotechnology EHS include requests for additional research and ongoing monitoring. Research within the insurance sector has indicated that there could be significant economic implications pertaining to the production and application of nanotechnology [18].

More than 60 countries are involved in national nanotechnology research and innovation programs [32], focusing on the use and application of nanotechnology worldwide. Figures 15.9 and 15.10 represent the growth of countries and continents in the field of nanotechnology, respectively. Regulation and oversight remain largely national, with limited international expansion. However, efforts for information exchange, harmonization, and standard setting in nanotechnology are underway at supranational and international levels, such as activities led by the OECD through its Working

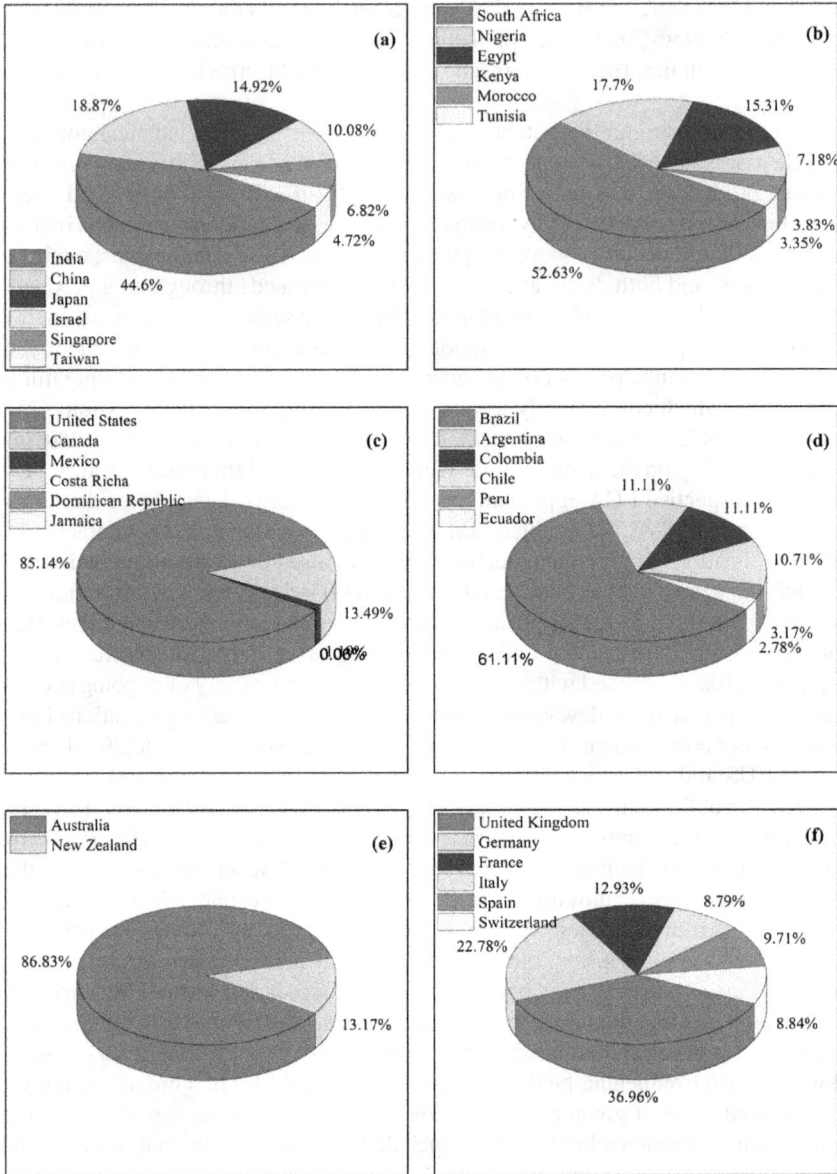

FIGURE 15.9 Representation of different countries working in the field of nanotechnology.

Parties on Nanotechnology [33] and Manufactured Nanomaterials. The implications of nanotechnology include variations in legislative and regulatory frameworks, which encompass both formal requirements and voluntary codes. These differences in governance structures involve the functions of agencies and industry, consumer organizations, and research institutions, and various entities in discussions regarding research in nanotechnology encompass commercialization, labeling, education, and regulation

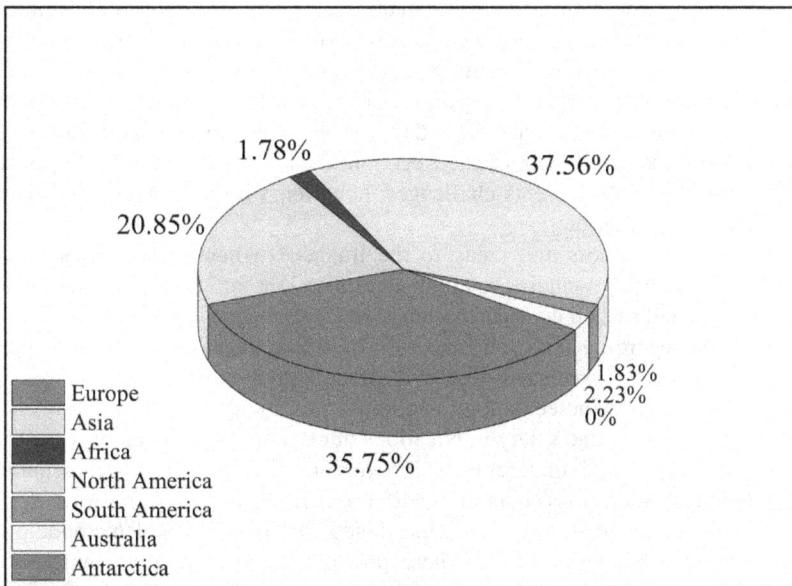

FIGURE 15.10 Representation of continents working in the field of nanotechnology.

[36]. Significant differences exist between developed and developing regions. The endeavors and competencies of developing countries in nanotechnology, as well as in the regulation and governance of this subject, are discussed in various sources [37].

Costs associated with applications in developing nations are a concern. For instance, several applications of green nanotechnology that selectively enhance agricultural productivity may have negative consequences, particularly for nations that depend on traditional approaches [38].

Anticipatory governance encourages a socially responsible framework for overseeing innovation by fostering the ability to anticipate, reflect on, and plan for future advancements. It emphasizes active participation from key stakeholders and meaningful public involvement [39]. Anticipatory governance methods may encompass foresight, scenario planning, technology integration, assessment, and various analytical techniques including LCA. Both the social and natural sciences should be included in research and innovation agendas. An approach to innovation that is socially responsible – also known as responsible research and innovation – incorporates openness and transparency, involves societal groups, and takes into account the effects of social, ethical, and environmental factors [40].

15.24 Concluding Remarks

Retrospective analyses of significant technological achievements often reveal their long-term economic impacts. The analysis is divided into three main categories: econometric analyses, survey-based studies, and case studies. According to Martin and Tang, economic and social impacts can emerge through several pathways, such as the advancement of knowledge, improvement of human capital, creation of new tools and techniques, establishment of networks and social connections, strengthened problem-solving capacity, the rise of new enterprises, and the spread of socially relevant knowledge.

It is important to note that these studies are retrospective in nature. The innovations are established, and an adequate duration has passed since their introduction for recognition and assessment of their economic and societal impacts. Conducting such studies effectively presents challenges; however, various methods and sources are available for utilization.

To prospectively assess and measure the impacts of new technologies, various methods are available, including technological foresight and forecasting techniques. However, especially when considering long-term projections, significant uncertainty is unavoidable regarding both the trajectory of technological development and the economic and societal conditions that will affect adoption and utilization. The economic assessment of nanotechnology is complicated by uncertainties regarding the environmental, health, and safety implications of certain nanotechnologies.

The complexities and uncertainties inherent in forecasting future technological developments should serve as motivation, rather than a deterrent, for assessing the potential economic impacts of green nanotechnology. Several methodological approaches are available to explore these potential impacts, such as multi-criteria evaluations, dynamic assessment methods, anticipatory life-cycle analyses, and scenario-based modeling. Advancing these and related methods will benefit from interdisciplinary collaboration that brings together expertise from both the natural and social sciences.

While the development and diffusion of green nanotechnologies may require more time than initially expected, they hold significant potential for addressing environmental challenges and promoting sustainable economic growth.

There is significant potential in nano-enabled applications, specifically in solar cells, photovoltaics, batteries, fuel catalysts, and water filtration. Additional nano-enabled applications may decrease operational energy requirements by providing similar or superior performance while being lighter, more durable, or more efficient. Attention must be directed throughout the life cycle to the energy and resource demands for the initial production of nanoscale materials, the types of energy (renewable or non-renewable) necessary for their large-scale manufacturing, the fate and disposal of nanomaterials during and post-use, as well as the related environmental, health, safety, and societal implications, alongside broader societal values and considerations. We propose that these considerations may be integrated via an anticipatory approach that models, deliberates on, and prepares for future developments. An anticipatory approach can be enhanced through a combination of measures and methods, enabling the modeling and exploration of various scenarios while prompting inquiries such as "what if" and "what about."

REFERENCES

[1] IEA: World energy outlook 2011 – Google Scholar, International Energy Agency, Paris (2020). https://scholar.google.com/scholar_lookup?title=World Energy Outlook&author=IEA&publication_year=2013%0Ahttps://scholar.google.com/scholar_lookup?title=World Energy Outlook 2012&author=IEA&publication_year=2012%0Ahttp://files/34/scholar_lookup.html.

[2] WHO/UNICEF, Progress on sanitation and DrWorld Health Organization UNICEF Joint Monitoring Programme for water supply and sanitation: Update 2010, WHO Libr. (2010) 1–55.

[3] M. Nasrollahzadeh, S.M. Sajadi, M. Sajjadi, Z. Issaabadi, Chapter 1 – An introduction to nanotechnology, in: M. Nasrollahzadeh, S.M. Sajadi, M. Sajjadi, Z. Issaabadi, M.B.T.-I.S. and T. Atarod (Eds.), *An introduction to green nanotechnology*, Elsevier (2019): pp. 1–27. https://doi.org/10.1016/B978-0-12-813586-0.00001-8.

[4] B. Ward, Development, synthesis and characterization of multifunctional nanomaterials (2014). www.researchgate.net/profile/Ward-Brullot/publication/263723744_Development_synthesis_and_characterization_of_multifunctional_nanomaterials/links/0f31753bbff1c9014b000000/Development-synthesis-and-characterization-of-multifunctional-nanomaterials

[5] S. Jagtap, P. Chopade, S. Tadepalli, A. Bhalerao, S. Gosavi, A review on the progress of ZnSe as inorganic scintillator, *Opto-Electronics Rev.* 27 (2019) 90–103. https://doi.org/10.1016/j.opelre.2019.01.001.

[6] L. De Almeida, A.T. Fujimura, M.L. Del Cistia, B. Fonseca-Santos, K.B. Imamura, P.A.M. Michels, M. Chorilli, M.A.S. Graminha, Nanotechnological strategies for treatment of leishmaniasis – a review, *J. Biomed. Nanotechnol.* 13 (2017) 117–133. https://doi.org/10.1166/jbn.2017.2349.

[7] O. Ali, *Nanotechnology market report: India*, AZoNano (2023).

[8] S. Bognár, P. Putnik, D. Šojić Merkulov, Sustainable green nanotechnologies for innovative purifications of water: Synthesis of the nanoparticles from renewable sources, *Nanomaterials.* 12 (2022). https://doi.org/10.3390/nano12020263.

[9] P. Singh, S. Pandit, C. Jers, A.S. Joshi, J. Garnæs, I. Mijakovic, Silver nanoparticles produced from *Cedecea* sp. exhibit antibiofilm activity and remarkable stability, *Sci. Rep.* 11 (2021) 12619. https://doi.org/10.1038/s41598-021-92006-4.

[10] E. Poorshakoor, M. Darab, Advancements in the development of nanomaterials for lithium-ion batteries: A scientometric review, *J. Energy Storage.* 75 (2024) 109638. https://doi.org/10.1016/j.est.2023.109638.

[11] *Nanotechnology market size, share & growth report, 2030*, (n.d.).

[12] *Shipbuilding market size, trends, report to 2030* (n.d.). https://straitsresearch.com/report/shipbuilding-market.

[13] N. Invernizzi, Nanotechnology between the lab and the shop floor: What are the effects on labor? *J. Nanoparticle Res.* 13 (2011) 2249–2268. https://doi.org/10.1007/s11051-011-0333-z.

[14] R. Clayton, *Bureau of labor statistics green jobs initiative BLS green jobs products* (n.d.).

[15] 이재흥, UNEP (2008) *Green jobs: Towards decent work in a sustainable, low-carbon world.* UNEP, ILO, IOE, and ITUC (2010).

[16] A.L. Porter, J. Youtie, How interdisciplinary is nanotechnology? *J. Nanoparticle Res.* 11 (2009) 1023–1041. https://doi.org/10.1007/s11051-009-9607-0.

[17] G. Tassey (2003), *Methods for assessing the economic impacts of government R&D, planning report 03–1*, National Institute of Standards and Technology, Gaithersburg, MD (n.d.).

[18] V.A. Basiuk, E.V. Basiuk, Green processes for nanotechnology: From inorganic to bioinspired nanomaterials, *Green Process. Nanotechnol. From Inorg. Bioinspired Nanomater.* (2015) 1–446. https://doi.org/10.1007/978-3-319-15461-9.

[19] M.C. Roco, C.A. Mirkin, M.C. Hersam, Nanotechnology research directions for societal needs in 2020: Summary of international study, *J. Nanoparticle Res.* (2011) 897–919. https://doi.org/10.1007/s11051-011-0275-5.

[20] P. Shapira, J. Youtie, The Economic Contributions of Nanotechnology to Green and Sustainable Growth. In: Basiuk, V., Basiuk, E. (eds) *Green Processes for Nanotechnology*. Springer, Cham (2015). https://doi.org/10.1007/978-3-319-15461-9_15.

[21] M.-A. Wolf, K. Chomkhamsri, M. Brandao, R. Pant, F. Ardente, D. Pennington, S. Manfredi, C.C. De, M. Goralczyk, *International reference life cycle data system (ILCD) handbook-general guide for life cycle assessment-detailed guidance*, Publications office of the European Union, Luxembourg, 2012.

[22] G. Rebitzer, T. Ekvall, R. Frischknecht, D. Hunkeler, G. Norris, T. Rydberg, W.-P. Schmidt, S. Suh, B.P. Weidema, D.W. Pennington, Life cycle assessment: Part 1: Framework, goal and scope definition, inventory analysis, and applications, *Environ. Int.* 30 (2004) 701–720.

[23] G. Finnveden, M.Z. Hauschild, T. Ekvall, J. Guinée, R. Heijungs, S. Hellweg, A. Koehler, D. Pennington, S. Suh, Recent developments in life cycle assessment, *J. Environ. Manage.* 91 (2009) 1–21.

[24] J.A. Isaacs, A. Tanwani, M.L. Healy, Environmental assessment of SWNT production, in: *Proc. 2006 IEEE Int. Symp. Electron. Environ. 2006* (2006) 38–41. https://doi.org/10.1109/ISEE.2006.1650028.

[25] A.E. Agboola, R.W. Pike, T.A. Hertwig, H.H. Lou, Conceptual design of carbon nanotube processes, *Clean Technol. Environ. Policy.* 9 (2007) 289–311.

[26] B.A. Wender, T.P. Seager, Towards prospective life cycle assessment: Single wall carbon nanotubes for lithium-ion batteries, in: *Proc. 2011 IEEE Int. Symp. Sustain. Syst. Technol.*, IEEE (2011) pp. 1–4.

[27] R.J. Aitken, K.S. Creely, C.L. Tran, *Nanoparticles: An occupational hygiene review*, HSE Books, London, 2004.

[28] P.A. Schulte, D. Trout, R.D. Zumwalde, E. Kuempel, C.L. Geraci, V. Castranova, D.J. Mundt, K.A. Mundt, W.E. Halperin, Options for occupational health surveillance of workers potentially exposed to engineered nanoparticles: State of the science, *J. Occup. Environ. Med.* 50 (2008) 517–526.

[29] V. Murashov, J. Howard, Health and safety standards, in: *Nanotechnology standards*, Springer (2011): pp. 209–238.

[30] S.R. Kanel, B. Manning, L. Charlet, H. Choi, Removal of arsenic (III) from groundwater by nanoscale zero-valent iron, *Environ. Sci. Technol.* 39 (2005) 1291–1298.

[31] Sigmund, W., El-Shall, H., Shah, D.O., & Moudgil, B.M. (Eds.). (2008). *Particulate Systems in Nano- and Biotechnologies (1st ed.).* CRC Press. https://doi.org/10.1201/9781420007534.

[32] N.C. Müller, B. Nowack, Nano zero valent iron – the solution for water and soil remediation, *Rep. Obs. Nano.* (2010) 1–34.

[33] S. Mahendra, H. Zhu, V.L. Colvin, P.J. Alvarez, Quantum dot weathering results in microbial toxicity, *Environ. Sci. Technol.* 42 (2008) 9424–9430.

[34] M. Bottrill, M. Green, Some aspects of quantum dot toxicity, *Chem. Commun.* 47 (2011) 7039–7050.

[35] P. Shapira, J. Youtie, The economic contributions of nanotechnology to green and sustainable growth BT – green processes for nanotechnology: From inorganic to bio-inspired nanomaterials, in: V.A. Basiuk, E. V Basiuk (Eds.), Springer International Publishing, Cham (2015): pp. 409–434. https://doi.org/10.1007/978-3-319-15461-9_15.

[36] J. Pelley, M.A. Saner, *International approaches to the regulatory governance of nanotechnology*, Regulatory Governance Initiative, Carleton University, 2009.

[37] A.P. Jayanthi, K. Beumer, M. Bhati, S. Bhattacharya, Nanotechnology: 'Risk governance' in India, *Econ. Polit. Wkly.* (2012) 34–40.

[38] G.P. Gruère, C.A. Narrod, L. Abbott, Agricultural, food, and water nanotechnologies for the poor: Opportunities, constraints, and role of the Consultative Group on International Agricultural Research, IFPRI Discussion Paper 01064 (2011).

[39] R. Karinen, D.H. Guston, Toward anticipatory governance: The experience with nanotechnology, *Gov. Futur. Technol. Nanotechnol. Rise an Assess. Regime.* (2010) 217–232.

[40] Research and innovation – European Commission (n.d.).

[41] Top Nanotechnology companies and suppliers in 2025 _ ensun, Data Accessed 11-05-2025. (n.d.).

Index

Note: *Italic* and **Bold** page numbers refer to *figures* and **tables**.

A

air pollution treatment, 107–110
  nanoadsorbents, air treatment, 109
  nanocatalysts, air purification, 108–109
  nanofilters and nanocoatings, air
    purification, 110
  nanosensors, air remediation, 107–108
aluminum oxide, 74, 215
AMNIVOR Medicare, 314
assessment objectives, business, 6
AVANSA Technology & Services, 329

B

batteries, 54–57
  lithium-ion batteries (LIBs), 54–55
  sodium-ion batteries (SIBs), 55–56
  solid-state batteries (SSBs), 56–57
bioactivity, actives, 70–71, 83, 272
biomass energy, 49–51
blood–brain barrier (BBB), 84, 85
breast implant-associated anaplastic large cell
    lymphoma (BIA-ALCL), 90
business potential, nanotechnology, 1–29
business scope, NT, 3–5, *4*

C

carbon-based anodes, 56
carbon-based nanomaterials, 1, 42, 43–45, *44*,
    93, 95, 100, 212, 261
carbon nanotubes (CNTs), 4–9, 14, 22, 27, 43,
    45, 49, 54, 73, 194
carbon quantum dots (CQDs), 133
catalytic activity, bacterial detection,
    310–311
circular economy, 66, 69, 138, 273, 305–306
climate change, 39, 40, 136, 236–246
  air pollution control, 305
  air quality monitoring, 236
  biodiversity loss, 236
  carbon capture and storage (CCS)
    technologies, 240, 245
  cleaning up oil spills, 244
  desalination, 126, 242, **262**, 305

  ecosystems, effects, 18, 23, 40, 66, 88, 93,
    174, 179, 182, 183, 238, 239, 244,
    245, 273, 276
  effects on environment, 238
  environment, monitoring, 89, 122n178,
    242–243, **263**
  filter media, 244
  greenhouse gases, sequestration, 239, **240**
  and greenhouse gases, 109, 236, 237, 239,
    240
  habitat destruction and agricultural impacts,
    238
  heavy metal removal, 231, 243
  infrastructure and human health, 238–239
  mitigation, sustainable solution, 236, 237,
    239, 245
  nanocatalysts for conversion, 241
  nanofiltration, 6, 100, 163, 242
  nanotechnology to combat, 239
  nanotechnology to reduce pollution, 243
  non-conventional energy resources
    (renewable energy), 239, 241
  scrubbers, nanocatalysts, 14, 244
  soil and water nanoremediation, 243
  sustainable architecture, 244
  water purification, 6, 11, 24, 94, 130, 138,
    163, 242, 256, 266, 277, 305, 310
  water quality sensors, 243
compliance, 291, 292
  mechanisms, 97, *99*, 128, 131, *135*, 185, 211,
    214–224, 285, 296
conducting polymer-based nanomaterials, 46,
    264
construction industries, nanomaterials, 194
  carbon nanotubes, 43, 44, 49, 73, 95–100,
    105, 109, 111, 163, 175, 181, 197–198
  cenosphere, 198
  challenges and recommendations, 201
  graphene and graphene oxide, 199, 214,
    223, 225
  nano-Al_2O_3, 199
  nano-cellulose, 194, 200, 225, 305
  nanoclays, 49, 55, 98, 104, 105, 196–197,
    216–225
  nano-Fe_2O_3, 198–199
  nano-SiO_2 and nano-TiO_2, 194–196

339

D

DaunoXome®, 84
dendrimers, 85, 93, 95, 96, 271
DepoCyt®, 84
display technology, 13
Doxil®, 84

E

eco-friendly alchemy, 151, 153, 155, 157, 159, 161
economic impact, indicators, 330
electrochemical biosensing, 134
energy flow, *40*
energy-generating devices, 134
enforcement, 291, 292
engineered nanomaterials (ENMs), 71, 93
environmental and health concerns, 21–22, 211
environmental quality, 134, 305
    air pollution control, 44, 305
    soil remediation, 103, 105, **106**, 305, 306
    water purification, 305–307
environmental remediation, 14, 88, 93–111
ethical implications, 19, 22

F

fire retardancy mechanisms, 211–215, 218, 219, 225
    barrier effect, 211, 214, 218–224
        compact char layer formation, 218–219
        heat and oxygen, physical shielding, 219
        volatile combustible gases, reduced escape, 215, 218, 219, 221
    catalytic charring, 215–218, 220
    gas phase action, 221
        radical scavenging effect, 221
        smoke suppression mechanism, 221
    heat absorption and energy dissipation, 214, 215, 218, 220
fire-retardant *nanomaterials, 213*, **220**, 221, 225
    coatings and surface treatments, 106, 221, 222
    nanocoating, textile industry, 24, 223
    paints and coatings, 199, 211, 224
fire retardants, 211–226
    boron-based nanomaterials, 212, 216
    carbon-based nanomaterials, 42–44, 93, 95, 100, 212–214, 261
    carbon nanofibers and fullerenes, 93–96, 177, 215, 245
    carbon nanotubes, 214–215
    clay-based nanomaterials, 212, 216

future prospects, nanomaterials, 225–226
graphene and graphene oxide, 199, 214, 223, 225
layered double hydroxide (LDH), 216, 217, 222, 224
metal oxide nanoparticles, 215
founders, nanotechnology, 81–82

G

gas-phase radical trapping, 217, 218
giant magnetoresistance (GMR) effect, 7
global regulatory frameworks, 18–21, 183, 284, 287
graphene, 45, 49
graphene–CNT composites, 57
graphene conducting polymers, 57
graphene–metal oxide hybrid composites, 57
Green economy, 281–299
    administering frameworks, nanomaterials, 289
    policy and regulatory approach, 283, 287, 289
    public engagement and transparency, 293, 294
    stakeholder engagement and public experience, 286–291
Green nanoparticles, 161
    agriculture, 165
    antimicrobial activities, 162
    biomedical applications, 160, 161, 182
    drug delivery systems, 6, 8, 29, 134, 161, 166, 273
    environmental applications, 18, 93–95, 107, 162, 163
    food industry, 164–165
    wastewater treatment process, 163–164
    water purification, 242
Green nanotechnology, 137, 253–254
    advancements, 137–138
    eco-friendly applications, 151
    eco-friendly synthesis using, 253–254
    nanoparticles synthesis
        microbial source, 255–256
        natural polysaccharides, 255
        plant extracts, 10, 27, 94, 97, 152–155, 165, 254
        yeast, 153, 165, 256
    next-generation nanomaterials, water treatment, 138
    sustainable economic development, 256
Green synthesis
    algae, 152
    bacteria, 162, **163**

challenges and future directions, 165–166
factors affecting synthesis, 156
fungi, 153, **154**
metal ions concentration, 157
methods, 137
of nanomaterials, 151–166, 212
pH, 255
plants, 154–155, **155**
resources for, 152
specific method approach, 156
temperature, 156
time, 156
yeasts, 153–154
Green-synthesized nanoparticles
characterization, 157
Fourier Transform Infrared (FTIR)
spectroscopy, 159
particle size and zeta potential, 161
scanning electron microscopy, 160
transmission electron microscopy, 160–161
UV–Vis spectroscopy, 157–158
X-ray diffraction (XRD), 159–160

H

healthcare industry, 81–91
healthcare nanotechnology, developments, 82
advancing with nanostructures, 82–84
biological system implications, awareness, 88
biosensor-based diagnostics advancement, 86
brain disorders, treating, 84–85
in cancer therapy, 84–87
carcinogenicity, 88–90
challenges, 87–88
dentistry, 85, 91
drug delivery, liposomal and polymeric
NPs, 84–87
environmental risks, 88–89
healthcare applications, trends, 90–91
liposomes, 84
in modern pharmaceuticals, 82–84
nanoparticles toxicity, 89–90
nanotoxicity and safety risks, 88–89
NP components, lack of understanding, 88
in ophthalmology, 85
polymeric nanoparticles, 84
risk management, 89
safety risk evaluation, 89
systemic toxicity, 89–90
targeted drug delivery, 82–84
toxicity level, variability, 88
toxicology, challenge, 87
heat release rate (HRR), 215, 219
human health impacts, nanomaterials

dermal exposure, 178
ingestion, 178
inhalation, 21, 176, 178, 181, 183, 245
systemic effects, 179
hybrid nanomaterials, 43, 46
hybrid supercapacitors, 58–59

I

Indian companies, nanotechnology, **315–326**
India's nanotechnology future, 329
India's present nanotechnology market, 312–314
industrial growth, 310
industry applications, 11–12, *12*
electronics and IT applications, 13–14
energy applications and environmental
remediation, 14, 27
future transportation benefits, 14–15
medical and healthcare applications, 12–13
international cooperation, 292–293
international harmonization, 282–283
challenges, achieving, 289
efforts toward, 283–284
enhancing, 284–285
initiatives toward, 285–286
lack of global standards and regulations,
nanomaterials, 282–283
iron oxide, 73

K

Kanopy Techno Solutions, 329

L

lab-on-a-chip (LOC) technology, 6
life cycle assessments (LCA), 64, 181, 184,
185, 290
in nanotechnology applications, 137
life cycle thinking (LCT), 64–76
balancing innovation and ecological impact,
65–66
end-of-life considerations, 65–66
operational phase, 65–66
lipid-based nanocarriers, 70
lithium-ion batteries (LIBs), 54–55, 311
enhancing performance with nanomaterials,
54–55
Lux Research projects, 5

M

magnetic random-access memory (MRAM), 13
market landscape, 11–15, *12*

metal-based nanomaterials, 45, *45*, 261–264
    structures of, *45*
metal nanoparticles, 45–46, 50, 111, 155, 158,
    258, **316**
metal oxide-based supercapacitors, 58
metal oxide nanoparticles, 93–95
microplastics detection, 132
modern industry updated systems, 5–6
modern NT systems, 6
Myocet®, 84

N

nano-alumina, 74, 193
nano-based remediation methods, 251
nanobioremediation, 106–107
nano-cargo transport, active filaments, 311–312
nano-cement particles, hydration, *71*
nanoclays, 49, 98, 104, 105, 196, 218–219, 220
nanocomposite electrolytes, 55, 56
nanocomposites, 95
nanoemulsions, 41, 164
nano-enabled cosmeceuticals, 70–71
nanoencapsulation, 69
nano-imaging technology, 6
nanomaterial-based biosensors
    aquaculture and environmental safety,
        applications, 133
    for water quality monitoring, 132–133
nanomaterial-based remediation
    adsorption, 128–129
    mechanisms, 128–129
    nanofiltration and size exclusion, 100
    photocatalysis, 99–100, 129–130
    pollutant-specific interactions, 100–101
    redox reactions and catalytic degradation,
        98–99
nanomaterial-based water treatment, 128
    adsorption mechanism, 128–129
    electrochemical processes, 130–131
    filtration techniques, 129
    mechanisms, 128
    photocatalysis, 99–100, 108, 128–130
nanomaterials (NMs), 193, 261, 286
    additives in concrete sector, 200–201
    as adsorbent, 101, **102**
    as antibacterial agents, 103, **103**
    in batteries, 54–57
    carbon nanotubes, occupational settings,
        181
    case studies, 180–182
    classification, *9*, 19, 42–43, 157, 277
    in construction industries, 194–200
    defined, 1, 7, 8, 86, 100, 174, 291, 331

in energy systems, types, 43–46, *44*
environmental fate of, 174–176
environmental health impacts, 179–180
    aquatic organisms, 179–182, *180*
    terrestrial organisms, 180
    trophic transfer, 176, 180, 182, **183**
environmental media, transformations,
        174–176
    biological transformations, 174, 176, 185
    chemical transformations, 175, **176**
    physical transformations, 175
factors influencing toxicity, 180
as fire retardants, 211–226
global regulatory frameworks, 18, *20*,
        183–184, 284
graphene-based materials, 59, 101, 182,
        183, 214
Green and clean energy production and
        storage, 39–60
Green synthesis of, 151–166
human health impacts, 178
management of, 287
mitigation strategies, 184–185, **185**, 288
MXenes, 182–183, *184*
persistence and bioaccumulation, 161, 176,
        182
as photocatalyst, 101–103, **103**
primary categories, 42
quantum dots, biomedical applications, 139,
        160, 161, 182
in renewable energy production, 46–54
risk assessment, challenges, 183
silver nanoparticles, wastewater treatment
        plants, 181
in solid-state electrolytes, 56
supercapacitors, 57–59
surface properties of, 309–310
for sustainable agriculture, 244–245, 266,
        267
for sustainable energy areas, *41*
synthesis methods, 96–97, 128, 137, 151,
        165, 269, 270
titanium dioxide nanoparticles, sunscreens,
        181
top-down and bottom-up approach, *152*
toxicity and environmental risk, 110–111,
        134–135
toxicological aspects and environmental
        concerns, 245
transport of, 176, 177, 181
    in air, 98, 100, 107, 109, 176
    in soil, 177, **178**
    in water, 177, *177*
types, 42, 95

in water treatment, 101, 125–132
zero-valent iron nanoparticles (nZVI),
    groundwater remediation, 94, 175,
    181–182
nano over bulk, superiority, 307, 308
nanoparticles, 54
    pesticides and hydrocarbons, degradation,
    105–106
nano-photonics, 6
nanorobotics, 6
nanoscale coatings, 11
nanoscale dimension advantages, 96
nanoscale innovations, 11, 261
Nanosentrix, 314
nano-silica, 72, *72*
    compressive strength, *72*, 194–196, **202**,
    204
Nanospan, 314–315
nanostructured anodes, 55, 56
nanostructured cathodes, 55, 56
nanostructured electrodes, 55
nanostructured electrolytes, 55
nanotechnological tools, challenges, 306–307
nanotechnology
    agri-food waste, sustainable valorization, 69
    for air pollution treatment, 107–110
    business growth and market entry,
        strategies, 27–28
    circular economy catalyst, sustainable
        industries, 66–67
    climate change mitigation, sustainable
        solution, 236–246
    commercialization challenges and
        opportunities, 17–18
    core principles and concepts, 7–8
    cost and scalability issues, 26–27
    creative founders in, 81–82
    eco-friendly and sustainable economic
        development, 249–261
    economic and business potential, 15–16
    emerging innovations in, 27
    environmental and health concerns, 21–22
    in environmental remediation, 93–111
    ethical implications, 19, 22–23
    fundamentals, 95, 128, 237, 261
    global regulatory frameworks, 18–21,
        183–184
    greater surface area, 237
    as Green alternative, 251–253
    in healthcare industry, 81–91
    heavy metals detection using, 133
    hybrid systems, integrating with
        bioremediation, 138
    incorporating into construction, 71–74, *76*

industrial growth of, 310
investment trends and funding sources,
    16–17
key industry players and startups, 24
life cycle thinking of, 64–76
market landscape and industry applications,
    11–15, *12*
market size and growth projections, 16
materials and processes, 8–10
net-zero goals, catalyzing, 67–68, 107, 130
next decade, predictions, 28–29
patents and intellectual property trends,
    24–25
policy and research directions, sustainable
    implementation, 138–139
quantum effects, 42, 54, 64, 96, 128, 237,
    308
regional and global market dynamics,
    25–26
regulation, ethical and social reflections,
    288
research and development, current trends,
    10–11
in soil remediation, 103–107, **106**, 305
strength and durability, 12, **202**, 237
sustainability and emerging risks, 71–74
sustainable development, pillar, 40–41
sustainable economy building, role,
    308–338
sustainable environmental performance and
    circular economy, path, 309
technical and manufacturing limitations, 26
technological revolutions, 6–7
in water treatment, 101–103, 125–139
    challenges and environmental
        implications, 134–137
    cost, scalability, and manufacturing
        challenges, 135–136
    fundamentals, 95, 128, 237, 261
    regulatory and safety considerations,
        136–137
non-conventional energy resources (renewable
    energy), 239, 241–242
biofuels, 49–51, 241
energy storage, 270
solar cells, 241
NoPo nanotechnologies, 314, **317**

O

one-dimensional (1D) nanomaterials, 42
organic photovoltaics (OPVs), 11, 93, 99,
    100–111, **102**, 129, 131, 250
organic pollutants detection, 133–134

P

peak heat release rate (PHRR), 211, 214–225
perovskite solar cells, 47, 241, 269, 270
point-of-use technology, 127
polymeric nanomaterials, 95

Q

quantum confinement, 47, 53, 54, 93, 270
quantum dots (QDs), 13, 47, 54
quantum dot solar cells, 47
quantum mechanics, 3, 7

R

rare earth metals, recycling, 306
regulatory challenges, 3, 287, 289
renewable energy production, 46–54
resource efficiency, 65–67, **263**, 305–306
risk assessment, 282, 286, 291
risk governance, 287

S

safer-by-design approaches, 64
safety testing, 282
Saint-Gobain Glass India, 329
Seebeck coefficient, 53–54
self-assembly, 5–8, 135
silica fumes, compressive strength, *72*
silver nanoparticles, 74
skin delivery, transforming, 70–71
smart nanomaterials, 237
sodium-ion batteries (SIBs), 55–56
    replacement for lithium-ion, 55–56
soil remediation, 103–107
    nanoclays, 105, **196**
    nZVI, 105
    physicochemical parameters, soil
        nanoparticle transport, 104–105
solar energy, 46–48, *47*
solid polymer electrolytes (SPEs), 51
solid-state batteries (SSBs), 56–57
    nanostructured electrodes for, 56, 270
stakeholder engagement, nanotechnology
        governance, 290
supercapacitors, 57–59
    graphene-based supercapacitor, 57, 59
    hybrid supercapacitors, 58–59
    metal oxide-based supercapacitors, 58
surface functionalization, 100
surface properties, 313–314
sustainable agriculture, 306

sustainable development goals (SDGs),
        266
    advancing, 261
    bottom-up approach, nanomaterial,
        269
    carbon-based nanomaterials, 43–46
    challenges and ethical considerations,
        nanotechnology, 276–277
    food quality and safety, nanotechnology,
        268
    hybrid and composite nanomaterials, 317
    medicine, nanotechnology, 271–273
    metal-based nanomaterials, 45
    nano-encapsulated fertilizers and
        herbicides, 268
    nanomaterials recovery, 273
        hydrometallurgy, 273, 274, *274*
        nanohydrometallurgy, 273, 274, *274*
        pyrometallurgical pathway, *276*
    nano-sensors, precision agriculture, 271
    nanotechnology and, 271
    plant growth and genetic improvement,
        nanoparticles, **263**
    polymer-based nanomaterials, 46
    soil and crop improvement, nano-
        agrochemicals, 268
    solar cells, nanomaterials, 270
        carbon nanomaterials, 56, 68, 128, 314
        energy storage and next-generation
            batteries, 274–275
        metal NPs, 98, 268, 269
        two-dimensional (2D) NMs, 138, 182
    sustainable agriculture, nanotechnology,
        266, *267*, 306
    synthesis and fabrication techniques,
        nanomaterial, 58, 268
    technology, 51, 261, 265
    top-down approach, nanomaterial, 151,
        265
        hybrid techniques, **320**
    water remediation, nanotechnology, 272
sustainable economy, 66, 282, 286, 288
sustainable nanotechnology innovation, 285
    regulation role, 291–292
sustainable valorization, 69
    agri-food waste, 69

T

techno-economic network paradigm, 24
technological readiness levels (TRLs), 29
technological revolutions, 6–7
technology, 51, 261, 265
three-dimensional (3D) nanomaterials, 43

titanium dioxide, 72–73, 108, 175
  self-cleansing activity, *73*
total heat release (THR), 211, 216
traditional flame retardants, 211, 215,
      216–218, 22
  notable nanomaterial–traditional retardant
      combinations, 217–218
  synergistic effects, nanomaterials, 215,
      216–218
traditional remediation techniques, 93
traditional water treatment processes,
      126
transition metal dichalcogenides (TMDs), 9,
      10, 270
transmaterialization, 40–41
triple junction catalysts, 68

two-dimensional (2D) nanomaterials, 42–43,
      133, 138, 182, 223, 270

W

waste valorization, 67–69, 305
wastewater treatment, 163, 164, 310, 314
water reuse technology, 126
water treatment, *101*, 101–103, 125–139
  next-generation nanomaterials, 138
wind energy, 14, 46, *48*, 48–49

Z

zero-dimensional (0D) nanomaterials, 42
zinc oxide, 74

For Product Safety Concerns and Information please contact our EU
representative GPSR@taylorandfrancis.com
Taylor & Francis Verlag GmbH, Kaufingerstraße 24, 80331 München, Germany

www.ingramcontent.com/pod-product-compliance
Lightning Source LLC
Chambersburg PA
CBHW060802220326
41598CB00022B/2519

9 781041 053798